Fundamentals of Turbomachinery

Fundamentals of Turbomachinery

Theory and Applications

Second Edition

Ryoichi S. Amano
The University of Wisconsin-Milwaukee
Milwaukee, WI

William W. Peng
California State University
Fresno, CA

WILEY

Published by John Wiley & Sons, Inc., Hoboken, New Jersey.
Published simultaneously in Canada.

Library of Congress Cataloging-in-Publication Data Applied for:

Hardback ISBN: 9781394273546

Cover design: Wiley
Cover image(s): © emel82/Shutterstock

Set in 9.5/12.5pt STIXTwoText by Straive, Chennai, India
Printed and bound by CPI Group (UK) Ltd, Croydon, CR0 4YY

C9781394273546_040925

I, Ryoichi S. Amano, dedicate this book to my wife, Jun, my daughter, Amy, my son, Kent, and my daughter, Naomi, for their love and patience during my writing and to my late parents, Kyusaburo and Fusako Amano.

I, William W. Peng, dedicate this book to my wife, Alice, and my son, Steven, for their love and patience while I was writing and to my late parents for their love and support of my education.

Contents

Preface

The second edition of this book has been written by modifying a significant portion of the first edition. Some chapters were changed to adjust to the current technological changes. Nonetheless, the book intends to serve turbomachinery in an undergraduate or a graduate curriculum. Readers are assumed to have completed basic fluid mechanics and thermodynamics courses, which engineering students normally take in their sophomore or junior year. Since the book covers application/selection, it can also be used as a supplement for courses such as thermal energy system design or heating, ventilation, and air conditioning (HVAC). The book can also be used by practicing engineers for self-studies. The uniqueness of this book is that (1) most turbomachines are covered, not just gas turbines, compressors, or pumps, and (2) more engineering applications of turbomachines are emphasized. The first two chapters serve as an introduction to general turbomachines. Basic fluid mechanics principles are reviewed. It is the starting point, leading to the derivation of energy transfer equations or the transport equations. They are the fundamentals of turbomachine analysis. The rest of the book is devoted to diverse types of turbomachines.

One chapter is on pumping devices, and four are on turbines—power-producing devices.

Flow-type classifications such as radial, mixed, or axial flow are also carefully grouped.

For each type of machine, the following topics are covered:

1) Basic principles
2) Preliminary design procedure
3) Ideal performance characteristics
4) Actual performance curves published by manufacturers
5) Application/selection of machine

The new chapters added are "Artificial Intelligence as a New Frontier in Turbomachinery Design and Applications" and "Computational Fluid Dynamics (CFD)." The book includes examples of working principles to illustrate the principles discussed. End-of-chapter problems are also included for the students and engineers to practice. In these problems, both the International System (SI) and English units are employed. Although conversion to SI is the trend, it will take another few decades for the U.S. industry to become fully SI. Hence, the current engineering students are required to be familiar with both systems.

Symbols

A	Flow passage area (in.^{2}, cm^2)
a	Acoustic speed (ft/s, m/s); constant
b	Blade height (in., cm); constant
C_d	Drag coefficient
C_L	Lift coefficient
C_p	Specific heat at constant pressure; pressure coefficient; power coefficient
C_v	Specific heat at constant volume [Btu/(lb$_m$-R)]
CV	Velocity coefficient
c	Chord length (in., cm); constant
D	Diameter (in., cm)
D_s	Dimensional specific diameter
d	Constant; diameter
e	Total energy of fluid per unit mass, $= u + V^2/2 + gZ$ (Btu/lb$_m$, kJ/kg)
F	Force (lb$_f$, N)
f	Constant; friction coefficient; function of
g	Gravitational constant (32.174 ft/s^2, 9.81 m/s^2)
g_c	Conversion factor (32.174 lb$_m$/slug, = 1 for SI system)
H	Pressure head, $= p/(\rho g)$ (ft, m)
h	Enthaply of fluid per unit mass (Btu/lb$_m$, kJ/kg)
i	Incidence angle (deg)
K	Cascade coefficient
k	Ratio of specific heats
$\mathbf{k}$	Unit vector in the axial direction
L	Length (in., cm)
M	Mach number
M_r	Relative Mach number
Mw	Molecular weight
m	Mass flow rate (lb$_m$/s, kg/s)
N	Rotating speed (rpm)
N_s	Dimensional specific speed
n	Polytropic constant; number
n_s	Number of stages
P_s	Shaft power (lb$_f$-ft/s, hp, kW)
p	Pressure (psi, Pa)
Δp	Pressure rise or change (psi, Pa)

Q	Volumetric flow rate (gpm, cfm, m^3/s)
q	Heat transfer rate (Btu/s, kJ/s)
R	Gas constant (Btu/lb_m-R, kJ/kg-K); degree of reaction
Re	Reynolds number
R_h	Reheat factor
r	Radius (in., cm); coordinate of radial direction
s	Entropy of fluid per unit mass (Btu/lb_m-R, kJ/kg-K); pitch (spacing between adjacent blades)
S	Dimensional suction-specific speed of the pump ($rpm(gpm)^{0.5}/ft^{0.75}$)
sp	Static pressure of fan (in. or mm of water)
T	Temperature (°F, R, °C, K)
t	Time (s); airfoil thickness; ratio of gas turbine inlet temperature ($= T_3/T_1$)
U, $\mathbf{U}$	Impeller peripheral velocity (ft/s, m/s)
u	Internal energy of fluid per unit mass (Btu/lb_m, kJ/kg)
$\mathbf{u}$	Unit vector
V, $\mathbf{V}$	Absolute flow velocity (ft/s, m/s)
v	Specific volume of fluid (ft^3/lb_m, m^3/kg)
W, $\mathbf{W}$	Relative flow velocity with respect to impeller
x_c	Distance of airfoil camber from the leading edge
Z	Elevation (ft, m); compressibility factor
Z_b	Number of blades
z	Coordinate of the axial direction
z_c	Camber of airfoil mean line

Greek Letters

α	Angle of attack (deg); absolute flow angle (deg)
β	Blade angle (deg); relative flow angle (deg)
γ	Stagger angle (deg); specific weight of fluid (lb_f/ft^3, N/m^3)
Δ_s	Nondimensional specific diameter
δ	Deviation angle at impeller discharge; ratio of drag and lift coefficients
ε	Flow deflection angle; utilization factor
η	Efficiency
θ	Blade deflection angle (deg)
λ	Compressor work done factor; wind turbine tip speed ratio ($= U_t/V$)
μ	Dynamic viscosity (lb_f-s/ft^2, N-s/m^2)
μ_s	Slip factor
υ	Kinematic viscosity (ft^2/s, m^2/s); hub–tip ratio
Π	Power coefficient
ξ	Rotor/stator loss coefficient
π	Nondimensional parameters; 3.14159
ρ	Fluid density (lb_m/ft^3, kg/m^3)
σ	Solidity
σ_s	Shear stress (psi, Pa)
τ	Torque (ft-lb_f, m-N)

ϕ	Flow coefficient
ψ	Pressure head coefficient; compressor stage loading factor
ω	Rotating speed (rad/s)
ω_s	Nondimensional specific speed

Subscripts

0 (zero)	Stagnation properties
1	Inlet of impeller/rotor for pumping devices; inlet of nozzle/stator for turbines; condition 1
2	Outlet of impeller/rotor for pumping devices; inlet of turbine impeller/rotor; condition 2
3	Outlet of diffuser/stator for pumping devices; outlet of turbine impeller/rotor
a	Axial direction
ac	Actual
ad	Adiabatic process
o/a	Overall
b	Blade
d	Drag
e	Impeller eye
f	Flow; frictional
h	Hub; hydraulic
i	Inlet; ideal condition; input; inertia
L	Lift; leakage
le	Leading edge
m	Meridional direction; mechanical; mean value; model
n	Nozzle
o	Outlet; output
p	Pump; polytropic; prototype
r	Radial direction; rotor
s	Stator; per stage; properties along isentropic process
t	Turbine; tip; tank
th	Theoretical; thermal
u	Tangential direction
v	Volumetric; viscous; dynamic head related to velocity

About the Companion Website

This book is accompanied by a companion website:

www.wiley.com/go/amano/turbomachinery2e

This website include:

- Solutions to problems

Scan this QR code to visit the companion website

1

Introduction

1.1 Definition

The word "turbo" comes from Latin, which implies spinning or whirling. A rotating blade row, a rotor, or an impeller varies the stagnation enthalpy of the fluid moving through it by doing either positive or negative work, depending upon the effect required of the machine. The turbomachine is an energy conversion device that converts mechanical energy to thermal/pressure energy or vice versa. The conversion is done through the dynamic interaction between a continuously flowing fluid and a rotating machine component. Both momentum and energy transfer are involved. Hence, positive displacement machines, such as piston-type or screw-type machines, which operate due to the static interaction between the fluid and mechanical components, are excluded.

Historically, the first turbomachines could be identified as water wheels, which appeared between the third and first centuries B.C. in the Mediterranean region. These were used throughout the medieval period, which began the first Industrial Revolution. A turbomachine has a rotating component that provides continuous interaction with a flowing fluid. Mechanical energy is delivered through this rotating element. Thermal/pressure energy in the flowing fluid can be in either kinetic energy or static enthalpy energy mode. These two energy modes can be converted in either direction through a diffuser or nozzle called stators, while rotating components are called rotors or impellers. Additional components are sometimes needed to direct the fluid in an appropriate direction.

1.2 Types of Turbomachines

Turbomachines can be classified according to

a) The direction of energy transfer, either from mechanical to thermal/pressure or vice versa.
b) Type of fluid medium handled, either compressible or incompressible.
c) The flow direction through the rotating impeller can be axial, radial, or mixed to the rotational axis.

A classification is presented in Table 1.1. In terms of the direction of energy transfer, the machine can be either a pumping device or a turbine. A pumping device converts mechanical energy into thermal/pressure energy. Such devices include liquid pumps, compressors, blowers, or fans. The gas-handling devices are classified based on their discharge pressure, which will be discussed in detail in later chapters. A turbine converts thermal/pressure energy to mechanical energy. Examples are hydraulic, wind, and gas or steam turbines.

Among these machines, the fluid medium handled by the liquid pump, hydraulic turbine, fan, and wind turbine can be treated as an incompressible fluid. Hence, the change of thermodynamic properties, other than pressure, of these fluids can be ignored. In machines that handle gas or steam, the variation of thermodynamic properties, such as temperature, pressure, and density, has to be incorporated into flow and energy transfer analysis.

Fundamentals of Turbomachinery: Theory and Applications, Second Edition. Ryoichi S. Amano and William W. Peng.

Companion website: www.wiley.com/go/amano/turbomachinery2e

Table 1.1 Classification of Turbomachines.[1]

	Fluid Machines → Turbomachines; Positive-displacement machines; Others
	Turbomachines → Pumping devices; Turbines
Direction of energy transfer	Pumping devices; Turbines
Type of fluid (liquid/gas)	Pump, fan, blower, compressor; Hydraulic, wind, gas, steam turbines
Flow direction	Axial-flow, mixed-flow, radial-flow
Mechanical arrangement	Horizontal- or vertical-axis pump, single- or double-suction pump/fan, single- or multistage pump/compressor, backward-, radial-, or forward-vane fan, full- or partial-admission turbine, horizontal- or vertical-axis wind turbine

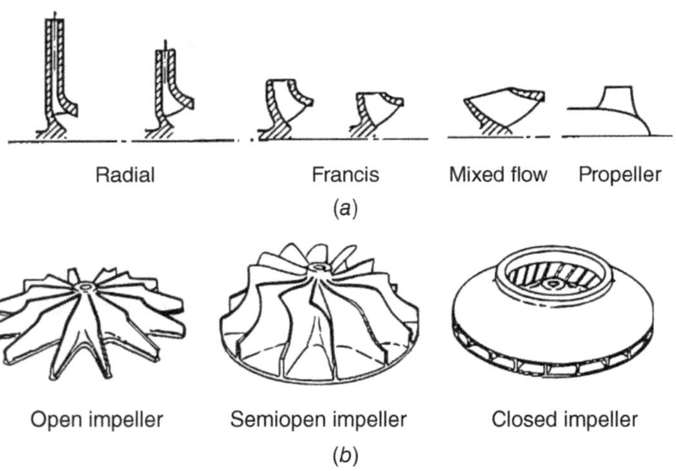

Figure 1.1 Types of turbomachines according to impeller type and flow direction through the impeller. ((*a*) Reproduced from Stepanoff, 1957/John Wiley & Sons;[1] (*b*) Reproduced from Gibbs, 1971/Ingersoll-Rand.[2])

Depending on the direction of flow in the impeller, concerning its rotating axis, the machines can be classified as radial-, mixed-, and axial-flow machines, as shown in Figure 1.1. The Francis turbine flows mostly in the mixed direction, except at discharge. In addition, the radial- and mixed-flow impellers can be closed, semi-open, or open-type, as shown.

Further classification of turbomachines according to their mechanical arrangement is also possible. This includes the basic single stage or the combinations of multistage, single suction or double suction, horizontal or vertical axis, and so on. Examples of the diverse types of arrangements are shown in Figure 1.2. These arrangements are chosen based on compactness or convenience of installation and maintenance.

Other classifications are based on inlet flow arrangements, such as full admission or partial admission, or the flow process in the rotor, either impulse (constant static enthalpy or pressure) or reaction machine. These classifications will be discussed in detail in later chapters when the individual types of machines are treated.

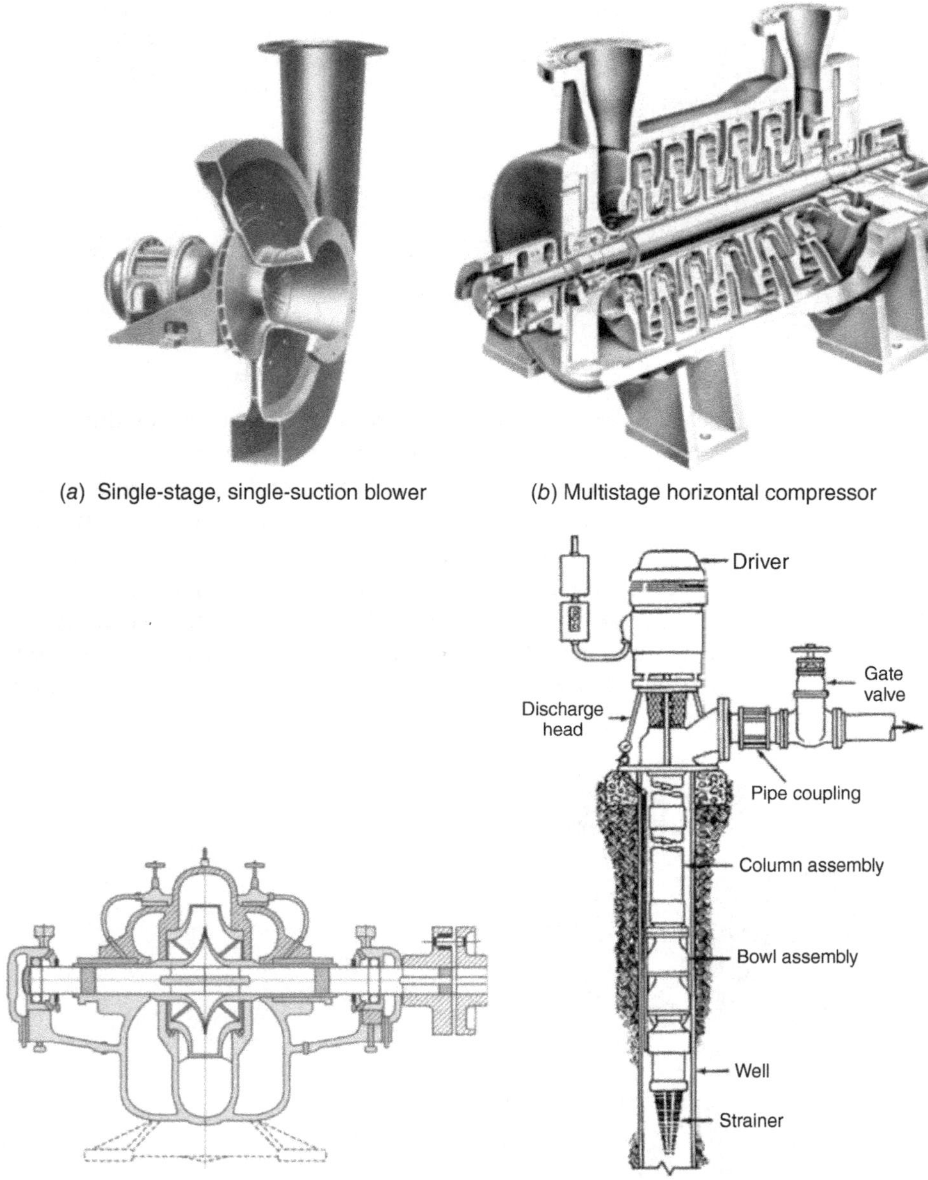

(*a*) Single-stage, single-suction blower
(*b*) Multistage horizontal compressor
(*c*) Double-suction pump
(*d*) Vertical pump

Figure 1.2 Types of turbomachines according to mechanical arrangements. ((*a*, *b*) Reproduced from Gibbs, 1971/Ingersoll-Rand; (*c*) Karassik et al., 1976/McGraw-Hill;[3] (*d*) Turbine Data Handbook, 1987/Weir Floway.[4])

1.3 Applications of Turbomachines

Turbomachines are widely used in power-generating and fluid-handling systems. In a typical central power plant, fossil, or nuclear, as shown in Figure 1.3,[5] the central component is a steam turbine, which is used to convert the thermal energy of steam into mechanical energy to drive an electric generator. Several pumps handle liquid water,

Figure 1.3 Typical central power plant with combined cycle. (Courtesy of Mechanical Engineering Power magazine, Nov. 1997, page 2; © Mechanical Engineering magazine, the American Society of Mechanical Engineers.)

including boiler-feed, condensate, and cooling-water circulating pumps. Turbomachines are also used in other energy-producing systems such as hydropower, wind power, and geothermal power installations.

The other major application of turbomachines is the gas turbine engines used in aircraft and industrial power plants. Multistage axial-flow gas turbines and compressors are exclusively used in high-power units. Centrifugal types are used in the smaller engines of propulsion systems for ground, marine, and air vehicles. A typical case is the automotive engine shown in Figure 1.4.[6] In the fluid-handling systems found in many industries, different types of pumps, fans, blowers, and compressors are employed to pressurize and transport the liquid or gas. Typical examples are the heating, ventilation, and air-conditioning (HVAC) systems shown in Figure 1.5[7] and water supply, water treatment, irrigation, oil production, oil refinery, gas transport, chemical process, and many other industries.

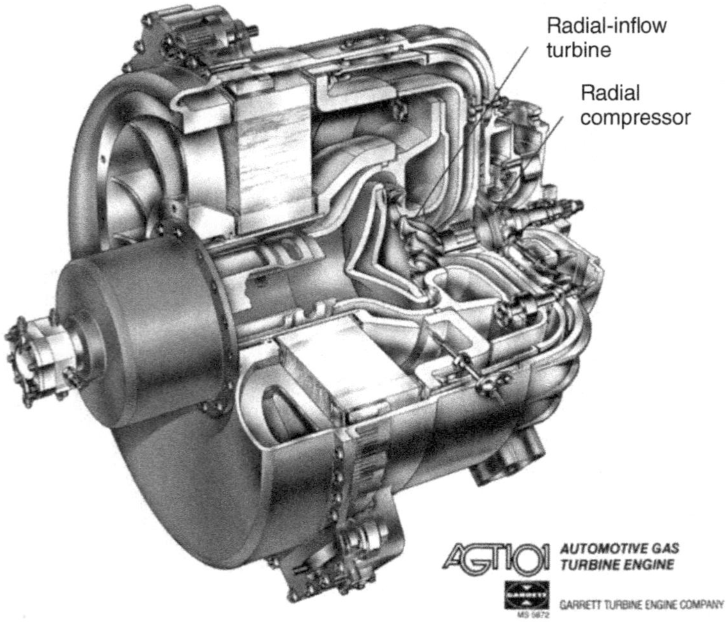

Figure 1.4 Automotive gas turbine engine. (Reproduced with permission from *Garrett/Ford AGT101 Advanced Gas Turbine Program Summary*, Garrett Turbine Engine Co., Honeywell Aerospace, Phoenix, AZ.)

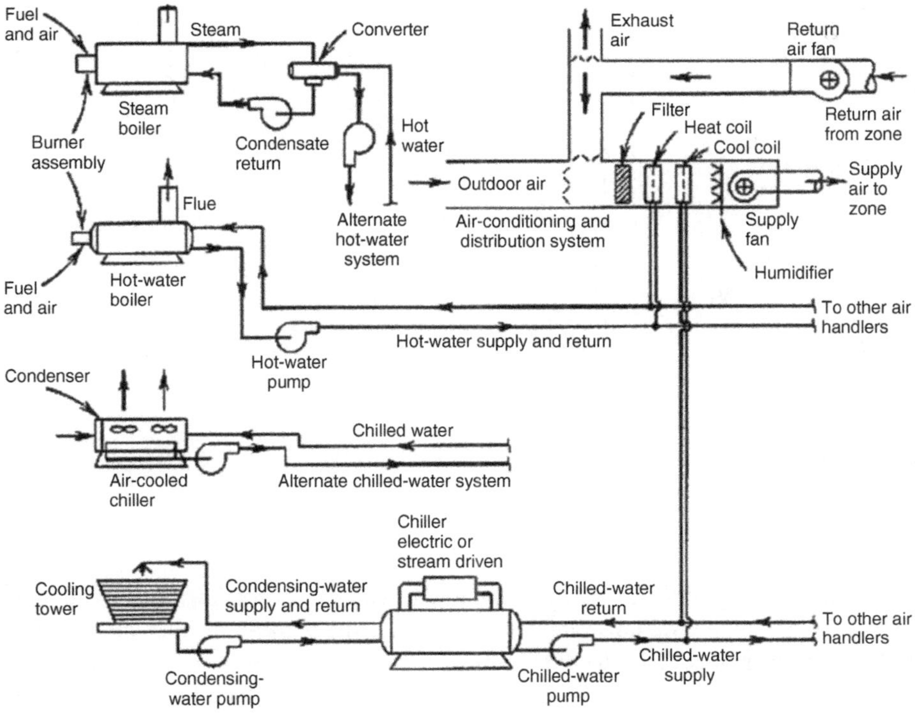

Figure 1.5 Turbomachines used in a typical commercial HVAC system. (Reproduced from McQuiston et al., 2005/John Wiley & Sons.)

1.4 Performance Characteristics

As an energy conversion device, a turbomachine is characterized by several parameters. These parameters and their relationship with machine geometry and dimensions based on the principles of fluid mechanics and thermodynamics are the main topics in this text.

The main parameters that characterize a turbomachine are input and output power, rotating speed, efficiency, through flow rate and inlet, outlet fluid properties, etc. In pumping devices such as liquid pumps, fans, or compressors, the output pressure is used to overcome the friction loss in the load, which is characterized by pressure loss versus flow rate. Hence, the performance of a typical pumping device is expressed in terms of the pressure rise Δp (or head rise H) versus the volumetric flow rate Q, or mass flow rate m, at a constant rotating speed N, as shown in Figure 1.6. The operating condition varies with a throttle valve at the discharge. In most cases, the input shaft power and efficiency are also included in this diagram. The overall efficiency is defined as the ratio of output power to input shaft power:

$$\eta = \frac{P_o}{P_s}, \tag{1.1}$$

where P_o is the output hydraulic power (product of volumetric flow rate and pressure rise), and P_s is the shaft power (product of angular velocity and torque of the shaft), that is,

$$P_o = Q\Delta p \quad \text{and} \quad P_s = \omega\tau.$$

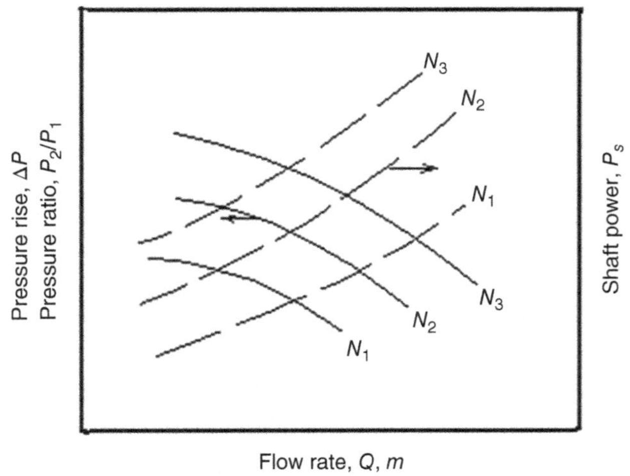

Figure 1.6 Typical pump, fan, and compressor performance curves at constant rotating speed.

For a fan or blower, the pressure rise is expressed in terms of the water head, either total or static head, and the flow rate is expressed in terms of the volumetric flow rate at the inlet since the density can vary slightly. In a compressor, the performance is normally expressed in the outlet–inlet pressure ratio p_2/p_1 versus the mass flow rate at a constant rotating speed. The adiabatic efficiency is expressed as the ratio of ideal enthalpy increase along the isentropic process over the actual enthalpy increase, $\eta_{ad} = \Delta h_s/\Delta h$.

At the high flow rate, the operation is limited by cavitation in pumps and choking due to shock waves in compressors. The low flow rate is limited by surging, a strong flow reversal at the inlet due to boundary layer separation. This problem is more severe in a compressor than in a pump.

The performance of turbines is also expressed in terms of head, rotating speed, output-shaft power, efficiency, and discharge flow rate. The loads, such as an electric generator or other mechanical machinery, are characterized by the input shaft power versus rotating speed. Hence, the basic turbine performance curve is plotted in terms of the output torque or shaft power versus the rotating speed, as shown in Figure 1.7. The head or inlet condition is usually fixed and is a function of the hydraulic installation or the combustion chamber condition for the gas/steam

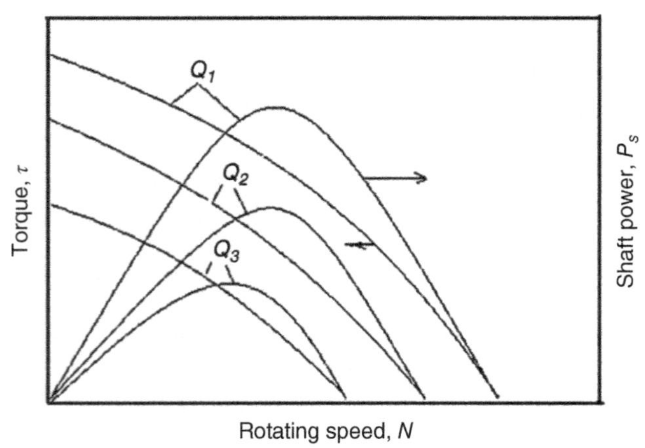

Figure 1.7 Typical basic performance curves of a turbine (torque and shaft power versus rotating speed at constant inlet condition).

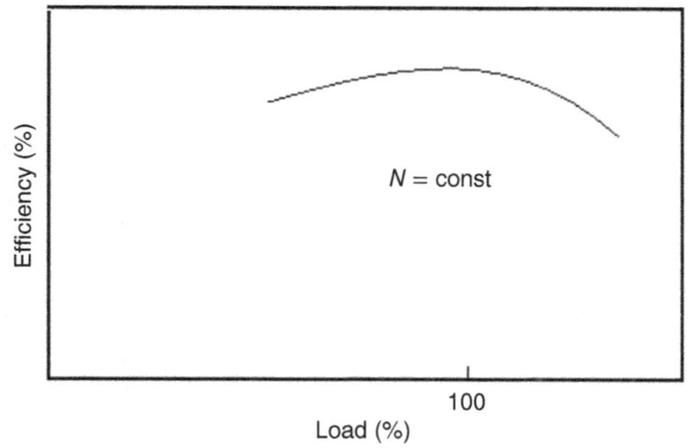

Figure 1.8 Turbine performance regarding efficiency versus load (with constant rotating speed).

turbine. The regulation is obtained by varying the flow rate using the gate or nozzle position. With a given inlet condition, the rotating speed varies by adjusting the load. In practice, sometimes, turbines are required to drive a constant-speed machine with a variable load. Hence, a performance diagram of efficiency versus load with a constant rotating speed is also frequently provided, as shown in Figure 1.8. A detailed discussion of these curves will be covered in later chapters.

1.5 Method of Analysis

The flow through a typical turbomachine is normally three-dimensional and turbulent and is either compressible or incompressible. The fluid medium can occasionally be a two-phase or two-component mixture of liquid, vapor, gas, and solid particles. Due to these complicated flow processes, especially inside the rotating impeller, analysis based on the first principles of three-dimensional fluid mechanics and thermodynamics is more difficult. In most cases, it requires a numerical method with a computer. In this entry-level treatment, the simplified one-dimensional or integral flow analysis form will emphasize the basic physical processes.

Dimensional analysis is widely used in the study of turbomachines. Specifically, dimensional analysis results can be applied to the correlation of experimental test data and to scale up the model test results to predict the prototype performance. Detailed procedures and the applications of dimensional analysis to turbomachines will be discussed in the next chapter.

For a completely new design, preliminary analysis must still be performed based on fluid mechanics and thermodynamics principles with some simplifications. The energy transfer equation, the so-called Euler equation, relates to the energy transfer rate between the flowing fluid and the rotating impeller of a turbomachine, which can be derived from the momentum equation of fluid mechanics. The momentum equation is expressed in integral form applied to the control volume enclosing the impeller, as shown in Figure 1.9. The Euler equation relates the rate of mechanical energy input or output of the shaft with the flow properties and geometric dimensions at the inlet and outlet of the impeller. The losses due to friction and three-dimensional effects through flow passages must be empirically estimated. These two types of analyses for various kinds of turbomachines will be discussed in detail in the next chapter. The flow processes in turbomachines can be treated as either the internal flow in a channel or the external flow over an airfoil, depending on the type of machine. The flow passages in radial- and mixed-flow machines are long, and internal flow models are used. In axial-flow machines, external flows over airfoils with the interference factor included are appropriate.

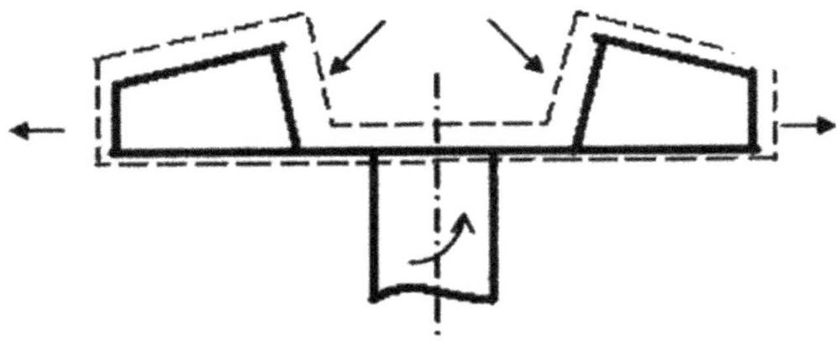

Figure 1.9 Control surface enclosing an impeller.

A quasi-three-dimensional flow analysis has been accomplished in more advanced analyses for the design operating condition. However, under off-design conditions, most analyses are still empirical or semiempirical. In recent years, computational fluid dynamics (CFD) software has become available and affordable for designing and analyzing various types of turbomachines.[8] Detailed flow analyses at different machine parts can be performed before the final design is fixed.

Example 1.1E A centrifugal pump is used to pump the oil with a specific gravity (s.g.) of 0.72. It requires 20.5 hp of shaft power when the flow rate is 385 gpm with an efficiency of 83%. Determine the pressure rise in terms of pounds per square inch, head of water, and head of oil pumped.

Solution
From

$$P_s = 20.5 \text{ hp} = 20.5 \times 550 = 11,275 \text{ ft-lb}_\text{f}/\text{s},$$

$$Q = 385 \text{ gpm} = \frac{385}{449} = 0.857 \text{ ft}^3/\text{s}, \quad \eta = \frac{Q\Delta p}{P_s},$$

we have

$$\Delta p = \frac{\eta P_s}{Q} = \frac{0.83 \times 11,275}{0.857 \times 144} = 75.8 \text{ psi}.$$

Also, from $\Delta p = \rho g \Delta H$, we have

$$\Delta H_w = \frac{75.8 \times 144}{62.4} = 175 \text{ ft of water},$$

$$\Delta H_\text{oil} = \frac{\Delta H_w}{\text{s.g.}} = \frac{175}{0.72} = 243 \text{ ft of oil}.$$

Example 1.1S A centrifugal pump is used to pump the oil with a specific gravity of 0.72. It requires 15.3 kW of shaft power when the flow rate is 87.4 m^3/h with an efficiency of 83%. Determine the pressure rise in kilopascals (kPa), the head of water, and the head of the oil pump.

Solution
From $P_h \equiv Q\,\Delta p = \eta P_s$, we have

$$\Delta p = \frac{\eta P_s}{Q} = \frac{0.83 \times 15.3 \times 1000}{87.4/3600} \text{ (N-m/s)/(m}^3\text{/s)}$$

$$= 523 \times 10^3 \text{ N/m}^2 = 523 \text{ kPa}.$$

Also from $\Delta p = \rho g\,\Delta H$, we have

$$\Delta H_w = \frac{523 \times 1000}{998 \times 9.81}(\mathrm{N/m^2})/[(\mathrm{kg/m^3})(\mathrm{m/s^2})],$$

$$= 53.4 \text{ m of water,}$$

$$\Delta H_{\text{oil}} = \frac{\Delta H_w}{\text{s.g.}} = \frac{53.4}{0.72} = 74.2 \text{ m of oil.}$$

Example 1.2E A centrifugal fan delivers air of 12,000 cubic feet per minute (cfm) measured at the inlet to an air duct. If the total resistance of the duct system is 2.5 in. of water at this flow rate and the total efficiency of the fan is estimated to be 85%, determine the shaft power input to the fan in horsepower (the discharge area of the duct back to ambient atmosphere is 3.5 ft^2).

Solution

The air flow velocity at discharge is $V = 12{,}000/(60 \times 3.5) = 57.14$ ft/s; hence the dynamic head can be calculated from

$$\frac{\rho_a V^2}{2} = \frac{0.0762 \times (57.14)^2}{32.2 \times 2} = 3.86 \ \mathrm{lb_f/ft^2} = \rho_w g H_v,$$

or $H_v = 0.062$ ft $= 0.74$ in. of water, or the total head $H_t = 2.5 + 0.74 = 3.24$ in. of water. So, we have the shaft power:

$$P_s = \frac{Q\Delta p_t}{\eta_t} = \frac{12,000 \times 3.24 \times 62.4}{60 \times 12 \times 0.85} = 3964.2 \text{ ft-lb}_\mathrm{f}/\mathrm{s} = 7.2 \text{ hp.}$$

Example 1.2S A centrifugal fan delivers 340 cubic meters per minute (cmm) of air measured at the inlet to an air duct. If the total resistance of the duct system is 6.35 cm of water at this flow rate and the total efficiency of the fan is estimated to be 85%, determine the shaft power input to the fan in kilowatts (kW) (the discharge area of the duct back to ambient atmosphere is 0.325 m^2).

Solution

The airflow velocity at discharge is $V = 340/(60 \times 0.325) = 17.4$ m/s. Hence, the dynamic head can be calculated from

$$\frac{\rho_a V^2}{2} = \frac{1.22 \times 17.4^2}{2}(\mathrm{kg/m^3})(\mathrm{m/s})^2 = 184.7 \ \mathrm{N/m^2} = 184.7 \text{ Pa.}$$

Converting static pressure, we have $\Delta p = \rho_w g H = 998 \times 9.81 \times 0.0635$ $(\mathrm{kg/m^3})(\mathrm{m/s^2})\mathrm{m} = 621.7$ Pa and the total pressure $\Delta p_t = 621.7 + 184.7 = 806.4$ Pa. So, we have the shaft power:

$$P_s = \frac{Q\Delta p_t}{\eta_t} = \frac{340 \times 806.4}{60 \times 0.85}$$

$$= 5376 \ (\mathrm{m^3/s})(\mathrm{N/m^2}) = 5.376 \text{ kW.}$$

Example 1.3E A hydropower site has a net head of 295 ft and an available water flow capacity of 148 ft^3/s. If a turbine rotating at 1800 rpm with an efficiency of 87% is to be installed, determine the total output power and the torque.

Solution

From

$$\Delta p = \rho g H = 62.4 \times 295 = 18,408 \ \mathrm{lb_f/ft^2} \quad \text{and} \quad P_s = \eta Q \Delta p,$$

we have

$$P_s = 0.87 \times 148 \times 18,408 = 2.37 \times 10^6 \text{ lb}_\text{f}\text{-ft/s} = 4309 \text{ hp},$$

$$\tau = \frac{P_s}{\omega} = \frac{2.37 \times 10^6}{1800 \times 2\pi/60} = 12.57 \times 10^3 \text{ ft-lb}_\text{f}.$$

Example 1.3S A hydropower site has a net head of 90 m and an available water flow capacity of 4.2 m^3/s. If a turbine rotating at 1800 rpm with an efficiency of 87% is to be installed, determine the total output power and the torque.

Solution
From $\Delta p = \rho g H = 998 \times 9.81 \times 90 = 881{,}134 \text{ N/m}^2$ and $P_s = \eta Q \Delta p$, we have

$$P_s = 0.87 \times 4.2 \times 881,134 = 3.22 \times 10^6 \text{ N-m/s} = 3220 \text{ kW},$$

$$\tau = \frac{P_s}{\omega} = \frac{3.22 \times 10^6}{1800 \times 2\pi/60} = 17.1 \times 10^3 \text{ N-m}.$$

1.6 Historical Evolution of Turbomachines

1.6.1 Hydraulic Pump

The development of modern turbomachines started in the eighteenth century. In 1705, Denis Papin published full descriptions of centrifugal blowers and pumps. However, crude centrifugal pumps were used in the United States until the early nineteenth century. In 1839, W. D. Andrews added a volute, and in 1875, a vaned diffuser was added and patented by Osborne Reynolds of England. Since then, it has been called a "turbine pump."

1.6.2 Blower/Compressor

In 1884, Charles Parsons patented an axial-flow compressor. Three years later, he produced a three-stage centrifugal compressor for ship ventilation. In 1899, he made an 81-stage axial-flow compressor with 70% efficiency. However, he had problems with the axial-flow machines in the next few years and returned to making the centrifugal machines in 1908. During this period, efforts on compressor development were also conducted by August Rateau in France. Continued work on compressor development was primarily in gas turbine engine development.

1.6.3 Gas/Steam Turbines

The Greek geometrician Hero devised the first steam turbine in 62 A.D. A simple closed, spherical vessel mounted on bearings discharges steam from a boiler with one or more pipes tangentially at the vessel's periphery, as shown in Figure 1.10. He called it an Aeolipile (wind ball). It is a pure reaction machine. Much later, in 1629, Giovanni de Branca in Italy developed an impulse-type steam turbine similar to a horizontal water wheel (also shown in Figure 1.10).

In 1791, John Barber of Britain was granted the world's first patent on the gas turbine, which consisted of all the elements of the modern gas turbine except the compressor, which was a reciprocating type.

Not until the early nineteenth century did steam turbines attract any interest in power generation. In 1831, William Avery in the United States produced Hero's steam turbine to drive circular saws. In 1848, Robert Wilson

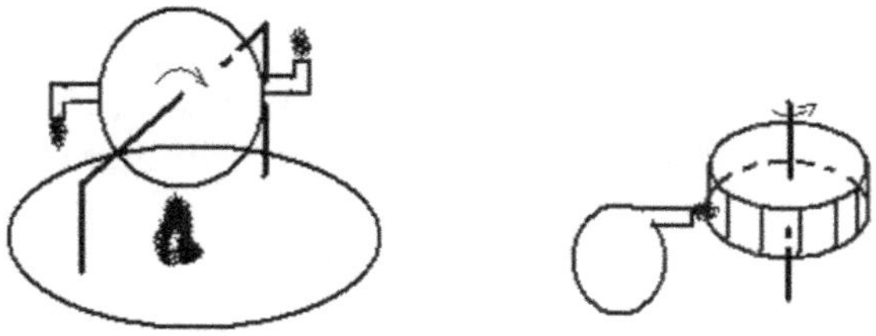

(a) Early reaction turbine (b) Early impulse steam turbine

Figure 1.10 Ancient steam turbines by Hero and Giovanni de Branca.

of Scotland patented a radial-inflow steam turbine. In 1875, Osborne Reynolds invented the turbine pump at the University of Manchester in England and made a multistage axial-flow steam turbine running at 12,000 rpm. In 1884, Charles Parsons, also of England, made a multistage axial-flow reaction turbine running at 18,000 rpm to produce 10 hp. He also tried but failed to produce a multistage radial-inflow turbine because of mechanical problems. In the following few years, he devoted his effort to further developing axial-flow machines. His machines were used for marine propulsion and electrical power generation.

In the early stage of gas turbine engine development, the failure was mostly due to the difficulty of designing an efficient compressor (pumping liquid water in a steam turbine engine is easier). To produce a positive net output power, the turbine output power must be greater than the power required by the compressor. This can be achieved by having a higher efficient compressor or a higher gas inlet temperature for the turbine.

In 1903, Aegidus Eilling, in Norway, constructed the world's first gas turbine with a net power output of 11 hp. His machine consisted of a 6-stage centrifugal compressor and a single-stage radial-inflow turbine. In France, August Rateau, in 1905, designed a gas turbine with a total power output of 400 hp. It consisted of a 25-stage centrifugal compressor with an intercooler and a 2-stage axial-flow turbine of impulse type.

With the further development and improvement of the gas turbine, the following milestones of aviation have been achieved:

1) In 1918, Sanford A. Moss, USA, an engineer with General Electric Co., came close to inventing a jet engine with his turbosupercharger, which used hot gases from the engine exhaust to drive a turbine that, in turn, drove a centrifugal compressor to supercharge the engine.
2) In 1920, Alan A. Griffith of England developed a theory of turbine design based on gas flow past airfoils rather than through passages. Griffith subsequently worked for many years for Rolls-Royce Ltd.
3) In 1930, Royal Air Force Lieutenant Sir Frank Whittle patented a jet-driven gas turbine engine with a single-stage centrifugal compressor and a single-stage turbine. His invention of the gas turbine engine powered the first British jet, the Gloster E.28/39, which made its first flight on May 15, 1941.
4) On August 27, 1939, the world's first jet engine-powered flight of Heinkel He 178 was completed in Germany.
5) On July 27, 1949, the world's first commercial jet airline, de Havilland Comet 1 of England, made its first flight.
6) On May 25, 1953, the world's first supersonic flight was made by the US Air Force F-100 fighter plane.
7) On December 31, 1968, the first commercial supersonic flight was made by the Russian TU-144, followed by the British French Concord flight on March 2, 1969. Only fourteen went into service, with the last of them retired in 2003; the Concorde proved to be a technical miracle but an economic disaster.

   In the past three decades, efforts have been made to increase the turbine inlet temperature with better materials and blade cooling. These efforts have resulted in thermal efficiency being increased from around 30 to 46% (GE's CF6-80E engine with a turbine inlet temperature of 1370°C in December 2003). Further improvement in thermal cycle efficiency can be achieved by combining the gas turbine and steam turbine in a combined-cycle plant.

1.6.4 Hydraulic Turbines

The Romans introduced the paddle-type water wheel, a pure impulse turbine, around 70 B.C. for grinding grain. The study of water wheels with systematic modeling was introduced by the British experimenter John Smeaton in the eighteenth century. He achieved maximum efficiency of 60%.

In France, where there are more rivers, the active development of water wheels was conducted in the early nineteenth century. In 1832, Benoit Fourneyron designed a radial-outflow machine to produce 50 hp with 85% efficiency. The activities moved to the United States when Uriah Boyden added a vaneless radial diffuser to this type of machine and achieved 88% efficiency. In 1851, James Francis designed a radial-inflow turbine, which is known as the Francis turbine today. During the same period, James Thomson in Britain worked on a more efficient radial-inflow turbine with a spiral inlet casing and adjustable inlet guide vanes.

1.6.5 Wind Turbine

Some simple versions of the windmill were used in Babylonia and China as early as 2000 B.C. Hero of Greece also described the horizontal-axis windmill with sails as aerodynamic surfaces. By the twelfth century, the windmill was introduced to Europe by both Arabs and the Crusaders returning from the Near East. In the nineteenth century, small multibladed windmills were exceedingly popular for grain grinding and water pumping in American farms.

In the 1970s, the US Energy Research and Development Administration (the predecessor of the Department of Energy) launched a series of research-and-development (R&D) projects on wind turbines, with the power output ranging from 100 to 2000 kW. The results of these projects have provided a foundation for the design, production, and operation of today's commercial wind turbines.

1.7 Organization of the Book

The types of turbomachines classified in Table 1.1 can also be depicted in a three-dimensional coordinate system as follows:

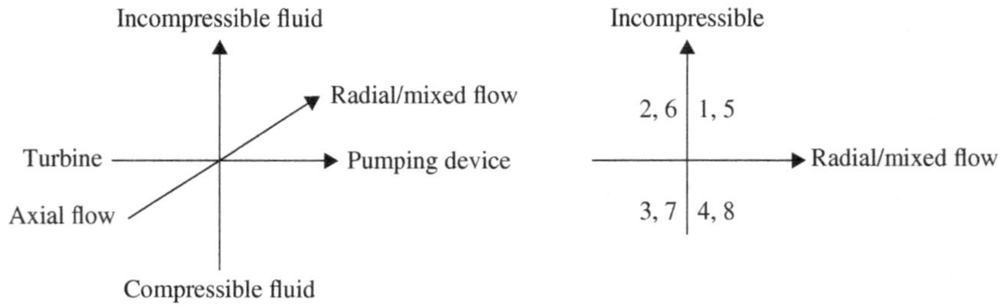

Compressible flow analysis requires the usage of thermodynamic parameters and processes. Flow through a radial/mixed-flow machine is similar to the channel flow, while that through an axial-flow machine can be treated as the external flow over the airfoils. The performance characteristics of a turbine are different from those of a pumping device.

In the diagrams above, each quadrant will represent a type of machine. Furthermore, each type of machine was developed by different groups of people and industries, and some parameters are unique for a certain type of machine. Hence, the presentation in a turbomachinery book is not straightforward and can be organized in several different ways, according to these three axes. Each way has its own merit.

Organization according to flow type is convenient for fluid dynamics researchers but is confusing for entry-level readers. They are more familiar with the machines according to the function (pumping devices or turbines) and the fluid medium handled (liquid or gas). So, in this book, we start with a general discussion of all turbomachines in the first chapter. After that, the readers should be comfortable with the different flow types. Dimensional Analysis, Similitude and Energy Transfer in Turbomachinery are provided in Chapter 2. Then, we proceed on to pumps and fans, both centrifugal and axial types, in Chapters 3 and 4 (quadrants 1, 2). With the low static pressure rise, axial-flow fans can be treated as incompressible fluid machines. Chapter 4 consists of Centrifugal and Axial flow Compressors.

The review of thermodynamics in Appendix A can be used as a reference. But it is recommended that at least the first five sections (A.1 to A.5) be reviewed before covering the centrifugal fan, blower, and compressor in Chapters 4 (quadrants 3, 4). In Chapter 5, the axial-flow gas turbine is covered (quadrant 7). It is integrated with a compressor in the gas turbine engine, which is a major topic in the study of turbomachinery. They are used in lower-power engines. Axial-flow steam turbines (quadrant 7) are covered in Chapter 6. Gas turbines are covered in Chapter 5 with the new topic of Manufacturing Challenges, Cooling Technologies, and Heat Transfer Sustainable Energy Systems.

Many concepts are similar to those of gas turbines. But some parameters and performance characteristics are different.

These topics have received renewed interest in recent years as renewable energy becomes an increasingly significant component of the global energy landscape, driven by concerns over global warming. Therefore, hydro turbines are discussed in Chapter 7, which covers common types such as impulse, reaction, and rim drive turbines. Chapter 8 focuses on wind turbines, including their applications and the challenges associated with icing. In response to the growing role of artificial intelligence (AI) in turbine design, a new Chapter 9 has been added, dedicated to AI in Turbomachinery Design and Applications. Additionally, due to the high demand for computational fluid dynamics (CFD) methods in turbine design, the former Appendix C has been expanded and elevated to Chapter 10.

For each type of machine, the following items are covered:

1) Theory based on the simplified fluid mechanics principles (one dimensional or integral form of equations)
2) Preliminary design procedure using basic theory and empirical formula/criteria (some of these sections can be skipped, depending on the instructor's and students' interest)
3) Ideal performance characteristics based on theory.
4) Actual performance characteristics with the modification due to loss mechanism and other flow processes (sample curves/tables published by the manufacturers are included)
5) Engineering applications and machine selection procedure (some of these sections can be read by the students themselves to save class time).

Since this book is primarily for entry-level readers, advanced topics on detailed design using a computer are not covered. However, some references are cited for those who want to pursue further studies on a particular machine or a design project. Also, some websites related to turbomachines are provided.

Both the International System (SI) and English system units are used in this book. Since turbomachinery is an applied subject, most of the information obtained from the industry is in the English system, although the trend is moving toward SI. Detailed discussions on the dimensions and units are given in basic engineering texts. They are also briefly discussed in Chapter 2. In the first three chapters, every example is worked out in both systems. After that, some are worked out in the English system and some in SI.

The prerequisite for using this book is a first course in fluid mechanics and thermodynamics at the undergraduate level. For some schools, if the basic turbomachinery principles are covered in fluid mechanics, Chapters 2 and 3 may be skipped or just briefly reviewed. Sections A.6 to A.10 in Appendix A are included for those students who plan to pursue more advanced studies on compressors and gas turbines.

References

1 Stepanoff, A. J., *Centrifugal and Axial Flow Pumps*, John Wiley & Sons, New York, 1957.
2 Gibbs, C. W. (Ed.), *Compressed Air and Gas Data*, 2nd ed., Ingersoll-Rand Co., Phillisburg, NJ, 1971.
3 Karassik, I. J., Krutzsch, W. C., Fraser, W. H., and Messina, J. P. (Eds.), *Pump Handbook*, McGraw-Hill, New York, 1976.
4 Weir Floway, Inc., *Floway Turbine Data Handbook*, 1st ed., Weir Floway, Fresno, CA, 1987.
5 Falcioni, J. G. (Ed.) *ASME, Mechanical Engineering Supplement*, American Society of Mechanical Engineers, New York, November 1997.
6 *Garrett/Ford AGT101 Advanced Gas Turbine Program Summary*, Garrett Turbine Engine Co., Phoenix, AZ, 1985.
7 McQuiston, F. C., Parker, J. D., and Spitler, J. D., *Heating, Ventilating & Air Conditioning*, 6th ed., John Wiley & Sons, New York, 2005.
8 CFD Software for turbomachine design: www.adapco.com; www.numeca.com; Concepts NREC.com; Fluent.com etc.

2

Basic Theories

2.1 Similitude

2.1.1 Dimensions and Dimensional Homogeneity

Dimensional analysis is a method for grouping several relevant parameters describing a certain physical phenomenon or process into a reduced number of nondimensional parameters. Experimental results can be more easily correlated with these nondimensional parameters. The test results from a model can be scaled up to predict the performance of a prototype machine. It is an important tool, especially for the study of turbomachinery, fluid mechanics, and heat transfer, because these subjects are more empirical in nature. Experimental studies are very important in understanding the processes or phenomena, such as turbulent flow, boundary layer separation, and flow with a two-phase mixture.

Although dimensional analysis is a powerful tool, it can only be used to group the relevant parameters into a smaller number of nondimensional parameters. The functional relationship still has to be obtained from the correlation of experimental test data.

The parameters describing a physical process are expressed in terms of some primary dimensions. If the process concerned is a mechanical system, the primary dimensions involved are either mass M, length L, and time t or force F, length L, and time t. If the process involves a system with heat transfer, then temperature T is also a primary dimension. The other dimensions expressed in terms of the combination of these primary dimensions are called secondary or derived dimensions. Examples are given in velocity (L/t), density (M/L^3), pressure (F/L^2), energy (ML^2/t^2), and so on.

In an equation describing a certain physical process or phenomenon such as

$$Q_1 = f(Q_2, Q_3, \ldots, Q_n), \tag{2.1}$$

the Q_i's are the relevant parameters. The dimensions of each term in the equation expressed as

$$f_1(Q_i) + f_2(Q_i) + \cdots = 0 \tag{2.2}$$

should be the same. This is called dimensional homogeneity. For example, in Bernoulli's equation, $p/\rho + V^2/2 + gZ = \text{const}$, we have the dimension of each term as

$$\frac{P}{\rho} = \frac{F/L^2}{M/L^3} = \frac{(ML/t^2)/L^2}{M/L^3} = \left(\frac{L}{t}\right)^2,$$
$$V^2 = \left(\frac{L}{t}\right)^2, \quad gZ = \left(\frac{L}{t^2}\right)L = \left(\frac{L}{t}\right)^2. \tag{2.3}$$

The resultant dimension of each term is $(L/t)^2$. In fact, this is also a dimension for energy per unit mass since it can be converted as $(L/t)^2 = (ML/t^2)(L/M) = (FL/M)$, where FL is the dimension for energy or work.

Fundamentals of Turbomachinery: Theory and Applications, Second Edition. Ryoichi S. Amano and William W. Feng.

Companion website: www.wiley.com/go/amano/turbomachinery2e

Another illustration can be shown with the ideal gas equation, $p = \rho RT$. The dimension on the right-hand side is $\rho RT = (M/L^3)[FL/(MT)](T) = (F/L^2)$, which is a dimension for pressure. This rule can be used to check the validity of an equation and associated conversion factors if there is any doubt.

To assign a numerical value for the dimensional parameter, a basic unit of measure has to be adopted. There are two systems of units commonly used: the English system (e.g., inch, pound) and the International System of Units (SI, Système International), which includes units such as meter, kilogram, and second. To avoid confusion in using the English system, it is important to remember that when relating parameters through Newton's second law, $F = Ma$, the units are related as $1\ \text{lb}_\text{f} = 1\ \text{slug} \times 1\ \text{ft/s}^2$, just like $1\ \text{N} = 1\ \text{kg} \times 1\ \text{m/s}^2$ in SI. The "slug" is a unit of mass, and one slug is equal to $32.2\ \text{lb}_\text{m}$. If the mass is given in pound mass, it is necessary to be converted to "slug." Some examples are given below:

1) To calculate dynamic pressure, $p_d = \rho V^2/2$, for air with a density of $0.075\ \text{lb}_\text{m}/\text{ft}^3$ moving at 120 ft/s, we have

$$\rho = 0.075\ \text{lb}_\text{m}/\text{ft}^3 = \frac{0.075}{32.2} = 0.00233\ \text{slug/ft}^3.$$

Hence,

$$p_d = \frac{0.00233 \times 120^2}{2} = 16.77\ (\text{slug/ft}^3)(\text{ft/s})^2$$
$$= 16.77\text{lb}_\text{f}/\text{ft}^2 = 0.116\text{lb}_\text{f}/\text{in.}^2 = 0.116\text{psi}.$$

2) To convert some thermal energy of air from temperature $T_0 = 600\ \text{R}$ to $T_1 = 550\ \text{R}$ into kinetic energy, the resulting air flow velocity can be calculated as

$$\frac{1}{2}V^2 = C_p(T_0 - T_1) = 0.24 \times 32.2 \times 778(600 - 550)$$
$$= 300{,}619(\text{Btu/slug-R})\ (\text{lb}_\text{f}\text{-ft/Btu})\ (\text{R})$$
$$= 300{,}619\text{lb}_\text{f}\text{-ft/slug} = 300{,}619\ (\text{ft/s})^2.$$

Hence, $V = 775.4\ \text{ft/s}$, where $C_p = 0.24\ \text{Btu/lb}_\text{m}\text{-R} = 0.24 \times 32.2\ \text{Btu/slug-R}$.

3) To convert the pressure of liquid with a specific gravity of 0.85 from 50 psi to the head of liquid, we use $p = \rho gH$ or

$$H = \frac{50 \times 144}{(0.85 \times 62.4/32.2) \times 32.2}$$
$$= 135.7(\text{lb}_\text{f}/\text{ft}^2)/[(\text{slug/ft}^3)(\text{ft/s}^2)] = 135.7\ \text{ft},$$

where $\rho = 0.85 \times 62.4 = 53.04\ \text{lb}_\text{m}/\text{ft}^3 = 1.647\ \text{slug/ft}^3$ is the liquid density and $g = 32.2\ \text{ft/s}^2$ is the gravitational acceleration.

4) To convert the dynamic pressure of air in the first example to water gage (wg), we use $p_d = \rho_w gh$ or

$$h = \frac{16.77}{(62.4/32.2) \times 32.2} = 0.269\ (\text{lb}_\text{f}/\text{ft}^2)/(\text{lb}_\text{f}/\text{ft}^3) = 0.269\ \text{ft}$$
$$= 3.23\ \text{in.wg},$$

where $\rho_w = 62.4/32.2 = 19.38\ \text{slug/ft}^3$ is the water density, and $\rho_w g = 62.4\ \text{lb}_\text{f}/\text{ft}^3$ is the water-specific weight (weight per unit volume). Manipulation of units as shown in these examples is important to assure a correct numerical result.

2.1.2 Buckingham Pi Theorem

The Buckingham pi theorem states that the number of independent nondimensional parameters needed to correlate the variables for a given process is $n - m$. Here, n is the number of relevant dimensional parameters involved, and m is the number of primary dimensions. As indicated in the previous section, there are three primary dimensions (M, L, t or F, L, t) in a mechanical system, and there are four primary dimensions (M, L, t, T or F, L, t, T) in a thermal-mechanical system with heat transfer involved. The procedure can be described as follows:

1) List n number of relevant dimensional parameters for the physical process based on experience and physical intuition, say $Q_1, Q_2, Q_3, \ldots, Q_n$.
2) Select a set of m (three or four) number of primary dimensions.
3) Select m (three or four) number of repeating parameters from Q_i's listed in step 1, say Q_{r1}, Q_{r2}, Q_{r3} (and Q_{r4} if applicable).
4) Express each Q_i in terms of the primary dimensions.
5) Express each of the nondimensional parameters π_i's in terms of one of the nonrepeating parameters and all repeating parameters, such as (if $m = 3$)

$$\pi_1 = Q_1 Q_{r1}^a Q_{r2}^b Q_{r3}^c, \quad \pi_2 = Q_2 Q_{r1}^d Q_{r2}^e Q_{r3}^f, \ldots, \quad \pi_{n-m} = Q_{n-m} Q_{r1}^x \ldots.$$

   There will be $m\ (n - m)$ number of undetermined exponents.
6) Group the same primary dimension in the expression of each π_i with $a, b, c, d, \ldots$ in the exponents.
7) Since the π_i's are nondimensional and the primary dimensions are independent of each other, the combined exponents for each primary dimension should always be equal to zero. Hence, $m\ (n - m)$ simultaneous equations can be obtained for the $m\ (n - m)$ undetermined $a, b, c, d, \ldots$.
8) Solve the simultaneous equations for $a, b, c, d, \ldots$.
9) Express the π_i's with these exponents numerically.
10) Make sure all π_i's are truly nondimensional.

The above procedures can be applied to the processes relevant to turbomachinery. The results will be used in later chapters.

2.1.2.1 Case 1

The drag force F_d (or lift force F_L) on a submerged object (such as an airfoil) due to a flowing fluid is a function of the fluid density ρ, the flow velocity V, the characteristic dimension of the object, D, and the fluid viscosity μ. In other words, the function can be expressed as $F_{d/L} = f(\rho, V, D, \mu)$. Thus, we have $n = 5$, $m = 3$, and we can expect two nondimensional parameters needed to correlate the test data for this problem. Next, the dimensions of these relevant parameters are expressed in terms of the primary dimensions as follows:

$$F_{d/L} = ML/t^2, \quad \rho = M/L^3, \quad V = L/t, \quad D = L, \quad \text{and} \quad \mu = M/(Lt). \tag{2.4}$$

Next, several repeating parameters equal to the number of primary dimensions are selected from the list of dimensional parameters. These repeating parameters should be expressible in terms of the primary dimensions, and no two repeating parameters can have the same net dimensions, differing only in the exponents. For example, do not include both length (L) and area (L^2) as the repeating parameters. The repeating parameters chosen may appear in all the nondimensional parameters obtained. Consequently, the dependent parameter should not be included in these repeating parameters.

In this example, ρ, V, and D are chosen as the repeating parameters. The nondimensional parameters can be formed in terms of these repeating parameters with the exponents to be determined, that is,

$$\pi_1 = \rho^a V^b D^c F_{d/L} = \left(\frac{M}{L^3}\right)^a \left(\frac{L}{t}\right)^b L^c \left(\frac{ML}{t^2}\right),$$
$$\pi_2 = \rho^d V^e D^f \mu = \left(\frac{M}{L^3}\right)^d \left(\frac{L}{t}\right)^e L^f \left(\frac{M}{Lt}\right). \tag{2.5}$$

By combining the like dimensions, we obtain $\pi_1 = M^{a+1} L^{-3a+b+c+1} t^{-b-2}$ and $\pi_2 = M^{d+1} L^{-3d+e+f-1} t^{-e-1}$.

Since both π_1 and π_2 are nondimensional, each expression of the exponents must be equal to zero. Hence, the following simultaneous equations are obtained:

$$\begin{aligned} &a+1=0, \quad -3a+b+c+1=0, \quad -b-2=0, \\ &d+1=0, \quad -3d+e+f-1=0, \quad -e-1=0. \end{aligned} \tag{2.6}$$

The above equations result in $a=-1, b=-2, c=-2$ and $d=-1, f=-1, e=-1$. Therefore, we have

$$\pi_1 = \frac{F_{d/L}}{\rho V^2 D^2}, \qquad \pi_2 = \frac{\mu}{\rho V D}. \tag{2.7}$$

Hence, the drag and lift due to the flow over an object can be expressed in terms of the nondimensional parameters as

$$\pi_1 = f(\pi_2) \qquad \text{or} \qquad \frac{F_{d/L}}{\rho V^2 D^2} = f\left(\frac{\mu}{\rho V D}\right). \tag{2.8}$$

Here, $C_{d/L} = F_{d/L}/(\rho V^2 D^2/2)$ is called the drag or lift coefficient, and $\text{Re} = 1/\pi_2 = \rho VD/\mu$ is the Reynolds number. It is shown from the test that under high Re with turbulent flow, the Re effect becomes insensitive. The forces mainly vary with the shape and orientation of the object with respect to the flow direction (i.e., angle of attack in an airfoil). Since shape and orientation are nondimensional parameters, they are not included in the dimensional analysis.

2.1.2.2 Case 2

For a centrifugal pump operating at a constant rotating speed with sufficient suction pressure, the discharge pressure Δp (or head H) will be a function of the flow rate Q, rotating speed N, fluid density ρ, fluid viscosity μ, and impeller outlet diameter D; that is, $\Delta p = f(Q, N, \rho, \mu, D)$.

Hence, we have $n=6$, $m=3$. Selecting ρ, D, and N as the repeating parameters, the nondimensional parameters can be expressed as

$$\begin{aligned} \pi_1 &= \rho^a D^b N^c \Delta p = \left(\frac{M}{L^3}\right)^a L^b \left(\frac{1}{t}\right)^c \left(\frac{ML}{t^2L^2}\right), \\ \pi_2 &= \rho^d D^e N^f Q = \left(\frac{M}{L^3}\right)^d L^e \left(\frac{1}{t}\right)^f \left(\frac{L^3}{t}\right), \\ \pi_3 &= \rho^g D^h N^i \mu = \left(\frac{M}{L^3}\right)^g L^h \left(\frac{1}{t}\right)^i \left(\frac{M}{Lt}\right). \end{aligned} \tag{2.9}$$

Following similar steps as those in the previous case, we obtain

$$\pi_1 = \frac{\Delta p}{\rho D^2 N^2} = \frac{gH}{(DN)^2}, \qquad \pi_2 = \frac{Q}{D^3 N}, \qquad \pi_3 = \frac{\rho D^2 N}{\mu}. \tag{2.10}$$

Here, $\psi = \pi_1$ is called the head coefficient, $\phi = \pi_2$ is the flow coefficient, and $\text{Re} = \pi_3$ is the Reynolds number for the rotating impeller.

The functional relationship can be written as $\psi = f(\phi, \text{Re})$. However, from laboratory tests, it is found that the effect of the Reynolds number is insensitive for the conventional vane-type impeller pumps, pumping clear water, where the Reynolds number is very high. The situation is similar to the case of flow in a stationary channel: When the Reynolds number is high, the friction coefficient is insensitive, as shown in a Moody diagram.

However, it was also found from experience that two similar machines of significantly different sizes did have different efficiencies at the corresponding operating conditions. This is due to the different relative roughness of the machine surfaces. For the same absolute surface roughness, the large machine has lower relative roughness and hence less pressure loss due to skin friction. An empirical formula to correlate this effect is suggested in Ref. 1 as $(1-\eta_1)/(1-\eta_2) = (\text{Re}_2/\text{Re}_1)^n$, where $n = 0.1$ to 0.25. If the shaft power P_s is the parameter that we are interested in, following the same procedure, we obtain the corresponding power coefficient:

$$\Pi = \frac{P_s}{\rho D^5 N^3}, \qquad \Pi = f(\phi, \text{Re}). \tag{2.11}$$

The overall efficiency, defined as $\eta = P_o/P_s = Q\,\Delta p/P_s$, can also be expressed in terms of the head, flow, and power coefficients, that is, $\eta = \phi\psi/\Pi$. Hence, it is also a nondimensional parameter.

The nondimensional parameters derived above can be interpreted as the ratio of two relevant parameters. The drag coefficient C_d is the ratio of drag force over the flow inertia force. The term $\rho V^2 D^2$ can be written as $(\rho VD^2)V$ or $(\rho VA)V$ or $\dot{m}V$, where A is a characteristic flow passage area, and m is the characteristic mass flow rate. For the Reynolds number Re, we can rewrite it in the form $\text{Re} = \rho VD/\mu \sim \rho V^2D^2/(VD)[(V/D)/\tau] \sim [F_i/(VD)][(V/D)/(F_v/D^2)] \sim F_i/F_v$, where τ is the shear stress, and F_i and F_v are the characteristic inertia and viscous forces, respectively. The flow coefficient ϕ can be written in the form $\phi = Q/(ND^3) \sim Q/(NDA) \sim V/U$, where $V = Q/A$ and $U \sim ND$ are the characteristic flow velocity and impeller tip linear velocity. The head coefficient ψ can be written in the form $\psi = gH/(DN)^2 \sim gH/(U^2/2)$, where gH is the flow energy change per unit mass and $U^2/2$ is the linear kinetic energy of the rotating impeller per unit mass. The parameter "head" is generally used in pumps and hydraulic turbines. For compressible fluid machines, compressors, or gas/steam turbines, pressure and enthalpy per unit mass are used.

2.1.2.3 Case 3

If the fluid medium handled by the turbomachine is compressible, such as the condition found in compressors and gas/steam turbines, we can expect an additional nondimensional parameter, the Mach number, to emerge from dimensional analysis. This number is the ratio of characteristic flow velocity and acoustic speed in the fluid and can also be interpreted as the ratio of flow inertia forces and the forces needed to compress the fluid (or elastic force).

In these machines, we can neglect fluid viscosity, but the temperature change is important due to the compressibility. So, the relevant parameters are $p_{01}, p_{02}, RT_{01}, RT_{02}, \dot{m}, N$, and D. Here, the combined RT_0 is selected instead of T_0 itself since we are concerned about the internal energy of the fluid medium, not the heat transfer effect in the process, so the number of primary dimensions is still three (M, L, t). From these seven relevant parameters, two nondimensional parameters can be formed immediately, p_{02}/p_{01} and T_{02}/T_{01}. Hence, the relevant parameters can be reduced by two, or only $p_{01}, RT_{01}, \dot{m}, N$, and D will be used to form the other two nondimensional parameters. Selecting $\dot{m}$, N, and D as the repeating parameters, since the number of primary dimensions is three, the nondimensional parameters can be expressed as

$$\pi_1 = \dot{m}^a N^b D^c p_{01} = \left(\frac{M}{t}\right)^a \left(\frac{1}{t}\right)^b L^c \left(\frac{M}{Lt^2}\right) = M^{a+1}L^{c-1}t^{-(a+b+2)},$$

$$\pi_2 = \dot{m}^d N^e D^f RT_{01} = \left(\frac{M}{t}\right)^d \left(\frac{1}{t}\right)^e L^f \left(\frac{L}{t}\right)^2 = M^d L^{f+2} t^{-(d+e+2)}. \tag{2.12}$$

By setting the exponents to be zero, we have $a = -1, b = -1, c = 1, d = 0, e = -2, f = -2$ or $\pi_1 = p_{01}D/(\dot{m}N)$, and $\pi_2 = RT_{01}/(ND)^2$. We can also manipulate these nondimensional parameters into forms such that

$$\pi_2' = \frac{1}{\sqrt{\pi_2}} = \frac{ND}{\sqrt{RT_{01}}} \sim \frac{N}{\sqrt{T_{01}}}, \qquad \pi_1'' = \frac{1}{\pi_1\pi_2'} = \frac{\dot{m}\sqrt{(RT_{01})}}{p_{01}D^2} \sim \frac{\dot{m}\sqrt{T_{01}}}{p_{01}}. \tag{2.13}$$

For a given machine and fluid medium, D and R are constants, so the parameters used in the compressible fluid machines are normally expressed in the functional relationship

$$f\left(\frac{p_{02}}{p_{01}}, \frac{T_{02}}{T_{01}}, \frac{\dot{m}\sqrt{T_{01}}}{p_{01}}, \frac{N}{\sqrt{T_{01}}}\right) = 0. \tag{2.14}$$

Since the adiabatic efficiency η_{ad} of the compressor and gas/steam turbine can be expressed in terms of the pressure ratio and temperature ratio to be shown in later chapters, T_{02}/T_{01} can be replaced by η_{ad} in the above set of relevant nondimensional parameters. So, typically, the performance curves of these machines are plotted in terms of p_{02}/p_{01} and η_{ad} versus $\dot{m}\sqrt{T_{01}}/p_{01}$ with $N/\sqrt{T_{01}}$ as parameters. Also, the nondimensional mass flow and rotational speed coefficients can be written in the forms shown in Eq. (2.15). They are called flow Mach number M_F and rotational Mach number M_R:

$$\frac{\dot{m}\sqrt{RT_{01}}}{p_{01}D^2} = \frac{\rho AV_1}{\rho\sqrt{RT_{01}}D^2} \sim \frac{V_1}{a_{01}} = M_F, \quad \frac{ND}{\sqrt{RT_{01}}} \sim \frac{U}{a_{01}} = M_R, \tag{2.15}$$

where V_1 is the characteristic flow velocity at the inlet, and U is the characteristic impeller tip velocity.

2.1.2.4 Case 4

The output shaft power P_s of a hydraulic turbine is a function of water density ρ, head H, flow rate Q, rotor diameter D, and rotating speed N. However, the head is always combined with the gravitational constant g since gH can be treated as the available hydraulic energy per unit mass of water. Hence, we have $P_s = f(\rho,Q,gH,D,N)$. These six parameters can be reduced to three nondimensional parameters:

$$\pi_1 = P_S\rho^a(gH)^bD^C = \frac{ML^2}{t^3}\left(\frac{M}{L^3}\right)^a\left(\frac{L^2}{t^2}\right)^b L^c = M^{1+a}L^{2-3a+2b+c}t^{-3-2b},$$

$$\pi_2 = Q\rho^d(gH)^eD^f = \frac{L^3}{t}\left(\frac{M}{L^3}\right)^d\left(\frac{L^2}{t^2}\right)^e L^f = M^dL^{3-3d+2e+f}t^{-1-2e},$$

$$\pi_3 = N\rho^g(gH)^hD^i = \frac{1}{t}\left(\frac{M}{L^3}\right)^g\left(\frac{L^2}{t^2}\right)^h L^i = M^gL^{-3g+2h+i}t^{-1-2h}. \tag{2.16}$$

By setting the exponents to zero, we have $a + 1 = 0$, $2 - 3a + 2b + c = 0$, $3 + 2b = 0$, $d = 0$, $3 - 3d + 2e + f = 0$, $1 + 2e = 0$, $g = 0$, $-3g + 2h + i = 0$, and $1 + 2h = 0$, or $a = -1$, $b = -1.5$, $c = -2$, $d = 0$, $e = -0.5$, $f = -2$, $g = 0$, $h = -0.5$, and $i = 1$. Hence, the nondimensional parameters can be expressed as

$$\pi_1 = P_s\rho^{-1}(gH)^{-1.5}D^{-2} = \frac{P_s}{\rho D^2(gH)^{1.5}} \equiv P_1,$$

$$\pi_2 = Q(gH)^{-0.5}D^{-2} = \frac{Q}{D^2\sqrt{gH}},$$

$$\pi_3 = N(gH)^{-0.5}D = \frac{ND}{\sqrt{gH}} \sim \frac{R\omega}{\sqrt{gH}} \equiv \lambda, \tag{2.17}$$

where P_1 and λ are defined as the unit power and speed ratio, respectively. So P_1 can be plotted versus λ with the flow rate Q as the parameter, which is discussed in more detail in Chapter 7. The head, flow, and power coefficients can also be derived from $\psi = 1/\pi_3^2\psi = 1/\pi_3^2\psi = 1/\pi_3^2\psi = 1/\pi_3^2\psi = 1/\pi_3^2\psi = 1/\pi_3^2 = gH/(ND)^2$, $\phi = \pi_2/\pi_3 = Q/(ND^3)$, and $\Pi = \pi_1/\pi_3^3\Pi = \pi_1/\pi_3^3\Pi = \pi_1/\pi_3^3\Pi = \pi_1/\pi_3^3\Pi = \pi_1/\pi_3^3\Pi = \pi_1/\pi_3^3 = P_s/(\rho N^3D^5)$, respectively.

2.1.2.5 Case 5

For a wind turbine, the output shaft power P_s is a function of rotor diameter D, rotating speed N, wind velocity V, and air density ρ and can be written as $P_s = f(D,N,V,\rho)$. These five parameters can be reduced to two nondimensional parameters.

Selecting ρ, D, and N as the repeating parameters, the nondimensional parameters can be expressed as

$$\pi_1 = D^a N^b \rho^c P_s = L^a \left(\frac{1}{t}\right)^b \left(\frac{M}{L^3}\right)^c \left(\frac{ML^2}{t^3}\right) = M^{c+1} L^{a-3c+2} t^{-b-3},$$

$$\pi_2 = D^d N^e \rho^f V = L^d \left(\frac{1}{t}\right)^e \left(\frac{M}{L^3}\right)^f \left(\frac{L}{t}\right) = M^f L^{d-3f+1} t^{-e-1}. \tag{2.18}$$

By setting the exponents to zero, we have $c+1=0$, $a-3c+2=0$, $b+3=0$ and $f=0$, $d-3f+1=0$, $e+1=0$ or $a=-5$, $b=-3$, $c=-1$, $d=-1$, $e=-1$, and $f=0$. Hence,

$$\pi_1' = \frac{P_s}{\rho N^3 D^5}, \qquad \pi_2' = \frac{V}{DN}. \tag{2.19}$$

These two parameters can be rearranged into

$$\pi_1 = \frac{\pi_1'}{\left(\pi_2'\right)^3} = \frac{P_s}{\rho D^2 V^3} \sim \frac{P_s}{\frac{1}{2}\rho A V^3}, \qquad \pi_2 = \frac{1}{\pi_2'} = \frac{DN}{V} \sim \frac{R\omega}{V} \tag{2.20}$$

which are defined as the power coefficient C_p and tip speed ratio λ, respectively, which are discussed in Chapter 8.

2.1.3 Other Nondimensional Parameters for Turbomachines

Although the total number of independent nondimensional parameters for a given machine is fixed, they can be combined and manipulated to form other new nondimensional parameters, as shown in the last section. Another important one is the specific speed $\omega_s = NQ^{1/2}/(gH)^{3/4}$. It can be derived from the head and flow coefficients by expressing $\omega_s = \phi^{1/2}/\psi^{3/4}$ to eliminate the characteristic dimension D. Hence, the specific speed is a parameter that characterizes the geometry (shape) of the machine and is independent of size. It is also called a shape number. Sometimes, this parameter is also interpreted as the rotating speed of a machine which delivers one unit of head at one unit of flow rate for the pumping devices. For turbines, the shaft power is more important, and hence, the power-specific speed

$$\omega_{sp} = \frac{\Pi^{1/2}}{\psi^{5/4}} = \frac{NP_s^{1/2}}{\rho^{1/2}(gH)^{5/4}} \tag{2.21}$$

is commonly employed. It is important to remember that specific speed is defined at the maximum efficiency point of the machine. It can be used to select the type of machine for a given duty or to correlate a certain parameter versus different types of machines.

In principle, specific speed is nondimensional. However, dimensional quantities are used in industry. When gpm or cfs is used for flow rate, feet for head, rpm for rotating speed, and hp for shaft power, specific speed can be expressed in dimensional form as follows:

$$N_s = \begin{cases} NQ^{1/2}/(H)^{3/4} & \text{rpm gpm}^{1/2}/\text{ft}^{3/4} & \text{for pumps,} \\ NQ^{1/2}/(\text{sp})^{3/4} & \text{rpm cfm}^{1/2}/(\text{in.wg})^{3/4} & \text{for fans,} \\ NQ^{1/2}/(H_{\text{ad}})^{3/4} & \text{rpm cfs}^{1/2}/(\text{ft}-\text{lb}_\text{f}/\text{lb}_\text{m})^{3/4} & \text{for compressors,} \end{cases}$$

$$N_{sp} = NP_s^{1/2}/(H)^{5/4} \quad \text{rpm hp}^{1/2}/\text{ft}^{5/4} \quad \text{for hydraulic turbines,} \tag{2.22}$$

where H is the water head for pumps and hydraulic turbines, sp is the static pressure rise in water gage for fans, and H_{ad} is the adiabatic head for compressors.

The type of impeller and average peak efficiencies versus specific speeds for the pumps are shown in Figure 2.1 (Ref. 2). It is clear that for the same inlet diameter, the radial-flow impeller has a larger outlet diameter and smaller

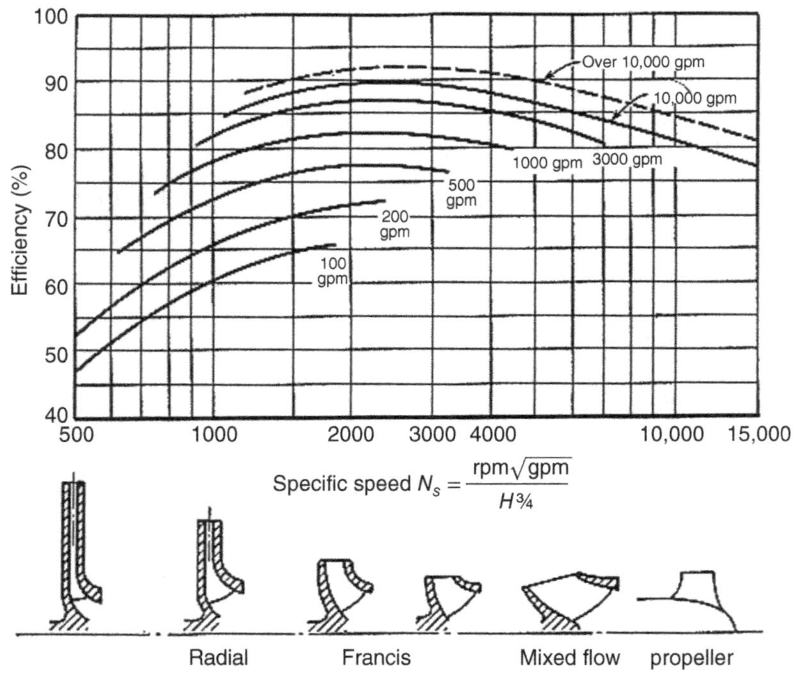

Figure 2.1 Variation of pump type and efficiency with specific speeds. (*Source:* Reproduced from Stepanoff[2]/John Wiley & Sons.)

flow passage area. So, it will deliver higher discharge pressure but can accommodate lower flow capacity for a given rotating speed. Hence, it has lower values of specific speed. The axial-flow type is just the opposite. Similar situations are found in other types of turbomachines, which are discussed later.

Another nondimensional parameter, also used in turbomachines, is specific diameter Δ_s, defined as

$$\Delta_s = \frac{\psi^{1/4}}{\phi^{1/2}} = \frac{D(gH)^{1/4}}{Q^{1/2}}, \tag{2.23}$$

or in dimensional form:

$$D_s = \begin{cases} DH^{1/4}/Q^{1/2} & \text{in.(ft)}^{1/4}/(\text{gpm})^{1/2} \quad \text{for pumps,} \\ D(sp)^{1/4}/Q^{1/2} & \text{in.(in.wg)}^{1/4}/(\text{cfm})^{1/2} \quad \text{for fans,} \\ D(H_{\text{ad}})^{1/4}/V_i^{1/2} & \text{ft(ft-lb}_\text{f}/\text{lb}_\text{m})^{1/4}/(\text{cfs})^{1/2} \quad \text{for compressors.} \end{cases} \tag{2.24}$$

Or SI system units can be used.

The specific diameter is a parameter to indicate the relative size of a machine for a given set of operating conditions. Or it can also be interpreted as the size of machine to deliver one unit of head at one unit of flow rate for the pumping devices. A correlation of Δ_s and ω_s developed by Csanady[3] for pumps and fans called the Cordier diagram is shown in Figure 2.2. It shows, again, that the radial-type impeller with a lower specific speed has a higher specific diameter and hence a larger diameter.

It should be remembered that in this diagram, the units for N, Q, H, and D have to be such that the resulting Δ_s and ω_s are nondimensional. With this correlation and that in Figure 2.1, the optimum size of a pump and its efficiency can be estimated for a set of operating requirements. The procedure using these correlations for the preliminary design and machine selection is discussed in later chapters.

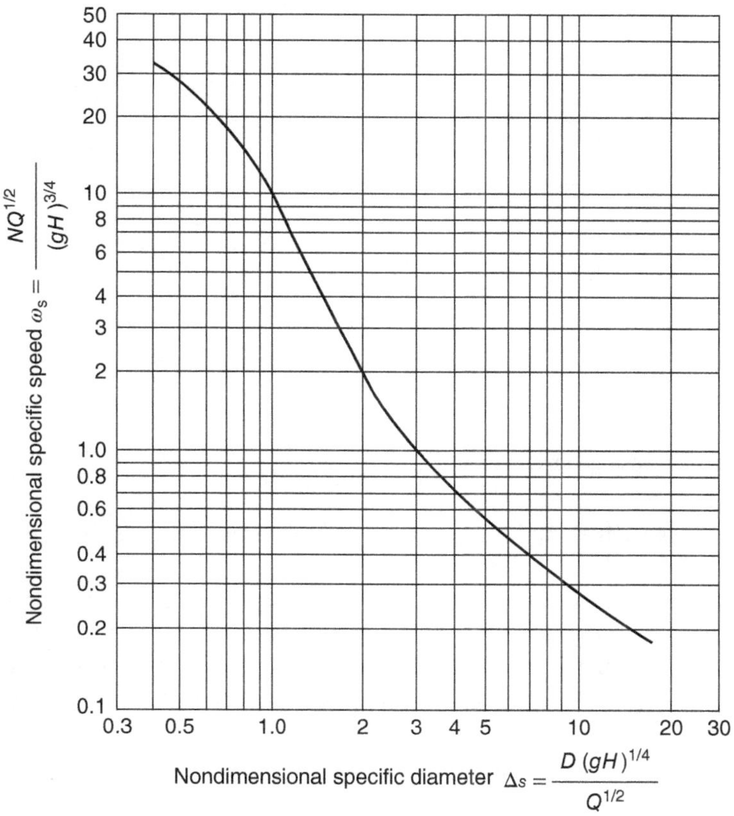

Figure 2.2 Cordier diagram for pumps and fans. (*Source:* Reproduced from Csanady[3]/McGraw-Hill.)

2.1.4 Similarity Laws

From the dimensional analysis illustrated in previous sections, the relevant nondimensional parameters for a given machine performance can be related through a certain function, that is, $\pi_1 = f(\pi_2, \pi_3, \ldots)$. The number of these nondimensional parameters is less than that of the corresponding dimensional parameters. Hence, it will be more convenient to correlate the test data in terms of the nondimensional parameters. Also, from these correlations, it will be possible to predict the performance of the same machine operating at different conditions or even to predict the performance of other geometrically similar machines operating at the corresponding conditions. These ideas will be demonstrated with centrifugal pumps in this chapter, although they also apply to other types of machines, which will be illustrated in later chapters.

From Case 2 in Section 2.2, it is shown that for a centrifugal pump, the head coefficient, flow coefficient, and Reynolds number are related through $\psi = f(\phi, \mathrm{Re})$. However, in practice, the Reynolds number effect is weak since flow through the pump is highly turbulent, and hence, we can write $\psi = f(\phi)$. Similarly, an expression for the power coefficient can be derived as $\Pi = g(\phi)$. If a pump is tested at a fixed rotating speed N, and the measured flow rate, discharge head, and shaft power are converted to flow coefficient, head coefficient, and power coefficient, a set of curves can be plotted, as shown in Figure 2.3. However, these curves will also be good for other rotating speeds of the same pump or other geometrically similar pumps. In fact, in exploration tests of a new pump, it is a good idea to run the machine at a few rotating speeds, if applicable. Then, a plot of these data in terms of the

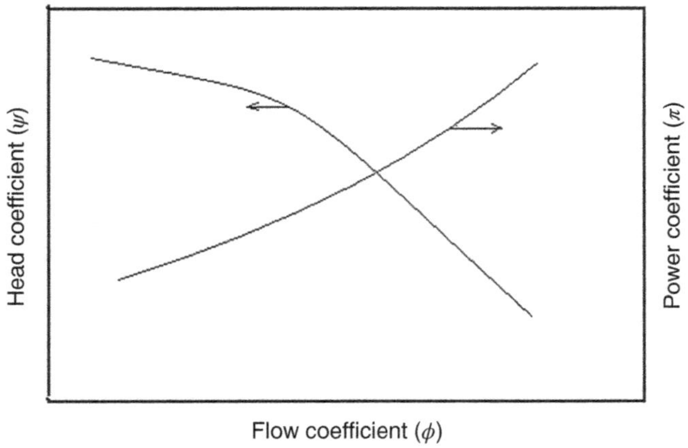

Figure 2.3 Typical pump performance curves in nondimensional parameters.

nondimensional parameters for different rotating speeds should collapse into a single curve. This can check the validity of tests.

Geometric similarity implies that all corresponding linear dimensions of two machines are in constant proportion, and all angular dimensions are the same for every component of the machines, such as the model and prototype shown in Figure 2.4. It also requires that the flow characteristics (or flow patterns) in two cases be the same, that is, kinematic similarity. The corresponding velocity diagrams should be similar, and no flow separation or cavitation is allowed in either case.

The relationships between model and prototype can be expressed explicitly as

$$\phi_p = \phi_m \quad \text{or} \quad \frac{Q_p}{D_p^3 N_p} = \frac{Q_m}{(D_m^3 N_m)},$$

$$\psi_p = \psi_m \quad \text{or} \quad \frac{gH_p}{(D_p N_p)^2} = \frac{gH_m}{(D_m N_m)^2},$$

$$\Pi_p = \Pi_m \quad \text{or} \quad \frac{P_{sp}}{(\rho_p D_p^5 N_p^3)} = \frac{P_{sm}}{\rho_m D_m^5 N_m^3}, \tag{2.25}$$

at the corresponding operating conditions of both machines. They are called similarity laws or affinity laws.

With these similarity laws, the test data of a small model machine can be used to predict the performance of a big prototype. Revision or correction on the design and the model can be made if necessary. Iterations may be needed

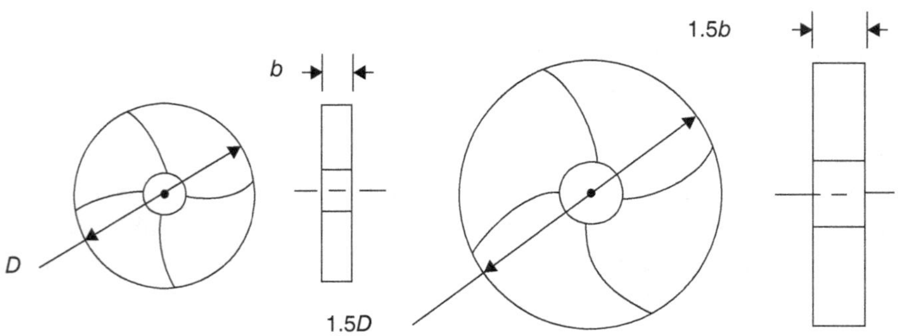

Figure 2.4 Example of model and prototype machines.

until a satisfactory design is achieved. This will save significant costs for new machine development. The following examples illustrate the associated calculations.

Example 2.1E A centrifugal pump operating at 1000 rpm is delivering 80 gpm of water at 150 psig and requires 8 hp of shaft power at the best efficiency point. Determine the flow rate, discharge pressure, required shaft power, and efficiency at the corresponding point when it is operating at 1500 rpm.

Solution

From $\phi_2 = \phi_1$ or $Q_2/(D^3N_2) = Q_1/(D^3N_1)$, we have

$$Q_2 = \frac{Q_1 N_2}{N_1} = \frac{80 \times 1500}{1000} = 120 \text{ gpm}.$$

From $\psi_2 = \psi_1$ or $\Delta p_2/(\rho D^2 N_2^2) = \Delta p_1/(\rho D^2 N_1^2)$, we have

$$\Delta p_2 = \Delta p_1 \left(\frac{N_2}{N_1}\right)^2 = 150\left(\frac{1500}{1000}\right)^2 = 337.5 \text{ psig}.$$

From $\Pi_2 = \Pi_1$ or $P_{s2}/(\rho D^5 N_2^3) = P_{s1}/(\rho D^5 N_1^3)$, we have

$$P_{s2} = P_{s1}\left(\frac{N_2}{N_1}\right)^3 = 8\left(\frac{1500}{1000}\right)^3 = 27 \text{ hp}.$$

The efficiencies are the same at the corresponding points, so

$$\eta_2 = \eta_1 = \frac{Q_1 \Delta p_1}{P_{s1}} = \frac{80 \times 0.00223 \times 150 \times 144}{550 \times 8} = 0.875 = 87.5\%.$$

Example 2.1S A centrifugal pump operating at 1000 rpm is delivering 18.2 m^3/h of water at 10.3 bars and requires 6 kW of shaft power at the best efficiency point. Determine the flow rate, discharge pressure, required shaft power, and efficiency at the corresponding point when it is operating at 1500 rpm.

Solution

From $\phi_2 = \phi_1$ or $Q_2/(D^3N_2) = Q_1/(D^3N_1)$, we have

$$Q_2 = Q_1\left(\frac{N_2}{N_1}\right) = 18.2\left(\frac{1500}{1000}\right) = 27.3\text{m}^3/\text{h}.$$

From $\psi_2 = \psi_1$ or $\Delta p_2/\left(\rho D^2 N_2^2\right) = \Delta P_1/\left(\rho D^2 N_1^2\right)$, we have

$$\Delta p_2 = \Delta p_1 \left(\frac{N_2}{N_1}\right)^2 = 10.3\left(\frac{1500}{1000}\right)^2 = 23.2 \text{ bars}.$$

From $\Pi_2 = \Pi_1$ or $P_{s2}/(\rho D^5 N_2^3) = P_{s1}/(\rho D^5 N_1^3)$, we have

$$P_{s2} = P_{s1}\left(\frac{N_2}{N_1}\right)^3 = 6\left(\frac{1500}{1000}\right)^3 = 20.2 \text{ kW}.$$

The efficiencies are the same at the corresponding points, so

$$\eta_2 = \eta_1 = \frac{Q_1 \Delta p_1}{P_{s1}} = \frac{(18.2/3600) \times 10.3 \times 10^2}{6} = 0.868 = 85.8\%.$$

Example 2.2E In order to design a pump to deliver a head of 320 ft at a flow rate of 350 ft^3/s (corresponding to the best efficiency point) when rotating at 600 rpm, a model is built and tested. If the model test flow rate at the best efficiency point is measured as 5 ft^3/s and hydraulic output power is 30 hp, determine the model rotating speed and the model/prototype scale ratio.

Solution

From the model hydraulic output power and measured flow rate, the discharge head can be determined as $\Delta p_m = 30 \times 550/5 = 3300\,\text{lb}_\text{f}/\text{ft}^2$. Hence, $H_m = 3300/62.4 = 52.88$ ft. From the similarity law, the ratio of discharge heads for the prototype and model is $H_p/H_m = (N_p/N_m)^2\,(D_p/D_m)^2 = 320/52.88 = 6.05$. So, $(N_p/N_m)(D_p/D_m) = 2.46$. Also, the flow rate ratio is $Q_p/Q_m = (N_p/N_m)(D_p/D_m)^3 = 350/5 = 70$. Therefore, we have

$$\frac{D_p}{D_m} = 5.33, \qquad \frac{N_p}{N_m} = 0.461, \qquad N_m = \frac{600}{0.461} = 1301\ \text{rpm}.$$

Example 2.2S In order to design a pump to deliver a head of 97.5 m at a flow rate of 9.9 m³/s (corresponding to the best efficiency point) when rotating at 600 rpm, a model is built and tested. If the model test flow rate at the best efficiency point is measured as 0.14 m³/s and hydraulic output power is 22.4 kW, determine the model rotating speed and the model/prototype scale ratio.

Solution

From the model hydraulic output power and measured flow rate, the discharge head can be determined as $\Delta p_m = 22.4/0.14 = 160\,\text{kN-m/(s-m}^3\text{/s)} = 160\,\text{kPa}$. Hence, $H_m = 160 \times 10^3/(998 \times 9.81) = 16.3$ m. From the similarity law, the ratio of discharge heads for the prototype and model is $H_p/H_m = (N_p/N_m)^2\,(D_p/D_m)^2 = 97.5/16.3 = 5.98$. So, $(N_p/N_m)(D_p/D_m) = 2.45$. Also, the flow rate ratio is $Q_p/Q_m = (N_p/N_m)(D_p/D_m)^3 = 9.9/0.14 = 70.7$.

Therefore, we have

$$\frac{D_p}{D_m} = 5.4, \qquad \frac{N_p}{N_m} = 0.46, \qquad N_m = \frac{600}{0.46} = 1315\ \text{rpm}.$$

Example 2.3E Select the type of pump to pump water of 250 gpm to overcome a resistance of 18 ft water head in the pipe if a motor of 1500 rpm is available. Also estimate its approximate size and efficiency.

Solution

From the dimensional specific speed $N_s = NQ^{0.5}/H^{0.75} = 1500 \times 250^{0.5}/18^{0.75} = 2714\,\text{rpm(gpm}^{0.5}\text{)/ft}^{0.75}$ and Figure 2.1, we select a Francis-type pump, and the efficiency is estimated to be $\eta = 74\%$. To find the approximate size, Figure 2.2 has to be used. The nondimensional specific speed is calculated from $\omega_s = \omega Q^{0.5}/(gH)^{0.75}$, where

$$\omega = \frac{N\pi}{30} = \frac{1500\pi}{30} = 157\ \text{rad/s}, \qquad Q = 250 \times 0.00223 = 0.5575\ \text{ft}^3/\text{s},$$

or

$$\omega_s = \frac{157 \times (0.5575)^{0.5}}{(32.2 \times 18)^{0.75}} = 0.992.$$

From Figure 2.2, it is obtained that

$$\Delta_s = \frac{D(gH)^{0.25}}{Q^{0.5}} = 3.1 \qquad \text{or} \qquad D = \frac{3.1 \times (0.5575)^{0.5}}{(32.2 \times 18)^{0.25}} = 0.472\ \text{ft} = 5.7\ \text{in}.$$

Example 2.4E A hydraulic turbine running at 1800 rpm with an estimated efficiency of 92% is to be designed for a hydropower site with a net head of 160 ft and a flow rate of 25 ft³/s. A model of 1:5 scale running at 3600 rpm is to be tested in the laboratory. Estimate its flow rate, head, and output power.

Solution

From $\phi = Q_m/(D_m^3 N_m) = Q_p/(D_p^3 N_p)$ and $\psi = gH_m/(D_m^2 N_m^2) = gH_p/(D_p^2 N_p^2)$, we have $Q_m = Q_p(D_m/D_p)^3 (N_m/N_p) = 25 \times (1/5)^3(3600/1800) = 0.4\ \text{ft}^3/\text{s}$ and $H_m = H_p(D_m/D_p)^2(N_m/N_p)^2 = 160 \times (1/5)^2 \times (3600/1800)^2 = 25.6$ ft.

Also, we have $P_s = \eta\ Q\rho\ gH = 0.92 \times 0.40 \times 62.4 \times 25.6/550 = 1.07$ hp since the efficiencies of both model and prototype should be the same.

Example 2.4S A hydraulic turbine running at 1800 rpm with an estimated efficiency of 92% is to be designed for a hydropower site with a net head of 50 m and a flow rate of 42 m^3/min. A model of 1:5 scale running at 3600 rpm is to be tested in the laboratory. Estimate its flow rate, head, and output power.

Solution

From $\phi = Q_m/(D_m^3 N_m) = Q_p/(D_p^3 N_p)$ and $\psi = gH_m/(D_m^2 N_m^2) = gH_p/(D_p^2 N_p^2)$, we have $Q_m = Q_p(D_m/D_p)^3(N_m/N_p) = 42 \times (1/5)^3(3600/1800) = 0.67$ cmm and $H_m = H_p(D_m/D_p)^2(N_m/N_p)^2 = 50 \times (1/5)^2 \times (3600/1800)^2 = 8$ m. Also, we have $P_s = \eta\ Q\rho\ gH = 0.92 \times (0.67/60) \times 998 \times 9.81 \times 8 = 804.6\ (m^3/s)(kg/m^3)(m/s^2)m = 804.6$ W since the efficiencies of both model and prototype should be the same.

2.2 Energy Transfer in Turbomachines

2.2.1 Review of Fluid Mechanics Related to Turbomachinery

Turbomachines are devices that transfer energy from or to flowing fluid. Hence, studies on turbomachinery are based on the physical laws governing flowing fluids. These laws and some basic flow processes related to turbomachines are reviewed briefly in this section. A more detailed discussion can be found in fluid mechanics texts.

If the flow passage concerned is long relative to the characteristic dimension normal to the flow direction, it can be treated as an internal flow (Figure 2.5), such as those in radial- and mixed-flow machines. Otherwise an external flow will be appropriate, such as those in axial-flow machines.

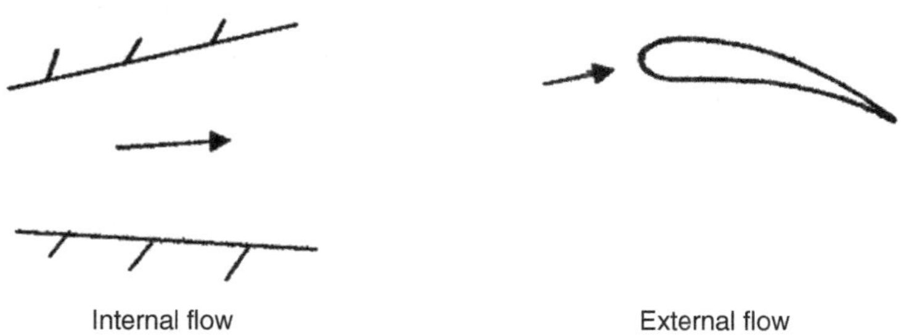

Figure 2.5 Typical flow types in turbomachines.

The fluid can be treated as either incompressible or compressible, depending on the characteristic Mach number of the flow. Some special phenomena of compressible flow are reviewed in Appendix A. In this book, all flows will be considered as steady-state flow, that is, the flow properties throughout the field do not change with respect to time. Complicated unsteady-state or transient-flow processes are beyond the scope of this book.

It is more convenient to apply the physical laws to a continuous flowing fluid in a fixed space in the field, called "control volume." A control volume is enclosed by several control surfaces. Across these surfaces, fluids may pass through, heat may transfer through, and mechanical work may be done to or extracted from the fluid through a rotor, as shown in Figure 2.6.

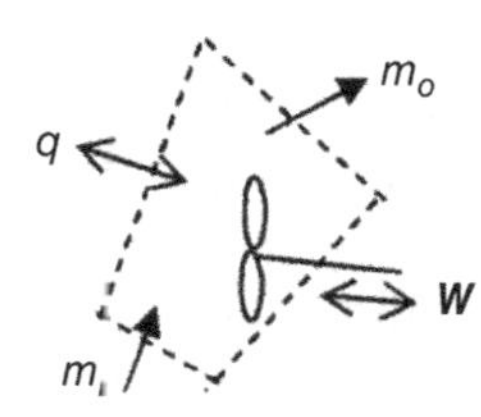

Figure 2.6 Control volume for flow analysis.

2.2.1.1 Conservation of Mass (Continuity Equation)

If there is no chemical reaction taking place (and no source or sink) inside the control volume, the total mass of fluid substance inside should be constant with respect to time. Fluid flows passing through the control surfaces should be balanced out, that is, $\int cs\ dm = 0$ or $\int cs\ \rho\ V{\cdot}dA = 0$, where ρ and V are the fluid density and velocity at the control surface with area dA, and m is the mass flow rate. The dot product between V and dA means only the component of V in the direction normal to the control surface will contribute to the product. Both V and A are the vector quantities, and the positive direction of A is directed away from the control volume. The term "cs" on the integral means the integration is carried over all control surfaces. In the simple case of a control volume with only two open control surfaces, that is, inlet and outlet A_i and A_o, the equation can be written as

$$\rho_i\ V_{ni}\ A_i = \rho_o V_{no} A_o. \tag{2.26}$$

2.2.1.2 Conservation of Momentum (Momentum Equation)

Newton's second law, $F = ma$, can also be described *as* $F = d(mV)/dt$, where mV is defined as the momentum or linear momentum, also a vector quantity. In the case of flowing fluid in a control volume, Newton's second law can be written as $\sum Fj = \int_{cs} Vdm = \int_{cs} V(\rho V\, dA)$, where $\sum Fj$ is the summation of all external forces acting on the fluid inside the control volume. The pressure forces between fluid elements are internal forces and are balanced out and hence are not included. The subscript j and $\sum$ sign indicate that there can be several discrete forces acting on the fluid.

2.2.1.3 Conservation of Energy (First Law of Thermodynamics)

From observation of physical processes, it is concluded that energy cannot be created or destroyed. It can only be converted from one mode to another. The modes of energy concerned in the study of turbomachinery are (1) mechanical energy (or work) related to force and displacement, that is, $w = F{\cdot}\Delta s$ for linear motion or $w = \tau{\cdot}\Delta\theta$ for rotational motion, where τ is the torque, and $\Delta\theta$ is the angular displacement; (2) thermal energy associated with heat and pressure, that is, $h = u + p/\rho$, where h and u are the enthalpy and internal energy of fluid medium; (3) kinetic energy related to the motion of fluid, that is, $\rho V2/2$; and (4) potential energy due to gravitation, that is, gZ, where g is the gravitational acceleration, and Z is the elevation with respect to a reference level.

Applying the equation to the fluid inside a control volume, we have

$$q - P_s = \int_{cs} \left(h + \frac{1}{2}V^2 + gZ\right) \rho V \cdot dA, \tag{2.27}$$

where positive q is the rate of heat transfer into the fluid inside the control volume, and positive P_s is the rate of work or power done by the fluid to the rotor and solid surfaces through shear friction. In the study of turbomachinery, q and shear friction are always neglected since they are small compared with the other terms in the equation.

2.2.1.4 Second Law of Thermodynamics

It is also concluded from observation of physical processes that the efficiency of any heat engine (to be defined later) is always less than 100%. And entropy always increases in a process without heat transfer across the control surfaces, that is, an adiabatic process. This is so due to the irreversible mechanism, such as friction, in any process. Applying this law to the fluid flowing through a control volume, we have $s_o > s_i$, that is, fluid entropy at the outlet is always greater than that at the inlet.

The physical laws discussed above are applied to fluid in a finite-size control volume. They are also applicable when the control volume is reduced to an infinitesimal size. A set of governing equations in differential forms will then be obtained. These differential equations can be solved analytically or numerically for various types of flow processes. They are the domain of fluid mechanics studies. For our purposes, only a few flow processes relevant to turbomachinery will be discussed.

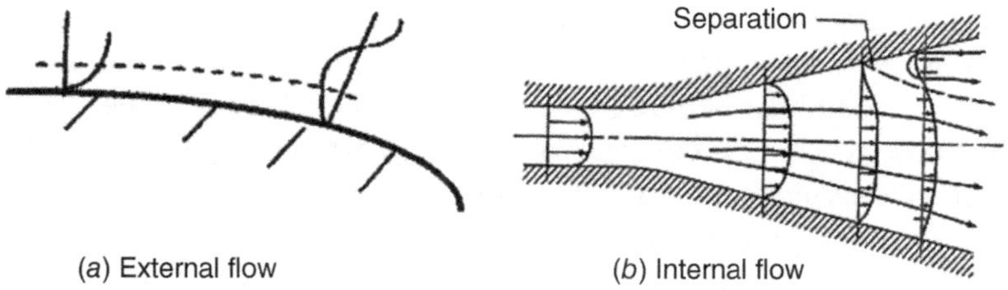

Figure 2.7 Boundary layer and its separation on solid surfaces. (*Source:* ASHRAE Handbook, Fundamentals[4]).

Due to the random motion of fluid particles, in either the microscopic or macroscopic scale, real fluid flow will suffer viscous shear forces between flow layers of different velocities and between fluid and adjacent solid surfaces. The magnitude of the shear force is proportional to the velocity gradient, that is, $\tau = \mu\, \partial U/\partial y$. The proportional constant μ depends on flow type, whether laminar or turbulent, and in general is governed by the Reynolds number. For laminar flow, μ varies with the type of fluid and its temperature. For turbulent flow, it varies significantly with the flow process and is more difficult to determine. In the vicinity of solid surfaces, to simplify the analysis, the flow field can be divided into the potential flow and boundary layer. Next to the solid surface, flow velocity increases rapidly from zero (in the no-slip condition) to the potential flow velocity within a thin layer. Most of the viscous effect is accounted for in this thin boundary layer. However, the boundary layer can separate from the solid surface under a certain adverse condition, as shown in Figure 2.7, and a change in potential flow pattern may result.

2.2.1.5 Typical Flow Types in Turbomachines

Flow Through a Channel The pressure loss due to viscous shear force in the flow through a channel can be evaluated with the expression $\Delta pf = (fL/D)\rho\, V^2/2$, where V is the average flow velocity, ρ is the fluid density, L is the length of the channel, and D is the characteristic dimension normal to flow passage. For a circular pipe, the diameter can be used, but for an arbitrary cross section, hydraulic diameter $Dh = 4A/P$ is used, where A and P are the area and perimeter of the cross section. If the cross-sectional area or geometry varies, an averaged value can be adopted.

Other than viscous shear force, the boundary layer separation from the internal wall surface is also a problem of main concern, especially for flow in a divergent channel, such as a diffuser, as shown in Figure 2.7. The adverse condition causing boundary layer separation is related to the rate of pressure increase in the flow direction. Hence, in designing a diffuser, a certain maximum divergent angle is imposed.

Flow Through a Bend Due to the centrifugal force and the interaction with boundary layers, flow through a bend will have a more complicated flow pattern (Figure 2.8). Some secondary flows normal to the main flow direction will occur. This will cause an extra pressure loss, which can be expressed as $\Delta p_f = K\rho V^2/2$, where K is the minor loss coefficient that generally depends on the radius ratio R/D, aspect ratio W/D for rectangular duct, angle of bend, and shape of bend. A vane installed inside the bend can reduce the loss since it tends to minimize the secondary flow. Numerical values have been determined empirically and can be found in fluid mechanics texts or a hydraulics handbook.

Figure 2.8 Flow through a bend.

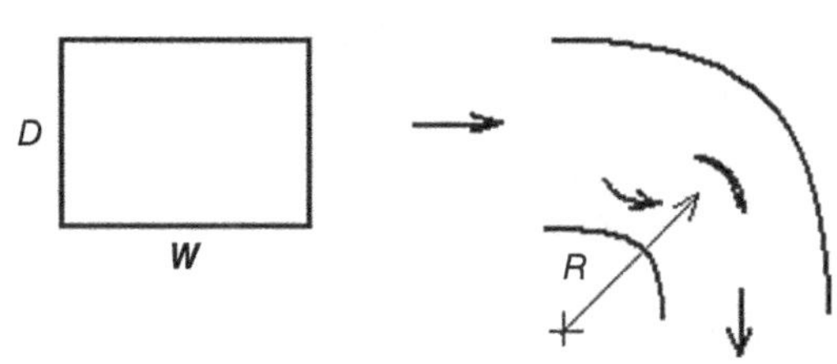

Flow Over Isolated Objects For flow over an isolated object, the viscous shear force between the fluid and object surfaces will result in a drag force on the object in the main flow direction. Due to the unbalanced fluid pressure distribution, additional forces on the object can result in any direction, depending on the geometry of the object and its orientation with respect to the main flow direction. If the object is moving with respect to the originally undisturbed stationary fluid, the main direction is the direction of object motion. The resultant force on the object can be decomposed into two directions. The one parallel to the main direction is called drag, and the other one perpendicular to the main direction is called lift. For most applications, lift is a useful force, while drag is a loss to be overcome with input power.

The boundary layer concept and boundary layer separation problems are also encountered in the external flow, as shown in Figure 2.9. A more detailed discussion of the flow over an airfoil is given in Appendix B.

Secondary Flow In the flow channel or flow field between airfoils cascade, as shown in Figure 2.10, sometimes flow in a direction normal to the mainstream direction or a local recirculation (eddy) occurs, especially when machine flow capacity is below the design condition. This is due to the curvature of the flow passage, similar to the flow through a bend and the rotational motion of the rotor. It incurs additional energy loss, which is accounted for in evaluating the hydraulic or adiabatic efficiency of a turbomachine.

Flow Around a Rotating Disk Fluid in the clearance between a rotor and stationary casing wall has a tangential velocity distribution profile, as shown in Figure 2.11. It increases from zero at the casing wall to $U = r\omega$ at the rotor wall because of the no-slip condition on both sides. Hence, a drag force is imposed on the rotor, and a disk friction loss occurs, expressed as $Pdf \sim \omega^3 D^5$. This loss is accounted for as part of mechanical loss in evaluating the mechanical efficiency of a turbomachine.

Also, a pressure distribution of the form $p = p_2 - \rho U_{av}^2 = p_2 - \rho r^2 \omega_{av}^2$ is expected, where ω_{av} is the average fluid rotating speed, and p_2 is the pressure at the rotor tip. Due to the different pressure distributions in the front and back shrouds of a rotor, an unbalanced axial thrust force occurs. Hence, thrust bearings or some special design features such as the balancing holes through the rotor will be required.

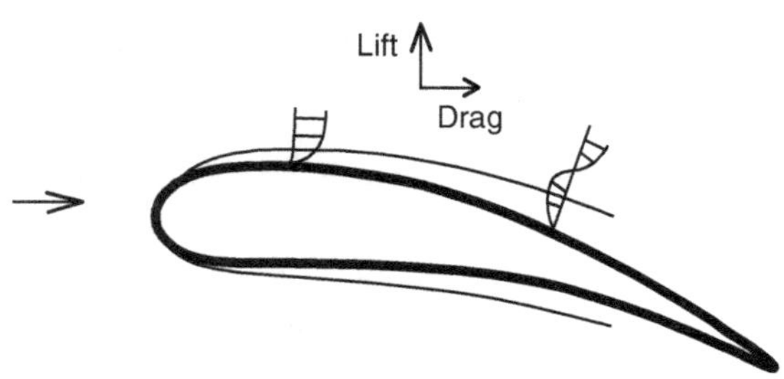

Figure 2.9 Flow over an airfoil to produce lift and drag.

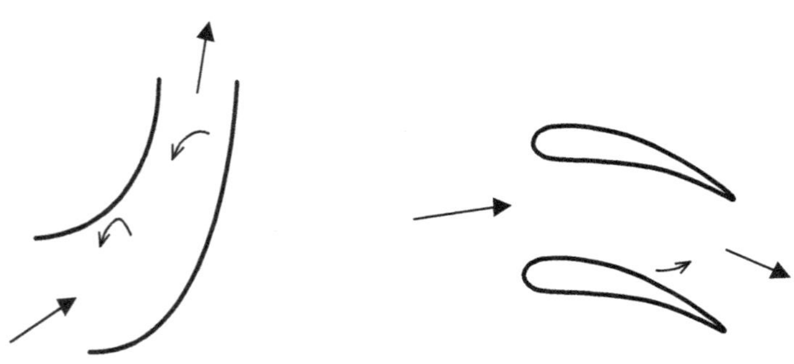

Figure 2.10 Secondary flows.

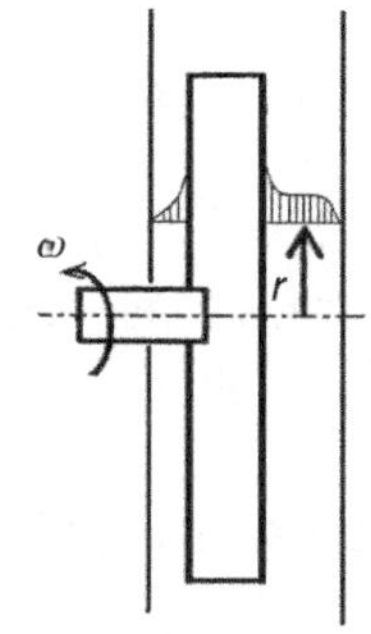

Figure 2.11 Flow around a rotating disk (different tangential flow patterns in the narrow space on the left side and the wide space on the right side).

Leakage Flow Due to the pressure difference between the inlet and outlet of an impeller, some fluid will leak through the clearance between the impeller and casing. Leakage can take place in one or several of the following places according to the type of machine: (1) between the casing and impeller in the impeller eye, (2) through the stuffing box in the back of the impeller, (3) through axial thrust balancing holes, and (4) between the adjacent stages in multistage machines. Since the leakage flow passage area is very small, the Reynolds number is low, and the leakage flow is normally laminar or Stokes flow, except in the very high rotating speed machine.

In turbomachine designs, some types of seals are installed to minimize these leakages if possible. But there is no 100% seal because they are dynamic seals. That means there is relative motion between the mating seal elements. Leakage losses are accounted for by the estimated volumetric efficiency of the machine.

2.2.2 Energy in Flowing Fluids

As discussed in the last section, energy associated with the rotating shaft is the mechanical power expressed as $Ps = \omega\tau$, where ω is the angular velocity of the shaft in radians per second and τ is the torque exerted on or by the shaft in Newton-meters (ft-lbf), with Ps in Newton-meters per second (lbf-ft/s). The energy associated with flowing fluids in general consists of internal energy u, flow work p/ρ, kinetic energy $V^2/2$, and gravitational potential energy gZ per unit mass of fluid. Hence, the total energy of the flowing fluid per unit mass is derived as

$$e = u + \frac{p}{\rho} + \frac{V^2}{2} + gZ = h + \frac{V^2}{2} + gZ, \tag{2.28}$$

where $h = u + p/\rho$ is the static enthalpy. The unit for each component of energy is Newton-meters per kilogram (lbf-ft/slug).

Within the flow passage shown in Figure 2.12, under a steady-state, uniform-flow condition, the energy equation (first law of thermodynamics) between positions 1 and 2 can be expressed as

$$\dot{m}\left(h + \frac{1}{2}V^2 + gZ\right)_1 + q = \dot{m}\left(h + \frac{1}{2}V^2 + gZ\right)_2 + P_s, \tag{2.29}$$

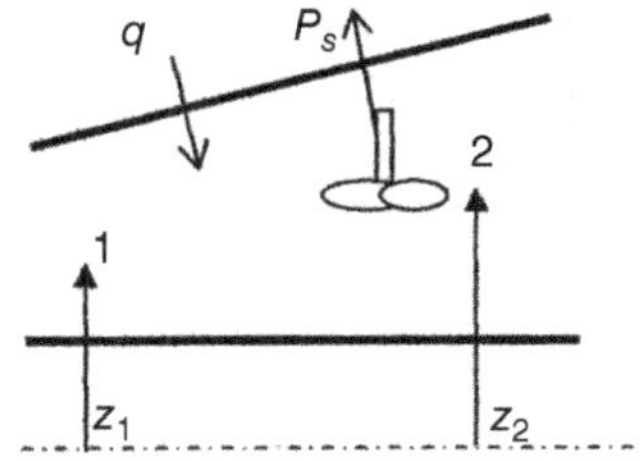

Figure 2.12 Energy in flowing fluid.

where m is the mass flow rate, positive q is the heat transfer rate into the fluid across the boundary, and positive Ps is the power done on the rotating impeller by flowing fluid. Under the simplified situation with neither heat transfer nor a rotating impeller, the familiar energy conservation equation results in the form

$$h_1 + V_1^2/2 + gZ_1 = h_2 + V_2^2/2 + gZ_2, \tag{2.30}$$

or

$$h + \frac{1}{2}V^2 + gZ = \text{const.} \tag{2.31}$$

In an isentropic process for incompressible fluids, where density ρ is constant, the change of enthalpy is equal to $\Delta p/\rho$. Hence, the above energy equation can be written in the form $p/\rho + V^2/2 + gZ = \text{const}$ or

$$gH + \frac{1}{2}V^2 + gZ = \text{const.} \tag{2.32}$$

This is the Bernoulli equation mentioned in fluid mechanics, where $H = p/(\rho g)$ is the head normally used in hydraulics.

2.2.3 Euler Equation

In the energy equation discussed in previous sections, the shaft power *Ps* is the energy transfer rate or the power between the rotating impeller and the flowing fluid. Its relationship with impeller geometry can be derived from the integral form of the momentum equation of fluid mechanics. To do this, several "velocities" at the impeller inlet and outlet regions have to be defined. Consider the flow around the rotating impeller shown in Figure 2.13, where

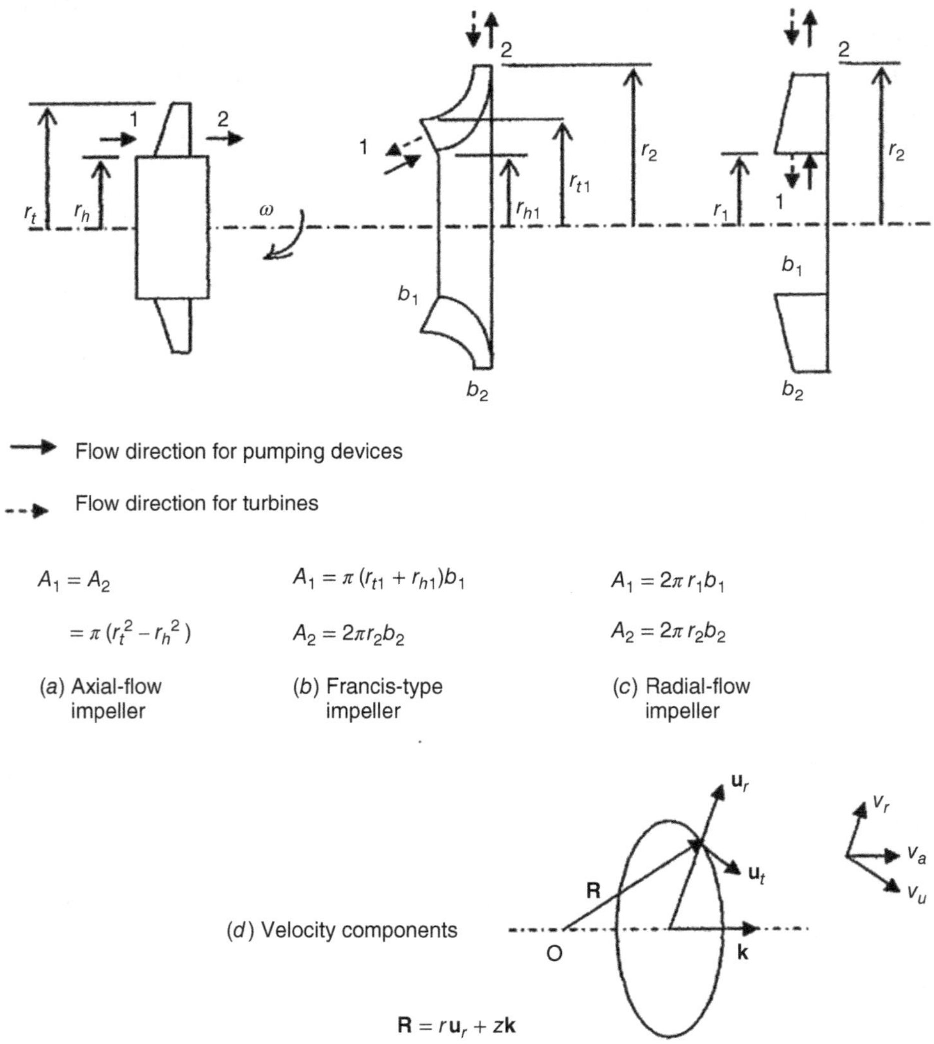

Figure 2.13 Geometries and flow velocities around impellers.

three types of impellers with their respective flow passage areas at the inlet and outlet are given. The flow velocity in general is three-dimensional. Using a cylindrical coordinate system, the flow velocity can be decomposed into axial, radial, and tangential components, represented by V_a, V_r, and V_u, respectively.

Consider a point fixed on the impeller blade defined with a position vector R which can be decomposed into radial and axial components, such as $R = ru_r + z_k$ (bold characters are used for vector quantities in this book). The tangential velocity at this point U due to the impeller rotation is $U = r\omega$. With respect to the impeller, the relative flow velocity can be defined with the vector eq. $W = V - U$. Or in a graphical form, the so-called velocity diagram (or velocity triangle) in a top or front view is shown in Figure 2.14. It can be drawn starting with either U or W as shown. The components of V and W in the direction of U are the tangential-flow velocities Vu and Wu, while those normal to U are axial, radial, or combined axial and radial (called meridional, with subscript m), depending on whether it is an axial-flow, radial-flow, or mixed-flow impeller. The volumetric flow rate through the impeller can then be expressed as $Q = V_m A$, where A is the flow passage area normal to the meridional direction.

Applying the steady-state integral form of the momentum equation to the control volume occupied by the impeller flow passage, we have the force exerted on the fluid passing through the impeller:

$$\mathbf{F} = \int_{cs} \mathbf{V}\ dm = \int_{cs} \mathbf{V}\rho\mathbf{V} \cdot d\mathbf{A}. \tag{2.33}$$

The integration carried over the control surface (cs) enclosing the impeller is the net momentum change of fluid. The corresponding moment of momentum equation can be written as

$$\boldsymbol{\tau} = \int_{cs} \mathbf{R} \times \mathbf{V}\rho\mathbf{V} \cdot d\mathbf{A} = \int_{A_2} \mathbf{R} \times \mathbf{V}\ dm - \int_{A_1} \mathbf{R} \times \mathbf{V}\ dm, \tag{2.34}$$

where A_1 and A_2 are the control surfaces at the impeller inlet and outlet, respectively, and m is the mass flow rate through the impeller. The cross-product $R \times V$ can be expanded into three components in terms of the unit vectors u_r, u_t, and k as

$$\begin{aligned} \mathbf{R} \times \mathbf{V} &= (r\mathbf{u}_r + z\mathbf{k}) \times (V_r\mathbf{u}_r + V_u\mathbf{u}_t + V_a\mathbf{k}) \\ &= -zV_u\mathbf{u}_r + (zV_r - rV_a)\mathbf{u}_t + rV_u\mathbf{k}, \end{aligned} \tag{2.35}$$

where $u_r \times u_t = k$, $u_t \times k = u_r$, $k \times u_r = u_t$, $u_r \times u_r = 0$, and so on. The torque is a vector quantity, but the shaft power is a scalar quantity, expressed as the dot product of ω and τ. Since ω is always in the k direction, only the z component of the torque, τ_z, will contribute to the power. The other two components of the vector equation can be ignored in deriving the energy transfer equation. Considering the average flow velocities at the impeller inlet and outlet, the above angular momentum equation in the z component can be written in the form $\tau z = m(r_2 V_{u2} - r_1 V_{u1})$, where the subscripts 1 and 2 designate the quantities evaluated at the impeller inlet and outlet, respectively.

Multiplying both sides of the equation with the impeller angular velocity, ω, we obtain $\omega\tau_z = \dot{m}(U_2 Vu_2 - U_1 V_{u1})$, or $-Ps = \dot{m}(U_2 V_{u2} - U_{1Vu1})$, where positive Ps is defined as the power done on the impeller by the flowing fluid in the last section. Expressed in terms of the energy transfer per unit mass of fluid, we have

$$\Delta E_p = -\frac{P_s}{\dot{m}} = U_2 V_{u2} - U_1 V_{u1} \tag{2.36}$$

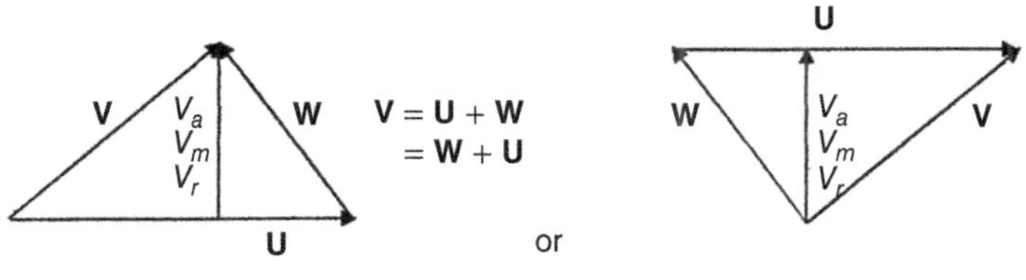

Figure 2.14 Velocity diagram (top view for axial-flow, front view for radial-flow impeller).

for pumps, fans, or compressors and

$$\Delta E_t = \frac{P_s}{\dot{m}} = U_1 V_{u1} - U_2 V_{u2} \tag{2.37}$$

for turbines. They are called ideal Euler equations. It is important to recognize that V_{u1} and V_{u2} in these equations are the actual average absolute flow velocities at the impeller inlet and outlet. How they are determined and related to the impeller geometry is discussed in detail in later sections.

The energy transfer per unit mass of fluid, ΔE_p or ΔE_t, can be related to the total enthalpy change per unit mass of fluid, as illustrated in Eq. (2.29), neglecting the heat transfer and elevation difference of the inlet and outlet of the machine. For incompressible fluid, they are the ideal total head change per unit mass of fluid. Hence, the Euler equation can be written in the forms:

$$\begin{aligned} g\,\Delta H_i &= U_2 V_{u2} - U_1 V_{u1} && \text{for pumps,} \\ \Delta h_i &= U_2 V_{u2} - U_1 V_{u1} && \text{for compressors,} \\ g\,\Delta H_i &= U_1 V_{u1} - U_2 V_{u2} && \text{for hydraulic turbines,} \\ \Delta h_i &= U_1 V_{u1} - U_2 V_{u2} && \text{for gas/steam turbines.} \end{aligned} \tag{2.38}$$

Example 2.5E The radial-flow impeller shown is rotating at 800 rpm and pumping water at 1750 gpm. Determine the radial-flow velocity and impeller tangential velocity at the inlet and outlet.

Solution

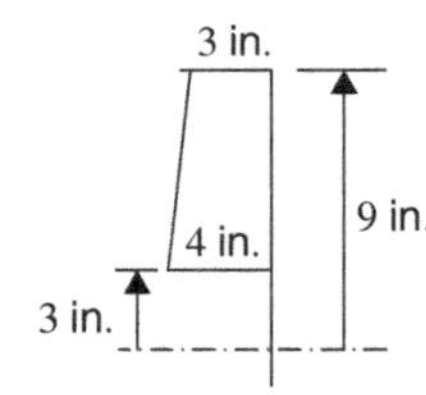

$$\omega = \frac{N\pi}{30} = \frac{800\pi}{30} = 83.7\ \text{rad/s},$$

$$U_1 = \omega r_1 = \frac{83.7 \times 3}{12} = 20.9\ \text{ft/s}, \quad U_2 = 62.7\ \text{ft/s},$$

$$A_1 = 2\pi r_1 b_1 = \frac{6\pi \times 4}{144} = 0.523\ \text{ft}^2, \quad A_2 = 1.18\ \text{ft}^2,$$

$$V_{r1} = \frac{Q}{A_1} = \frac{1750 \times 0.00223}{0.523} = 7.47\ \text{ft/s}, \quad V_{r2} = 3.27\ \text{ft/s}.$$

Example 2.5S The radial-flow impeller shown is rotating at 800 rpm and pumping water at 397 m^3/h. Determine the radial-flow velocity and impeller tangential velocity at the inlet and outlet.

Solution

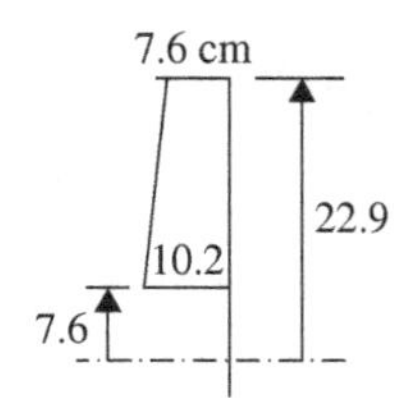

$$\omega = \frac{N\pi}{30} = \frac{800\pi}{30} = 83.7\ \text{rad/s},$$

$$U_1 = \omega r_1 = \frac{83.7 \times 7.6}{100} = 6.36\ \text{m/s}, \quad U_2 = 19.2\ \text{m/s},$$

$$A_1 = 2\pi r_1 b_1 = 2\pi \times 7.6 \times 10.2 = 487\ \text{cm}^2, \quad A_2 = 1093.5\ \text{cm}^2,$$

$$V_{r1} = \frac{Q}{A_1} = \frac{397/3600}{487/10^4} = 2.26\ \text{m/s}, \quad V_{r2} = \frac{V_{r1} A_1}{A_2} = 1.01\ \text{m/s}.$$

Example 2.6E In Examples 2.5E and 2.5S, the inlet flow is swirl free, and the relative flow direction decreases by 10° with respect to the tangential direction from the inlet to the outlet through the impeller passage. Determine all flow velocities at both the inlet and outlet and plot the velocity diagrams.

Solution

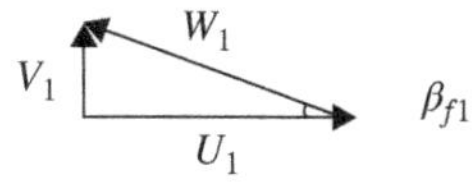

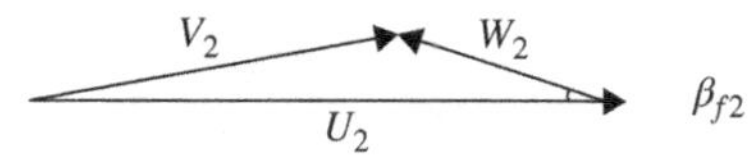

$$\tan\,\beta_{f1} = \frac{V_{r1}}{U_1} = \frac{7.47}{20.9} = 0.357.$$

So, $\beta_{f1} = 19.6°$ and

$$V_1 = V_{r1},\quad W_1 = \left(U_1^2 + V_{r1}^2\right)^{0.5} = (20.9^2 + 7.47^2)^{0.5} = 22.19\text{ft/s},$$

$$\beta_{f2} = 19.6 - 10 = 9.6°,\quad W_{u2} = \frac{V_{r2}}{\tan\beta_{f2}} = \frac{3.27}{\tan 9.6°} = 19.3\text{ft/s},$$

$$V_{u2} = U_2 - W_{u2} = 62.7 - 19.3 = 43.4\text{ft/s},$$

$$W_2 = \left(W_2^{u2} + V_{r2}^2\right)^{0.5} = (19.3^2 + 3.27^2)^{0.5} = 19.57\text{ft/s},$$

$$V_2 = \left(V_{u2}^2 + V_{r2}^2\right)^{0.5} = (43.4^2 + 3.27^2)^{0.5} = 43.52\text{ft/s}.$$

Example 2.6S Again, the inlet flow is swirl free and the relative flow direction decreases by 10^{c} with respect to the tangential direction from the inlet to the outlet through the impeller passage. Determine all flow velocities at both the inlet and outlet and plot the velocity diagrams.

Solution

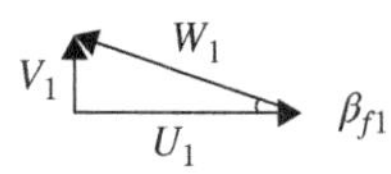

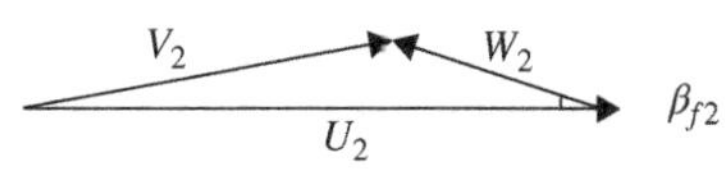

$$\tan\,\beta_{f1} = \frac{V_{r1}}{U_1} = \frac{2.26}{6.36} = 0.356.$$

So, $\beta_{f1} = 19.6°$ and

$$V_1 = V_{r1},\quad W_1 = \left(U_1^2 + V_{r1}^2\right)^{0.5} = (6.36^2 + 2.26^2)^{0.5} = 6.75\text{ m/s},$$

$$\beta_{f2} = 19.6 - 10 = 9.6°,\quad W_{u2} = \frac{V_{r2}}{\tan\beta_{f2}} = \frac{1.01}{\tan 9.6°} = 5.97\text{ m/s},$$

$$V_{u2} = U_2 - W_{u2} = 19.2 - 5.97 = 13.23\text{ m/s},$$

$$W_2 = \left(W_{u2}^2 + V_{r2}^2\right) = (5.97^2 + 1.01^2)^{0.5} = 6.05\text{ m/s},$$

$$V_2 = \left(V_{u2}^2 + V_{r2}^2\right) = (13.23^2 + 1.01^2)^{0.5} = 13.27\text{ m/s}.$$

2.2.4 Equations for Axial-Flow Machines

For axial-flow machines (either pumping devices or turbines), flow through the rotor will be predominantly in the axial and tangential directions. In other words, there is no change in the radial distance from the inlet to the outlet, as shown in Figure 2.15. In one-dimensional flow analysis, average flow conditions at the mean radius will be treated. Hence, the following conditions prevail:

$$U_2 = U_1 = U_m = r_m\omega,\quad A_1 = A_2 = \pi\left(r_t^2 - r_h^2\right), \tag{2.39}$$

where $r_m = \left[\left(r_t^2 + r_h^2\right)/2\right]^{1/2}$ is the mean radius, and r_t and r_h are the tip and hub radii, respectively. In some of the literature, the mean radius is defined as $r_m = (r_t + r_h)/2$. Ideally, the relative flow velocity is tangent to the blade mean line at both the inlet and outlet, that is, $\beta_f = \beta_b$, as shown in Figure 2.16. Under ideal (design) operating conditions, the velocity diagrams at the inlet and outlet for a

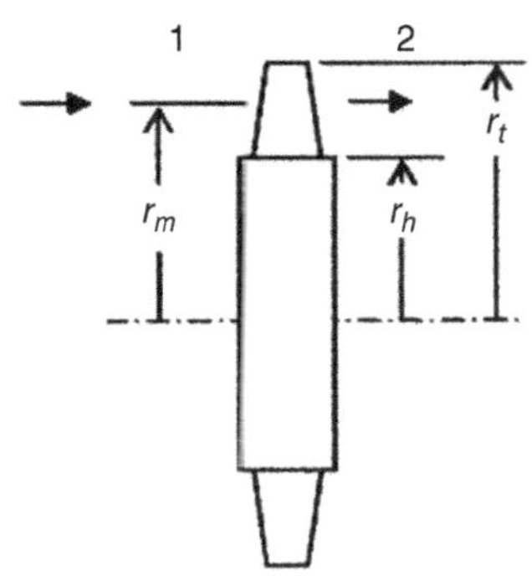

Figure 2.15 Axial-flow impeller (meridional view).

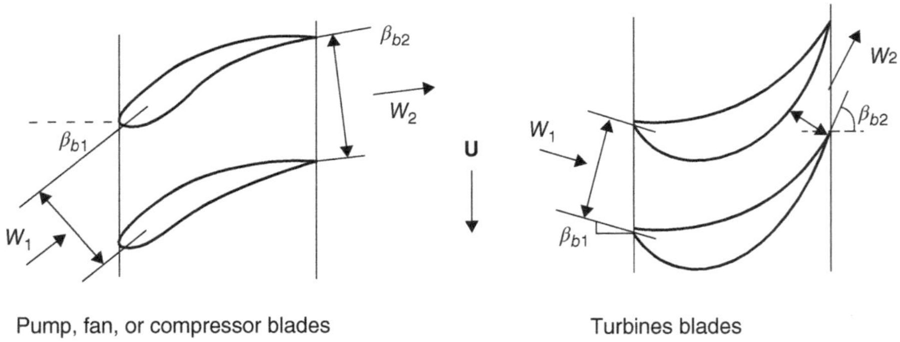

Figure 2.16 Blade profiles for axial-flow machines (top view).

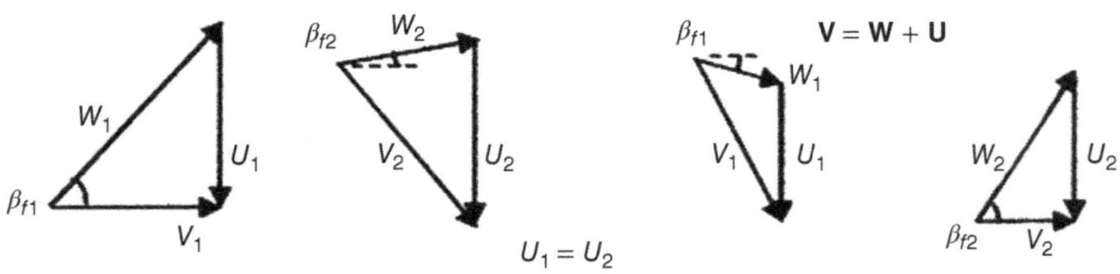

(*a*) Pump, fan, or compressor (*b*) Turbine
(All β angles are measured with respect to the axial direction for axial–flow machines)

Figure 2.17 Velocity diagrams for axial-flow machines (top view).

pump, fan, or compressor are as illustrated in Figure 2.17*a* and those for turbines as in Figure 2.17*b* to have maximum energy transfer without whirl at the inlet or outlet. Hence, the corresponding Euler equations are $\Delta E_p = U_m V_{u2}$ and $\Delta E_t = U_m V_{u1}$, respectively. In reality, there are some deviations between the relative flow direction and the blade profile at both the inlet and outlet. This is discussed in later chapters. The absolute flow direction can be controlled with inlet guide vanes if necessary. In a pump, fan, or compressor, the flow passage diverges from the inlet to the outlet, similar to a diffuser, and hence, relative flow decelerates and pressure increases. For turbines, the flow passage converges, similar to a nozzle, and hence, relative flow accelerates and pressure decreases in the direction of flow.

Example 2.7E An axial-flow fan with tip and hub radii of 24 and 10 in., respectively, is rotating at 1250 rpm. If it delivers an ideal total head of 2 in. wg at a flow rate of 53,000 cfm, estimate its leading and trailing edge blade angles at the mean radius.

Solution

Mean radius

$$\text{Mean radius } r_m = \left[\frac{1}{2}(24^2 + 10^2)\right]^{0.5} = 18.4 \text{ in.} = 1.53 \text{ ft},$$

$$U_m = \frac{1250 \times \pi}{30} \times 1.53 = 200.5 \text{ ft/s}, \ A = \frac{\pi(24^2 - 10^2)}{144} = 10.4 \text{ ft}^2.$$

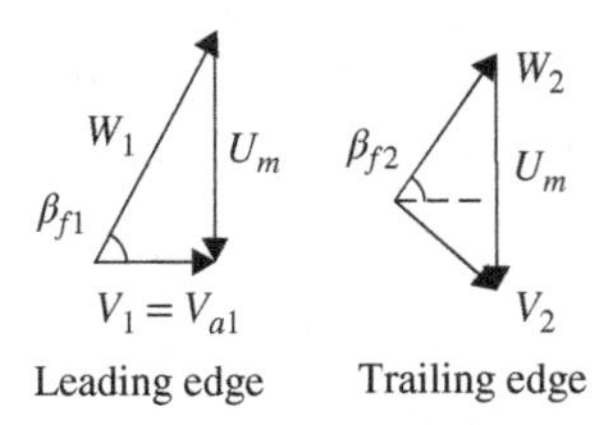

So,

$$V_{a1} = V_{a2} = \frac{Q}{A} = \frac{53,000}{10.4 \times 60} = 84.9 \text{ ft/s},$$

$$\beta_{b1} = \beta_{f1} = \tan^{-1}\left(\frac{U_m}{V_{a1}}\right) = 67°$$

with respect to the axial direction. From the ideal Euler equation, $gH_a = (\rho_w/\rho_a)gH_w = U_m V_{u2} = U_m(U_m - V_{a2}\tan\beta_{f2})$, where H_a is the total head of air, and H_w is the total head of water, or $(62.4/0.0762)\times 32.2\times 2/12 = 200.5(200.5–84.9\tan\beta_{f2})$. So $4395 = 200.5(200.5–84.9\tan\beta_{f2})$. Hence, we have $\beta_{b2} = \beta_{f2} = 64.5°$ with respect to the axial direction. The velocity diagrams are shown below:

Example 2.7S An axial-flow fan with tip and hub radii of 0.6 and 0.25 m, respectively, is rotating at 1250 rpm. If it delivers an ideal total head of 5 cm H_2O at a flow rate of 1500 cmm, estimate its leading and trailing edge blade angles at the mean radius.

Solution

Mean radius

$$r_m = \left[\frac{1}{2}(0.6^2 + 0.25^2)\right]^{0.5} = 0.46\ \text{m},$$

$$U_m = \left(\frac{1250\times\pi}{30}\right)\times 0.46 = 60.2\ \text{m/s}, \quad A = \pi(0.6^2 - 0.25^2) = 0.934\ \text{m}^2.$$

So,

$$V_{a1} = V_{a2} = \frac{Q}{A} = \frac{1500}{0.934\times 60} = 26.7\ \text{m/s},$$

$$\beta_{b1} = \beta_{f1} = \tan^{-1}\left(\frac{U_m}{V_{a1}}\right) = 66°$$

with respect to the axial direction. From the ideal Euler equation, $gH_a = (\rho_w/\rho_a)gH_w = U_m V_{u2} = U_m(U_m - V_{a2}\tan\beta_{f2})$, where H_a is the total head of air, and H_w is the total head of water, or $(998/1.22)\times 9.8\times 5/100 = 60.2(60.2–26.7\tan\beta_{f2})$. So $400.8 = 60.2(60.2–26.7\tan\beta_{f2})$. Hence, we have $\beta_{b2} = \beta_{f2} = 63.5°$ with respect to the axial direction. The velocity diagrams are shown below:

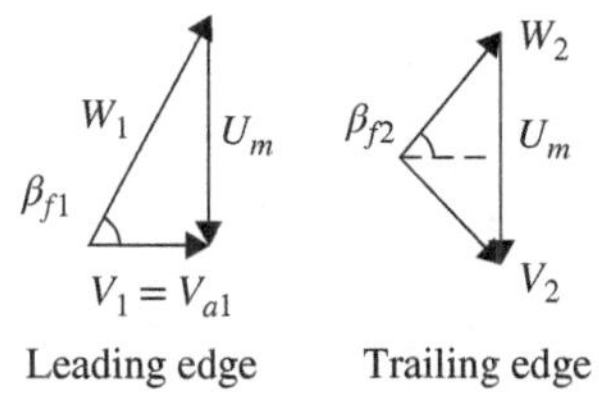

2.2.5 Equations for Mixed- and Radial-Flow Machines

For mixed- and radial-flow machines, the inlet and outlet radii are different. For a pump, fan, or compressor, the outlet radius is larger than the inlet radius. For turbines, it is reversed to take advantage of the centrifugal force in converting energy. The flow passage areas are

$$A_1 = 2\pi r_1 b_1, \qquad A_2 = 2\pi r_2 b_2 \qquad \text{for radial flow},$$

$$A_1 = \pi(r_{t1} + r_{h1})b_1, \qquad A_2 = \pi(r_{t2} + r_{h2})b_2 \qquad \text{for mixed flow},$$

where b_1 and b_2 are the blade heights at the inlet and outlet, as shown in Figure 2.18. And for the mixed-flow impeller, the mean radii at both the inlet and outlet can be defined as $r_m = \left[\left(r_t^2 + r_h^2\right)/2\right]^{1/2}$. It is defined such that the flow passage area is equally divided into two parts. The expressions for A and r_m for the mixed-flow impeller will become those for the axial- and radial-flow cases with $b = r_t - r_h$ and $r_t = r_h$, respectively.

The velocity diagrams are shown in Figures 2.19*a* and *b*. They are opposite to each other when applied to the same machine rotating in the opposite direction for a pumping device or turbine. Here, we have $U_2 > U_1$ for a pump, fan, or compressor and $U_2 < U_1$ for turbines. The corresponding Euler equations are the same as those for axial-flow machines. The relative flow directions are close to the blade profiles, as shown in Figure 2.20, where the dashed-line vectors are for turbines.

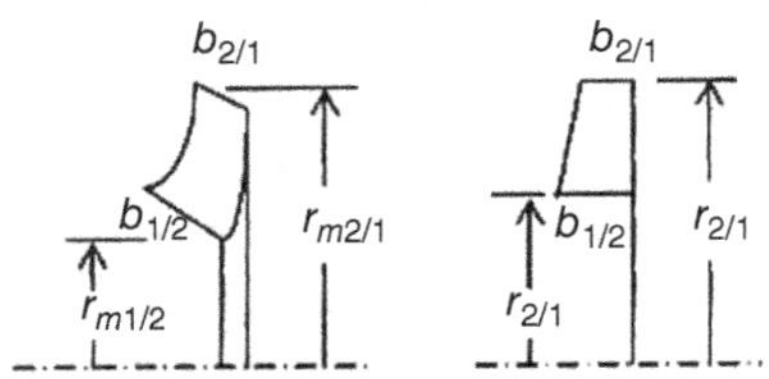

Figure 2.18 Mixed- and radial-flow impeller for pumping devices and turbines (meridional view, subscript 1 is inlet, subscript 2 is outlet).

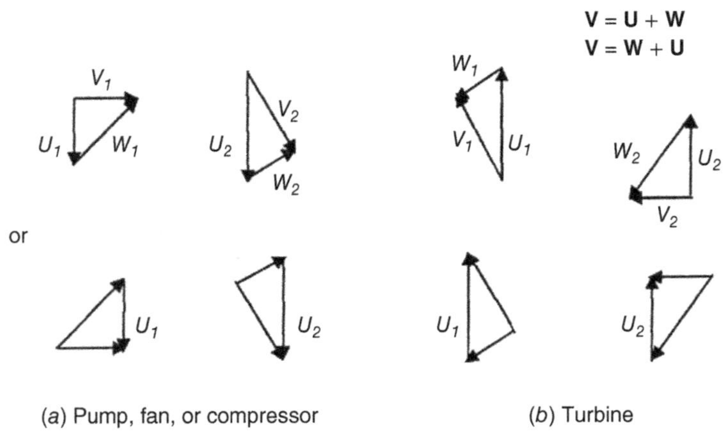

Figure 2.19 Velocity diagrams for mixed- and radial-flow machines (front view at 3 O'clock position).

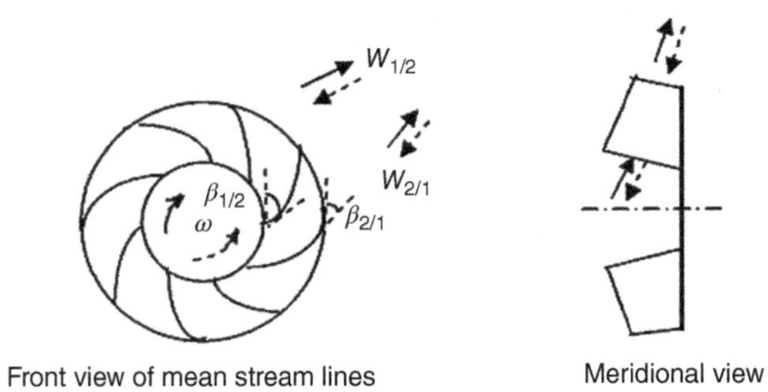

Figure 2.20 Blade profiles for mixed- and radial-flow machines (dashed line vectors are for turbines).

It is noted that in axial-flow machines, the flow angles and blade angles are referred with respect to the axial direction, while in mixed- and radial-flow machines, they are referred with respect to the tangential direction. The minor angles, less than 90°, are always specified. To draw these velocity diagrams, some basic guidelines can be followed for convenience:

a) Start with the impeller tangential velocity U.
b) Add the relative flow velocity W in the direction parallel to the blade mean line tangent, with the adjustment for incidence at the leading edge or flow deviation at the trailing edge for actual flows.
c) If the volumetric flow rate and flow passage area are given or can be determined, the radial, axial, or meridional component of flow velocity can be used to complete the velocity triangle since $Q = V_{rA}$ or V_{aA} or V_{mA}.

Example 2.8E A Francis-type inflow hydraulic turbine with the dimensions shown is rotating at 120 rpm. Determine the ideal output shaft power and the flow rate at design condition.

Solution

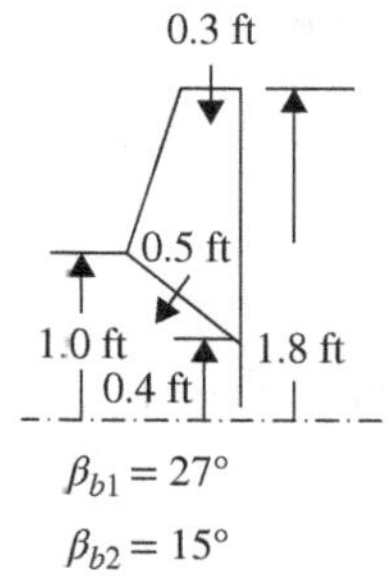

$$\omega = \frac{N\pi}{30} = \frac{120\pi}{30} = 12.56 \text{ rad/s},$$

$$U_1 = r_1\omega = 1.8 \times 12.56 = 22.6 \text{ ft/s},$$

$$U_2 = r_{m2}\omega = \left[\frac{1}{2}(1.0^2 + 0.4^2)\right]^{0.5} \times 12.56$$

$$= 9.56 \text{ ft/s},$$

$$A_1 = 2\pi r_1 b_1 = 2\pi \times 1.8 \times 0.3 = 3.39 \text{ ft}^2,$$

$$A_2 = \pi(r_{t2} + r_{h2})b_2 = \pi(1.0 + 0.4) \times 0.5 = 2.20 \text{ ft}^2.$$

Assume swirl-free flow at discharge, that is, $V_{u2} = 0$ and $\beta_{f1} = \beta_{b1}$, $\beta_{f2} = \beta_{b2}$. So, $V_{m2} = U_2$ tan 15° = 2.56 ft/s, and hence,

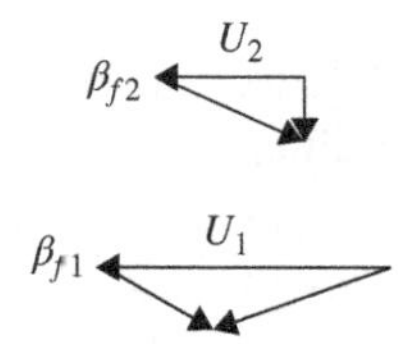

$$Q = V_{m2}A_2 = 2.56 \times 2.20 = 5.63 \text{ ft}^3\text{/s} = V_{r1}A_1,$$

$$V_{r1} = 1.66 \text{ ft/s}, \quad W_{u1} = \frac{V_{r1}}{\tan \beta_{f1}} = 3.26 \text{ ft/s},$$

$$V_{u1} = U_1 - W_{u1} = 19.3 \text{ ft/s},$$

$$\Delta E_t = U_1 V_{u1} = 22.6 \times 19.3 = 437.1 \text{ (ft/s)}^2,$$

$$P_s = m\,\Delta E_t = \left(\frac{62.4}{32.2}\right) \times 5.63 \times 437.1 \quad (\text{slug/ft}^3)(\text{ft}^3\text{/s})(\text{ft/s})^2$$

$$= 4769 \text{ ft-lb}_f\text{/s} = 8.67 \text{ hp}.$$

Example 2.8S A Francis-type inflow hydraulic turbine with the dimensions shown is rotating at 120 rpm. Determine the ideal output shaft power and the flow rate at design condition.

Solution

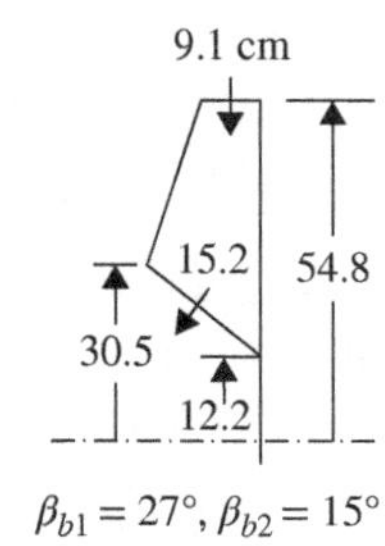

$$\omega = \frac{N\pi}{30} = \frac{120\pi}{30} = 12.56 \text{ rad/s},$$

$$U_1 = r_1\omega = \frac{54.8 \times 12.56}{100} = 6.9 \text{ m/s},$$

$$U_2 = r_{m2}\omega = \left[\frac{1}{2}(30.5^2 + 12.2^2)\right]^{0.5} \times 12.56 = 2.92 \text{ m/s},$$

$$A_1 = 2\pi r_1 b_1 = \frac{2\pi \times 54.8 \times 9.1}{10,000} = 0.313 \text{ m}^2,$$

$$A_2 = \pi(r_{t2} + r_{h2})b_2 = \frac{\pi(30.5 + 12.2) \times 15.2}{10,000} = 0.204 \text{ m}^2.$$

Assume swirl-free flow at discharge, that is, $V_{u2} = 0$ and $\beta_{f1} = \beta_{b1}$, $\beta_{f2} = \beta_{b2}$. So, $V_{m2} = U_2$ tan 15° = 0.782 m/s, and hence,

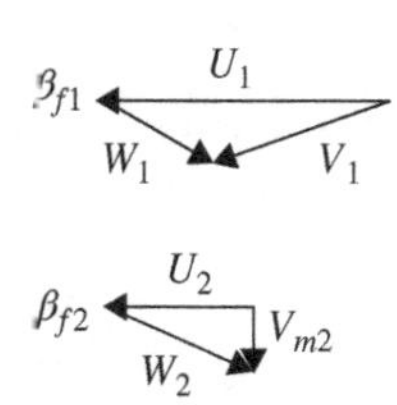

$$Q = V_{m2}A_2 = 0.782 \times 0.204 = 0.159 \text{ m}^3\text{/s} = V_{r1}A_1,$$

$$V_{r1} = 0.508 \text{ m/s}, \quad W_{u1} = \frac{V_{r1}}{\tan \beta_{f1}} = 0.997 \text{ m/s},$$

$$V_{u1} = U_1 - W_{u1} = 5.92 \text{ m/s},$$

$$\Delta E_t = U_1 V_{u1} = 6.9 \times 5.92 = 40.85 (\text{m/s})^2,$$

$$P_s = m\Delta E_t = 998 \times 0.782 \times 0.204 \times 40.85 (\text{kg/s}) \ (\text{m/s})^2$$

$$= 6503 \text{ W} = 6.503 \text{ kW}.$$

2.2.6 Degree of Reaction

The Euler Eqs. (2.36) and (2.37) derived in Section 2.2.3 can be converted into another form to illustrate the physical meaning of energy transfer between the flowing fluid and the rotor. From the velocity diagram, we have

$$V_u^2 = V^2 - V_m^2, \qquad V_m^2 = W^2 - (U - V_u)^2, \tag{2.40}$$

and hence, $V_u^2 = V^2 - W^2 + \left(U^2 - 2UV_u + V_u^2\right)$ or $UV_u = (U^2 + V^2 - W^2)/2$. The Euler equations can therefore be written in the forms

$$\Delta E_p = \frac{1}{2}\left[\left(U_2^2 - U_1^2\right) + \left(V_2^2 - V_1^2\right) + \left(W_1^2 - W_2^2\right)\right], \tag{2.41}$$

$$\Delta E_t = \frac{1}{2}\left[\left(U_1^2 - U_2^2\right) + \left(V_1^2 - V_2^2\right) + \left(W_2^2 - W_1^2\right)\right]. \tag{2.42}$$

Hence, the energy transfer can be divided into three terms: (1) centrifuging associated with the change of rotation at different radii, (2) change of flow kinetic energy, and (3) diffusion or expansion associated with the change of flow passage area. From the conservation of energy, the summation of kinetic energy and static enthalpy is constant throughout the stator (diffuser or nozzle). They can only be converted from one to the other, while the other two terms are related to the static enthalpy change in the impeller.

In axial-flow machines, there is no energy transfer due to centrifuging since $U_1 = U_2$. In radial- and mixed-flow machines, the flow must be radially outward for a pumping device and radially inward for a turbine so that the centrifuging effect will predominantly contribute to the needed energy transfer. In pumping devices, centrifuging and diffusion cause static enthalpy and pressure rise in the direction of flow; hence, these machines are likely to suffer from a boundary layer separation. There is no such problem in turbines since the flow accelerates and the pressure drops in the direction of flow in most cases. However, if we try to minimize this problem associated with centrifuging and diffusion by increasing the kinetic energy change term, the burden will be placed on the diffuser downstream of the impeller. That means more kinetic energy has to be converted into pressure rise in the diffuser. A parameter used to designate the relative magnitude of these energy transfer components is called degree of reaction or simply reaction. Reaction is defined as the ratio of static enthalpy change across the rotor (impeller) over the complete stage of rotor and stator (impeller–diffuser or nozzle–impeller), that is,

$$R_p = \frac{h_2 - h_1}{\Delta h_s} = \frac{h_2 - h_1}{\Delta E_p} \cong \frac{\Delta p/\rho}{\Delta E_p} \qquad \text{or} \qquad R_t = \frac{h_1 - h_2}{\Delta h_s} = \frac{h_1 - h_2}{\Delta E_t} \cong \frac{\Delta p/\rho}{\Delta E_t}. \tag{2.43}$$

Here, the subscript s is used to designate the "stage" and $\Delta h_s \cong \Delta h_{os} = \Delta E_p/t$, as the velocities at the inlet and outlet of a stage are almost the same.

Since the energy transfer takes place in the rotor, we have

$$\Delta E_p = h_{o2} - h_{o1} = h_2 - h_1 + \frac{1}{2}\left(V_2^2 - V_1^2\right),$$

$$\Delta E_t = h_{o1} - h_{o2} = h_1 - h_2 + \frac{1}{2}\left(V_1^2 - V_2^2\right), \tag{2.44}$$

with negligible heat transfer across the machine. Hence, from Eqs. (2.41) and (2.42), the reaction can also be expressed in the forms

$$R_p = \frac{\frac{1}{2}\left(U_2^2 - U_1^2\right) + \frac{1}{2}\left(W_1^2 - W_2^2\right)}{\Delta E_p} \quad \text{for pumping device,} \tag{2.45}$$

$$R_t = \frac{\frac{1}{2}\left(U_1^2 - U_2^2\right) + \frac{1}{2}\left(W_2^2 - W_1^2\right)}{\Delta E_t} \quad \text{for turbine.} \tag{2.46}$$

If the fluid enters and leaves the rotor at different radii, that is, radial- or mixed-flow impeller, zero degree of reaction means the relative flow velocity also changes with an equal amount in the same direction. Then all

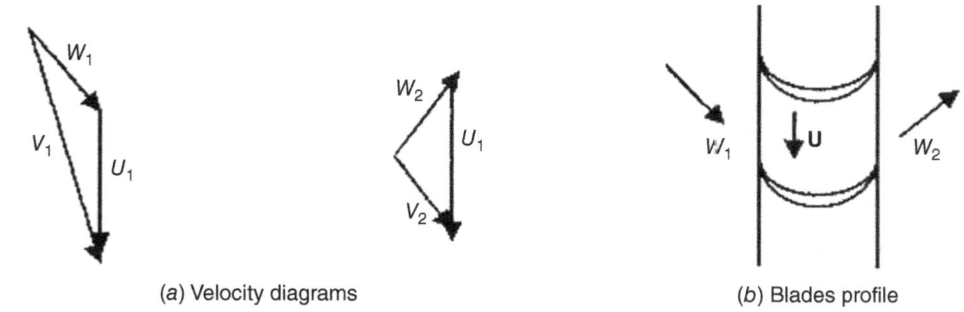

Figure 2.21 Axial-flow impulse turbine.

the burden of pressure or static enthalpy change has to be on the stator. This kind of design is rarely employed. A more detailed discussion of the reaction of the compressor and gas turbine is given in later chapters. Although the reaction is defined for both pumps and turbines, it is more commonly used to classify the types of turbines. The value may range from zero to unity.

For axial-flow turbines, $R_t = 0$ implies $W_1 = W_2$ since $U_1 = U_2$ automatically. This type of turbine is called an impulse turbine. Its velocity diagram is shown in Figure 2.21*a*, and the blade profile should be symmetric, as shown in Figure 2.21*b*. For the turbine with a reaction of $R_t = 1$, the fluid of high pressure entering the rotor is converted into kinetic energy. It is discharged at the outlet with high relative flow velocity but results in $V_2 = V_1$. A sprinkler with a velocity diagram shown in Figure 2.22 is an example. A popular axial-flow reaction turbine, however, has a reaction $R_t = 0.5$. Its velocity diagram and blade profile are shown in Figure 2.23. Detail of these turbines is discussed in later chapters.

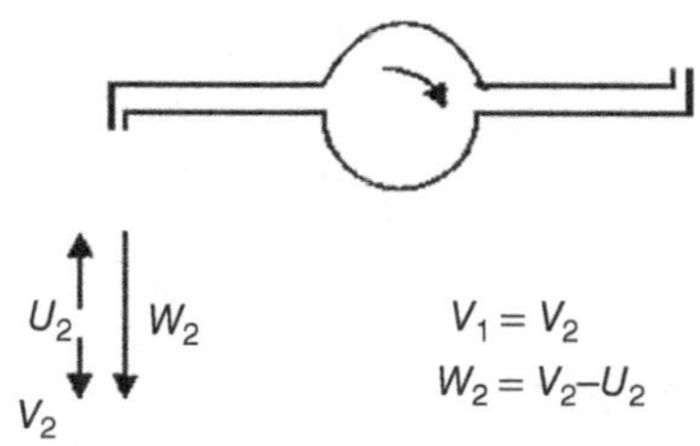

Figure 2.22 Sprinkler (a radial type reaction turbine with reaction, $R_t = 1$).

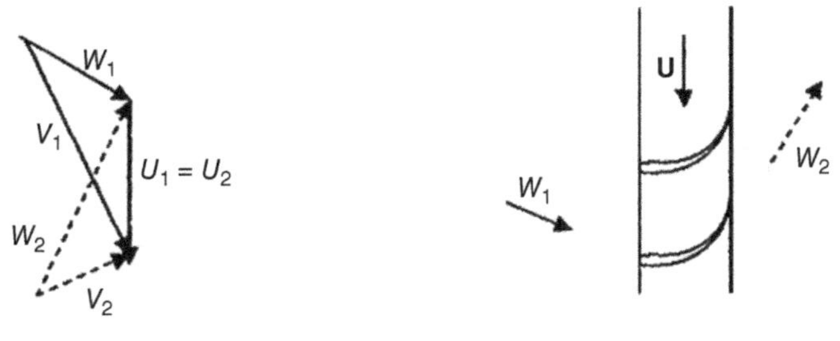

Figure 2.23 Axial-flow reaction turbine, $R_t = 0.5$.

Example 2.9E Determine the degree of reaction for the hydraulic turbine in Example 2.8E.

Solution

From Example 2.8E, we have $\Delta E_t = 437.1\ (\text{ft/s})^2$, $U_1 = 22.6$ ft/s, $U_2 = 9.56$ ft/s, and $V_2 = V_{m2} = 2.56$ ft/s. Also, we have

$$W_1 = \left(V_{r1}^2 + W_{u1}^2\right)^{0.5} = (1.66^2 + 3.26^2)^{0.5} = 3.66\ \text{ft/s},$$
$$W_2 = (9.56^2 + 2.56^2)^{0.5} = 9.9\ \text{ft/s},$$
$$V_1 = \left(V_{r1}^2 + V_{u1}^2\right)^{0.5} = (1.66^2 + 19.63^2)^{0.5} = 19.37\ \text{ft/s}.$$

So,

$$R_t = \frac{0.5\left[\left(U_1^2 - U_2^2\right) + \left(W_2^2 - W_1^2\right)\right]}{\Delta E_t}$$
$$= \frac{0.5[(22.6^2 - 9.56^2) + (9.9^2 - 3.66^2)]}{437.1} = 0.576.$$

Example 2.9S Determine the degree of reaction for the hydraulic turbine in Example 2.8S.

Solution

From Example 2.8S, we have $\Delta E_t = 40.85\ (\text{m/s})^2$, $U_1 = 6.9\,\text{m/s}$, $U_2 = 2.92\,\text{m/s}$, and $V_2 = V_{m2} = 0.782\,\text{m/s}$. Also, we have

$$W_1 = \left(V_{r1}^2 + W_{u1}^2\right)^{0.5} = (0.508^2 + 0.997^2)^{0.5} = 1.12\ \text{m/s},$$
$$W_2 = (2.92^2 + 0.782^2)^{0.5} = 3.02\ \text{m/s},$$

So,

$$R_t = \frac{0.5\left[\left(U_1^2 - U_2^2\right) + \left(W_2^2 - W_1^2\right)\right]}{\Delta E_t}$$
$$= \frac{0.5[(6.9^2 - 2.92^2) + (3.02^2 - 1.12^2)]}{40.85} = 0.575.$$

Example 2.10E Air with a density of 0.085 lbm/ft^3 enters a reaction axial-flow turbine at 600 ft/s in the direction of 60° with respect to the axis. If the rotor has dimensions $r_t = 1.2$ ft, $r_h = 0.7$ ft and is rotating at 4800 rpm, determine the blade angles at the mean radii of both the inlet and outlet and estimate its output shaft power.

Solution

The velocity diagram is similar to that in Figure 2.23. With

$$r_m = \left[\frac{1}{2}(1.2^2 + 0.7^2)\right]^{0.5} = 0.982\ \text{ft}, \qquad A = \pi(1.2^2 - 0.7^2) = 2.98\ \text{ft}^2,$$
$$U = \frac{N\pi r_m}{30} = \frac{4800\pi \times 0.982}{30} = 493.6\ \text{ft/s},$$

we have

$$V_{a1} = V_1 \cos\, 60° = 300\ \text{ft/s},$$
$$V_{u1} = V_1 \sin\, 60° = 519.6\ \text{ft/s}.$$

From $\tan\beta_{f1} = (V_{u1} - U)/V_{a1} = 0.0867$, we have $\beta_{b1} = \beta_{f1} = 4.95°$. Also $\beta_{b2} = \beta_{f2} = \alpha_1 = 60°$. From the velocity diagram, we have $V_{u2} = W_{u1} = V_{u1} - U = 26$ ft/s. Hence,

$$\frac{P_s}{m} = U(V_{u1} + V_{u2}) = 493.6 \times (519.6 + 26)$$
$$= 2.69 \times 10^5\ \ (\text{ft/s})^2,$$

where $\dot{m} = \rho\, V_{a1}\, A = (0.085/32.2) \times 300 \times 2.98 = 2.36$ slugs/s. So, we have $P_s = 2.36 \times 2.69 \times 105/550 = 1154$ hp.

Example 2.10S Air with a density of 1.36 kg/m^3 enters a reaction axial-flow turbine at 183 m/s in the direction of 60° with respect to the axis. If the rotor has dimensions $r_t = 0.36$ m, $r_h = 0.21$ m and is rotating at 4800 rpm, determine the blade angles at the mean radii of both the inlet and outlet and estimate its output shaft power.

Solution

The velocity diagram is similar to that in Figure 2.23. With

$$r_m = \left[\frac{1}{2}(0.36^2 + 0.21^2)\right]^{0.5} = 0.295\text{ m}, \qquad A = \pi(0.36^2 - 0.21^2) = 0.27\text{ m}^2,$$

$$U = \frac{N\pi r_m}{30} = 4800\pi \times 0.295/30 = 148.3\text{ m/s},$$

we have

$$V_{a1} = V_1 \cos 60° = 91.5\text{ m/s}, V_{u1} = V_1 \sin 60° = 158.5\text{ m/s}.$$

From $\tan \beta_1 = (V_{u1} - U)/V_{a1} = 0.111$, we have $\beta_{b1} = \beta_{f1} = 6.35°$. Also $\beta_{b2} = \beta_{f2} = \alpha_1 = 60°$. From the velocity diagram, we have $V_{u2} = W_{u1} = V_{u1} - U = 10.2$ m/s. Hence,

$$\frac{P_s}{\dot{m}} = U(V_{u1} + V_{u2}) = 148.3 \times (158.5 + 10.2)$$
$$= 2.5 \times 10^4\ (\text{m/s})^2,$$

where $\dot{m} = \rho V_{a1} A = 1.36 \times 91.5 \times 0.269 = 33.5$ kg/s. So, we have $P_s = 33.5 \times 2.5 \times 104 = 83.7 \times 104$ W $= 837$ kW.

Example 2.11E Air with a density of 0.095 lbm/ft^3 flows through the axial-flow impulse turbine stage as shown. If the impeller is rotating at 4500 rpm, draw the velocity diagrams at the mean radii of both the inlet and outlet of the rotor under design conditions. Also, estimate the flow rate and shaft power output from this stage (assuming the ideal case).

Solution

The velocity diagram under design conditions is shown below.

The annular flow area can be calculated as

$$A = \pi\left(r_t^2 - r_h^2\right) = \pi(5.2^2 - 3.5^2) = 46.46\text{ in.}^2 = 0.322\text{ ft}^2$$

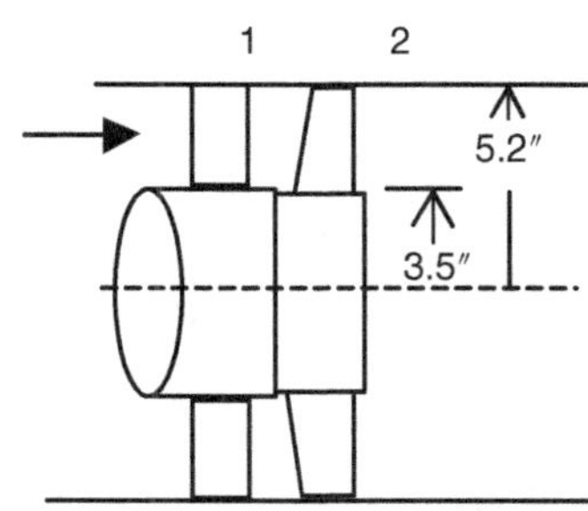

$\beta_{sb1} = 55°$, $\beta_{rb1} = \beta_{rb2} = 30°$

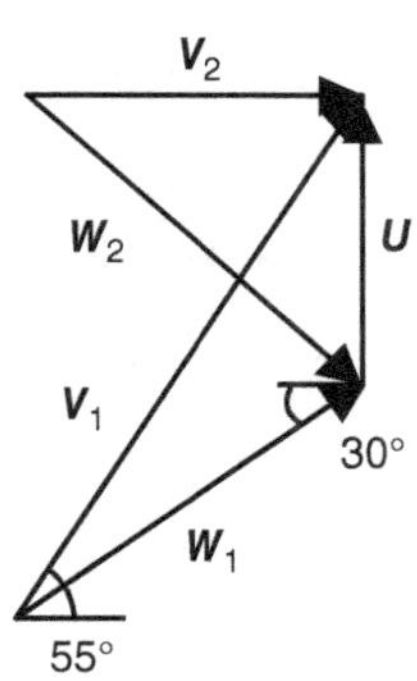

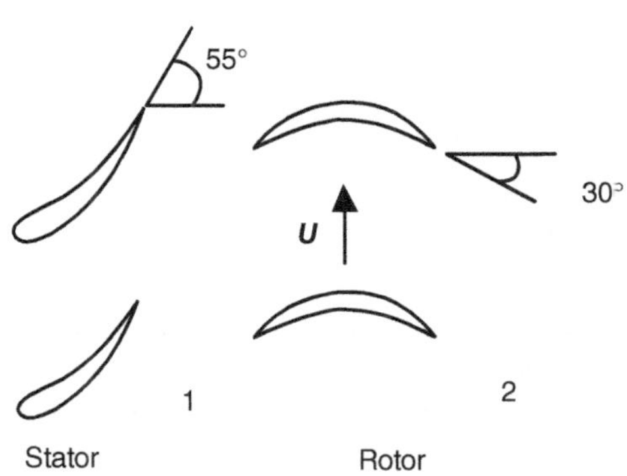

and the mean radius is

$$r_m = \left[\frac{1}{2}\left(r_h^2 + r_t^2\right)\right]^{0.5} = \left[\frac{1}{2}(3.5^2 + 5.2^2)\right]^{0.5} = 4.43 \text{ in} = 0.369 \text{ ft}.$$

Hence, the peripheral velocity of the blade at the mean radius is $U_m = \pi\, N_{rm}/30 = \pi \times 4500 \times 0.369/30 = 174$ ft/s. From the inlet velocity triangle, we have

$$\frac{V_1}{\sin(\beta_1 + \pi/2)} = \frac{U_m}{\sin(\alpha_1 - \beta_1)}.$$

Hence,

$$V_1 = 174 \times \frac{\sin\ 120°}{\sin\ 25°} = 356.5 \text{ ft/s},$$
$$V_a = V_1 \cos\ \alpha_1 = 356.5\ \cos\ 55° = 204.5 \text{ ft/s}.$$

So, we have the flow rate $m = \rho V_a A = 0.095 \times 204.5 \times 0.322 = 6.25$ lbm/s. Also from $\Delta E = U_m V_{u1}$, we have the output shaft power $P_s = m\,\Delta E = (6.25/32.2) \times 174 \times 356.5 \sin 55° = 9862.7$ ft-lbf/s $= 17.9$ hp.

Example 2.12 Draw the velocity diagrams for the air flow around a propeller-type horizontal-axis wind turbine as shown, at both upstream and downstream, in both the frontal and meridional views.

Solution

The airfoil sections are viewed from the top of the rotor, and the relative flow angles are close to the corresponding blade as shown.

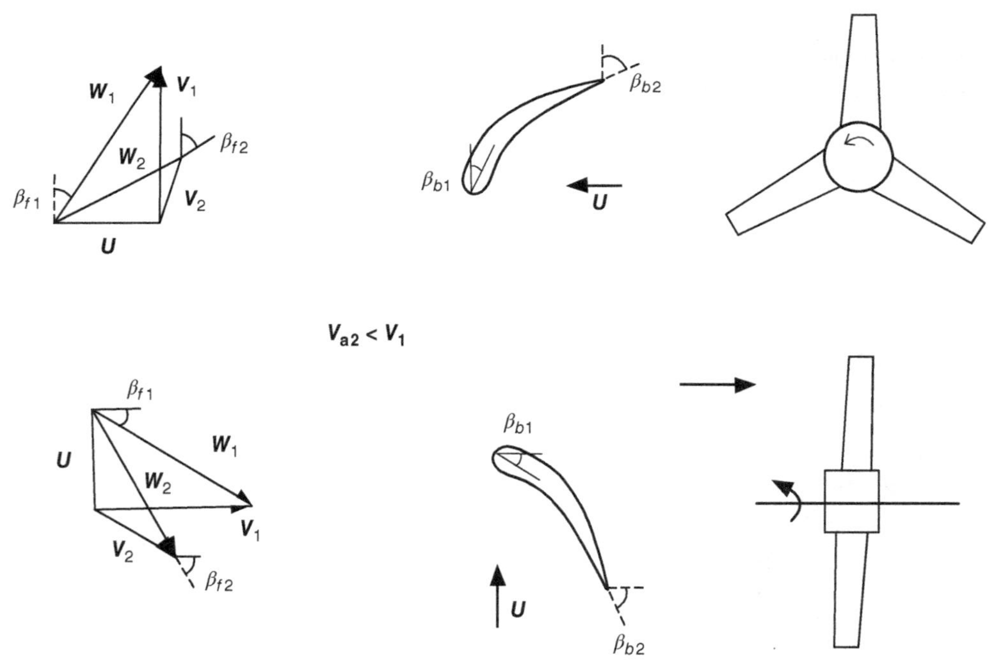

References

1 Davis, C. V. (Ed.), *Handbook of Applied Hydraulics*, McGraw-Hill, New York, 1942.
2 Stepanoff, A. J., *Centrifugal and Axial Flow Pumps*, John Wiley & Sons, New York, 1957.
3 Csanady, G. T., *Theory of Turbomachines*, McGraw-Hill, New York, 1964.
4 *ASHRAE Handbook, Fundamentals Volume*, American Society of Heating, Refrigerating and Air-Conditioning Engineers, Atlanta, GA, 2005, https://shop.iccsafe.org/media/wysiwyg/material/8950P203-toc.pdf.

Problems

Problems from 2.1

2.1 Determine the model-to-prototype wheel diameter ratio for a water turbine which will produce 30,000 hp at 100 rpm with a head of 16 m while the model produces 55 hp under a head of 5 m. What is the speed of the model turbine?

2.2 Use the Cordier diagram shown in Figure 2.2 to estimate the rotor diameter of a pump which will produce a head of 10 m at 17,000 Lpm while running at 1000 rpm. Estimate the efficiency of the pump.

2.3 A model centrifugal fan is tested to deliver 15,000 L/min of air at standard conditions with a pressure rise of 6 cm of water at a speed of 1800 rpm. While the prototype fan will produce the same pressure at 1200 rpm, determine the diameter ratio and the flow rate of the prototype fan.

2.4 The thrust T of a propeller is known to be dependent on axial-flow velocity V, fluid density, ρ viscosity μ, rotating speed N in rpm, and diameter D. Determine the appropriate nondimensional parameters by means of dimensional analysis.

2.5 A centrifugal pump is selected to pump the oil with a specific gravity of 0.7. The pressure rise required is 180 kPa at 100 L/min. If the pump performance given in the catalog is $Q = 160$ L/min, $\Delta P = 250$ kPa at $N = 3600$ rpm for water, by how much must the impeller be trimmed and what is the rotating speed?

2.6 A centrifugal fan rotating at 1200 rpm delivers 5.3 kg/s of air at 20 °C through a duct system, and the motor requires 4.5 kW. If the air temperature changes to 50 °C, what should the fan speed and motor power be in order to deliver the same mass flow rate?

2.7 A jet fan used in highway tunnel ventilation is an axial-flow fan with a cylindrical casing. Its performance is expressed in terms of the static thrust T versus shaft power P_s. These two parameters are a function of impeller diameter D, rotating speed ω, and air density ρ. Determine the corresponding nondimensional parameters.

2.8 A model propeller 0.5 m in diameter rotating at 1800 rpm is tested in a wind tunnel with an air velocity of 60 m/s. The thrust and torque measured are 90 N and 18 N.m, respectively. Determine the expected performance of the prototype and its rotating speed if its diameter is 2.5 m and air velocity is 100 m/s. Air viscosity and compressibility can be neglected.

2.9 Repeat problem 2.4, but now the air compressibility effect is considered but the viscosity effect can be neglected in the high-speed propeller.

2.10 Calculate the efficiency of the following ump test data:

Q (gpm)	H (ft)	τ (lb_f-ft)	N (rpm)	η(%)
80	25	3.4	1800	
160	23	4.0	1800	
240	14	4.8	1800	

2.11 A multistage boiler-feed pump is designed to pump water for a total head of 25 m at 0.10 m^3/s when rotating at 25 rad/s. If the impeller diameter is selected to be 70 cm for the optimum design, how many stages are required?

2.12 A hydraulic turbine model tested with a head of 40 m delivers 4700 kW with 15 m^3/s while rotating at 600 rpm. If the prototype is double the model in size, what are the corresponding rotating speed, flow rate, and output power with the same available head?

2.13 A centrifugal pump model one-tenth the size of the prototype, running at 1500 rpm, delivers a head of 2 m at a flow rate of 0.2 m^3/s and requires an input shaft power of 9.8 kW. Determine the expected flow rate, head, and shaft power for the prototype running at the same speed.

2.14 A multistage pump is required to deliver water to a head of 50 m at a flow rate of 0.1 m^3/s. A motor rotating at 1800 rpm is available. Determine the optimal number of stages for the pump.

2.15 A hydraulic turbine operating under a head of 40 m will deliver an optimal shaft power of 4700 kW when running at 600 rpm with a flow rate of 15 m^3/s. If the head is changed to 60 m, determine its rotating speed, flow rate, and output shaft power under the optimal condition.

2.16 A wind turbine is rotating at 50 rpm and producing 6.5 hp under a wind speed of 32 km/h. Determine its rotating speed and output power when the wind speed is 50 km/h.

2.17 If a scale-up wind turbine to produce 50 hp of output power under a wind speed of 25 mph is needed, determine the scale-up factor.

Problems from 2.2

The ideal case of inviscid flow and zero blade thickness is assumed in the following problems.

2.18 A centrifugal pump impeller rotating at 1800 rpm has the dimensions shown. Assuming a radial inlet flow, draw the velocity diagrams at the inlet and outlet; also determine the rated flow capacity and the ideal head at design condition.

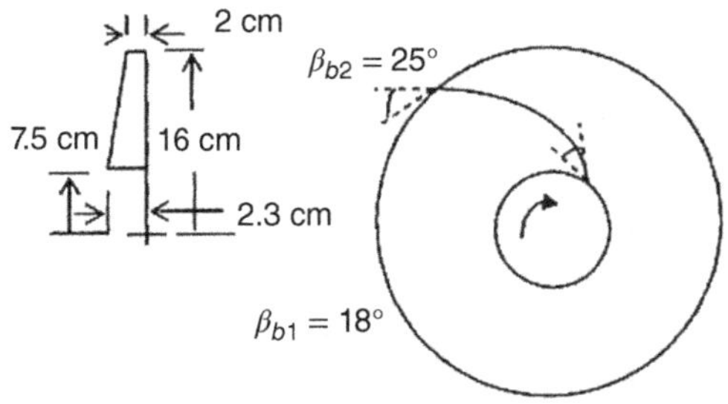

2.19 The axial-flow compressor shown is rotating at 2400 rpm and receives air in a pure axial direction. If the air density change across the single stage is negligible, draw the velocity diagrams at both the inlet and outlet. Also determine the ideal power input required if the axial-flow area is 0.15 m^2 and the air density is 1.20 kg/m^3.

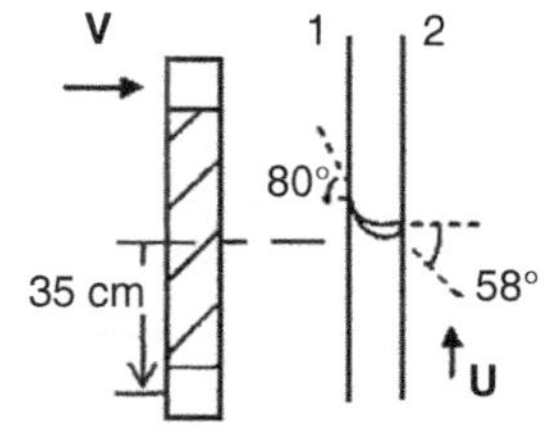

2.20 The impeller of a centrifugal oil pump with inlet and outlet radii of 5 and 15 cm, respectively, is rotating at 4800 rpm. Oil with a specific gravity of 0.81 enters the pump at a rate of 45 lps (liters per second). The blade angle at the impeller outlet is pure radial, and the constant blade height is 4 cm. Draw the velocity diagrams at both the inlet and outlet at the design condition. Determine the ideal head rise and input power required.

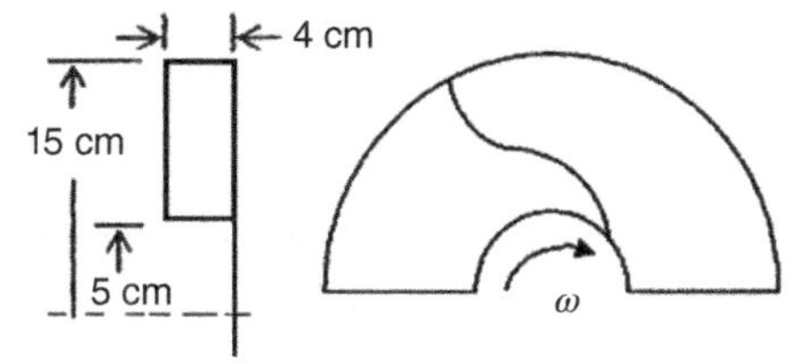

2.21 An axial-flow gas turbine of 8 in. mean radius with the blade profile shown is rotating at 2400 rpm. Draw the velocity diagrams at both the inlet and outlet at the design condition. If the mass flow rate of air is 3.5 kg/s, what is the ideal shaft power output in horsepower?

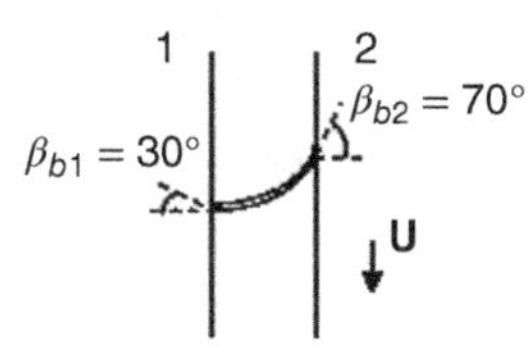

2.22 Air leaves the nozzle of an axial-flow turbine stage at 30° with respect to the tangential direction, as shown, at an axial flow velocity of 180 m/s. If the rotor blade speed at the mean radius is also 600 fps, find the rotor blade angles β_{b1} and β_{b2} if the reaction of the stage is to be 50%. Draw the velocity diagrams at both the inlet and outlet of the rotor.

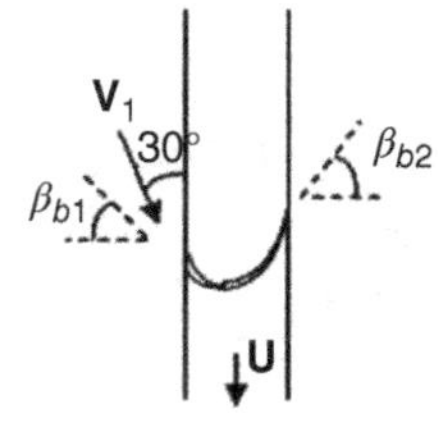

2.23 An inward-flow turbine has a 60% reaction. The blade speed at the inlet is 40 fps, and the meridional velocity of the flow is constant at 8 fps. The rotor diameter at the inlet is twice that at the outlet. Find the blade angle at both the inlet and outlet assuming there is no whirl at the outlet.

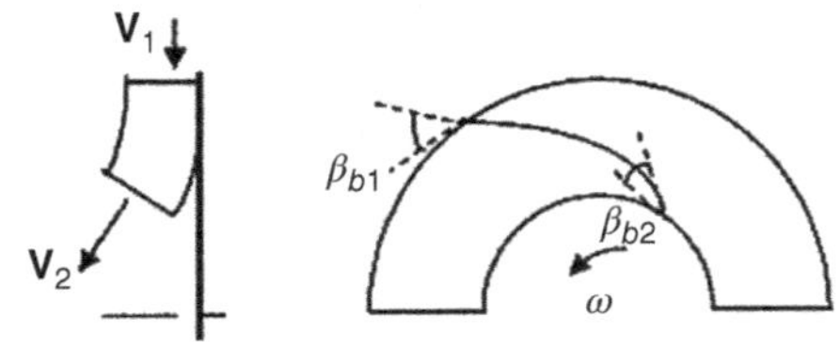

2.24 The axial-flow hydraulic turbine impeller shown is rotating at 1800 rpm. Draw the velocity diagrams of both the inlet and outlet at the mean radius and determine the design flow rate and ideal output horsepower.

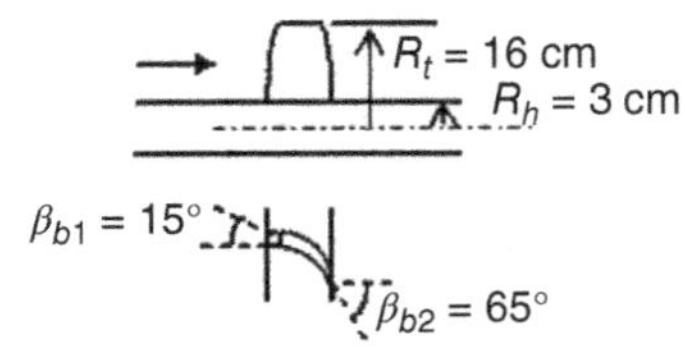

2.25 If the turbine impeller in problem 2.23 rotates in the reverse direction with the same speed as a pump, what is the ideal head rise when the meridional flow velocity is 5 ft/s?

2.26 Draw the velocity diagrams at the impeller inlet and outlet for an axial-flow impulse turbine with fixed blade angles at three different rotating speeds with a fixed flow rate.

2.27 The axial-flow fan with hub/tip radii as shown is rotating at 690 rpm and delivers 4 m^3/s with a discharged static pressure of 60 Pa. If the exit area is 0.8 m^2 and the hydraulic efficiency based on total pressure is 87%, determine the blade angles at mean radii of leading and trailing edges. Neglect the incidence at the leading edge and the flow deviation at the trailing edge.

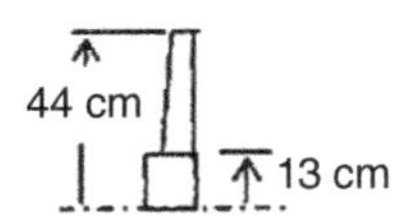

2.28 The Francis-type pump impeller with inlet/outlet dimensions shown is rotating at 1750 rpm. Determine the flow rate and ideal discharge head under the design condition.

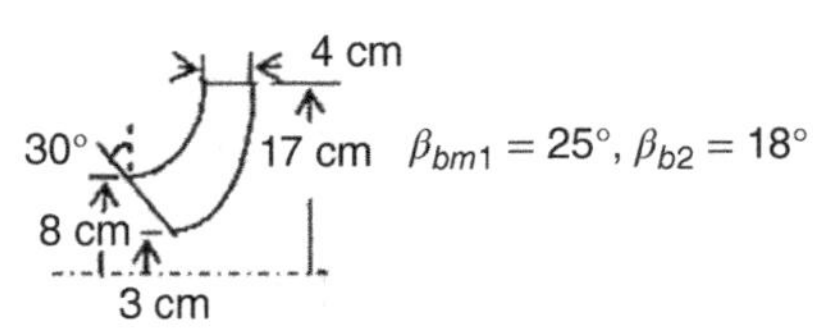

2.29 Air at 101.3 kPa and 15°C enters and leaves an axial-flow compressor stage with a velocity of 150 m/s without swirl. The stage has a tip radius of 56 cm and a hub radius of 35 cm and rotates at 8000 rpm. The air is turned through 12° as it passes through the rotor with negligible incidence and deviation from the blades. (a) Determine the blade angles at the leading and trailing edges of both the rotor and stator at the mean radii. (b) Calculate the mass flow rate and the power required for each stage.

3

Pumps

3.1 Centrifugal Pumps

The centrifugal pump is a common machine in modern civilization. Its application can be found in every industry, household, and daily life wherever there is liquid to be transported or processed. Sometimes the liquid can be very hot or even hazardous.

The construction of a basic centrifugal pump is shown in Figure 3.1.[1] The rotating components, consisting of an impeller, shaft, and bearings, are enclosed in a housing with the associated seals in the back and a wear ring in the front. This is the so-called end-suction, overhung, single-stage pump with a radial-type impeller.

Pumps can be classified according to (a) impeller type, (b) casing type, (c) single or double suction, (d) single or multistage, (e) orientation of axis, either horizontal or vertical, and (f) direction of casing split, either axially or radially. The types of impellers are shown in Figure 1.1 and in Figure 3.2[2] again with the corresponding specific speeds. The ratio of discharge diameter to the inlet eye is smaller for the Francis type compared to the radial impeller type. Flow discharge from the impeller is collected with a volute and then exits through a diffuser, as shown in Figure 3.1.

Types of casings are shown in Figure 3.3. The double volute is designed to reduce the radial thrust force due to the unbalanced pressure distribution around the impeller. The concentric casing is designed to provide generous space for the solid particles to pass through when pumping solid–liquid mixtures. The detailed classifications are given in the Hydraulic Institute Standards.[3] The schematic front view, meridional plane view, isometric view, and associated terminology for a Francis impeller are shown in Figure 3.4.

3.1.1 Basic Working Principles

The centrifugal pump is a machine used to convert the mechanical energy from a prime mover into the hydraulic energy of a flowing fluid. The ideal energy transfer rate without considering any loss was discussed in Chapter 2. From Equation (2.38), the energy transfer per unit mass of fluid, $\Delta E_p = U_2 V_{u2} - U_1 V_{u1}$, is related to the head rise, designated as $gH_i' = P_s/m = \Delta E_p$. The swirl velocity at inlet V_{u1} can be neglected at the design flow rate or around the best efficiency points (b.e.p.). Hence the ideal head rise can be written in the form.

$$H_i' = \frac{U_2 V_{u2}}{g} = \left(\frac{U_2}{g}\right)\left(U_2 - \frac{Q}{A_2 \tan \beta_{f2}}\right), \tag{3.1}$$

Fundamentals of Turbomachinery: Theory and Applications, Second Edition. Ryoichi S. Amano and William W. Peng.

Companion website: www.wiley.com/go/amano/turbomachinery2e

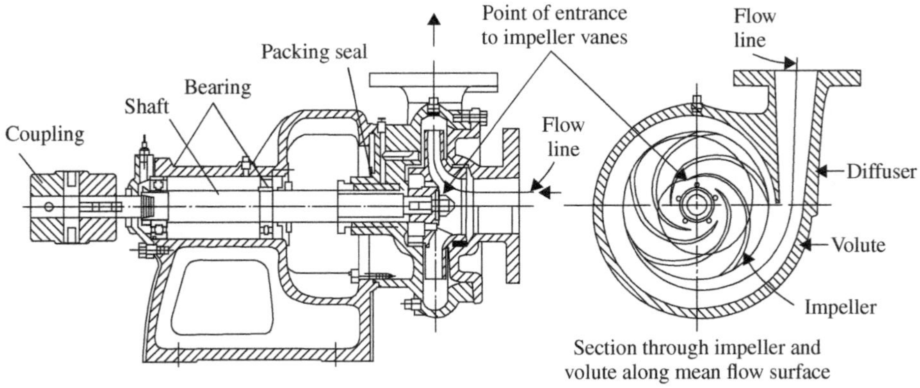

Figure 3.1 Construction of a basic centrifugal pump (single suction, overhung, horizontal axis with radial impeller). (Reprinted by permission from Karassik, I. J. et al. (Eds.), *Pump Handbook*, McGraw-Hill, Inc., New York, 1976.)

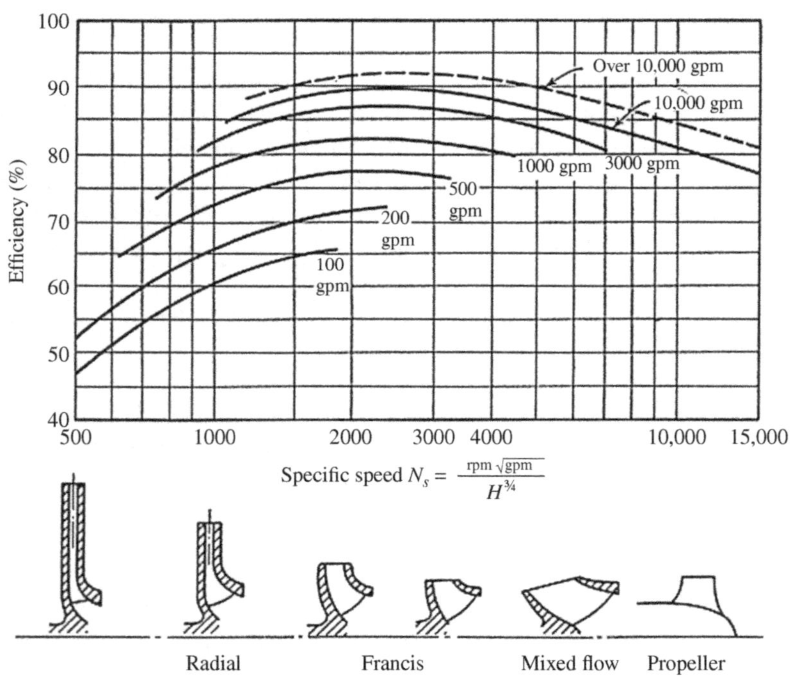

Figure 3.2 Impeller types and efficiencies as a function of specific speed. (Reprinted by permission from Stepanoff, A. J., *Centrifugal and Axial Flow Pumps*, 2nd ed., John Wiley & Sons, New York, 1957.)

where A_2 is the impeller exit flow area and β_{f2} is the relative flow angle, which is equal to the blade angle β_{b2} at the impeller exit in the ideal case, as shown in Figure 3.5. Hence for a given impeller rotating at a fixed speed, that is, constant U_2, the ideal head rise decreases with an increase in the flow rate if the blade angle is backward (or $\beta_{b2} < 90°$) because tan β_{f2} in Equation (3.1) is a positive value. Otherwise the head rise will increase or be constant when the flow rate increases, as shown in Figure 3.6*a*. In practice, most pump impellers are designed with exit blade angle less than 90° to have a stable operation in the hydraulic system.

Flow through a pump will never be ideal. It is generally turbulent and three-dimensional or even unsteady. Hence the actual relative flow direction at the impeller exit is different from the blade angle. In other words, β_{f2} is

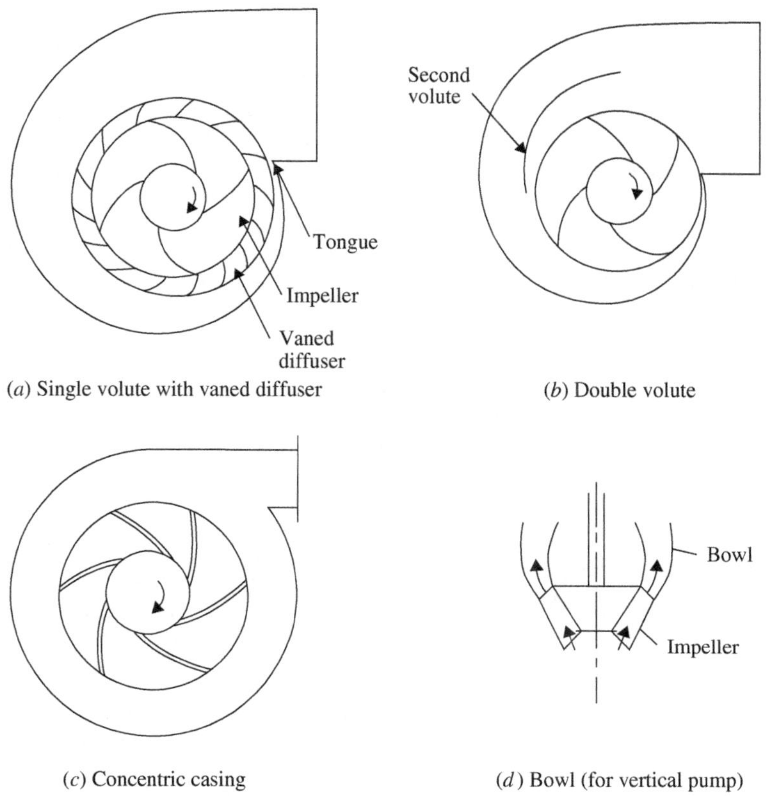

Figure 3.3 Types of casing for centrifugal pumps.

always less than β_{b2}, as shown by the dashed lines in Figure 3.5. And some pressure (or head) losses due to friction or secondary flow in the flow passage from the pump inlet to discharge should be expected. Hence the actual head rise can be written as

$$H_a = \eta_h H_i = \eta_h \mu_s \left(\frac{U_2}{g}\right)\left(U_2 - \frac{Q}{A_2 \tan \beta_{b2}}\right), \tag{3.2}$$

where μ_s, the slip factor, is defined as the ratio of actual tangential flow velocity and the ideal one at the impeller exit, while η_h is the hydraulic efficiency to account for friction and secondary flow losses of the through flow. Both of these parameters are empirical in nature and can only be determined from tests or correlations in general. They are not constant as the pump through flow rate changes, and hence the head–flow rate characteristic is a curve instead of a straight line, as shown in Figure 3.6*b*. The slip coefficient μ_s can be expressed as $\mu_s = V_{u2}/V'_{u2} = 1 - \Delta V_s/V'_{u2}$, where ΔV_s, called the slip velocity and related as $\Delta V_s = V'_{u2} - V_{u2} = W_{u2} - W'_{u2}$, is proportional to the flow circulation at impeller discharge. Or we have $\Delta V_s \sim Nd$, where N is the impeller rotation speed and d is the effective flow passage dimension at the discharge shown. Hence, we have

$$\frac{\Delta V_s}{V'_{u2}} \sim \frac{Nd}{V'_{u2}} \sim \frac{r_2 Nd}{r_2 V'_{u2}} \sim \frac{U_2 d}{s Z_b V'_{u2}} \sim \frac{\sin \beta_{b2}}{Z_b},$$

where $s = 2\pi r_2/Z_b$ is the arc length between two adjacent blades, $\sin \beta_{b2} \sim d/s$, Z_b is the number of blades, and V'_{u2} is proportional to U_2.

A simple correlation for μ_s cited in the literature (Ref. 1) is Stodola's formula, $\mu_s = 1 - \pi \sin \beta_{b2}/Z_b$. It is intuitively reasonable since a smaller value of β_{b2} and a higher number of blades, Z_b, will make the flow channels longer and

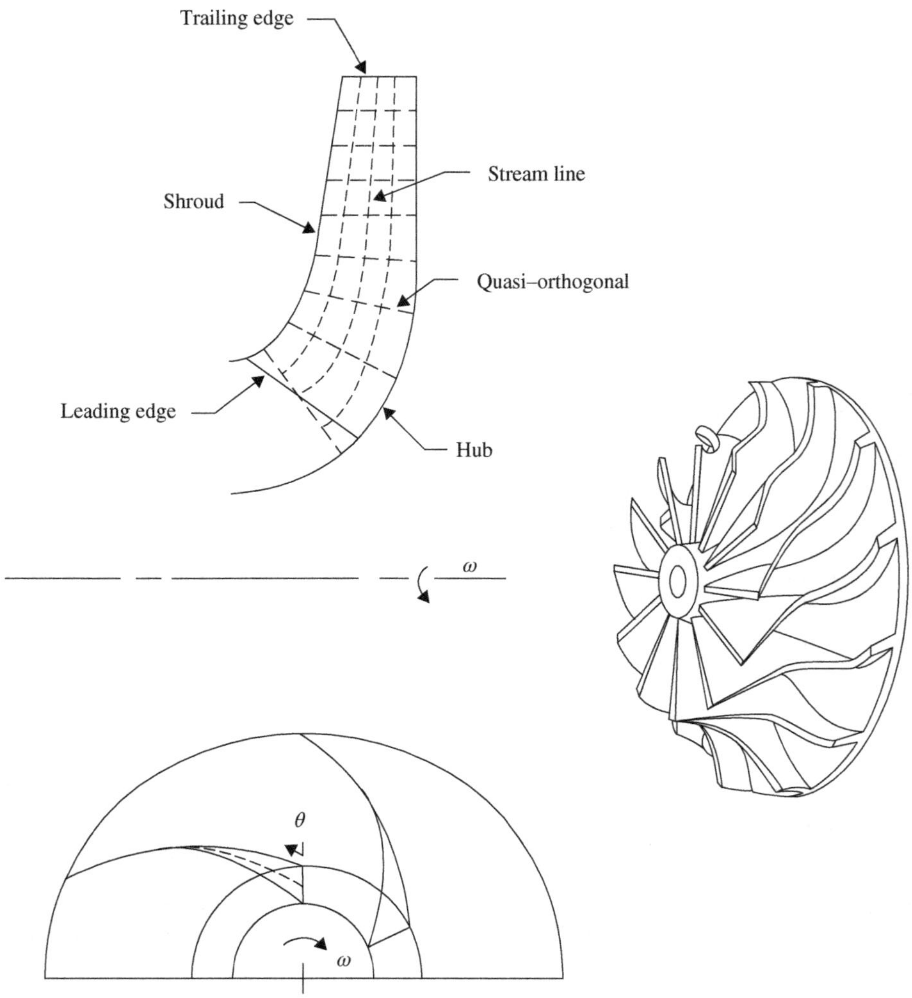

Figure 3.4 Isometric, meridional, and front view of a Francis type impeller.

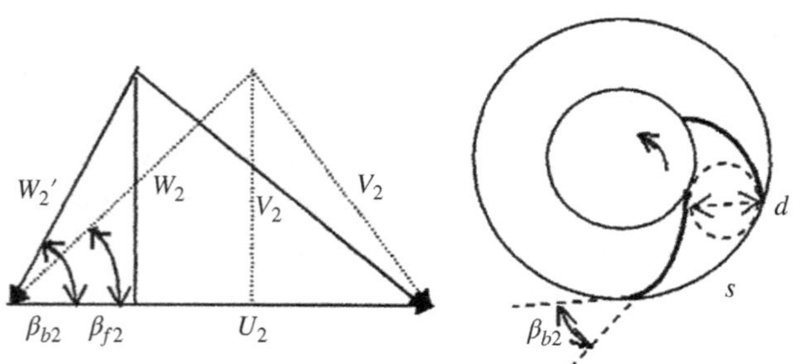

Figure 3.5 Velocity triangle with flow slip at impeller discharge.

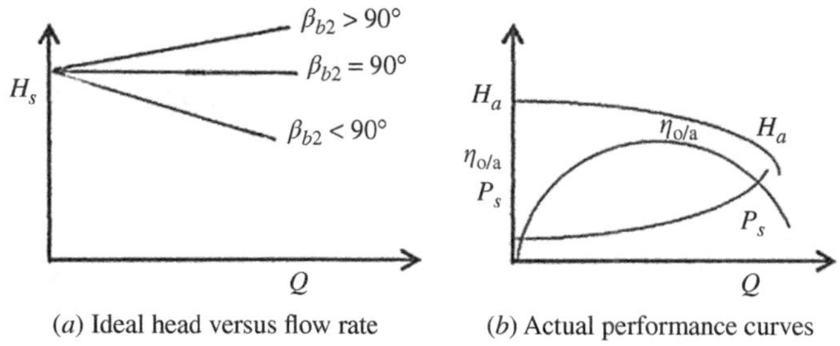

Figure 3.6 Basic pump performance curves.

narrower. Hence it results in better guidance for the flow, less flow slip, and a value closer to unity for μ_s. Of course, that will cause an increase in flow friction loss.

Also, a simple empirical formula for η_h is cited in Ref. 1, $\eta_h = 1 - 0.8/Q^{0.25}$, where Q is the design flow rate in gallons per minute (gpm). More rigorous procedures to determine friction loss in the flow passage are available, but they are beyond the scope of this text, or flow analysis with CFD software is needed.

The total shaft power required (also called brake horsepower) will include some extra power for pumping leakage flow (designated as Q_L) from the impeller exit back to the inlet through the clearance and some power to overcome the friction in bearings and seals (designated as P_m) and disk friction due to impeller shrouds (designated as P_{df}). These losses are accounted for by the volumetric efficiency and mechanical efficiency, defined as

$$\eta_v = \frac{Q}{Q+Q_L}, \quad \eta_m = \frac{P_{imp}}{P_s} = \frac{P_s - P_m - P_{df}}{P_s},$$

where P_{imp} is the power delivered by the impeller to the fluid and is expressed as $P_{imp} = \rho(Q+Q_L)gH_i$. The overall efficiency of a centrifugal pump (not including the prime mover's efficiency), defined as $\eta_{o/a} = P_a/P_s = \rho QgH_a/P_s$, can be expressed in the form

$$\eta_{o/a} = \frac{\rho QgH_a}{P_s} \times \frac{Q+Q_L}{Q+Q_L} \times \frac{H_i}{H_i} = \frac{H_a}{H_i} \times \frac{Q}{Q+Q_L} \times \frac{\rho(Q+Q_L)gH_i}{P_s} = \eta_h\eta_v\eta_m.$$

Or the total shaft power can be related to the power components as $P_s = P_h + P_f + P_L + P_m + P_{df}$, where $P_h = \rho QgH_a$ is the useful output power, $P_f = \rho Qg(H_i - H_a)$ is the loss due to flow friction, and $P_L = \rho Q_L gH_i$ is the loss due to leakage.

The pressure drop due to flow friction can also be written as $\Delta H_f = kQ^2$, similar to that in the stationary channel flow, in the vicinity of the design point. Hence the actual head H_a and its derivative with respect to Q can be written as

$$H_a = \mu_s\left(\frac{U_2}{g}\right)\left(U_2 - \frac{Q}{A_2\tan\beta_{b2}}\right) - kQ^2, \tag{3.3}$$

$$\frac{\partial H_a}{\partial Q} = -\mu_s\frac{U_2}{gA_2\tan\beta_{b2}} - 2kQ, \tag{3.4}$$

where μ_s is assumed to be constant. Using these expressions, it will be possible to determine the parameters μ_s and k from the test performance curves of a given pump at the design flow rate. And the hydraulic efficiency η_h can be calculated from $\eta_h = H_a/(H_a + kQ^2)$. The detailed procedure is given in Refs. 4 and 5. This procedure can be used to check the formula for μ_s and η_h just mentioned if test data are available.

The similarity laws discussed in Section 2.1.4 are also extensively used in predicting the pump performance from one speed to another or between two similar pumps. They are listed here for convenience, that is,

$$Q \sim ND^3, \quad H_a \sim N^2D^2, \quad P_s \sim N^3D^5.$$

Example 3.1 The centrifugal pump with a radial impeller shown is tested at 1750 rpm. If the result is given as $H_a = 60$ m at the design flow rate $Q = 4.5$ L/s, $P_s = 3.2$ kW, and $\partial H_a/\partial Q = -2.5$ m/(L/s). Volumetric efficiency due to leakage is estimated to be 0.95. Determine the slip factor μ_s, hydraulic efficiency η_h, and mechanical efficiency η_m.

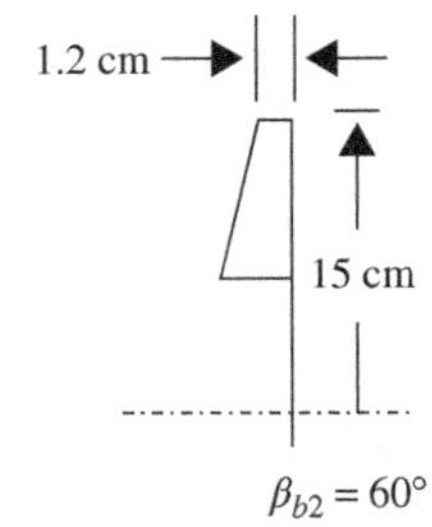

Solution

With

$$U_2 = \left(\frac{N\pi}{30}\right) r_2 = \left(\frac{1750\pi}{30}\right) \times 0.15 = 27.5 \text{ m/s},$$

$$A_2 = 2\pi r_2 b_2 = 2\pi \times 0.15 \times 0.012 = 0.0113 \text{ m}^2,$$

Equations (3.3) and (3.4) can be written as

$$\left(\frac{27.5}{9.8}\right)\left(27.5 - \frac{4.5}{1000 \times 0.0113 \tan 60°}\right)\mu_s - \left(\frac{4.5}{1000}\right)^2 k = 60,$$

$$\left(\frac{27.5}{9.8 \times 0.0113 \tan 60°}\right)\mu_s + \left(2 \times \frac{4.5}{1000}\right) k = 2.5 \times 1000,$$

or $76.52\mu_s - 2.025 \times 10^{-5}k = 60$ and $1.434\mu_s + 9 \times 10^{-5}k = 25$. Solve for μ_s and k. We obtain

$$\mu_s = 0.852, \quad k = 2.642 \times 10^5,$$

$$\eta_h = \frac{H_a}{H_a + kQ^2} = 0.918, \quad \eta_{o/a} = \frac{\rho Q g H_a}{P_s} = 0.825, \quad \eta_m = 0.946.$$

3.1.2 Performance Characteristics

The parameters of head, shaft power, and overall efficiency versus flow rate for a typical centrifugal pump with the impeller rotating at a constant speed are shown in Figure 3.6*b*. However, in industrial applications, for convenience of pump selection, manufacturers generally publish performance curves for their pump models in the form shown in Figure 3.7*a*.[4] Several impellers trimmed to different diameters fitted for a given size of casing are presented in a single chart. The shaft power, efficiency, and net positive suction head required (NPSHR to prevent cavitation, to be discussed in the next section) are plotted in constant-contour forms. Performance ranges of different sizes of the same pump model are plotted in the composite form shown in Figure 3.7*b*. Detailed procedure for reading these charts will be illustrated in the examples later.

The various types of impellers illustrated in Figure 3.2 have different distinguished performance curves when plotted on a percentage basis, as shown in Figure 3.8. It should be realized that this is only true in a general qualitative nature. Depending on the detail design, there is variation from one pump to another within the same type. It is noted that low-specific-speed pumps with radial-flow impellers have flat characteristic curves, while higher-specific speed impeller pumps have steeper head curves.

The corresponding shaft power increases to a peak at the shutoff condition. This is due to severe flow recirculation at the impeller's inlet and outlet regions at low through-flow rates, as shown in Figure 3.9. Recirculation flow is a complicated phenomenon. It causes severe vibration and temperature rise due to the extra power consumption. More details on this topics can be found, for example, in Refs. 5–7.

Example 3.2 Test data of an end-suction water pump operating at 1350 rpm are given as

Q (gpm)	285	435	540	785	920	1275
H (ft)	200	195	190	186	172	130
P_s (hp)	31	36	42	44	49	58

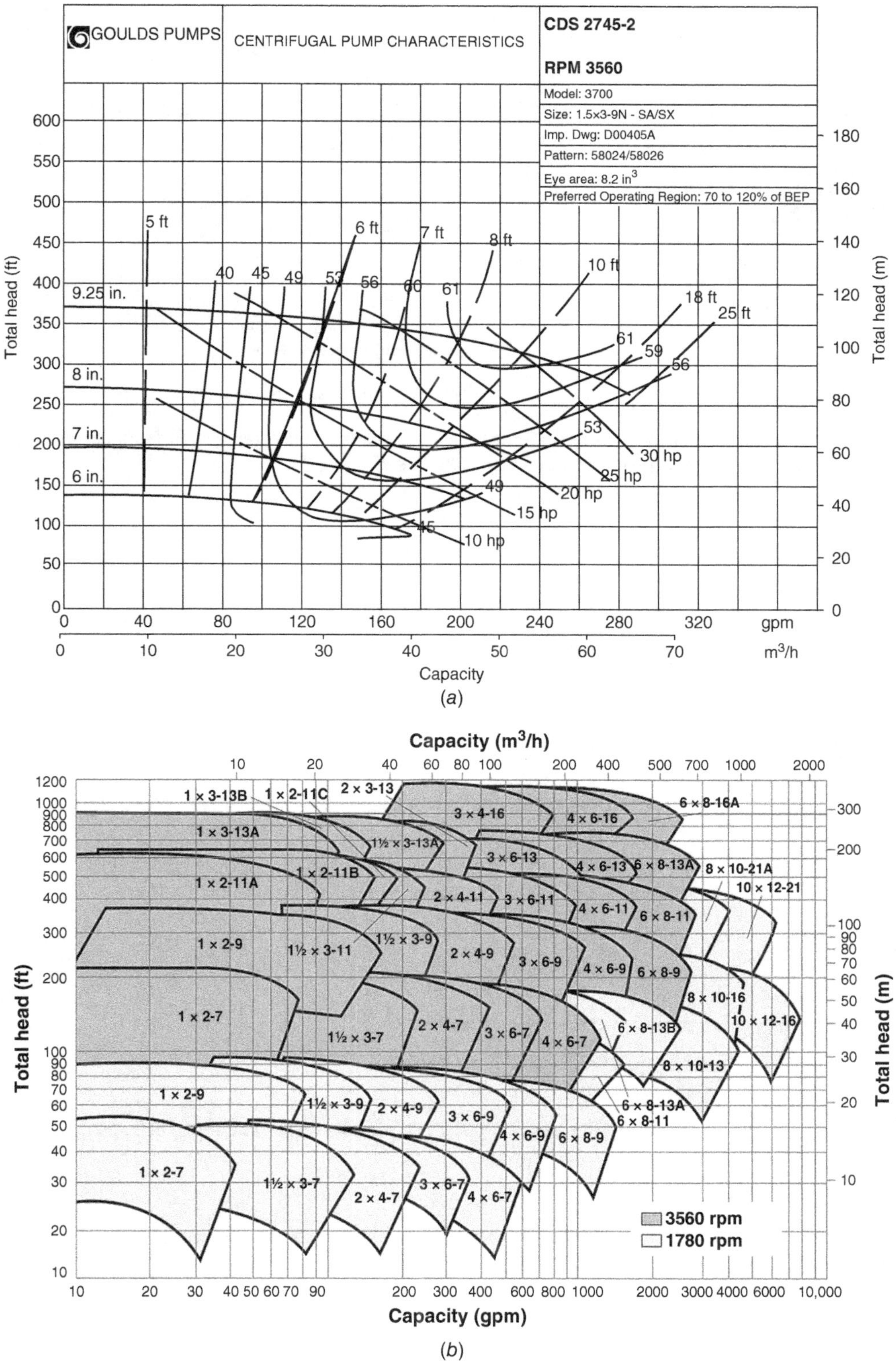

Figure 3.7 (*a*) Performance curves of the pump (size 1.5 × 3–9N) with impeller trimmed. (*b*) Composite performance chart of different sizes of a pump model. (Courtesy of Goulds Pumps, ITT Corp., Seneca Falls, NY.)

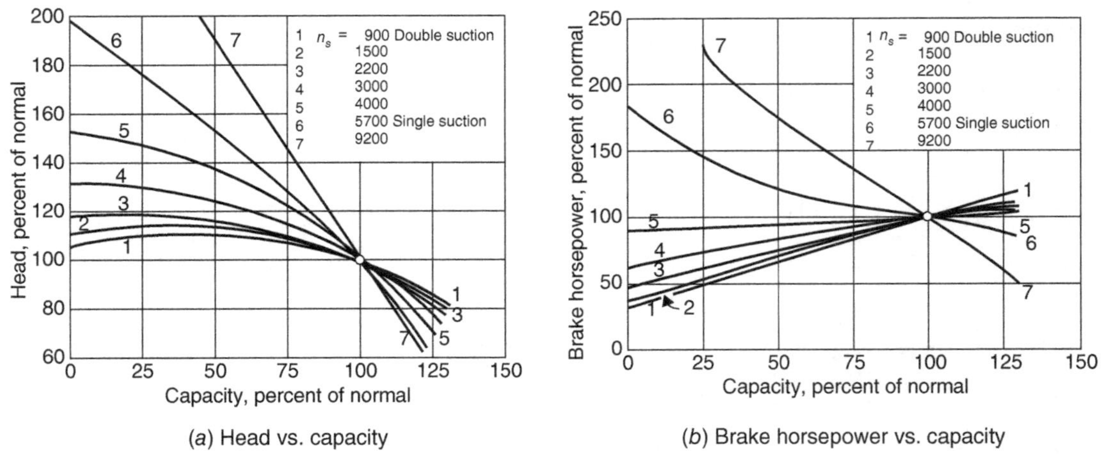

Figure 3.8 Effect of specific speed (impeller type) on pump performance curve shapes. (Reprinted by permission from Stepanoff, A.J., *Centrifugal and Axial Flow Pumps*, 2nd ed., John Wiley & Sons, New York, 1957.)

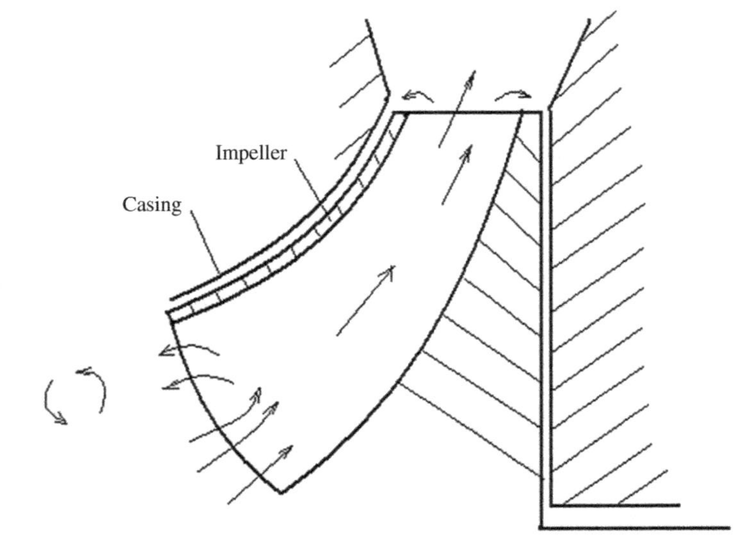

Figure 3.9 Flow recirculation at impeller inlet and outlet, for partial through flow rate.

Plot the complete performance curves including the efficiency with a spread sheet (such as Excel). Also, predict this pump's performance for $N = 1850$ rpm.

Solution

The relevant formulas are given as

$$\eta = \frac{\rho QgH}{P_s} = \frac{62.4 \times 0.00223QH}{550P_s},$$

$$Q_2 = \frac{Q_1N_2}{N_1}, \quad H_2 = H_1\left(\frac{N_2}{N_1}\right)^2, \quad P_{s2} = P_{s1}\left(\frac{N_2}{N_1}\right)^3, \quad \eta_2 = \eta_1.$$

Figure 3.10 Performance deterioration due to cavitation.

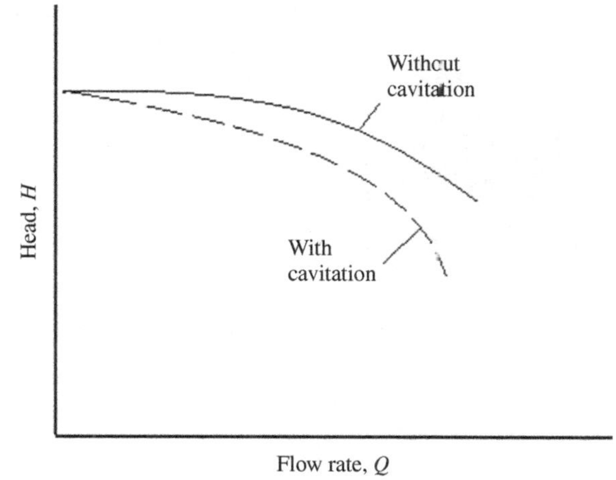

Note: H and P_s at $Q = 0$ are obtained with extrapolation

Q_1 (gpm)	H_1 (ft)	P_{s1} (hp)	η_1 (%)	Q_2 (gpm)	H_2 (ft)	P_{s2} (hp)	η_2 (%)
0	205	28	0	0	384.99	72.044	0
285	200	31	46.5194	390.45	375.6	79.763	46.5168
435	195	36	59.6131	595.95	366.21	92.628	59.6099
540	190	42	61.8043	739.8	356.82	108.066	61.8009
785	186	44	83.9558	1075.45	349.308	113.212	83.9512
920	172	49	81.7035	1260.4	323.016	126.077	81.6991
1275	130	58	72.3013	1746.75	244.14	149.234	72.2974

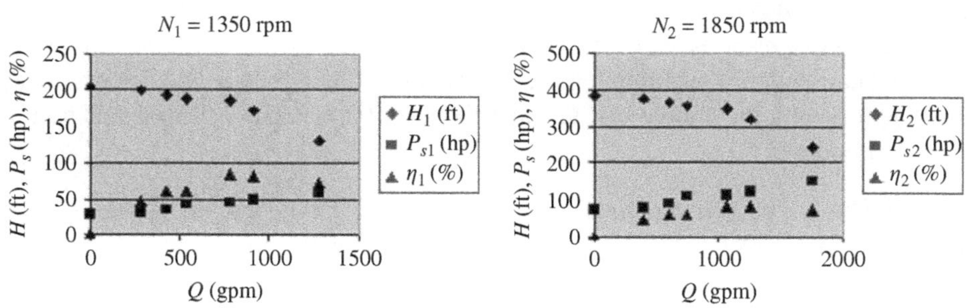

3.1.3 Cavitation

Cavitation is a phenomenon that may occur in all hydraulic machines and systems. When the local pressure reaches a certain low value close to the vapor pressure corresponding to the fluid temperature, part of the fluid will vaporize and cause a change in the flow pattern. In a centrifugal pump, this happens in the vicinity of the impeller inlet. The occurrence of cavitation causes noise, vibration, and deterioration of performance, as shown in Figure 3.10. In the long run, it will also damage the pump components because of the impinging forces resulting from vapor bubbles collapsing at the higher-pressure regions in the flow passage. This is called cavitation erosion. In some high-quality pumps, an inducer such as shown in Figure 3.11[8] is installed upstream of the impeller to improve the cavitation performance.

Figure 3.11 Inducers to improve cavitation performance. (Courtesy of Ingersoll Rand Company, Phillipsburg, NJ.)

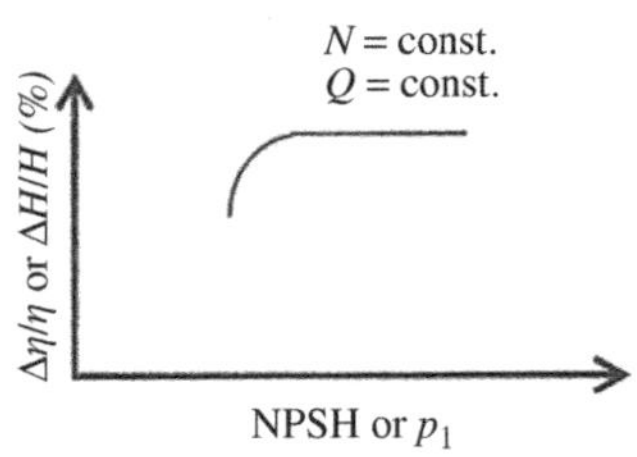

Figure 3.12 Typical cavitation test result.

In order to prevent the occurrence of cavitation, a certain minimum pressure must be imposed at the pump inlet. A parameter called the net positive suction pressure (NPSP) or net positive suction head (NPSH = NPSP/ρg) is used to designate this minimum inlet pressure. It is defined as NPSP = $p_{01} - p_v$, where p_{01} is the total pressure at the pump suction flange and p_v is the vapor pressure corresponding to the pumped fluid temperature, which can be obtained from steam tables in the case of water. For each pump, the required NPSP also varies with respect to the flow rate, as shown in Figure 3.7*a*. It is called NPSPR. Other parameters are also used in industry, such as the Thomas parameter $\sigma = \text{NPSH}/H$, where H is the pump discharge head, the cavitation parameter $\tau = \text{NPSP}/\rho U_1^2$, or the suction specific speed $S = NQ^{0.5}/(\text{NPSH})^{0.75}$ in the dimension of rpm(gpm)$^{0.5}$/ft$^{0.75}$. The values of these parameters for each pump have to be determined from laboratory tests or empirical correlations. The onset of cavitation always takes place at the impeller inlet tip, where the relative flow velocity is maximum, but the detailed flow pattern is difficult to determine unless CFD software is used.

A typical cavitation test result is shown in Figure 3.12. Detailed discussion on the test facility and procedure will be given in a later section. The required NPSP is usually set at a 3% drop in efficiency. The general range of suction specific speeds with the dimension indicated above for the meticulously designed pumps is estimated by G. F. Wislicenus[9] as follows:

Single-suction pumps with overhung impeller	$7500 < S < 10{,}000$
Single-stage pumps with shaft running through impeller inlet	$6500 < S < 9000$
Single-suction high-pressure multistage pumps	$5500 < S < 7500$

The lower value of suction specific speed means poor cavitation performance since it requires higher NPSH.

In selecting a pump for a hydraulic system, the pressure at the pump inlet should be such that its corresponding available NPSP be greater than the required NPSP specified by the pump manufacturer, that is, NPSPA > NPSPR. The available NPSP or NPSPA can be calculated according to the system condition. Two examples are illustrated in Figure 3.13. If the pump is required to pump water from an open tank, the available NPSP at the pump inlet is

$$\text{NPSPA} = p_{01} - p_v = p_1 + \frac{1}{2}\rho V_1^2 - p_v = p_a \pm \rho g Z - p_f - p_v,$$

where the minus sign is for the lift when the pump centerline is above the tank water level and the plus sign is for the head when the pump centerline is below the tank water level. While *the pressure drops are* due to friction in the suction pipe, p_a is the atmospheric pressure, which will be lower at higher altitude. Or $p_a = p_t$ is the tank pressure for an enclosed tank.

To improve a pump's cavitation performance, that is, to have a lower required NPSP, it is important to streamline the flow at the suction area. For the high-specific-speed pumps (mixed- or axial-flow impeller), a higher number

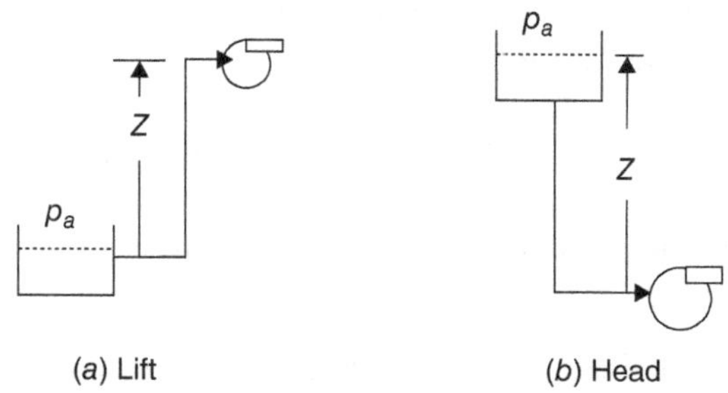

Figure 3.13 Available NPSP with system lift and head.

of vanes will help. For the low-specific-speed pumps (radial-flow impeller), removal of some vanes at the leading edge to increase the flow passage area can generally improve cavitation performance. Also, to reduce the noise and vibration due to cavitation, a small amount of air can be introduced to reduce the collapsing forces of vapor bubbles.

Example 3.3 Plot the performance curves of the pump with 8-in. impeller in Figure 3.7*a* in the typical form shown in Figure 3.6*b*.

Solution

The expression $P_s = \rho QgH/\eta = 62.4 \times 0.00223QH/\eta$ is used to calculate the shaft power for all finite flow rate conditions. For the shutoff condition, since η is zero, P_s has to be extrapolated. Also, $0.1H$ is plotted, instead of H in the same chart, because its order of magnitude is higher than the others.

Q (gpm)	$H/10$ (ft)	eff (%)	NPSHR (ft)	P_S (hp)
40	27	30	5	9.108
80	26.5	42.5	5.5	12.62024
120	25.5	52	6.1	14.88808
160	24	56.7	7.3	17.13439
200	21.5	57	11.5	19.08596
220	20	54.5	16	20.42569

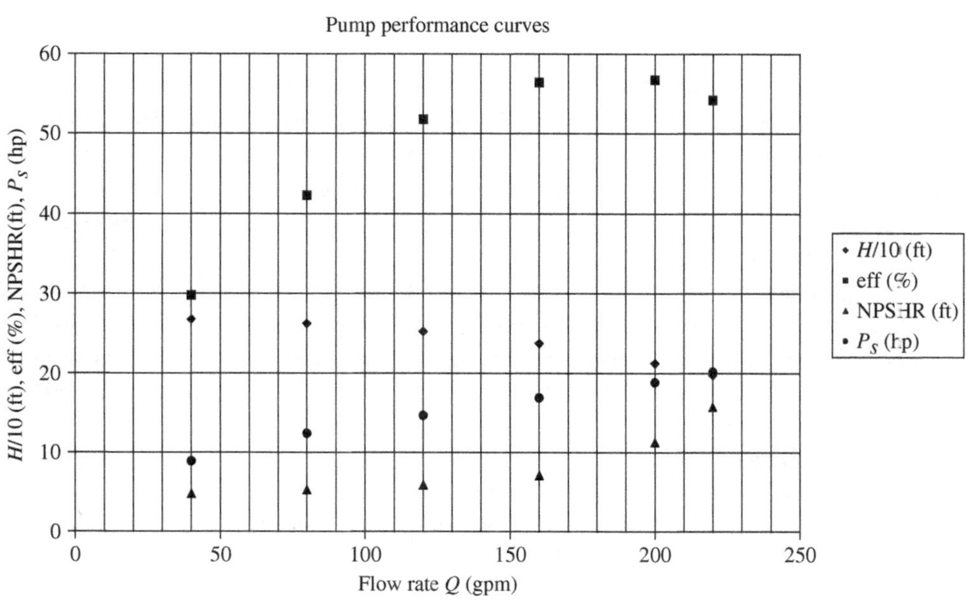

Note: The efficiency and NPSHR are read from the constant contours with interpolation and P_s is calculated with the above equation.

Example 3.4 The pump with impeller trimmed to 8-in. in Figure 3.7*a* is selected for the system shown. Read the head, flow rate, and NPSHR at the b.e.p. Determine the minimum height Z required if the tank pressure is 20 psia, friction loss in the suction line is 1.5 psi at 50 gpm, and water temperature is 150°F.

Solution

From Figure 3.7*a*, $H = 235$ ft, $Q = 170$ gpm at b.e.p. of 57.5%. The corresponding NPSHR is 8.7 ft. From the steam table, we have $P_v = 3.73$ psia at $T = 150$°F and NPSPA $= p_t - p_f + \rho g Z - p_v \geq$ NPSPR. So we have

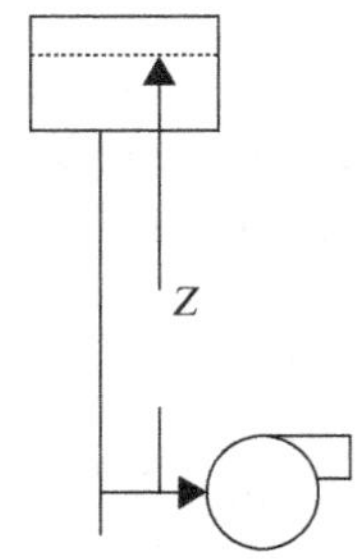

$$Z \geq \text{NPSHR} - \frac{p_t - p_f - p_v}{\rho g} = 8.7 - \left[20 - 1.5 \times \left(\frac{170}{50}\right)^2 - 3.73\right]\left(\frac{144}{62.4}\right),$$

or $Z \geq 8.7 + 2.5 = 11.2$ ft.

3.1.4 Performance Modifications

Typical performance curves published by manufacturers such as those of Figure 3.7 are based on tests of pumping clear water at room temperature. Sometimes the pumps are used to pump other fluids with peculiar physical properties, such as fluids of high viscosity and two-phase mixtures of liquid with solid particles or gaseous bubbles. The performance characteristics can be changed when pumping these fluids, normally toward reduced efficiency, head, and capacity.

3.1.4.1 Effects of Viscosity

Corrections for pumping viscous liquids are published in Hydraulic Institute Standards[3, 10] and are expressed in charts and mathematical equations. These corrections are averages from tests of conventional single-stage vaned pumps handling petroleum oils. They are not exactly for all pumps.

A special type of impeller pump for pumping highly viscous liquids is shown in Figure 3.14. The impeller consists of a set of disks, instead of blades. The dynamic interaction between the impeller and fluid is through the viscous force. Although the efficiency of this type of pump is low in general, it is less expensive to make.

3.1.4.2 Effects of Solid Particles

Pumps designed specifically for pumping solid–liquid mixtures are called slurry pumps. These pump designs should have an excessive flow passage area to prevent solid-particle clogging, minimize particle erosion on pump components, and minimize performance degradation. Two typical designs are shown in Figure 3.15. They are inherently of lower efficiency. In the vortex pump, the impeller is used to produce a vortex in the chamber to suck the flow in from the inlet and discharge it to the outlet. Most of the flow does not pass through the impeller. Detailed flow analysis for this type of pump is given in Ref. 11.

The slurry mixture parameters affecting performance degradation are (a) particle concentration, (b) specific gravity of the particles, (c) particle size and size distribution, and (d) the geometry of the particles. Typical slurry pump performance characteristics are published in Ref. 12 and shown in Figure 3.16. It is shown that when solid-particle concentration increases, the head and efficiency decrease, while the shaft power required increases.

3.1.4.3 Effects of Gaseous Bubbles

The effects of gaseous bubbles in liquid on pump performance are similar to those due to solid particles. But gaseous bubbles have variable density, possible mass transfer or phase change between gas/vapor with liquid, and

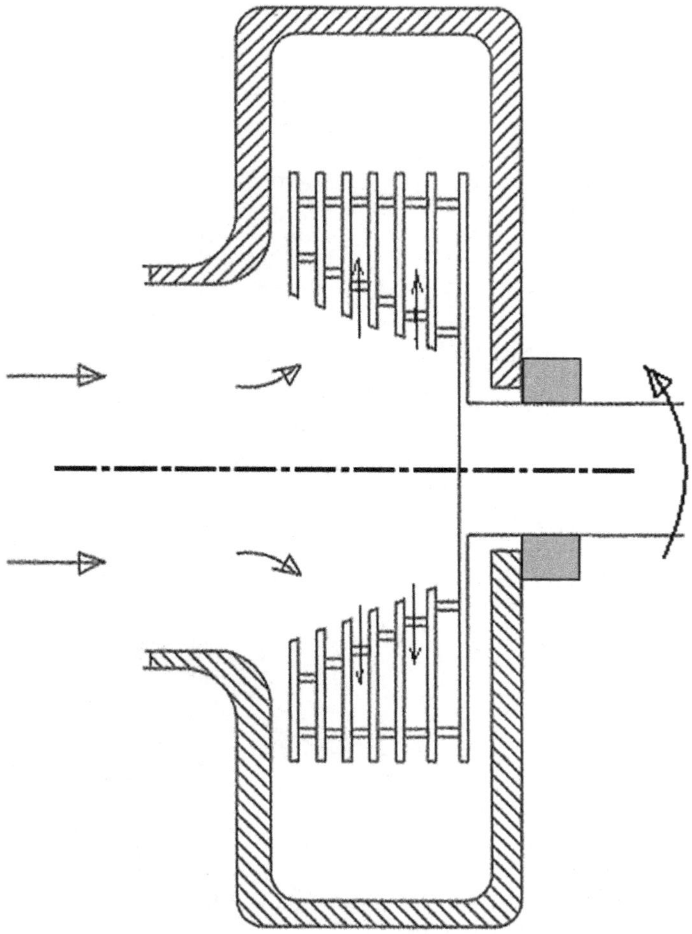

Figure 3.14 Disk-flow impeller pump for pumping viscous liquids.

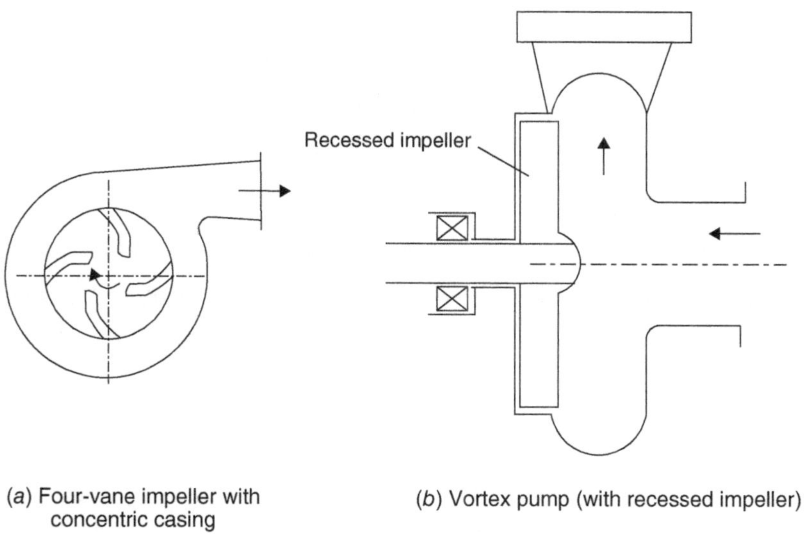

Figure 3.15 Typical slurry pumps.

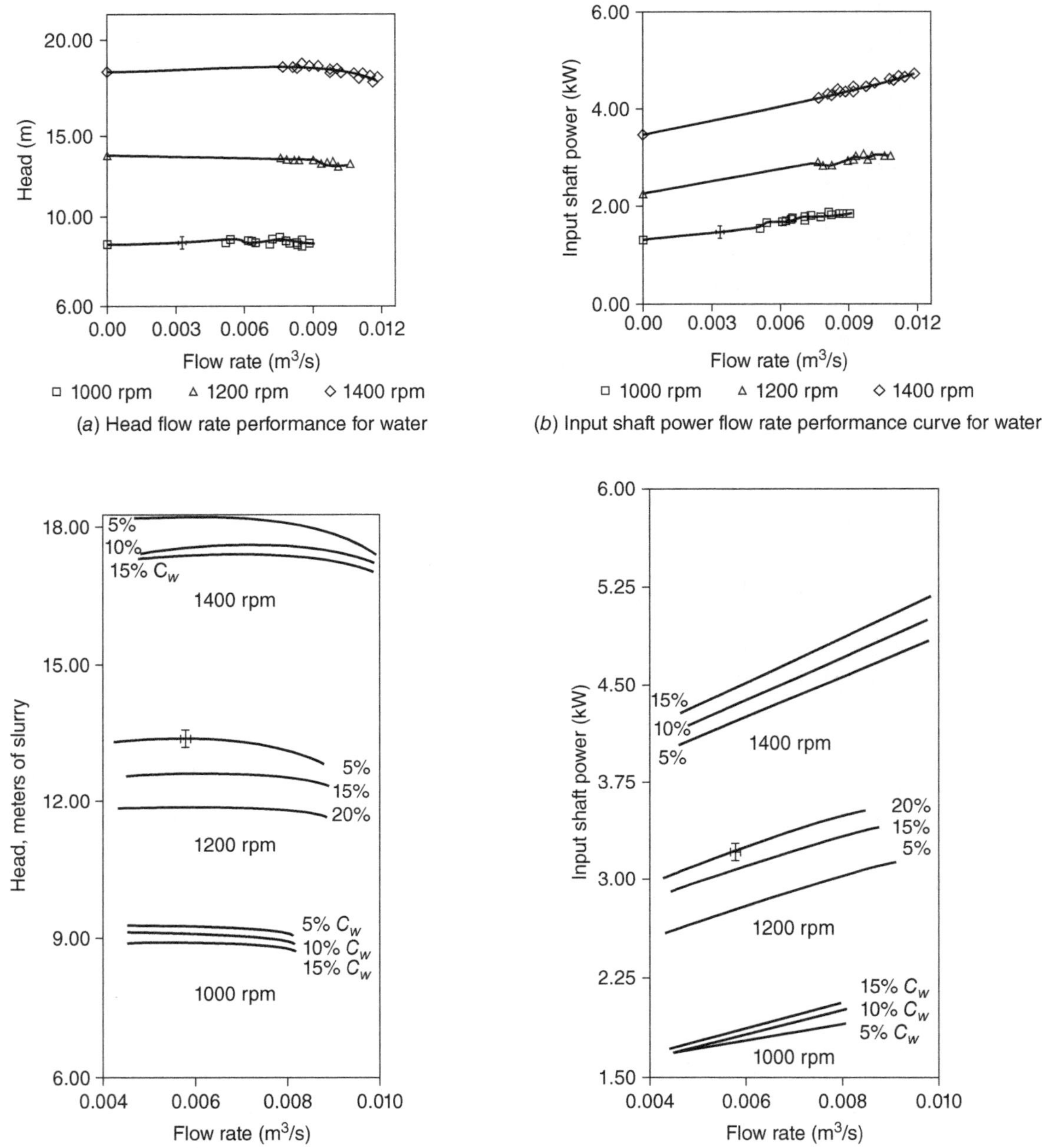

(*a*) Head flow rate performance for water

(*b*) Input shaft power flow rate performance curve for water

(*c*) Head flow rate performance for sand slurry

(*d*) Input shaft power flow rate performance curve for sand slurry

Figure 3.16 Characteristics of slurry pump with sand slurry. (Reprinted by permission from Sheth, K. K., Morrison, G. L., and Peng, W. W., Slip Factors of Centrifugal Slurry Pumps, *J. Fluids Eng. ASME Trans.* Vol. 109, pp. 313–318, Sept. 1987. © the American Society of Mechanical Engineers.)

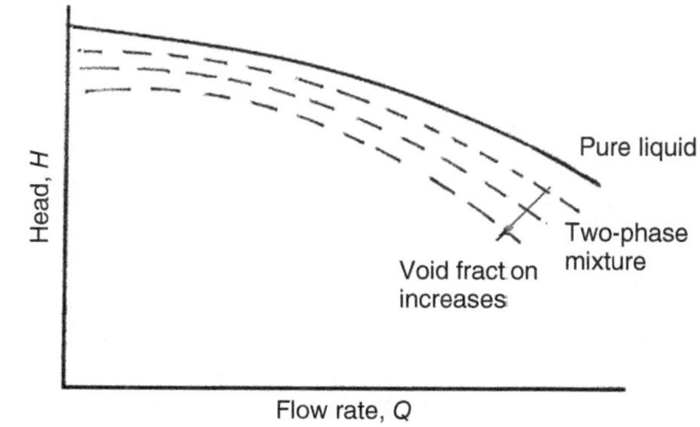

Figure 3.17 Typical pump performance with gas/liquid two-phase mixture.

more different flow regimes. A typical performance is shown in Figure 3.17 qualitatively. As shown, the head and flow rate decrease as the void fraction of the gaseous phase increases. This is due to the decrease of average fluid density and the extra pressure loss of two-phase interaction.

3.1.4.4 Effects of Fluid Properties on Cavitation

Experience has indicated that the NPSP required by a pump pumping hydrocarbon fluids or water at hot temperature is generally less than that for pumping cold water. This is so because when vapor is formed from liquid, the latent heat must be supplied by the internal energy of the liquid in the vicinity, and hence local liquid temperature drops. This is called a thermal cavitation effect and is expressed in terms of NPSPR reduction, or ΔNPSP. A chart to determine ΔNPSP was prepared by the Hydraulic Institute[3] and is reproduced in Figure 3.18. To use this chart, we start with the desired temperature on the horizontal axis and proceed vertically upward to the type of fluid and then read the inclined NPSP reduction scale to the right. Since the chart was based on a limited number of test data, it is recommended only for a limited range. If the resulting ΔNPSP is more than 50% of the original NPSPR, the reduction should be limited to 50%. The example on the chart is to determine ΔNPSP for the propane at 56°F. It shows that $\Delta NPSP = 9.5$ ft.

Some liquids, especially hydrocarbon liquids, can dissolve a significant amount of air or other no condensable gases. The solubility in general is proportional to the ambient pressure. Hence in the cavitation process, some dissolved gas will be released and mixed with the vapor. The dissolved gas in the liquid has an adverse effect on the required NPSP. More detailed discussion on this topic can be found in Ref. 13.

Example 3.5 Read the best efficiency points (b.e.p.) of all four impeller trims in Figure 3.7*a* and show that they do not follow the similarity laws since the trimmed impellers do not satisfy geometric similarity.

Solution

The b.e.p. of all four trims are read as follows:

Dia. (in.)	*Q* (gpm)	*H* (ft)	P_s (hp)	Efficiency (%)
9.25	220	330	29	62.5
8	170	235	17	57.5
7	155	170	12.5	53.5
6	130	120	8	49.5

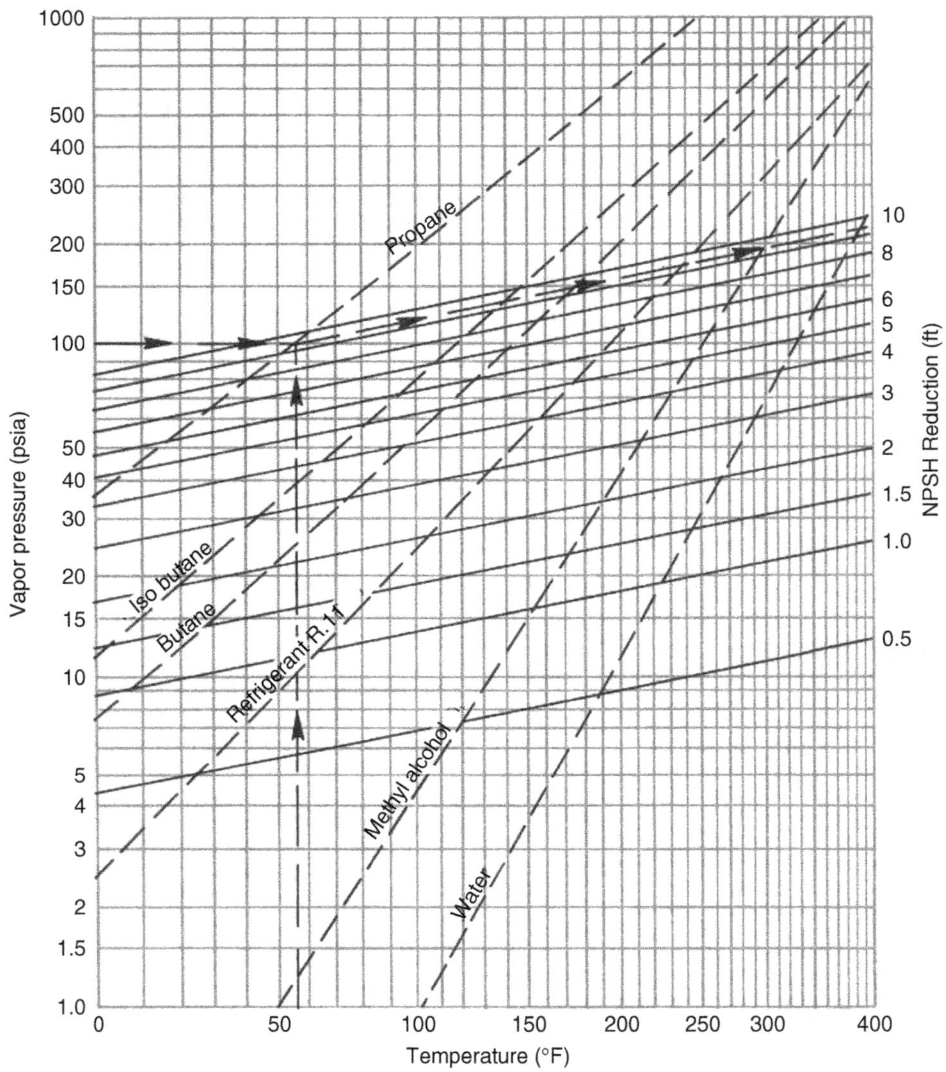

Figure 3.18 NPSP reduction for pumps handling hydrocarbon liquids and high-temperature water. (Reprinted by permission from *Hydraulic Institute Vertical Pump Design and Application–2000*, Hydraulic Institute, Parsippany, NJ, 1997–2005, www.pumps.org.)

From the similarity laws, $Q \sim D$, $H \sim D^2$, $P_s \sim D^3$, the performance should be

Dia. (in.)	*Q* (gpm)	*H* (ft)	P_s (hp)	Efficiency (%)
9.25	220	330	29	62.5
8	190	246	18.7	62.5
7	166	189	12.5	62.5
6	143	139	7.9	62.5

Example 3.6 The pump with impeller trimmed to 8-in. in Figure 3.7a is selected to pump hot water of 90°C in the enclosed tank shown. Determine the minimum tank pressure in kilopascals required to avoid cavitation at the pump inlet under the b.e.p. condition. The estimated pressure loss due to friction in the suction line is 10 kPa.

Solution

From Figure 3.7a, we have NPSHR = 8.7 ft = 2.65 m at Q = 170 gpm for the b.e.p. condition of the 8-in. impeller. From Figure 3.18, we have ΔNPSH = 0.6 ft = 0.18 m and p_v = 69 kPa at T = 90°C (194°F) for water. The total NPSPA is equal to $p_t - \rho gZ - p_f - p_v$, and the total NPSPR is equal to ρg(NPSHR − ΔNPSH). To avoid cavitation at the pump inlet, it is required to have NPSPA > NPSPR. Hence it is required that $(p_t - \rho gZ - p_f - p_v) > \rho g(\text{NPSHR} - \Delta\text{NPSH})$ or $p_t > p_f + p_v + \rho g(Z + \text{NPSHR} - \Delta\text{NPSH}) = 10 + 69 + 998 \times 9.81(2 + 2.65 - 0.18)/1000 = 122.7$ kPa.

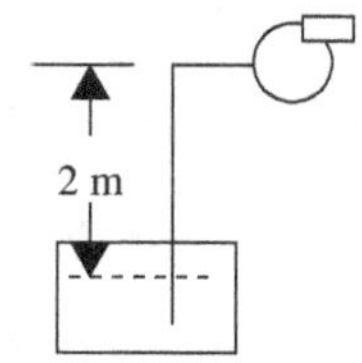

3.1.5 Preliminary Design Procedure

There is no rigorous procedure to design a pump. In most cases, it depends on the information available and the designer's experience. When test data of similar types of pumps are available, similarity laws can be used to check whether the new pump can be designed by scaling up or down from these existing pumps. Primary parameters such as hydraulic efficiency and slip factor can be determined using the procedure suggested in Section 3.1.2. Otherwise, a brand new design must be performed. A few basic rules and guidelines for designing a new pump impeller will be discussed in this section. The basic radial-flow impeller shown in Figure 3.19 will be considered for simplicity.

3.1.5.1 Impeller Type

A new impeller design starts with the selection of impeller type based on the required flow rate Q, discharge head H, and rotating speed N at the design point. From these design conditions, specific speed N_s is calculated as $N_s = NQ^{0.5}/H^{0.75}$ rpm(gpm)$^{0.5}$/(ft)$^{0.75}$. An appropriate impeller type is then selected from Figure 3.2, and some empirical constants correlated concerning N_s can be determined. Sometimes, double-suction or multistage arrangements can be considered, if necessary, to accommodate very high flow rates or high head requirements. The impeller inlet and outlet configurations, blade profile, and blade numbers or solidity can be determined with an impeller type selected.

3.1.5.2 Inlet Configuration

The inlet geometry is determined by the minimum shaft size to transmit the torque required and to accommodate the suction flow properly to meet the NPSHR and minimum-flow requirements. From the design flow rate and discharge head, the output hydraulic power $P_h = Q\Delta p$ and the required shaft power (or brake horsepower) $P_s = P_h/\eta_{o/a}$

Figure 3.19 Radial-flow impeller dimensions.

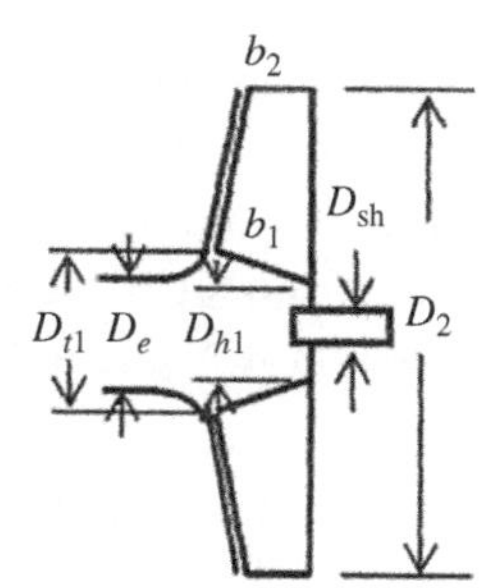

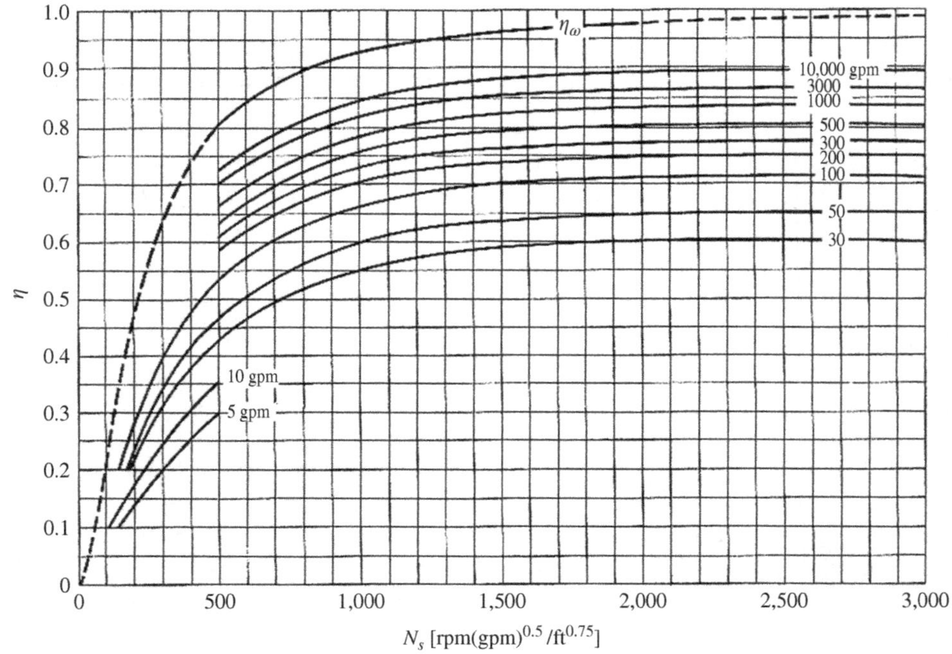

Figure 3.20 Pump efficiency as a function of specific speed and capacity. (Reprinted by permission from Karassik, I. Y. et al. (Eds.), *Pump Handbook*, McGraw-Hill, Inc., New York, 1976.)

can be determined from the overall pump efficiency $\eta_{o/a}$ estimated from Figure 3.20. The shaft diameter required to transmit torque is then obtained from $\sigma_s = \tau r/J$, with $J = \pi D_{sh}^4/32$, or $D_{sh} = (16\tau/\pi\sigma_s)^{1/3}$, where $\tau = P_s/\omega$ is shaft torque, and σ_s is the maximum allowable shear stress of the shaft. The hub diameter D_{h1} is then selected to be $D_{h1} = D_{sh} + \left(\frac{5}{16} \sim \frac{1}{2} \text{ in.}\right)$ if the blade leading edge is inclined and the shaft is extended into the eye.

The low NPSHR and low minimum-flow requirements always compete. In general, a large suction area, a small leading-edge blade angle, and low suction flow velocity are good for low NPSHR but bad for minimum flow (low efficiency, too) and vice versa. A compromise design is to set the leading-edge blade angle at around 17°.[1] Other empirical criteria are used by different designers, such as setting the eye flow coefficient $\varphi_e = V_e/(r_e\omega)$ to be between 0.25 and 0.30. More detailed configurations are discussed in Refs. 1 and 2.

The other dimensions and blade angles are sized such that the meridional (or radial) component of the flow velocity increases from the eye area to avoid boundary layer separation, that is, $V_{m1} = (1.05 - 1.20)V_e$, where $Q = V_e\pi D_e^2/4 = V_{m1}\pi(D_{t1} + D_{h1})b_1/2$, with V_e being around 14 ft/s. The incidence angle between the blade and the relative flow $i = \beta_{b1} - \beta_{f1}$ is chosen to be between 2° and 3° to avoid a negative angle for the flow rates slightly higher than the design condition. The flow area at the leading edge can be calculated as $A_1 = \pi(r_{t1} + r_{h1})b_1\varepsilon_1$, where ε_1 is the contraction factor due to blade thickness t and boundary layer. It can be expressed as $\varepsilon_1 = 1 - Z_b t/(2\pi r_{m1} \sin \beta_{b1})$, where Z_b is the number of blades and r_{m1} is the mean radius as defined earlier. The contraction factor can be estimated as 0.8–0.9 and checked later when Z_b and t are selected. The blade thickness is selected primarily based on strength and manufacturing.

3.1.5.3 Outlet Configuration

The impeller outlet dimensions are mainly determined by the required discharge head at the design point. The diameter D_2 can be determined from the Cordier diagram (Figure 2.2) or Figure 3.36 later in the chapter, where the specific diameter is correlated with the specific speed. Then we have $U_2 = D_2\omega/2$. From the Euler

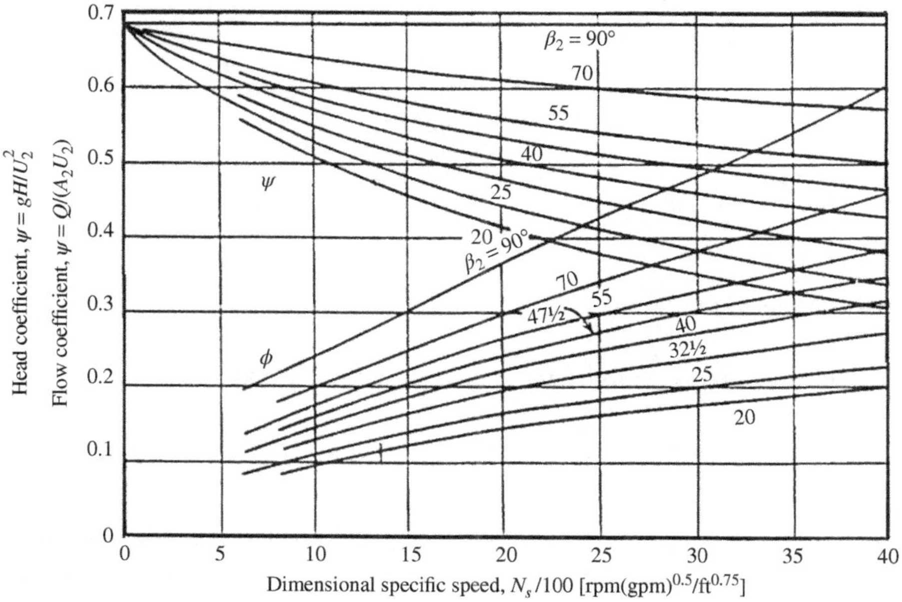

Figure 3.21 Head and flow coefficients versus specific speed. (Reprinted by permission from Stepanoff, A. J., *Centrifugal and Axial Flow Pumps*, 2nd ed., John Wiley & Sons, New York, 1957.)

equation at the design flow rate, $H_a = \eta_h \mu_s (U_2/g)[U_2 - Q/(A_2 \tan \beta_{b2})]$, the outlet area A_2 and blade angle β_{b2} are related. These parameters can be determined with the following steps:

1. As discussed in Section 3.1.2, hydraulic efficiency can be estimated from[1] $\eta_h = 1 - 0.80/(\text{GPM})^{0.25}$, or $\eta_h = \eta_{o/a}/(\eta_m \eta_v)$ (with η_m and η_v estimated from the empirical formula).
2. The blade angle β_{b2} can be set as values of 22.5° ~ 27.5° for optimum efficiency.[2]
3. The slip factor is determined from empirical formulas such as Stodola's expression mentioned earlier, $\mu_s = 1 - \pi \sin \beta_{b2}/Z_b$. The blade number Z_b can be set as $Z_b = 6.5[(1 + \nu)/(1 - \nu)] \sin[(\beta_{b1} + \beta_{b2})/2]$, where $\nu = D_1/D_2$. This implies that more blades are needed for the high values of ν and β_b since the blade length is shorter under these conditions. Other formulas and a detailed discussion of the slip factor can be found in the literature (e.g., Refs. 1, 2, 14).
4. Then the discharge area A_2 can be calculated from the above Euler equation.
5. The meridional velocity V_{m2} can be calculated from $V_{m2} = Q/A_2$.
6. The relative flow velocity W_2 is calculated as $W_2^2 = V_{m2}^2 + (U_2 - V_{u2})^2$ from the velocity diagram.
7. The diffusion factor should be checked such that $1 > (D_f = W_2/W_1) > 0.70$ to ensure there is no boundary layer separation. Otherwise, a new outlet diameter and/or blade angle should be selected, and new calculations should be repeated starting with step 2.
8. If the diffusion factor is satisfactory, the contraction factor at discharge can be calculated similar to the one at the inlet, $\varepsilon_2 = 1 - Z_b t/(2\pi r_2 \sin \beta_{b2})$. The blade height b_2 can then be set as $b_2 = A_2/(\varepsilon_2 \pi D_2/2)$.

Other procedures to determine the outlet configuration are available. Among them, a set of head and flow coefficients versus specific speed and discharge blade angle are suggested in Ref. 2. They are reproduced in Figure 3.21.

3.1.5.4 Blade Profile

The detailed blade profile can be determined from an appropriate blade angle distribution with the impeller inlet and outlet dimensions fixed. Blade angle is related to the blade mean line profile as $\tan \beta_b = dr/(r\, d\theta)$ for a pure

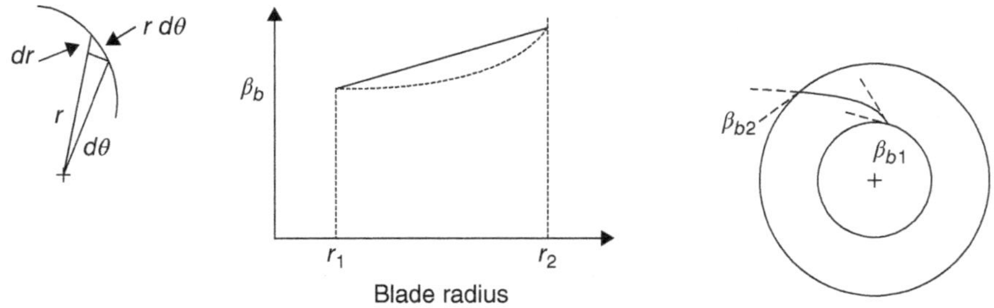

Figure 3.22 Blade angle distribution for preliminary design.

radial impeller. So the mean line coordinate (r, θ) can be related in the form $\Delta\theta = (180/\pi)dr/(r \tan \beta_b)$, or

$$\theta = \frac{180}{\pi} \int_{r1}^{r} \frac{dr}{r \tan \beta_b}$$

where θ is expressed in degrees.

A simple linear distribution of β_b, as shown in Figure 3.22, can be adopted for the preliminary design. Other blade angle distributions have been suggested for advanced design procedures to avoid or delay boundary layer separation in the flow passage. A numerical integration procedure can be used to calculate the blade profile step by step.

The total length of the blade, L, can also be calculated from $L = \int dL$, where $dL = [(r\, d\theta)^2 + (dr)^2]^{0.5}$ and θ is in radians. It can be used to calculate a parameter called solidity, defined as $\sigma = Z_b L/(\pi D_2)$. Some designers suggest that the optimum solidity should be between 2.5 and 3.0. If the solidity is too high, the flow may suffer excessive friction; if it is too low, the flow may not be properly guided.

The blade profile will be three-dimensional and more complicated for a mixed-flow impeller. The blade mean surface must be described with a set of space lines in terms of (r, θ, z). It will require sophisticated graphical construction or computer-aided graphics to describe these lines. The isometric, front, and meridional plane views of a Francis-type impeller are shown in Figure 3.4.

3.1.5.5 Diffuser and Volute

Downstream of the impeller, some diffuser is needed to convert the kinetic energy into pressure rise, and a volute is used to collect the liquid circumferentially before discharging it through the outlet. Sometimes, the volute also serves as a diffuser or is followed with a conical diffuser.

A vaned diffuser immediately downstream of the impeller can improve efficiency, but it tends to limit the stable operating range. Because the flow discharged from the impeller will be mismatched with the vane leading-edge angle under the off-design flow rate. The vaned diffuser is used only when operation can be limited to a narrow range. This component will be discussed in more detail in Chapter 4, Section 4.1 on centrifugal compressors.

The basic geometry of a volute is shown in Figure 3.23. Assuming constant angular momentum of flow, that is $r_c V_\theta = C r_2 V_{u2} = CL$ and $A_\theta V_\theta = Q\theta/(2\pi)$, the theoretical value of the volute cross-sectional area A_θ can be expressed as

$$\frac{A_\theta}{r_c} = \frac{Q\theta}{2\pi CL} \tag{3.5}$$

It increases from the cut water at $\theta = 0$ to the volute outlet area A_t. Here r_c is the centroid radius of the volute cross-sectional area, θ is measured in radians, and L is the angular momentum of flow at the impeller outlet, that is, $L = r_2 V_{u2}$ and $C \cong 0.95$ to account for the friction loss. The average flow velocity at the volute outlet is related to the pump head H as $V_3 = K_3\sqrt{2gH}$, where the volute constant K_3 is an empirical parameter correlated with the specific speed, as shown in Figure 3.24 along with other volute parameters, the volute angle α_v, and the diameter

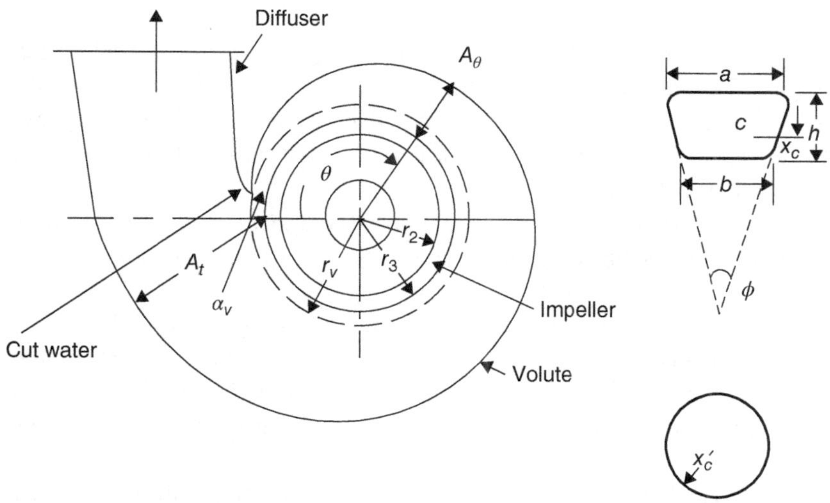

Figure 3.23 Terminology and parameters of centrifugal pump casing with volute and diffuser.

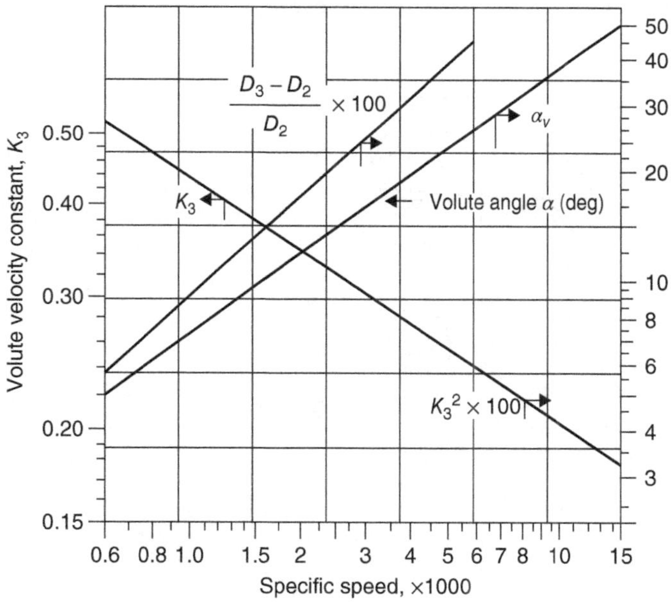

Figure 3.24 Volute constants correlated with specific speed, N_s[rpm(gpm)$^{0.5}$/ft$^{0.75}$]. (Reprinted by permission from Stepanoff, A. J., *Centrifugal and Axial Flow Pumps*, 2nd ed., John Wiley & Sons, New York, 1957.)

of the volute basic circle, D_3. The volute profile at each section can be determined as follows: For a circular volute, we have $A_\theta = \pi x_c^2$ and

$$r_c = r_3 + x_c = r_3 + \sqrt{\frac{A_\theta}{\pi}} \tag{3.6}$$

so A_θ and r_c can be solved from Equations (3.5) and (3.6). For a trapezoid volute as shown, $A_\theta = h(a + b)/2$, $\tan(\varphi/2) = (a - b)/2h$ and $x_c = h(2a + b)/[3(a + b)]$. With φ being selected (normally 30°) and b being set as $b \cong (1.5 \sim 2.0)b_2$, the parameters A_θ, r_c, a, and h at each section can be determined.

3.1.5.6 Determination of Shaft Power

The overall efficiency and input shaft power estimated from Figure 3.20 are average values. The actual magnitude will depend on the detailed design of each pump. For example, the volumetric efficiency depends on the seal design and type of impeller, whether an open, semi-open, or closed impeller. Mechanical efficiency depends on the bearing arrangement and type of seal (with or without mechanical seal). A typical distribution of power consumption is given in Figure 3.25 for double-suction pumps.

It is shown that disk friction loss is significant in the low-specific-speed pump, and casing loss is significant in the high-specific-speed pump, while mechanical loss due to bearings and leakage loss is constant with respect to specific speed.

The disk friction loss in horsepower for both sides can be estimated from the formula $P_{df} = K\rho D^2U^3$, where K is a numerical coefficient plotted against the Reynolds number $\mathrm{Re} = UR/\nu$, as shown in Figure 3.26, $U = R\omega$ is the tangential velocity of the disk periphery in feet per second, R is the disk radius in feet, D is the disk diameter in feet, ν is the fluid kinematic viscosity in square feet per second, and ρ is the fluid density in pound mass per cubic feet.

It is also shown that disk friction loss varies with the clearance between the disk and casing wall (B/D). This is understandable since the wider clearance means more space for the flow recirculation generated by the rotating disk. When the clearance is small, the flow recirculation is restricted, and only the shear viscous force will contribute to disk friction loss. As shown, the optimum value of B/D corresponds to the minimum disk friction loss.

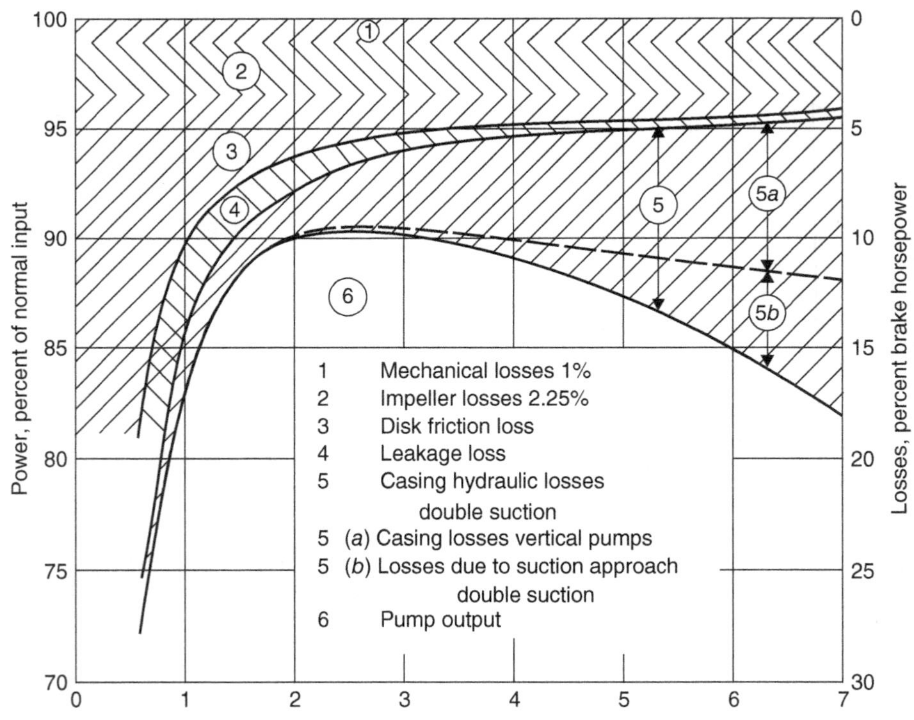

Figure 3.25 Power balance for double-suction pumps at best efficiency point. (Reprinted by permission from Stepanoff, A. J., *Centrifugal and Axial Flow Pumps*, 2nd ed., John Wiley & Sons, New York, 1957.)

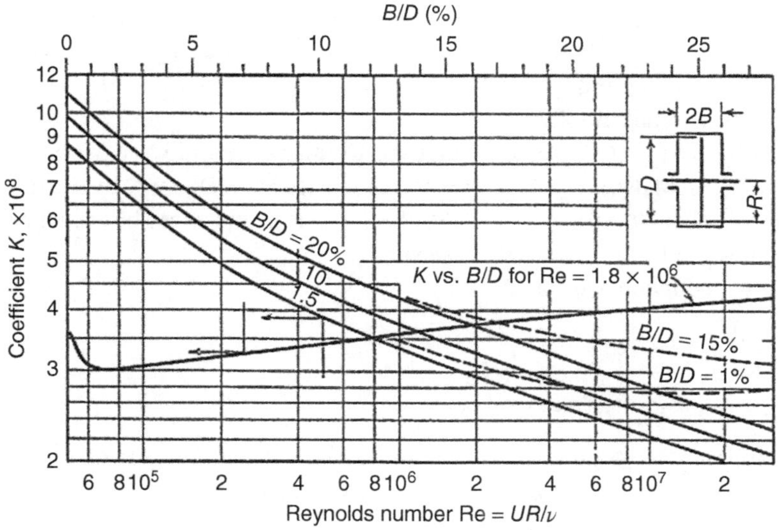

Figure 3.26 Disk friction coefficient versus Reynolds number and clearance (full lines for smooth disk, dashed lines for rough disk). (Reprinted by permission from Stepanoff, A. J., *Centrifugal and Axial Flow Pumps*, 2nd ed., John Wiley & Sons, New York, 1957.)

Example 3.7 Design a water pump impeller with the design point at $Q = 1200$ gpm, $H = 450$ ft, and $N = 2400$ rpm.

Solution

a) Start with a calculation of specific speed:

$$N_s = \frac{NQ^{0.5}}{H^{0.75}} = \frac{2400(1200)^{0.5}}{450^{0.75}} = 851 \text{ rpm(gpm)}^{0.5}/\text{ft}^{0.75}.$$

From Figure 3.20, we obtain $\eta_{o/a} \approx 0.76$. So

$$P_s = \frac{Q\Delta p}{\eta_{o/a}} = \frac{1200 \times 0.00223 \times 62.4 \times 450}{0.76} = 98,871 \text{ ft-lb}_f/\text{s}.$$

From Figure 3.2, it is shown that a radial impeller is appropriate.

b) To determine the inlet configuration: With $\omega = N\pi/30 = 2400\pi/30 = 251$ rad/s, we have $\tau = P_s/\omega = 393$ lb$_f$-ft. Hence the shaft diameter D_{sh} can be determined as

$$D_{sh} = \left(\frac{16\tau}{\pi\sigma_s}\right)^{1/3} = \left(\frac{16 \times 393}{\pi \times 10,000 \times 144}\right)^{1/3} = 0.1116 \text{ ft} = 1.34 \text{ in.},$$

where $\sigma_s = 10{,}000$ psi (for 316 stainless steel, including a safety factor of 1.5) is used. A pure radial inlet configuration is selected for simplicity, that is, $D_{t1} = D_{h1} = D_1$ (see Figure 3.19). The impeller eye area and leading edge diameters can be calculated from the following equations:

$$U_1 = \frac{1}{2}\omega D_1, \quad Q = A_1V_1 = A_eV_e, \quad V_1 = 1.1V_e,$$

$$\beta_{f1} = \tan^{-1}\left(\frac{V_1}{U_1}\right)$$

where $A_1 = \varepsilon_1 \pi D_1 b_1, \; A_e = \pi \; D_e^2/4$.

With $\varphi_e = Q/(A_e U_e) = Q/\left(\pi r_e^3 \omega\right) = 0.27$, we obtain $r_e = 2.8$ in., and selecting $r_1 = 3.2$ in., the final results are obtained as $V_e = Q/\left(\pi D_e^2/4\right) = 15.6$ ft/s, $V_1 = 1.1V_e = 17.2$ ft/s, $A_1 = 22.4$ in.^{2}, $b_1 = 1.24$ in., with $\varepsilon_1 = 0.9$ assumed. Also $U_1 = 66.9$ ft/s, $\beta_{f1} = 14.4°$, $\beta_{b1} = 17°$, and $W_1 = 69.1$ ft/s.

c) The impeller discharge configuration is determined as follows:
   1) With $\omega_s = \omega Q^{0.5}/(gH)^{0.75} = 251(2.676)^{0.5}/(14{,}490)^{0.75} = 0.311$ and the Cordier diagram, we have $\Delta_s = 9.0 = D_2(gH)^{0.25}/Q^{0.5}$ and $D_2 = 9(2.676)^{0.5}/(14{,}490)^{0.25} = 1.341$ ft $= 16.1$ in. Hence $U_2 = D_2\omega/2 = 168.5$ ft/s.
   2) Selecting $\beta_{b2} = 27.5°$ and $Z_b = 6.5[(1+0.4)/(1-0.4)]\sin[(17+27.5)/2] = 5.74 \approx 6$, μ_s and η_h can be estimated from $\mu_s = 1 - \pi \sin 27.5°/6 = 0.76$ and $\eta_h = 1 - 0.8/(1200)^{0.25} = 0.864$.
   3) Substituting these values into the Euler equation, we obtain

$$V_{u2} = \mu_s\left(U_2 - \frac{V_{r2}}{\tan\beta_{b2}}\right) = \frac{gH}{\eta_h U_2} = 99.6 \text{ ft/s},$$

$$V_{r2} = 19.4 \text{ ft/s}, \quad W_2 = \left[V_{r2}^2 + (U_2 - V_{u2})^2\right]^{0.5} = 71.5 \text{ ft/s}.$$

   Hence $W_2/W_1 > 1$, which is not appropriate. A different D_2 has to be selected, say $D_2 = 15$ in. So we have $U_2 = 156.9$ ft/s, $V_{u2} = 106.9$ ft/s, $V_{r2} = 8.5$ ft/s, and $W_2 = 50.7$ ft/s. Hence $W_2/W_1 = 0.74$, which is acceptable.
   4) The outlet dimensions are calculated as $A_2 = Q/V_{r2} = 0.315$ ft^2 $= 45.3$ in.^{2} $= \varepsilon_2 \pi D_2 b_2$, where $\varepsilon_2 = 1 - Z_b t/(\pi D_2 \sin\beta_{b2}) = 1 - 6 \times 0.25/(\pi \times 15 \sin 27.5°) = 0.93$. Hence $b_2 = 1.0$ in.
   5) To check $\varepsilon_1 = 1 - 6 \times 0.25/(\pi \times 6.4 \sin 17°) = 0.74$, b_1 has to be adjusted to $b_1 = 1.3 \times 0.9/0.74 = 1.58$ in.

d) The blade profile is constructed by assuming a linear blade angle distribution,

$$\beta_b = \beta_{b1} + \frac{(\beta_{b2} - \beta_{b1})(r - r_1)}{r_2 - r_1} = 17 + 2.44(r - 3.2),$$

where β_b is in degrees and r is in inches. Also, $\Delta\theta$ can be expressed as $\Delta\theta = (180/\pi)\Delta r/(r \tan\beta_b)$.

If we select $\Delta r = 0.43$ in., then β_b, $\Delta\theta$, θ, dL, and L can be calculated and tabulated as follows:

r (in.)	*β*	*dθ*	*θ* (deg)	*dL*	*L* (in.)	*r* (in.)	*β*	*dθ*	*θ* (deg)	*dL*	*L* (in.)
3.2	17	25.188	0	1.4709	0	3.2	17	25.19	0	1.471	0
3.63	18.05	20.831	25.188	1.388	1.471	3.63	17.6	21.4	25.2	1.422	1.471
4.06	19.1	17.528	46.018	1.3143	2.859	4.06	18.2	18.46	46.6	1.377	2.893
4.49	20.15	14.957	63.546	1.2484	4.173	4.49	18.8	16.12	65	1.334	4.27
4.92	21.2	12.913	78.502	1.1892	5.422	4.92	19.4	14.22	81.2	1.295	5.604
5.35	22.25	11.259	91.415	1.1358	6.611	5.35	20	12.65	95.4	1.257	6.899
5.78	23.3	9.8993	102.67	1.0873	7.747	5.78	20.6	11.34	108	1.222	8.157
6.21	24.35	8.768	112.57	1.0431	8.834	6.21	21.2	10.23	119	1.189	9.379
6.64	25.4	7.8157	121.34	1.0026	9.877	6.64	21.8	9.279	130	1.158	10.57
7.07	26.45	7.006	129.16	0.9655	10.88	7.07	22.4	8.456	139	1.129	11.73
7.5	27.5	6.3116	136.16	0.9314	11.85	7.5	23	7.74	147	1.101	12.85

The table shows that with $\beta_{b2} = 27.5°$, the blade length $L = 11.85$ in. Hence, the solidity is $\sigma_s = 6 \times 11.85/(\pi \times 15) = 1.51$, which is too low. Hence the new $\beta_{b2} = 23°$ and $Z_b = 9$ are selected so that we have $\sigma_s = 9 \times 12.85/(\pi \times 15) = 2.45$, which is acceptable. The final values of the impeller outlet dimensions are revised:

$$A_2 = 0.194\ \text{ft}^2 = 27.9\ \text{in.}^2, \quad b_2 = 0.67\ \text{in.} \quad \text{with} \quad \mu_s = 0.86, \quad \varepsilon_2 = 0.88,$$
$$V_{r2} = 13.8\ \text{ft/s}, \quad W_2 = 51.8\ \text{ft/s}.$$

Example 3.8 Design a volute for the pump designed in Example 3.7.

Solution

From Example 3.7, we have $Q = 1200$ gpm, $H = 450$ ft, $N = 2400$ rpm, $N_s = 851$ rpm(gpm)$^{0.5}$/ft$^{0.75}$, $D_2 = 15$ in., $b_2 = 0.67$ in., $U_2 = 156.9$ ft/s, $\beta_{b2} = 23°$, $\mu_s = 0.86$, $V_{r2} = 13.8$ ft/s, and $V_{u2} = 106.9$ ft/s. Hence we have the angular momentum $L = r_2 V_{u2} = 15 \times 106.9/(2 \times 12) = 66.8$ ft^2/s. From Figure 3.24, we have $K_3 = 0.46$ for $N_s = 851$. Hence the volute exit velocity $V_t = K_3\sqrt{2gH} = 0.46\sqrt{2 \times 32.2 \times 450} = 78.3$ ft/s and $A_t = Q/V_t = 1200 \times 0.00223/78.3 = 0.034$ ft^2 $= 4.92$ in.^{2} Also, from Figure 3.24, we have $(D_3 - D_2)/D_2 = 0.08$, or $r_3 = 7.5(1 + 0.08) = 8.1$ in. is selected.

If the trapezoid with an included angle of 30°, as shown, is selected for the basic geometry of the volute cross-section, the dimensions at each section can be calculated from

$$A_\theta V_\theta = \frac{Q\theta}{2\pi}, \tag{E3.8.7}$$

$$r_c V_\theta = CL, \tag{E3.8.8}$$

$$A_\theta = \frac{1}{2}h(a + b), \tag{E3.8.9}$$

$$\frac{a - b}{2h} = \tan 15°, \tag{E3.8.10}$$

$$x_c = \frac{h(2a + b)}{3(a + b)}, \tag{E3.8.11}$$

$$r_c = r_3 + x_c, \tag{E3.8.12}$$

where $b = 1.8b_2 = 1.206$ in. is selected.

The equations are solved with iterations and performed with a spreadsheet. The initial velocity is selected as $V_\theta = V_t$. Then A_θ is calculated from (1), a, and h from (3) and (4), and x_c and r_c from (5) and (6). Then, the improved V_θ is obtained from (2), and the process is repeated until the discrepancies of the repeated values are small enough. In the final design, the trapezoid is modified by rounding off the corners, and the area may be increased by 5–10% to accommodate the blockage due to the boundary layer and some secondary flows.

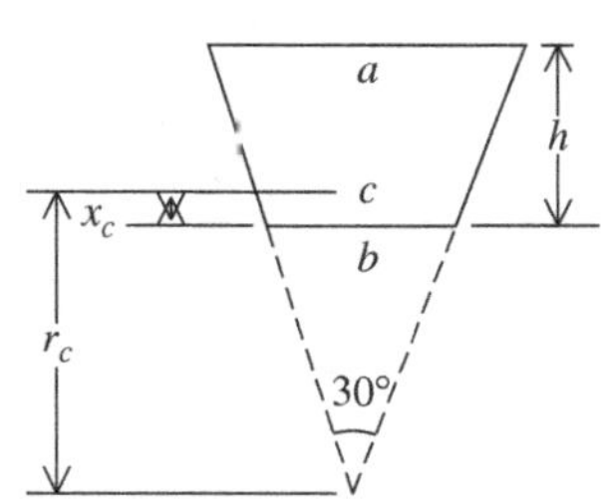

3.1.6 Pump Performance Tests

A pump can be evaluated in a laboratory test facility or in the field to investigate its hydraulic performance. The purpose of the test could be for new pump development, production verification, or troubleshooting. A typical closed-loop test setup in a laboratory is shown in Figure 3.27. The tank is used to control the fluid temperature and pressure, especially the suction condition for cavitation tests. The straightening vanes are installed to minimize the secondary flow or swirl at the pump suction. An open-loop setup is also possible if fluid temperature control is not necessary.

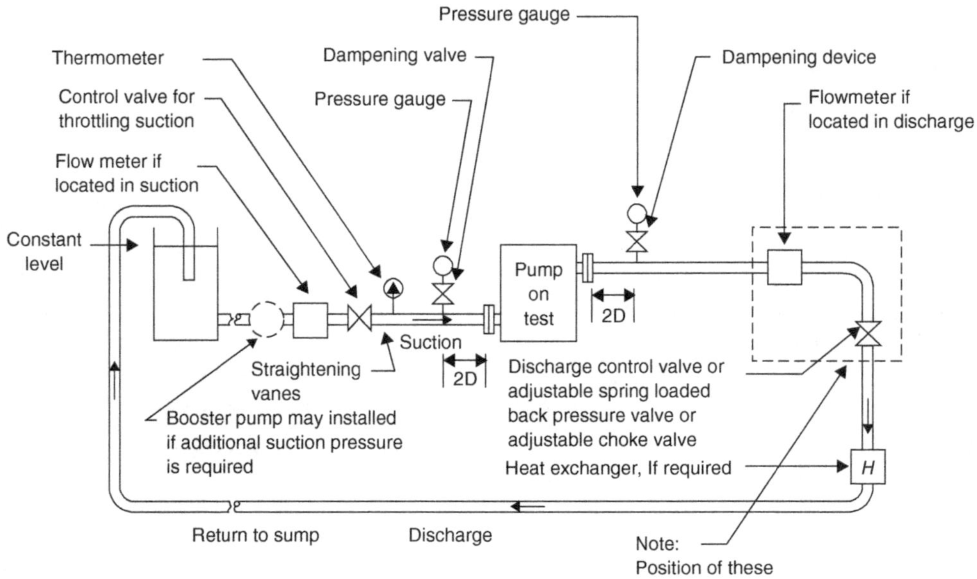

(*a*) Horizontal pump (with open or closed tank)

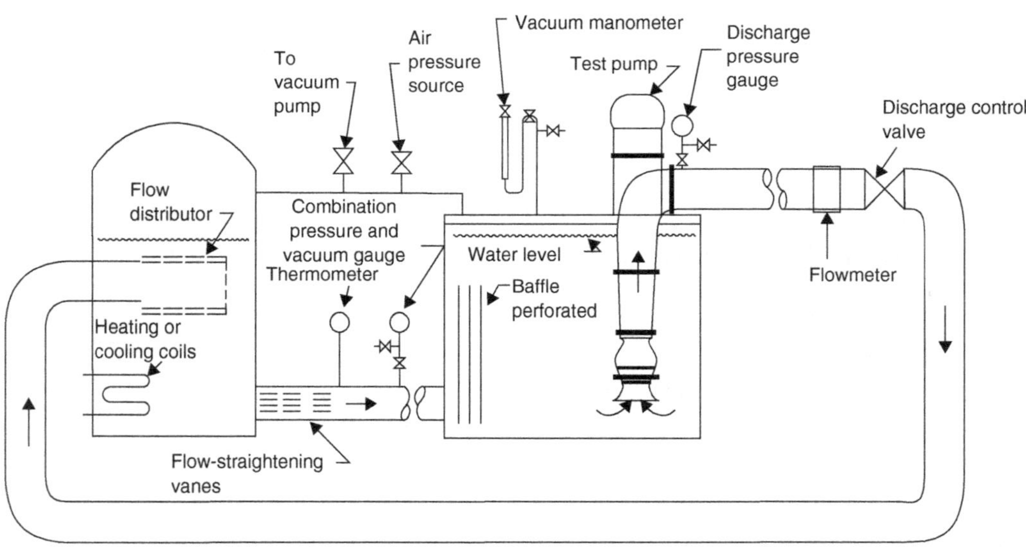

(*b*) Vertical pump (with closed loop)

Figure 3.27 Typical setup for pump tests. (Reprinted by permission from Hydraulic Institute *Centrifugal Pump Tests–2000*, and *Vertical Pump Tests and Application–2000*, Hydraulic Institute, Parsippany, NJ, www.pumps.org.)

The main parameters to be measured with instrumentation are (1) volumetric flow rate, (2) suction and discharge pressures (heads), (3) rotating speed of the impeller, (4) input shaft torque or power, and (5) fluid temperature at pump suction. Other parameters may be needed in development and troubleshooting tests.

Flow rate can be measured with (a) venturi, (b) orifice plate, (c) pitot tube traverse, and (d) weighing tank or other head meter depending on the magnitude of the flow rate and the accuracy required. For a very high flow rate, a weir in an open channel is used. In an irrigation field, the distance of water ejecting from a pipe can be used to estimate the flow rate. Pressure or head can be measured with (a) a water manometer, (b) a mercury monometer, (c) a Bourdon gage, or (d) a pressure transducer if the computer data acquisition system is employed. The input shaft power can be determined with (a) a transmission dynamometer, (b) a torsion dynamometer, (c) strain gage, or (d) electrical input power and motor calibration if a motor is used as a driver. The impeller rotating speed can be measured with (a) a tachometer, (b) a photocell and timer, or (c) a stroboscope. Many other new devices are being developed and used, especially in automation facilities.

During the tests, the rotating speed may fluctuate. For the purpose of plotting, the flow rate, head, and shaft power should be corrected from measured values at the test speed to the rated speed. The corrections are made according to similarity laws such that

$$Q_0 = \frac{Q_1 N_0}{N_1}, \quad dp_0 = dp_1 \left(\frac{N_0}{N_1}\right)^2, \quad P_{s0} = P_{s1}\left(\frac{N_0}{N_1}\right)^3, \tag{3.7}$$

where subscript 1 is for the test values and subscript 0 is for the values at the rated speed.

Before the test, the system has to be primed, that is, to get rid of all air in the system by venting it from the highest point of the system. The noncavitation test normally starts with the throttle valve at discharge fully open, that is, the highest flow rate. Then, the valve is closed to reduce the flow rate. For the cavitation test, the fluid must be circulated through the system for a few minutes below atmospheric pressure (with the vacuum pump shown in the loop). Thus, the dissolved air in the fluid can be purged. The test will then be run at several constant flow rates while the suction pressure (or NPSH) is reduced with the suction valve or by lowering the tank pressure. A typical result is shown in Figure 3.12 of Section 3.1.4. More detailed pump test procedures are discussed in Refs. 1 and 3.

In production tests of small pumps, air can be used instead of liquid water to facilitate the process. Of course, a correlation between water test and air test performance has to be established first.

Example 3.9 A pump driven with an electrical motor is tested in the laboratory. The test result and motor calibration curves are given below. Convert these raw data into the performance curves at the rated speed of 1750 rpm. Include the pressure rise, shaft power, and pump efficiency versus flow rate. Here, Q_1 is the flow rate, dp_1 is the total pressure rise, P_e is the input electrical power to the motor, and N_1 is the impeller rotating speed.

Q_1 (gpm)	dp_1 (psi)	P_e (kW)	N_1 (rpm)
40	77	5	1775
80	76	5.4	1772
120	70	5.7	1765
140	66	6.3	1760
160	64	6.5	1755
180	62	6.7	1753
200	55	7	1750
220	48	7.2	1747

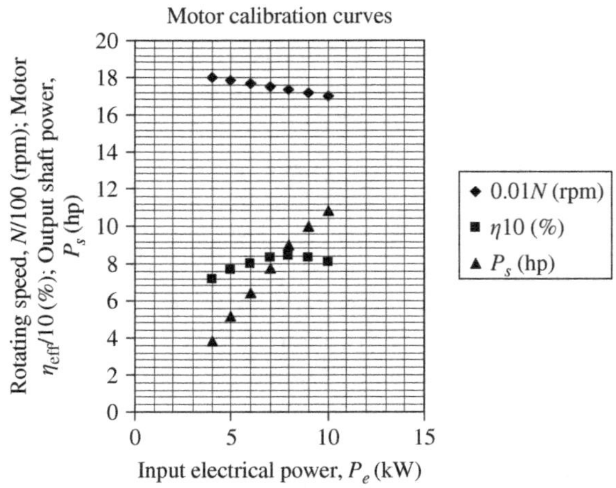

Solution

1) Use the motor calibration curves to convert the input electrical power to shaft power for the corresponding rotating speed. The result is provided below.

Q_1 (gpm)	dp_1 (psi)	P_{s1} (hp)	N_1 (rpm)
40	77	5.2	1775
80	76	5.6	1772
120	70	6	1765
140	66	6.5	1760
160	64	7.2	1755
180	62	7.6	1753
200	55	7.75	1750
220	48	7.8	1747

2) Convert each operating condition of flow rate, pressure rise, and shaft power from the test rotating speed to the rated speed of 1750 rpm. They are related as $Q_0 = Q_1(1750/N_1)$, $dp_0 = dp(1750/N_1)^2$, and $P_{s0} = P_{s1}(1750/N_1)^3$.] Calculate the efficiency for each operating condition with the equation.

$$\eta = \left(\frac{Q\ (\text{gpm})}{449}\right)\frac{dp\ (\text{psi}) \times 144}{550P_s} = \frac{Qdp}{1715P_s}.$$

The results are tabulated and plotted below.

Q_0 (gpm)	dp_0 (psi)	$10p_s$ (hp)	η_0 (%)
39.4366	74.8463	47.9169	35.9128
79.0068	74.1246	52.0136	65.6415
118.98	68.8153	55.5591	85.9158
139.205	65.2521	61.9322	85.5067
159.544	63.6359	64.446	91.8449
179.692	61.788	66.6566	97.1087
200	55	70	91.6143
220.378	48.165	72.3716	85.5067

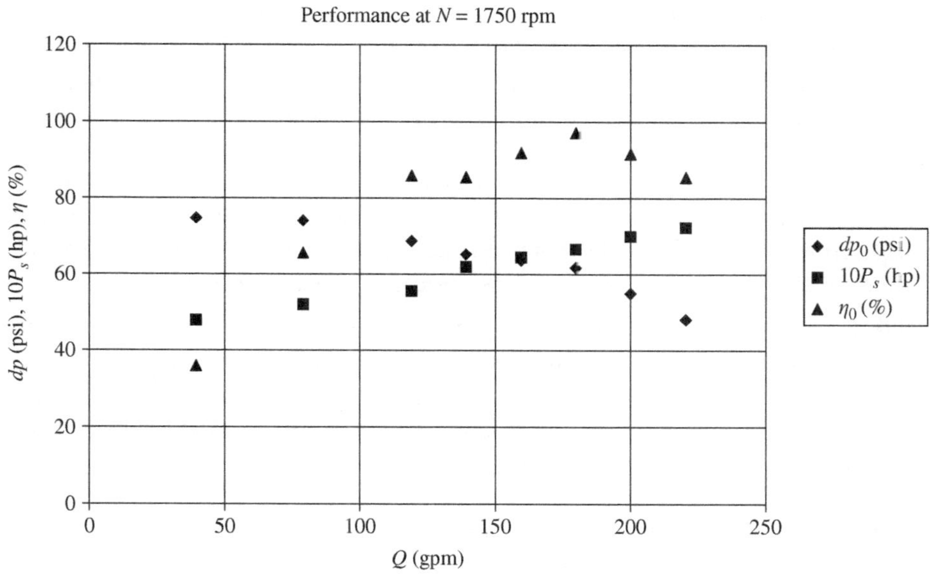

Example 3.10 A centrifugal pump is tested for its required NPSH at the design condition of $Q = 220$ gpm, $dp = 70$ psi, and $N = 3500$ rpm. The flow rate is controlled to be constant by adjusting the inlet and outlet valves. With the following data obtained, determine its required NPSH and the corresponding suction-specific speed. Assume that the inlet water temperature is 80°F and the suction pipe inside diameter is 5 in.

p_1 (psia)	12	10	9	8	7.5	7.2	7.0
dp (psi)	70	70	69	68.5	68	67.5	67

Solution

The dynamic pressure at the inlet is calculated from

$$V_1 = \frac{Q}{A} = \frac{4Q}{\left(\pi D_i^2\right)} = \frac{4 \times 220 \times 0.00223}{\pi \times 25/144} = 3.60 \text{ ft/s},$$

$$P_{d1} = \frac{\rho V_1^2}{2} = \frac{62.4 \times 3.60^2}{2 \times 32.2 \times 144} = 0.087 \text{ psi}.$$

The vapor pressure of water at 80°F is obtained from steam tables as $p_v = 0.507$ psia. The required NPSH is designated at the condition such that the discharge pressure drops by 3% from the noncavitation condition. Hence from the above test data, the corresponding $dp' = 0.97 \times 70 = 67.9$ psi and $p_1' = 7.2 + (7.5 - 7.2) \times (67.9 - 67.5)/0.5 = 7.44$ psia can be obtained. Also we have $\text{NPSPR} = p_1' + p_{d1} - p_v = 7.44 + 0.087 - 0.507 = 7.02$ psia, or $\text{NPSHR} = 2.307 \times 7.02 = 16.2$ ft of water, and

$$S = \frac{NQ^{0.5}}{\text{NPSHR}^{0.75}} = \frac{3500 \times 220^{0.5}}{16.2^{0.75}} = 3500 \times 14.8/8.07$$

$$= 6419 \text{ rpm(gpm)}^{0.5}/\text{ft}^{0.75}.$$

3.1.7 Pumping Systems

A pumping system consists of one or several pumps to energize the flowing fluid through a hydraulic system. A hydraulic system typically consists of several components (such as heat exchangers and/or other processing units) connected with the pipes and accessories (such as valves, fittings, and reservoir tanks). The pump characteristic curves for a constant rotating speed imply that all the conditions of the head and flow rate on the curves are possible operating conditions. However, the actual operating condition of the pump depends on the flow resistance in the hydraulic system, that is, system curve.

The flow resistance due to viscous force in the system, including pipes, piping components, and equipment, in terms of head loss can be related as $H_s = kQ^2$, where Q is the volumetric flow rate. For the high flow rate with high Reynolds number, the loss coefficient k is independent of the flow rate. The k values for the standard pipes and fittings can be obtained from fluid mechanics texts or hydraulics handbook (e.g., Ref. 15). For the components, these data are provided by manufacturers. The combined loss coefficients due to several components can be obtained from $k_t = \sum k_i$ if they are connected in series or $1/k_t = \left(\sum 1/\sqrt{k_i}\right)^2$ if they are connected in parallel.

In a closed-loop or open system with the same free surface levels in the tanks of both ends, as shown in Figure 3.28*a*, the viscous force is the only flow resistance. Hence a parabolic curve representing the system curve can be plotted on the H–Q chart. The intersection point, A, of the pump and system curves will determine the pump operating condition of the head and flow rate. The shaft power and efficiency corresponding to this flow rate can then be determined.

When the throttle valve installed at the pump discharge is closed, additional flow resistance will be imposed. It results in increased system resistance. Hence the k value increases and the system curve becomes steeper as shown. The pump operating point shifts from point A to B. A lower flow rate with a higher head result.

In the open systems, if the free surface levels in the tanks of both ends are different, such as those shown in Figures 3.28*b* and *c*, a constant flow resistance (positive or negative due to gravitation) will be added to the variable flow resistance. Hence the system curves are shifted upward or downward. In the extreme case with most of the resistance due to gravitation, as shown in Figure 3.28*d*, the system curve becomes fairly flat. Or if the surface levels change during pumping, the system curve will change accordingly, as shown in Figure 3.28*e*. Of course, in the cases shown in Figures 3.29*b*, *d*, and *e*, a check valve is needed if the reversed flow is to be prevented. In Figure 3.29*c*, no pump is needed for a flow rate below Q_0.

In other situations, if two or more pumps are combined, the resultant pump performance curve will be modified. As shown in Figure 3.29*a*, two pumps are connected in parallel, and the flow rate through the hydraulic system will be the sum of those for both pumps. That means, at any head, the combined pump curve will have a flow rate of $Q_c = Q_a + Q_b$. If two pumps are connected in series, as shown in Figure 3.29*b*, at any flow rate, the combined pump head will be the sum of both pumps, that is, $H_c = H_a + H_b$.

However, in these systems, when one pump fails, the other may be overloaded if no check valve or some control system is installed. In the case of parallel connection, if pump B fails, part of the flow from pump A will be recirculated through pump B back to suction. Hence the flow resistance due to the failed pump B is connected with the hydraulic system in parallel, and the resulting system curve is changed from 01 to 02, and the operating point is shifted from 1 to 2, as shown in Figure 3.30*a*. Similarly, in the case of series connection, if pump B fails, its flow resistance is added to the hydraulic system in series. Hence the operating point is shifted from point 1 to 2, as shown in Figure 3.30*b*. The possibility of pump A being overloaded will depend on its characteristic curve of brake horsepower, as illustrated in Figure 3.8*b*.

3.1.7.1 Stability of Pumping System

A pumping system, like many other physical systems, when subject to an external disturbance, may become unstable. That means the disturbance may grow in magnitude and eventually destroy the complete system.

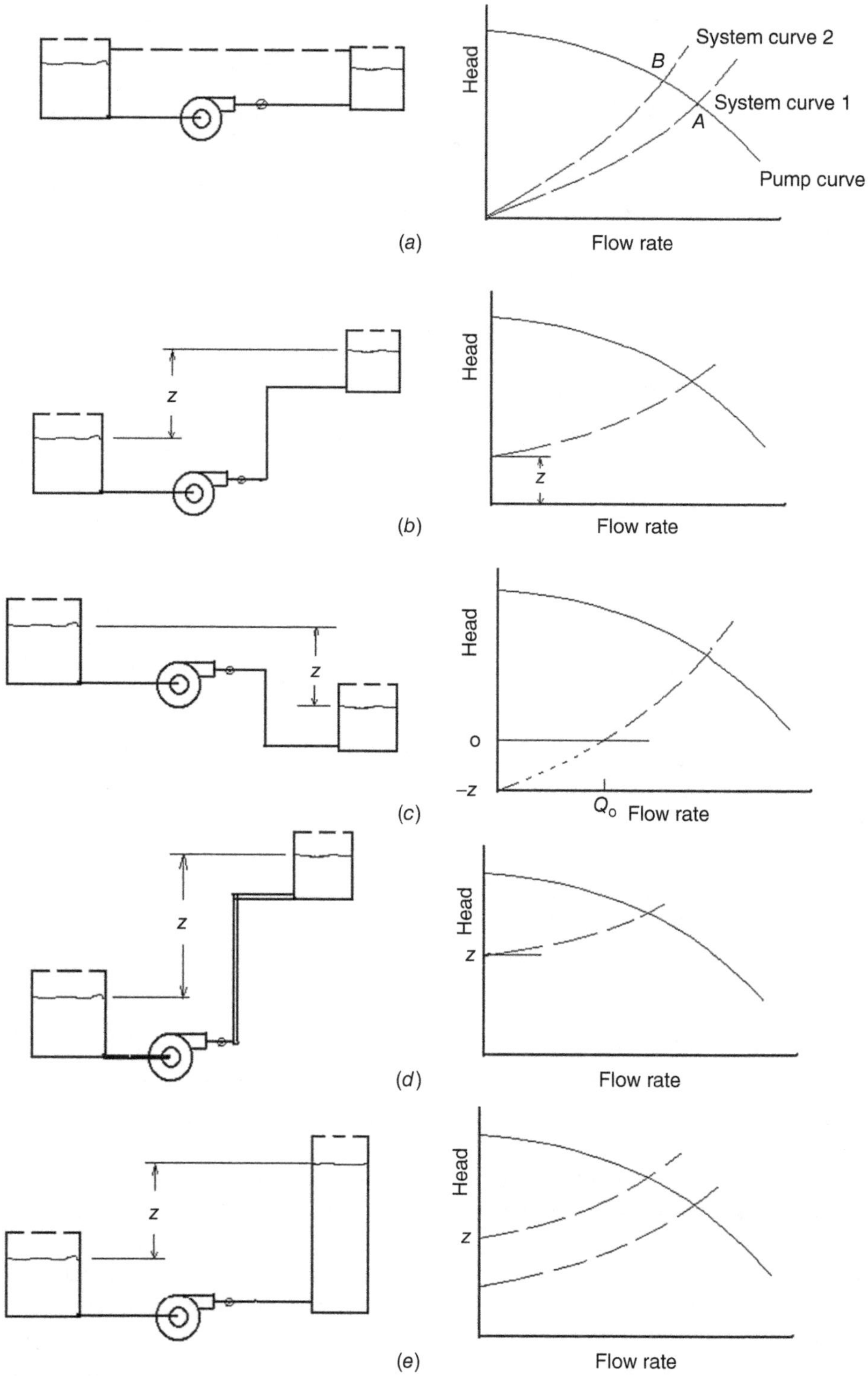

Figure 3.28 Hydraulic systems with flow resistance due to viscous force and gravitation.

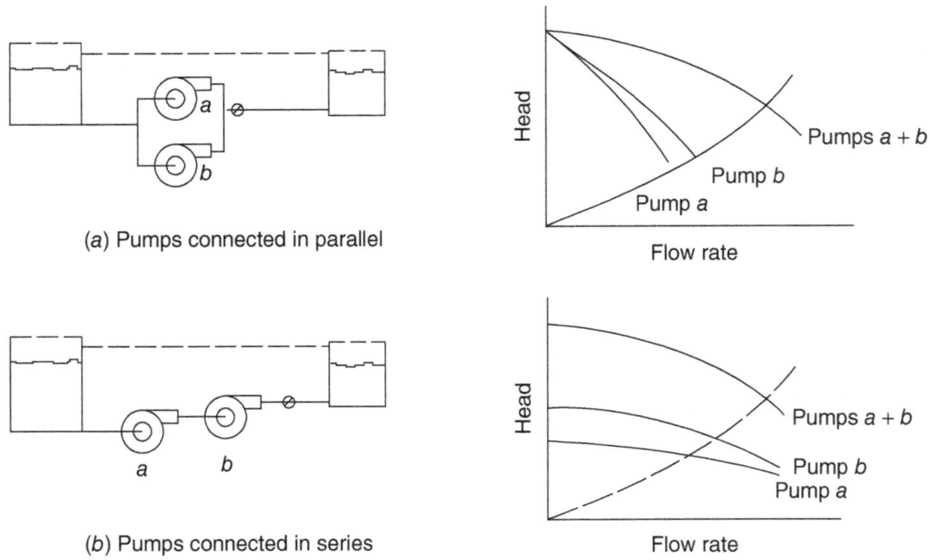

Figure 3.29 Performance curves of combined pumps and system.

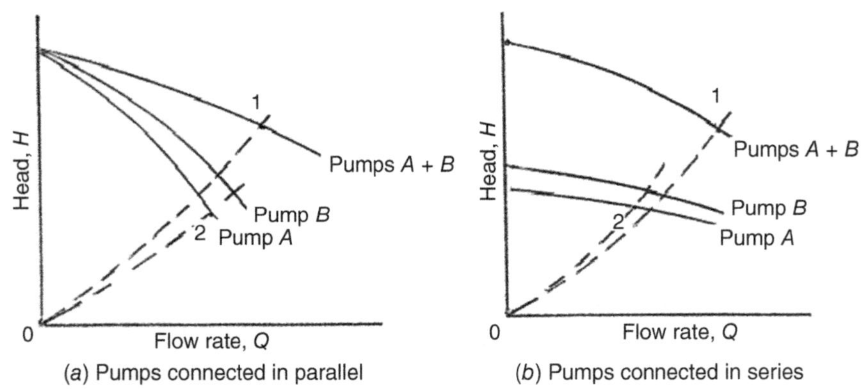

Figure 3.30 Change of the operating point due to failure of one pump in a two-pump system.

The likelihood that this will happen depends on the system characteristics. In a pumping system, instability involves the fast variation of flow rate. It may originate locally and then propagate to the whole system and induce mechanical vibration.

The flow surge can be associated with the local flow recirculation at the inlet and outlet of a pump, as mentioned in Section 3.1.3. Another type of flow surge is associated with the pump performance characteristics and the system.

For a pump with part of the head–capacity curve decreasing to the shut-off condition (or positive H–Q gradient in the left of maximum head), as shown in Figure 3.31, unstable operation is possible. If the hydraulic system is subject to a disturbance with additional resistance, the system curve will change from $0A$ to $0B$. The flow rate through the pump decreases from Q_a to Q_b and results in a lower pump head H_b. Since this head is insufficient to overcome the increased system resistance, the flow through the system will further decrease until the system resistance is lower than the pump head. Due to the system inertia, the flow rate will change back and forth. This kind of flow surge is especially possible in hydraulic systems with free surfaces in the tanks of both ends.

Figure 3.31 Unstable operation of pump with positive head-capacity gradient.

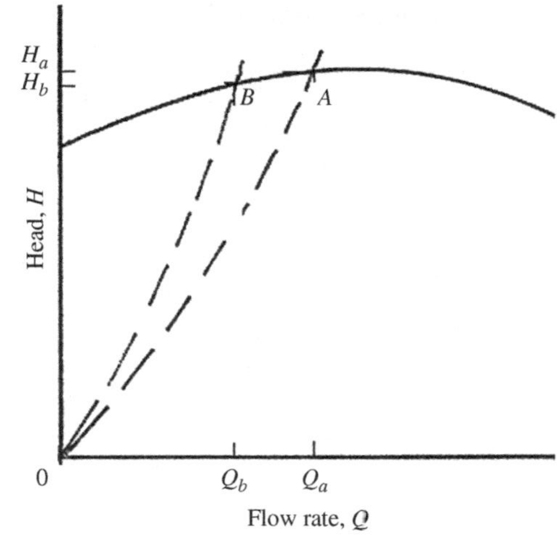

Example 3.11 The pipe component shown has been tested at a flow rate of 25 gpm and a pressure drop of 10 psi due to flow friction is obtained. If its inlet and outlet have an elevation difference of 3 ft, (a) plot the system curve in a head–flow rate diagram; (b) if two of the identical components are combined in parallel, derive an expression for the combined system curve in the form $H_c = f(Q_c)$.

Solution

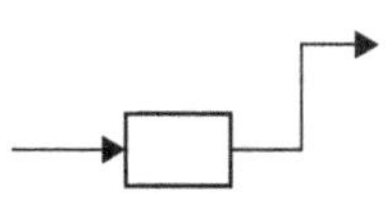

a) With $H = kQ^2 + 3$, the constant k can be obtained by substituting $H = 10 \times 144/62.4 = 23.07$ ft and $Q = 25$ gpm into the expression. That is, from $23.07 = k \times 25^2 + 3$, we obtain $k = 0.0321$ ft/(gpm)^2. Then the H/Q relationship can be tabulated and plotted as follows:

Q (gpm)	0	5	10	15	20	25	30
H (ft)	3	3.8	6.21	10.22	15.84	23.07	31.9

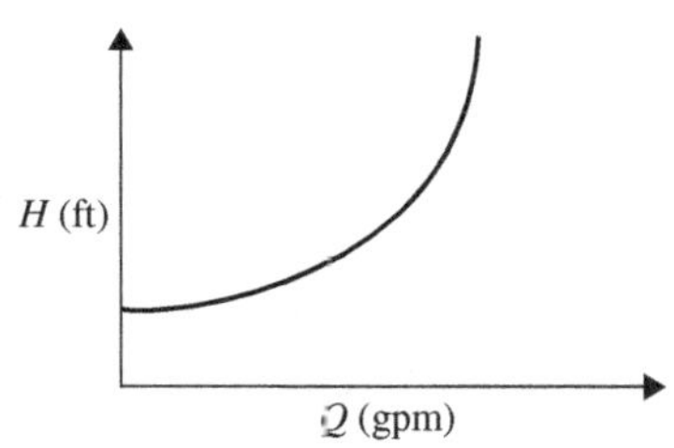

b) Rewrite $H = 0.0321Q^2 + 3$ into $Q = [(H - 3)/0.0321]^{0.5} = 5.581(H - 3)^{0.5}$. From $Q_c = 2Q$ and $H_c = H$, we have $Q_c = 11.16(H_c - 3)^{0.5}$ or $Q_c^2 = 124.6(H_c - 3)$ or $H_c = 0.00802Q_c^2 + 3$. It is the expression for the combined system.

Example 3.12 Plot the combined performance curves for two of the pumps given below: (a) two connected in series and (b) two connected in parallel.

Solution

The single-pump performance is given as follows:

Q (gpm)	0	70	130	172	186	190	195	200
H (ft)	1600	1500	1275	920	785	540	435	285

a) Two connected in series:

Q (gpm)	0	70	130	172	186	190	195	200
H_1 (ft)	3200	3000	2550	1840	1570	1080	870	570

b) Two connected in parallel:

Q (gpm)	0	140	260	344	372	380	390	400
H_2 (ft)	1600	1500	1275	920	785	540	435	285

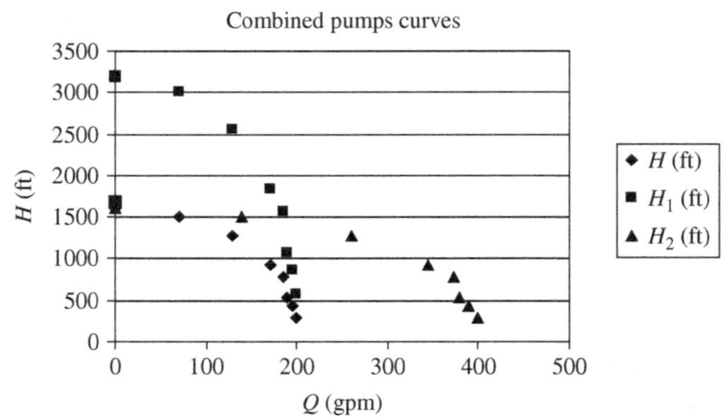

3.1.8 Pump Applications

As mentioned earlier, pumps are widely used in industry and the household. Wherever there is liquid to be transported or to be processed, there will be some types of pumps and very likely they will be the centrifugal pumps. Some typical pumps used in various industries are illustrated below. Two photographs shown in Figures 3.32 and 3.33 are taken from Refs. 10 and 18.

1) Water supply:

Low-lift pumps	Used in transporting water from the source
Carbon slurry pumps, fluoride pumps, sampling pumps, washwater pumps, plant water pumps, high-service pumps	Used in the water treatment plant
In-line booster pumps	Used in the distribution system
Vertical turbine pumps, submersible pumps	Used in the groundwater well source

Figure 3.32 Typical vertical pump for farm irrigation. (Courtesy of Weir Floway, Inc., Fresno, CA.)

2) Sewage systems:

Nonclog pumps	Used to pump solid–liquid slurry mixtures in collection system, treatment plant, and effluent system. The nonclog pumps can be either a radial impeller with 2 or 3 blades in a concentric casing or a recessed impeller to provide a generous clearance for slurry mixture to pass through, as shown in Figure 3.15.

3) Irrigation systems:

Low-head axial-flow pumps	Used in the surface source (lakes, ponds, dams)
Volute pumps, deep-well turbine pumps, submersible turbine pumps (shown in Figure 3.32)	Used in the subsurface source (well)
Nonclog pumps	Used in the source from sewage treatment effluents

4) Petroleum industry:

Nonclog pumps	Used in drilling
Vertical turbine pumps, submersible turbine pumps	Used in lifting crude and secondary recovery
In-line booster pumps	Used in transferring crude and refined products
Various types of centrifugal pumps with single or multistages	Used in refining (shown in Figure 3.33)

The effects of specific gravity, viscosity, temperature, air content, and gas–liquid mixtures are important considerations in these applications. The American Petroleum Institute (API) standard 610 is widely used in designing and selecting these pumps[16].

5) Thermal power plants: High-pressure boiler feed pumps (shown in Figure 3.34), cooling-water circulating pumps, condensate pumps, heater drain pumps, and reactor coolant pumps are used. Some special requirements for boiler feed pumps include (a) stable head–capacity characteristics for the whole operating range, (b) low required NPSP, (c) protection from overheating at low or zero capacity, and (d) special construction material for the feed water chemistry and pressure/temperature condition.

Figure 3.33 Centrifugal pump in a refining plant. (Courtesy of Ingersoll Rand Company, Phillipsburg, NJ.)

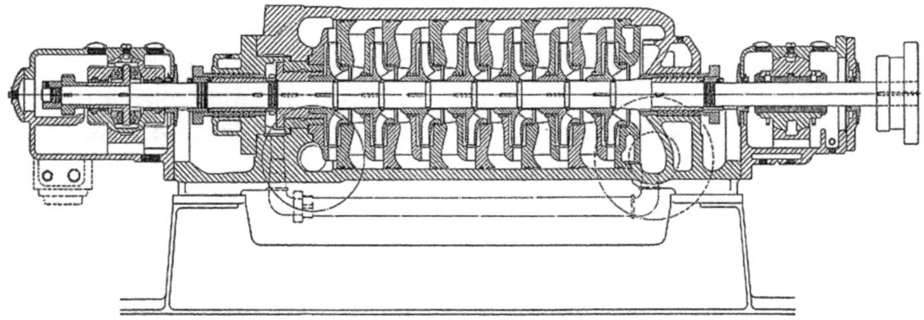

Figure 3.34 Multistage high-pressure centrifugal boiler-feed pump for thermal power plants or process plants. (Reprinted by permission from Stepanoff, A. J., *Centrifugal and Axial Flow Pumps*, 2nd ed., John Wiley & Sons, New York, 1957.)

6) HVAC systems: Hot- and chilled-water circulating pumps, condenser cooling-water circulating pumps, and refrigerant circulating pumps are used.
7) Others: Aircraft fuel pumps, liquid fuel rocket engine pumps, marine pumps, and many other special pump types are used in the chemical industry, food and beverage processing, and hydraulic transport of solids.

3.1.8.1 Variable-Speed Pumps

The centrifugal pump operating at a fixed rotating speed has only one flow rate corresponding to the most efficient condition. When the system requirement changes, it is necessary to change the throttle valve setting to the new operating point with a lower efficient condition of the pump. In some cases, this drop can be drastic.

To achieve the best efficient condition for different system requirements, the pump can be operated with a variable-speed drive. As shown in Figure 3.35, the normal design operating condition is at 1, with the pump operating at speed N_a. When a lower flow rate Q_2 is required, instead of throttling the system curve to S_2', the pump speed can be decreased to N_b and the system is throttled to S_2, where $\eta_2 > \eta_2'$. Also, the extra Δp between 2 and 2′ can be saved. Or if a lower discharge pressure Δp_3 is required, the pump speed can be decreased to N_c and the valve opened to S_3, instead of to S_3' with constant pump speed N_a and bypassing some extra flow rate. For more complicated system requirements, a multiple-pump system connected in series and/or in parallel can be designed.

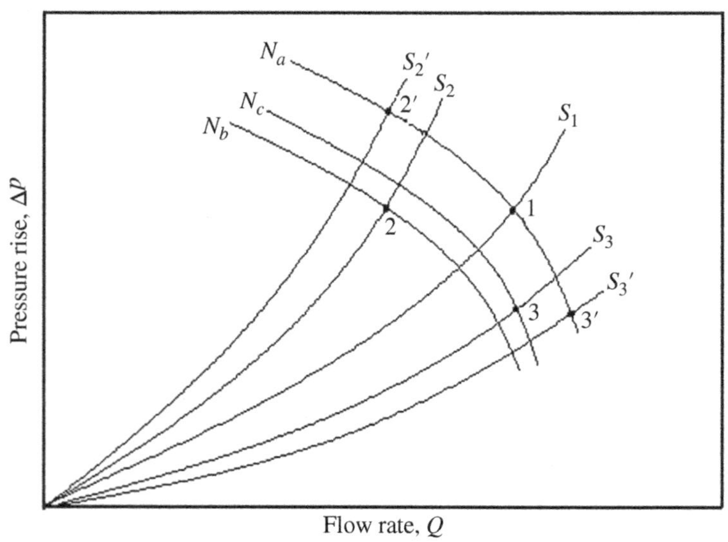

Figure 3.35 Variable-speed pump performance.

3.1.9 Pump Selection

For any application, picking the proper pump from the multitude of available types and sizes can be a difficult job for the user or the contracting engineer unless it is a small to medium-size pump to be used in a simple system. Pump selection requires careful analysis of the hydraulic system, the fluid to be pumped, and the pump's location and function within the system. With knowledge of pump capabilities and limitations, a step-by-step procedure will direct the engineer to the least cost of equipment and operation. The economic analysis of total cost should include the cost of equipment, space, installation, and the operating cost of energy, maintenance, and down time. To compare the alternatives, a simple calculation of the total cost, including the initial and equivalent present value of operating cost, can be performed, that is, $C_t = C_i + C_{po}$. The present value of operating cost can be expressed as $C_{po} = C_o[1 - (1 + i)^{-n}]/i$, where C_o is the periodic operating cost, i is the prevailing interest rate, and n is the expected equipment life. Of course, these parameters are difficult to predict. They must be determined based on experience.

A detailed listing of fluid characteristics is usually the first step in the selection process. These include the chemical identity of the fluid, its viscosity, specific gravity, and vapor pressure at the temperature during normal operation. If fluid viscosity is above 500 SSU (Saybolt Universal Unit) or there are more than 5% entrained gases in normal operation, a positive-displacement pump is probably the better choice. When handling slurries with a centrifugal pump, the impeller speed should be 1500 rpm or less. When solid volume is over 3%, a special slurry pump may be required.

In the second step, the hydraulic system should be studied. If the required head varies widely with a relatively constant capacity in the system, a reciprocating pump operated at a constant speed will be preferred. But if the pulsation of the head is objectionable, a rotary pump can be selected. However, this type of pump is limited to low and intermediate head. A system curve with the effect of throttling, as discussed in Section 3.1.3 and shown in Figure 3.28*a*, can be established, including the available NPSH. If a remarkably high flow rate or high head is required, a double-suction or multistage pump can be employed. If space is permitted, pumps arranged in parallel or in series as shown in Figures 3.29*a* and *b* are preferred, because some units can be closed or bypassed to save energy when operating conditions change. Margins should be included for the wear and potential future system changes.

From the system curve, a normal operating point is selected. And with a rotating speed, the specific speed of the required pump can be calculated. In general, a higher rotating speed will mean the pump is of higher efficiency and smaller size. However, more problems associated with cavitation and vibration will be likely during operation. The maximum speed can be estimated with the definition of suction specific speed if the available NPSH is known. The suction-specific speeds $S = NQ^{0.5}/(\text{NPSH}^{0.75})$ will vary only slightly within a limited range for a specific type of pump, as discussed in Section 3.1.4. An average value can be selected if the type of pump is contemplated.

When a centrifugal pump is selected, a diagram based on the Cordier diagram specifically for the pumps in dimensional specific speed and specific diameter, shown in Figure 3.36, can be used to select the appropriate type and size. Efficiency can be read from Figure 2.1 or Figure 3.20. Similar curves are also given in Ref. 17.

The values from these charts are agreeable, except for the low flow rate. If the double-suction or multistage pump is considered, the specific speed should be calculated based on single-suction, single-stage flow rate and head.

Based on the piping layout and space available for pump installation, different types of pump constructions can be selected:

a) An overhung or cantilever pump is selected for perpendicular inlet/outlet piping.
b) An in-line pump is selected for continuous piping layout.
c) A vertical pump has to be selected for limited space or when the fluid is pumped from a well. It has some advantages in certain applications in terms of installation, maintenance, and NPSH requirements. Typical installations are shown in Figure 3.37.

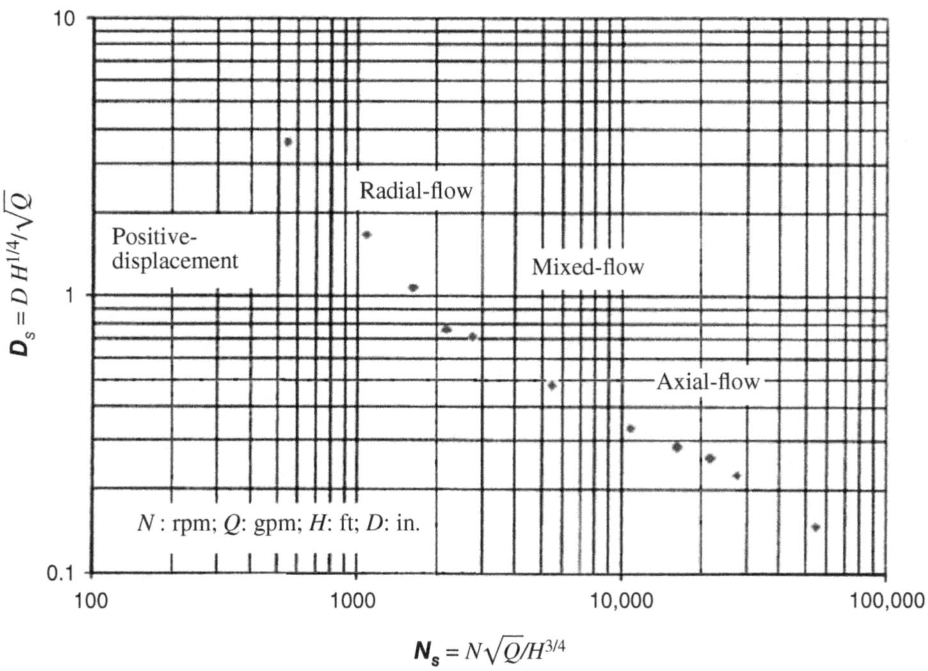

Figure 3.36 $N_s - D_s$ curves for single-stage pumps (converted from Cordier diagram).

Each vertical pump consists of four basic components: (1) driver, (2) discharge head, (3) column and shaft assembly, and (4) bowl assembly. The configuration and dimensions of each component can be varied to suit a particular application.

A particular model can then be selected from the composite chart as shown in Figure 3.7*b*, and detailed performance curves as shown in Figure 3.7*a* can be obtained from the vendors.

A detailed discussion of the pump materials and their resistance to various environments and pumped fluids may be found in Hydraulic Institute standards.[3] A few examples are listed below:

a) Bronze or iron for cold water with less than 3% salt
b) Austenitic steel for water with more than 3% salt
c) Nickel-base alloys or high-silicon cast iron for sulfuric acid

It should be kept in mind that the manufacturer's standard materials should be used whenever possible.

Selection of the driver depends on the type of energy available. In most cases, if electrical power is readily available, an electrical motor will be used, especially for the small to medium power range. Otherwise, an internal combustion or diesel engine will be used for power up to 50 hp. A gas turbine or steam turbine will be employed for higher horsepower. Occasionally, if compressed air or high-pressure fluid is readily available, an air motor or hydraulic motor can be considered. For small to medium power, the variable-speed drive is preferred if the initial cost is justifiable.

A technical specification has to be prepared for prospective vendors. The applicable codes and standards [such as the American National Standards Institute (ANSI) or API] should be referenced. The final selection from several technically acceptable candidates submitted by manufacturers is normally based on economic considerations, including initial and operating costs, maintenance, and other business conditions.

In some applications, such as marine and aerospace pumps, size and weight are important considerations, while in other applications noise and vibration can be more important considerations.

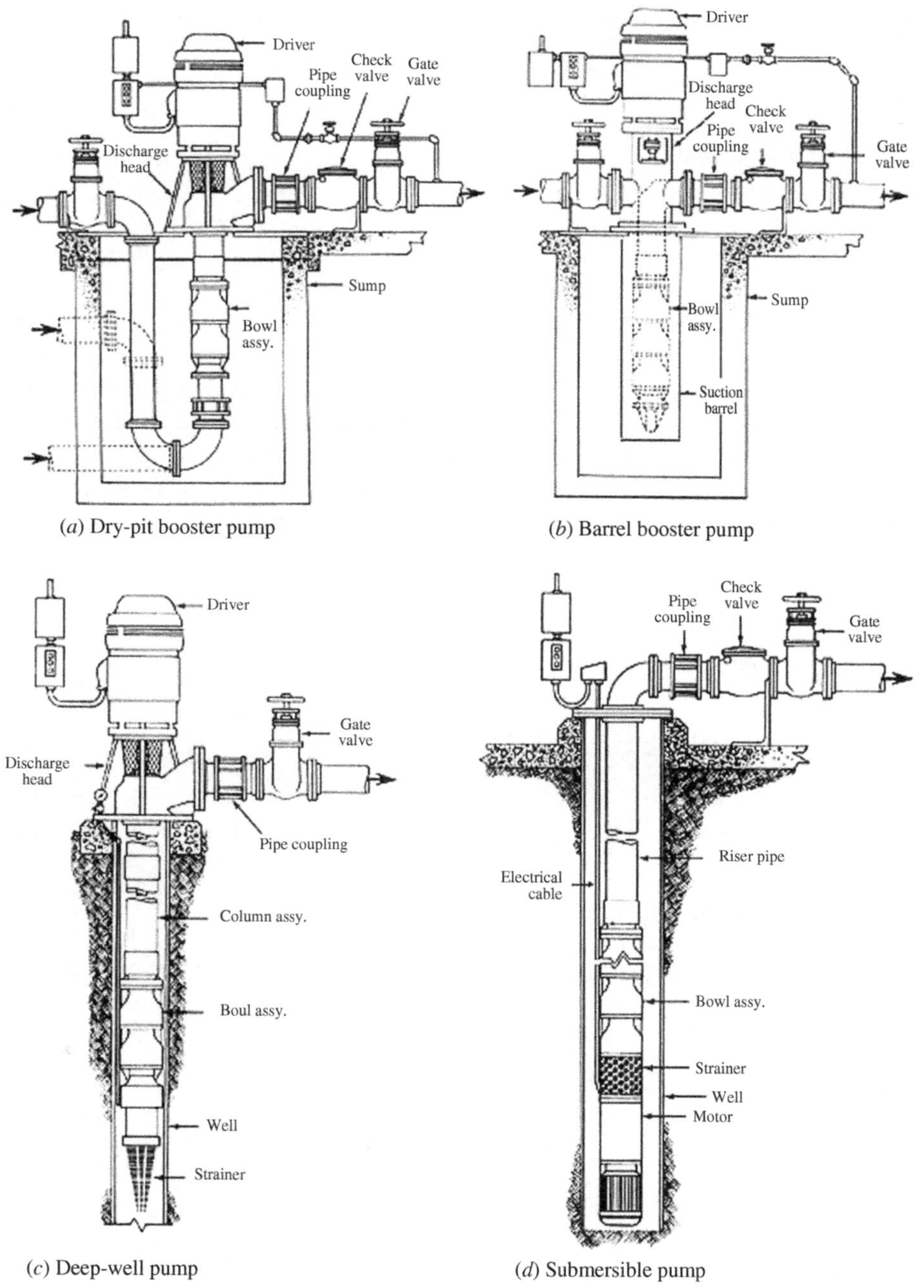

Figure 3.37 Typical vertical pump installations. (Courtesy of Weir Floway, Inc., Fresno, CA.[32])

Pump data supplied by vendors should cover at least the following items:

1) Model number and type
2) Number of stage
3) Impeller diameter
4) Suction diameter and type of connection
5) Discharge diameter and type of connection

6) Horse power requirements at (a) shutoff condition, (b) normal operation condition (b.e.p.), and (c) maximum horsepower condition
7) Weight
8) Type of coupling with driver
9) Type of shaft seal (packing or mechanical seal)
10) Type of bearings and lubricating instruction
11) Bed plate (cast or fabricated)
12) Performance curves including NPSHR
13) List of materials
14) Pump drawing with all salient dimensions, including all external connections.

Example 3.13 A pump is needed to supply 500 gpm of water to a hydraulic system with a total system friction loss of 350 ft. Three pump arrangements are proposed, all driven with an alternating current (AC) motor of 90% efficiency. Option *A* is a pump directly connected to the motor running at 3600 rpm. Option *B* is a pump geared up to 4320 rpm with a gear efficiency of 95%. Option C employs two smaller pumps connected in series running at 3600 rpm. Estimate the electricity consumption of each option and the approximate impeller diameters.

Solution

The specific speeds for all three options are calculated as follows:

$$N_{sa} = \frac{3600 \times 500^{0.5}}{350^{0.75}} = 995 \text{ rpm} \times \text{gpm}^{0.5}/\text{ft}^{0.75},$$

$$N_{sb} = \frac{4320 \times 500^{0.5}}{350^{0.75}} = 1194, \quad N_{sc} = 3600\left(\frac{500^{0.5}}{175^{0.75}}\right) = 1673.$$

From Figures 3.20 and 3.36, we obtain $\eta_a = 0.75$, $D_{sa} = 1.7$, $\eta_b = 0.78$, $D_{sb} = 1.4$, $\eta_c = 0.80$, $D_{sc} = 1.1$, and the diameters $D_a = D_{sa}Q^{0.5}/H^{0.25} = 1.7(500^{0.5}/350^{0.25}) = 8.8$ in., $D_b = 7.2$ in., and $D_c = 6.8$ in. The output hydraulic power can be calculated as

$$P_h = Q\rho gH = 500 \times 0.00223 \times 62.4 \times 350 = 24,352 \text{ ft-lb}_f/\text{s} = 44.3 \text{ hp} = 33 \text{ kW}.$$

Hence the electrical power required $P_e = P_h/(\eta_{\text{pump}} \times \eta_{\text{gear}} \times \eta_{\text{motor}})$ can be calculated for the three options as

$$P_a = \frac{33}{0.75 \times 0.9} = 48.9 \text{ kW},$$

$$P_b = \frac{33}{0.78 \times 0.95 \times 0.9} = 49.5 \text{ kW},$$

$$P_c = \frac{33}{0.80 \times 0.9} = 45.8 \text{ kW}.$$

3.2 Axial-Flow Pumps and Fans

3.2.1 Introduction

Axial-flow pumps are used at specific speeds ranging from 10,000 to 15,000 rpm(gpm)$^{0.5}$/(ft)$^{0.75}$, as shown in Figures 3.2 and 3.36. They are all single-stage machines. Multistage axial-flow pumps are not commercially valuable. Axial-flow fans can be found at specific speeds ranging from 100,000 to 200,000 rpm(cfm)$^{0.50}$/(in water)$^{0.75}$, as shown in Figure 3.38, where the flow rate is in cubic feet per minute static pressure is in inches of water, and diameter is in inches. It is important to note that the dimensional-specific speeds are defined differently for different types of turbomachines. The axial-flow compressors to be discussed in Chapter 4, Section 4.2 are

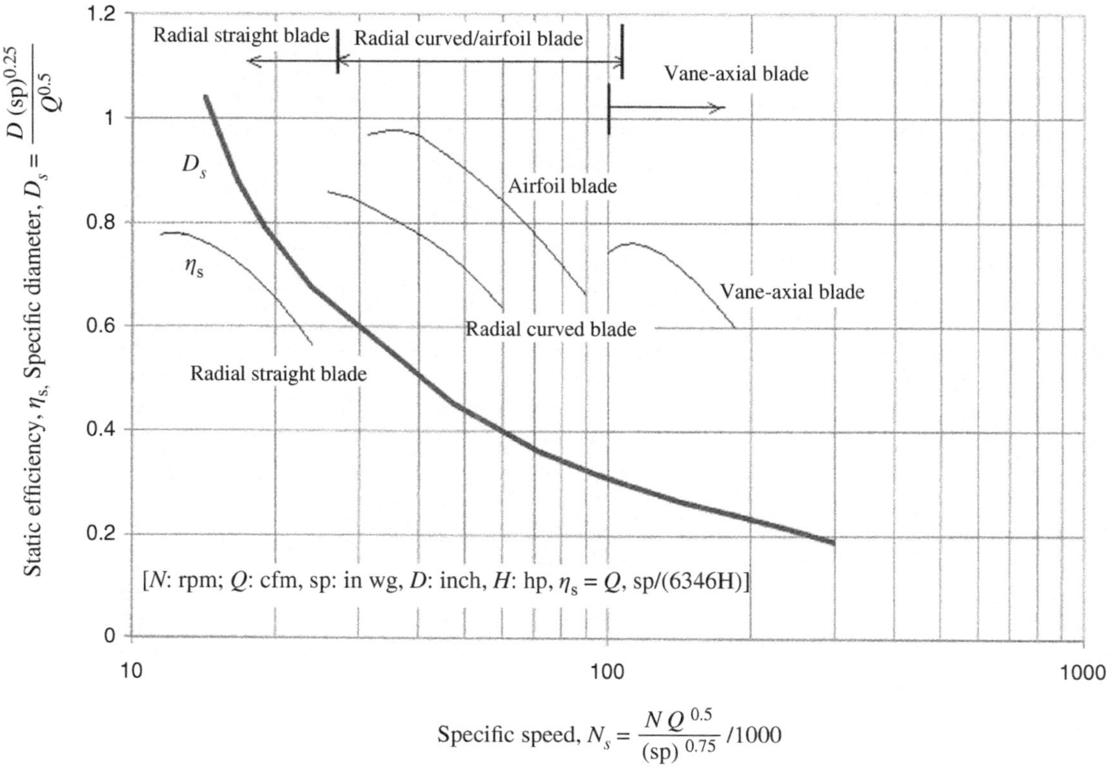

Figure 3.38 N_s–D_s curve (converted from Cordier diagram to dimensional parameters) and estimated static efficiency (from Ref. 18) for fans and blowers.

normally built in multistage to achieve a high-pressure ratio to compete with centrifugal machines in price, size, and efficiency. The fan static efficiency defined in terms of discharge static pressure is normally used, where the static pressure is related to the total pressure in the form $\text{tp} = \rho V^2/2 + \text{sp}$.

Flow analysis through an axial-flow machine is treated differently from that of a centrifugal machine because the flow passage in the former is short compared to that in the latter. Therefore, two types of analysis for designing axial flow machines exist: (a) relating the lift and drag acting on an isolated airfoil, including interference effects from adjacent ones, with the flow momentum change and (b) using the cascade test data to relate the performance in terms of fluid deflection and pressure loss, to be accounted for in the Euler equation.

It is recalled from Section 2.2.3 that the mean radius of an axial-flow impeller is defined as $r_m = \left[\left(r_t^2 + r_h^2\right)/2\right]^{0.5}$. This is the radius at which the first step of flow analysis will be performed. On cylindrical development, blades appear equally spaced at a distance $s = 2\pi r_m/Z_b$, which is called pitch. The ratio of the blade chord length to the pitch, called blade solidity (c/s), is an important design parameter for the axial-flow impeller.

3.2.2 Flow Over Isolated Airfoil

An object immersed in a flow field will be subjected to forces due to both viscous force and unbalanced pressure distribution around the object. The magnitudes of these forces are related to the geometry and orientation of the object with respect to the free stream flow direction. The net resultant force can be resolved into two components. The one parallel to the mean flow direction is called drag, while the one normal to the flow direction is called lift. A streamlined airfoil, properly orientated, can produce a maximum lift and a minimum drag. A typical airfoil

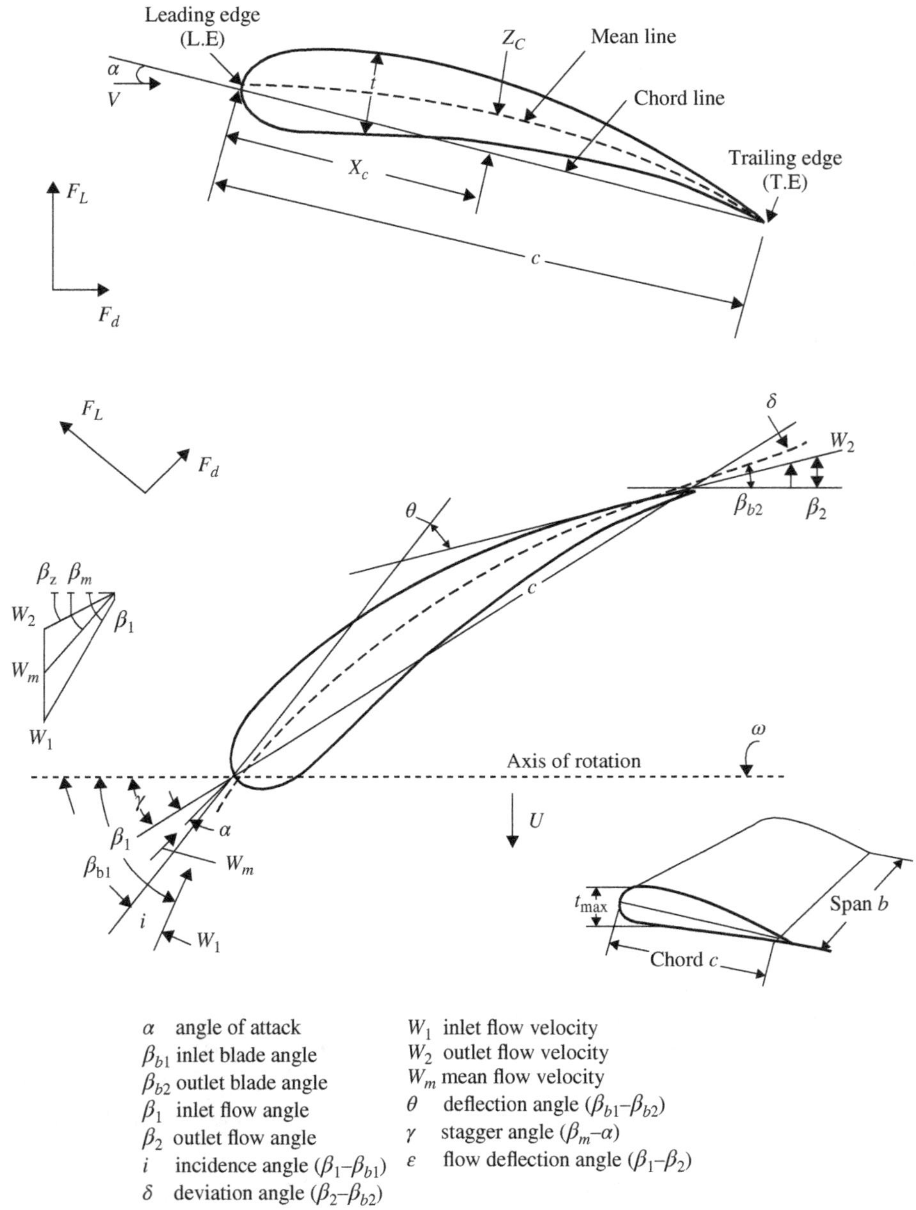

Figure 3.39 Nomenclature of airfoil section.

with the relevant nomenclature is shown in Figure 3.39. The mean line curvature is designed to change the flow direction, the vane thickness is for strength, and the streamlined shape is to delay the onset of boundary layer separation on the surface. The resultant forces in terms of lift and drag can be expressed in terms of the nondimensional parameters C_L and C_d, as discussed in Chapter 2,

$$F_L = C_L bc\left(\frac{1}{2}\rho W_m^2\right), \quad F_d = C_d bc\left(\frac{1}{2}\rho W_m^2\right) \tag{3.8}$$

where b is the span, c is the chord length of the airfoil ($A_p = bc$ is the plane form area, which is the maximum projected area of the airfoil), C_L and C_d are the lift and drag coefficients, respectively, and W_m is the mean flow

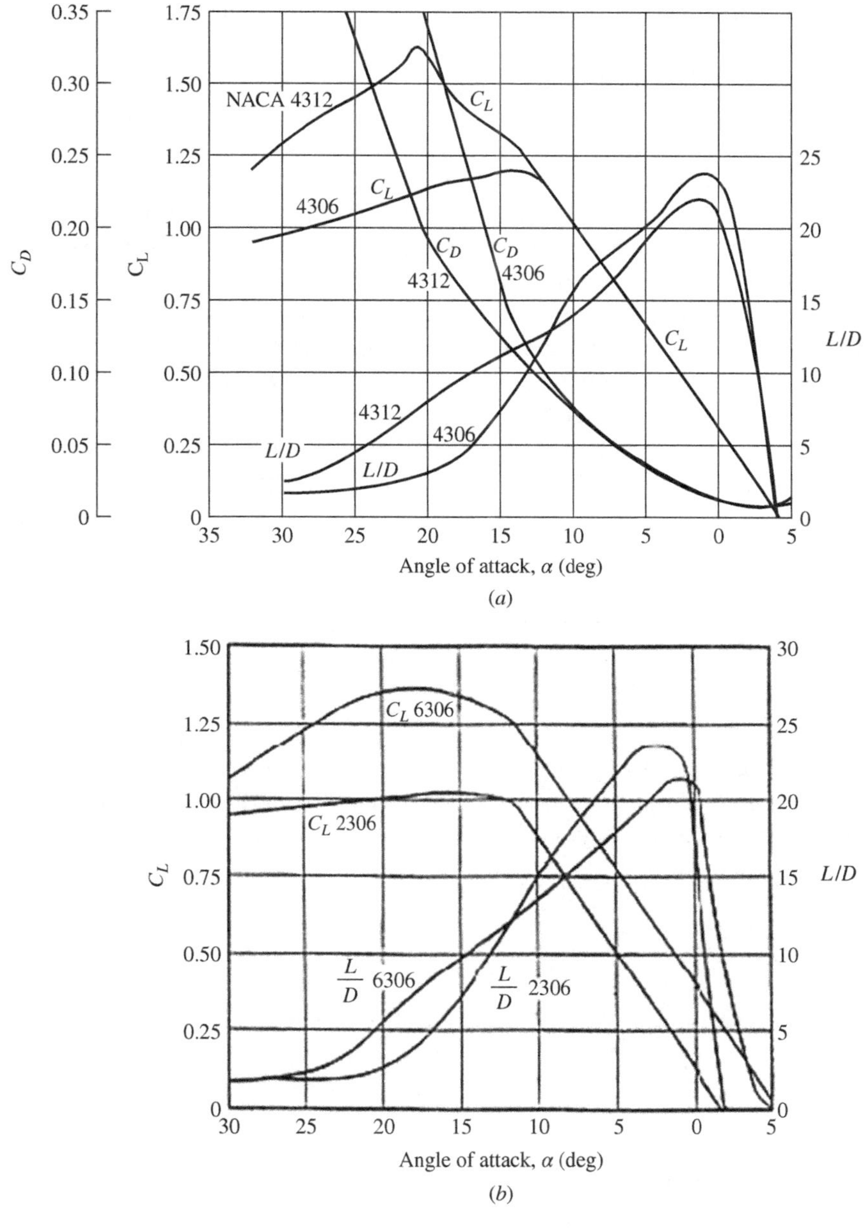

Figure 3.40 (*a*) Effect of vane thickness on airfoil performance (*b*) Effect of vane camber on airfoil performance. (Reprinted by permission from Stepanoff, A. J., *Centrifugal and Axial Flow Pumps*, 2nd ed., John Wiley & Sons, New York, 1957.)

velocity, to be defined later. Lift is used to support the weight of an airplane, but in the rotating impeller of an axial-flow machine, it is used to convert energy from the shaft to the fluid, or vice versa, while drag is part of the energy losses due to friction.

The National Advisory Commission on Aeronautics (NACA) tested and published a series of different airfoils. One type of NACA airfoil profile is assigned with a four-digit number, such as 4312. The first number indicates the camber of the mean line in percent of the chord (z_c/c); the second indicates the location of the camber from the leading edge in tenths of the chord length (x_c/c); the last two digits show the maximum vane thickness in percent of the chord t_{max}/c. Other types of airfoils include the five-digit and the series 6 sections. In five-digit code, the design lift coefficient can be obtained from the multiplication of the first digit with 0.15. The series 6 airfoils are designed for the laminar boundary layer on the surfaces.

Some typical NACA data are shown in Figure 3.40 to illustrate the effect of vane thickness and vane camber. These data are based on two-dimensional flow without end effects. For a finite-span airfoil, the trailing vortex will be generated at the tip due to the pressure difference between the lower and upper surfaces. This trailing vortex causes a downwash of the flow at the trailing edge and hence an effective decrease of the angle of attack. It can be estimated with the expression $\Delta\alpha = -C_L/(\pi\ ar)$, where ar is the aspect ratio defined as the span over the chord (b/c). More detailed discussion of the airfoil is given in Appendix B.

The axial-flow machine performance parameters are related to the inlet and outlet flow velocity and direction. Hence it will be useful to relate these parameters with the lift and drag coefficients of the airfoil, although the isolated airfoil data must be modified to account for the interference effect due to the adjacent airfoils in the cascade, to be discussed later. Consider the control volume of unit height (in the span direction, or $b = 1$) with an isolated airfoil enclosed, as shown in Figure 3.41, where s is the width of the control volume to enclose one airfoil section, which is equivalent to the pitch of an impeller, defined as $s = 2\pi r_m/Z_b$. The mean flow angle β_m of W_m, as shown in Figure 3.39, is expressed as

$$\tan\beta_m = \frac{1}{2}(\tan\beta_1 + \tan\beta_2), \tag{3.9}$$

and the resultant force per unit length of span in the peripheral direction expressed in terms of lift and drag is

$$F_u = F_L\cos\beta_m + F_d\sin\beta_m = \left(\frac{1}{2}\rho W_m^2\right)c(C_L\cos\beta_m + C_d\sin\beta_m), \tag{3.10}$$

where $W_m = W_1\cos\beta_1/\cos\beta_m$, as shown in Figure 3.41.

From the integral form of the momentum equation in the tangential direction, with the velocity triangles shown in Figure 3.42, the force on the fluid due to the airfoil is just the opposite to the force on the airfoil due to the fluid and is equal to the gain of flow momentum. Hence F_u can be expressed as

$$\begin{aligned} F_u &= m(V_{u2} - V_{u1}) = m(W_{u1} - W_{u2}) = \rho W_a s(W_a\tan\beta_1 - W_a\tan\beta_2) \\ &= \rho W_a^2 s(\tan\beta_1 - \tan\beta_2), \end{aligned} \tag{3.11}$$

where $m = \rho W_a s$ is the mass flux.

Also, with $W_a = W_m\cos\beta_m$, by equating (3.10) and (3.11) and neglecting C_d, which is small compared with C_L, we have

$$C_L = 2\left(\frac{s}{c}\right)(\tan\beta_1 - \tan\beta_2)\cos\beta_m. \tag{3.12}$$

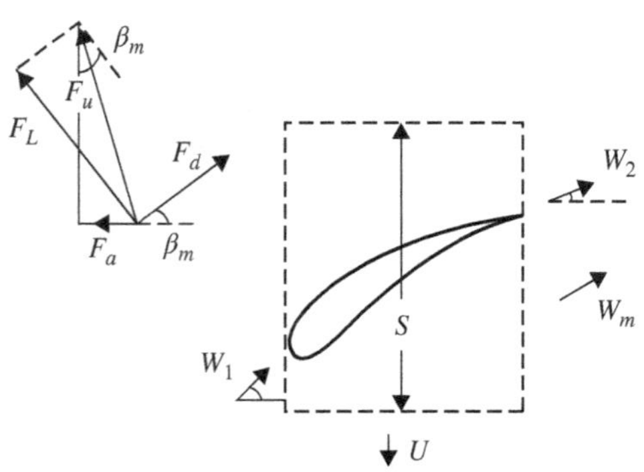

Figure 3.41 Control volume enclosing an isolated airfoil.

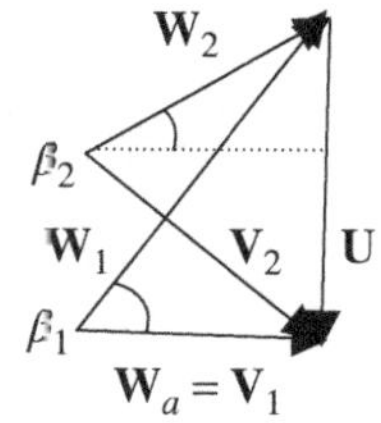

Figure 3.42 Velocity diagrams at inlet and outlet of axial-flow impeller pump or fan.

Neglecting the blade blockage effect, the axial-flow velocity is assumed to be constant, so we have $F_a = (p_2 - p_1)s$ from the momentum equation in the axial direction. Hence, the total pressure loss can be related to the forces from the momentum equation as

$$\frac{\Delta p_0}{\rho} = \frac{p_1 - p_2}{\rho} + \frac{W_1^2 - W_2^2}{2} = \frac{-F_a}{\rho s} + \frac{(W_{u1} + W_{u2})(W_{u1} - W_{u2})}{2}$$
$$= \frac{-F_a}{\rho s} + \frac{(W_{u1} + W_{u2})F_u}{2\rho s W_a} = \frac{-F_a}{\rho s} + \left(\frac{F_u}{\rho s}\right)\tan\beta_m. \tag{3.13}$$

So it can be written as

$$\Delta p_0 s\cos\beta_m = -F_a\cos\beta_m + F_u\sin\beta_m = F_d = \frac{1}{2}C_d c\rho W_m^2,$$

or substituting $W_a = W_m\cos\beta_m = W_1\cos\beta_1$ and assuming $\cos^3\beta_m \approx \cos^2\beta_1$ (by Howell[19]), we have

$$C_d = \left(\frac{s}{c}\right)\left(\frac{\Delta p_0}{\rho W_1^2/2}\right). \tag{3.14}$$

It is concluded that, with W_1, β_1 and the other geometric parameters given or specified, β_2, β_m, and Δp_0 can be expressed in terms of C_L and C_d from Equations (3.9), (3.12), and (3.14).

The pressure rise across the blades can also be related to the lift and drag coefficients from the momentum equation in the axial direction applied to the control volume depicted in Figure 3.41. Since the axial component of the flow velocity from the inlet to the outlet is unchanged as mentioned, we can write the momentum equation as $(p_2 - p_1)s = F_a = F_L\sin\beta_m - F_d\cos\beta_m$, or

$$p_2 - p_1 = \left(\frac{\rho W_m^2}{2}\right)\left(\frac{c}{s}\right)(C_L\sin\beta_m - C_d\cos\beta_m). \tag{3.15}$$

These equations relating the turbomachine parameters with the airfoil parameters will be used for axial-flow machine design and analysis.

Example 3.14 An axial-flow fan impeller with the dimension indicated is designed with a NACA 4312 airfoil. If it delivers 11,560 cfm at the design flow rate when rotating at 3500 rpm, estimate the static pressure rise across the rotor and stator at the mean radius (assuming efficiency across the stator to be 85%).

Solution

From Figure 3.40a, we have $C_L = 0.36$, $C_d = 0.017$, and $\alpha = 1.8°$ at maximum L/D. Also, we have

$$A = \pi\left(r_t^2 - r_h^2\right) = 518.4\ \text{in.}^2 = 3.6\ \text{ft}^2,$$

$$r_m = \left[\frac{1}{2}\left(r_t^2 + r_h^2\right)\right]^{0.5} = 9.65\ \text{in.} = 0.804\ \text{ft}.$$

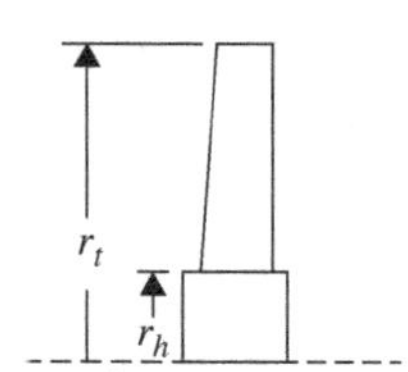

$Z_b = 7$, $\gamma = 70°$
$r_t = 13.25$ in., $c_m = 3.5$ in.
$r_h = 3.25$ in.,

So

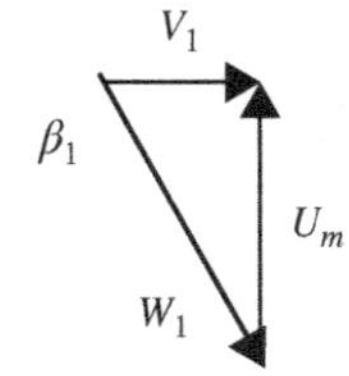

$$U_m = \left(\frac{N\pi}{30}\right) r_m = 294.7\ \text{ft/s}, \quad s = \frac{2\pi r_m}{Z_b} = 8.66\ \text{in.},$$

$$V_a = V_1 = \frac{Q}{A} = \frac{11,560}{60 \times 3.6} = 53.5\ \text{ft/s},$$

$$\beta_1 = \tan^{-1}\left(\frac{U_m}{V_1}\right) = 79.7°, \quad \beta_m = \gamma + \alpha = 71.8°.$$

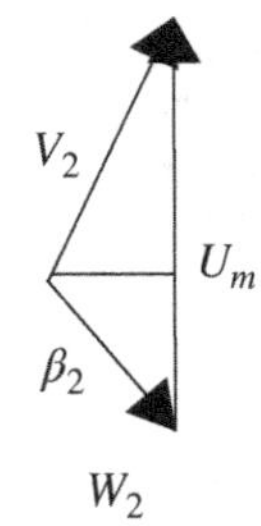

From $\tan \beta_m = (\tan \beta_1 + \tan \beta_2)/2$, $\beta_2 = 30.1°$; Then $W_m \cos \beta_m = V_a$, so $W_m = 171.3$ ft/s. Hence across the rotor, we have

$$\Delta p_r = \left(\frac{\rho W_m^2}{2}\right)\left(\frac{c_m}{s}\right)(C_L \sin \beta_m - C_d \cos \beta_m)$$

$$= \left(\frac{0.0762 \times 171.3^2}{2 \times 32.2}\right)\left(\frac{3.5}{8.66}\right)(0.36 \sin 71.8° - 0.017 \cos 71.8°)$$

$$= 14.03 \times 0.337 = 4.73\ \text{lb}_\text{f}/\text{ft}^2 = 0.0328\ \text{psia} = 0.91\ \text{in. wg.}$$

Across the stator, from the velocity diagram, we have $W_{u2} = V_a \tan \beta_2 = 53.5 \times \tan 30.1° = 31.0$ ft/s and $V_2 = \left[V_a^2 + (U_m - W_{u2})^2\right]^{0.5} = 269.0$ ft/s. So, assuming $V_3 = V_a$,

$$\Delta p_s = \left(\frac{\eta_s \rho}{2}\right)(V_2^2 - V_a^2) = \frac{[0.85 \times 0.0762/(2 \times 32.2)](269^2 - 53.5^2)}{144}$$

$$= 0.485\ \text{psia} = 1.12\ \text{ft} = 13.44\ \text{in. of } H_2O.$$

3.2.3 Axial-Flow Cascade

A typical axial-flow impeller and its relevant dimensions are shown in Figure 3.43, where s is the pitch or spacing mentioned in the previous section and can be determined as $s = 2\pi r/Z_b$. Here, s/c is the pitch/chord ratio, while c/s is the solidity for the axial-flow impeller. The aspect ratio is defined as the ratio of the span and chord (b/c), where $b = r_t - r_h$.

To discuss the interaction between the flow and impeller, it is convenient to show the vanes on the development of several cylindrical sections, such as tip, hub, and mean radius, which is defined as $r_m = \left[\left(r_t^2 + r_h^2\right)/2\right]^{0.5}$. A typical development is shown in Figure 3.44. This is called a cascade, which simulates the axial-flow impeller better than the isolated airfoil. A stationary cascade can be tested in a wind tunnel to provide data for an axial-flow machine design. The measurements in terms of the total pressure loss Δp_0 and the flow angle deflection $\varepsilon = \beta_1 - \beta_2$ can be taken along the leading and trailing edges. These data can be averaged and converted into lift and drag coefficients with Equations (3.12) and (3.14). Their variations versus the incidence angle, $i = \beta_1 - \beta_{b1}$, are shown in Figure 3.45 qualitatively. It is shown that ε and C_L increase with increasing incidence until the stall point, where $\varepsilon = \beta_1 - \beta_2 = \beta_{b1} + i - (\beta_{b2} + \delta) = \theta + i - \delta$. Hence the flow deviation δ can be determined. The angles are depicted in Figure 3.46. The stalling point ε_s is defined as the condition when total pressure loss is double the minimum value, and the nominal deflection selected for design is set at $\varepsilon^* = 0.8\varepsilon_s$.

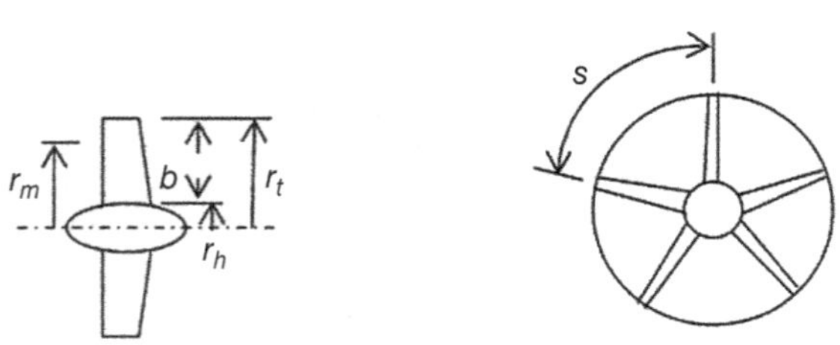

Figure 3.43 Basic dimensions of an axial-flow impeller.

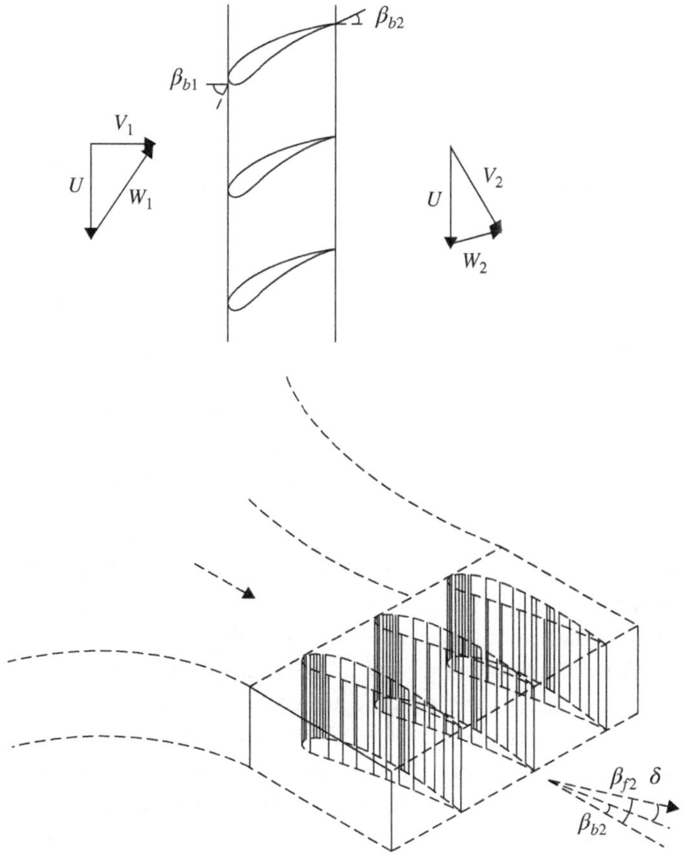

Figure 3.44 Cascade of axial-flow impeller blades.

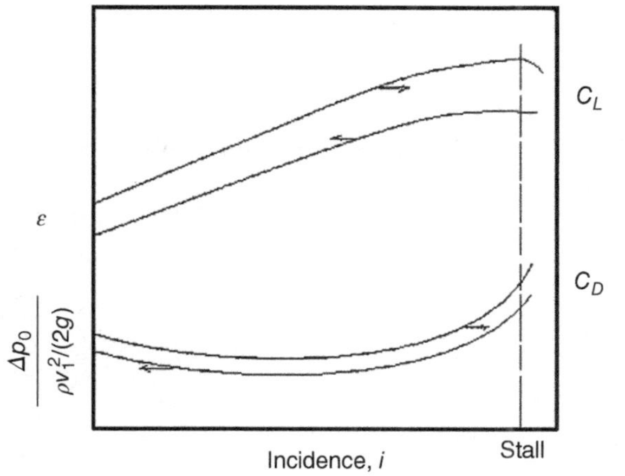

Figure 3.45 Typical cascade test results in terms of C_L, C_d, ε, and total pressure loss coefficient.

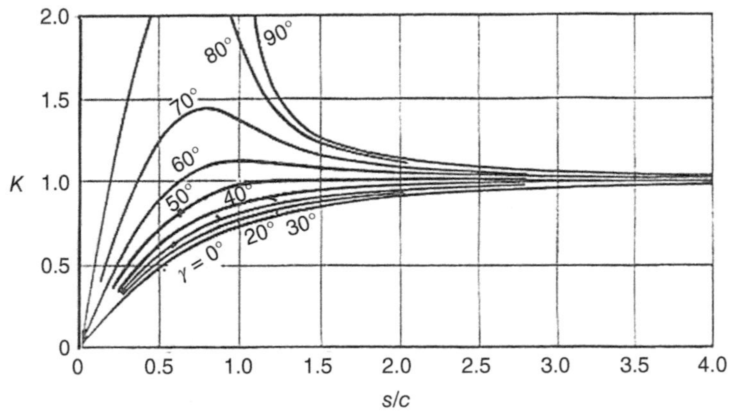

Figure 3.46 Variation of cascade coefficient K for circular arc airfoil with pitch–chord ratio and stagger angle. (Reprinted by permission from Lakshminarayana, B., *Fluid Dynamics and Heat Transfer of Turbomachinery*, John Wiley & Sons, Inc., New York, 1996.)

Although the type of cascade data shown in Figure 3.44 can provide information for axial-flow machine design, it requires extensive tests for one type of machine design. Hence another type of cascade test is devised to supplement the existing isolated airfoil data such that the lift coefficient of a cascade is related to the isolated airfoil lift coefficient by $C_L = KC_{Li}$, where the cascade coefficient K is a function of the pitch/chord ratio (s/c) and the stagger angle γ of a cambered blade, as shown in Figure 3.46[20] (the stagger angle γ shown is concerning the axial direction). It is noted that, as s/c increases, the cascade coefficient approaches unity since the interference diminishes. As s/c decreases to zero, K approaches zero. In the range of practical applications, where $s/c > 0.5$, K is generally less than 1. Although it decreases with decreasing s/c (i.e., increasing blades number), the total effectiveness of energy transfer still increases with increasing the number of blades within a reasonable limit. The interference effect on C_d is negligible.

A more detailed discussion of cascade test data will be given in Chapters 4 and 5 when the axial-flow compressor and turbine are covered.

3.2.4 Preliminary Design Procedure

The design procedure for both pumps and fans of axial-flow type are similar except for a few empirical correlations since the gas flowing through the fans can also be treated as incompressible.

The design starts with the specification of flow rate Q, head rise H (or static pressure rise SP for a fan), and rotating speed N in rpm or ω in radians per second. From these parameters, the specific speed can be calculated as

$$N_s = \begin{cases} \dfrac{N\sqrt{\text{GPM}}}{H^{0.75}} & \text{for pumps,} \\ \dfrac{N\sqrt{\text{CFM}}}{(\text{SP})^{0.75}} & \text{for fans,} \end{cases} \tag{3.16}$$

or

$$\omega_s = \frac{\omega\sqrt{Q}}{(gH)^{0.75}} \tag{3.17}$$

The Cordier diagram shown in Figure 2.2 (or Figures 3.36 and 3.38) can be used to determine the specific diameter and hence the impeller tip diameter D_t. An appropriate hub–tip ratio (D_h/D_t) generally falls in the range of

0.3–0.7, and its relationship with N_s and solidity c/s at the impeller periphery for both pumps and fans is shown in Figure 3.47.[21, 22] Or the relationship shown in Figure 3.47*a* can be expressed approximately as

$$\frac{c}{s} = \frac{10}{(D_h/D_t)(N_s/1000)^{1.5}}. \tag{3.18}$$

Some alternative designs with an appropriate value of c/s to avoid a stall in an axial fan have been suggested in Ref. 23. The solidity calculated should be in the range of 0.4–1.1; otherwise, a different hub–tip ratio should

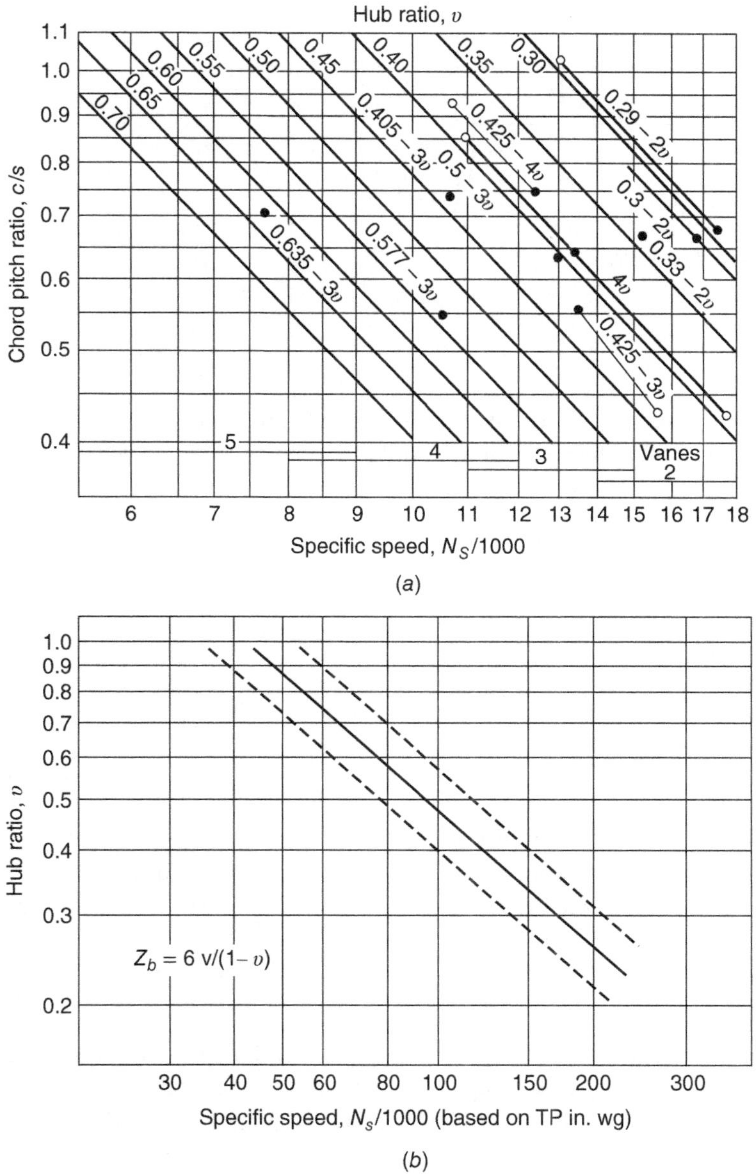

Figure 3.47 (*a*) Hub–tip ratio, blade number and chord pitch ratio for axial-flow pump (*b*) Optimum hub–tip ratio for axial-flow fan (*c*) Mean head and flow coefficients for axial-flow fans. (Reprinted by permission from (*a*) Stepanoff, A. J., *Centrifugal and Axial Flow Pumps*, 2nd ed., John Wiley & Sons, New York, 1957; (*b*, *c*) Jorgensen, R., *Fan Engineering*, 9th ed., Howden Buffalo Company, Buffalo, NY, 1999.)

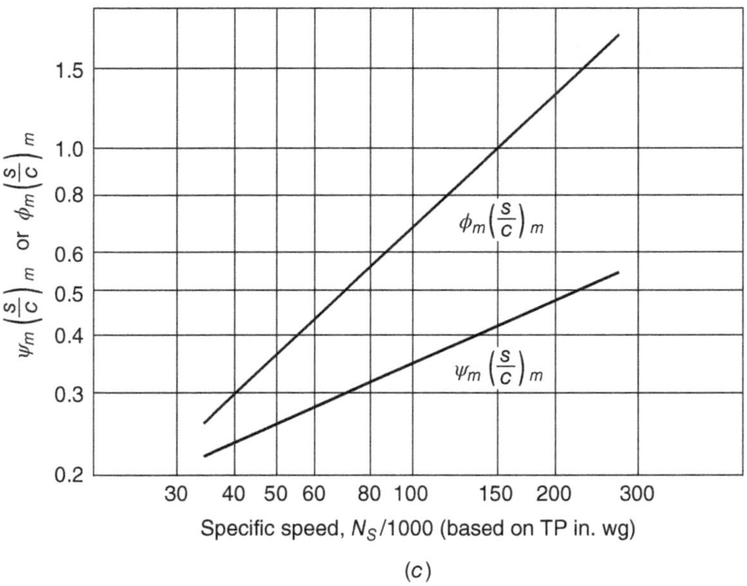

Figure 3.47 *(Continued)*

be chosen. For fans, the solidity can be chosen from the related cascade data to match the required flow angles. The chord c is selected to provide an aspect ratio, that is, blade height to chord (b/c), of 3 or more. For pumps, this ratio should be lower for mechanical strength.

The inlet and outlet flow angles at the mean diameter are then determined from the velocity triangles using the following equations:

$$D_m = \left[\frac{1}{2}\left(D_t^2 + D_h^2\right)\right]^{0.5}, \quad U = \frac{1}{60}N\pi D_m,$$
$$V_a = \frac{4Q}{\pi\left(D_t^2 - D_h^2\right)}, \quad \beta_1 = \tan^{-1}\left(\frac{U}{V_a}\right),$$
$$V_{u2} = \frac{gH}{\eta_h U}, \quad \beta_2 = \tan^{-1}\left(\frac{U - V_{u2}}{V_a}\right), \tag{3.19}$$

where the hydraulic efficiency η_h can be estimated first (with the overall efficiency from Figure 3.36 or 3.38), then checked with the airfoil or cascade data later. For axial-flow pumps and fans, η_h runs from 0.85 to 0.90. With flow angles β_1, β_2 and the solidity c/s determined, the corresponding required lift coefficient can be calculated from Equation (3.13):

$$C_L = 2\left(\frac{s}{c}\right)(\tan\beta_1 - \tan\beta_2)\cos\beta_m. \tag{3.20}$$

The angle of attack or incidence angle can then be determined from the appropriate airfoil or cascade test data, such as those illustrated in Figures 3.40, 3.45, or 3.46. The stagger angle or the inlet blade angle can be calculated from $\gamma = \beta_m - \alpha$ or $\beta_{b1} = \beta_1 - i$. Hence the setting of the selected blade section concerning the axial direction is fixed at the mean diameter.

Hydraulic efficiency, sometimes called stage efficiency in the literature, including rotor and stator estimated earlier, can now be compared with the value calculated from the expression.

$$\eta_h = \frac{\Delta p}{\Delta p_i} = \phi\left[\frac{R_r - \phi\delta_r}{\phi + \delta_r R_r} + \frac{1 - R_r - \phi\delta_s}{\phi + \delta_s(1 - R_r)}\right], \tag{3.21}$$

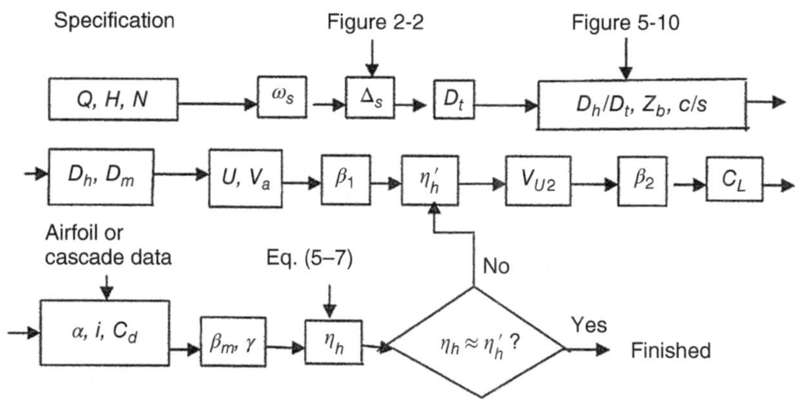

Figure 3.48 Flowchart for axial-flow impeller design.

where $\phi = V_a/U$, $R_r = W_{mu}/U$, and $\delta_{r/s} = \tan^{-1}(C_d/C_L)$ for the rotor and stator. The detailed derivation for this equation is given below. If a significant difference is found, the design process should be repeated with a new hydraulic efficiency until they match closely (5% or less).

Additional pressure losses due to the annulus loss associated with the hub and casing surfaces, secondary flow loss, and tip clearance leakage loss can be accounted for by adding the corresponding loss coefficients to the profile drag coefficient obtained from the airfoil or cascade data. Howell gives these empirical loss coefficients[19] as follows:

$$C_{da} = \frac{0.02s}{b}, \quad C_{ds} = 0.018C_L^2, \quad C_{dt} = 0.29\left(\frac{c_t}{b}\right)C_L^{1.5}, \tag{3.22}$$

where s is the pitch, b is the blade height, and c_t is the tip clearance. The design procedure described above can be briefly depicted in the flowchart in Figure 3.48.

Equation (3.21) is derived as follows: From the force balance across the control volume enclosing the rotor shown in Figure 3.49, we have

$$(p_2 - p_1)_r s = F_a = F_L \sin\beta_m - F_d \cos\beta_m. \tag{3.23}$$

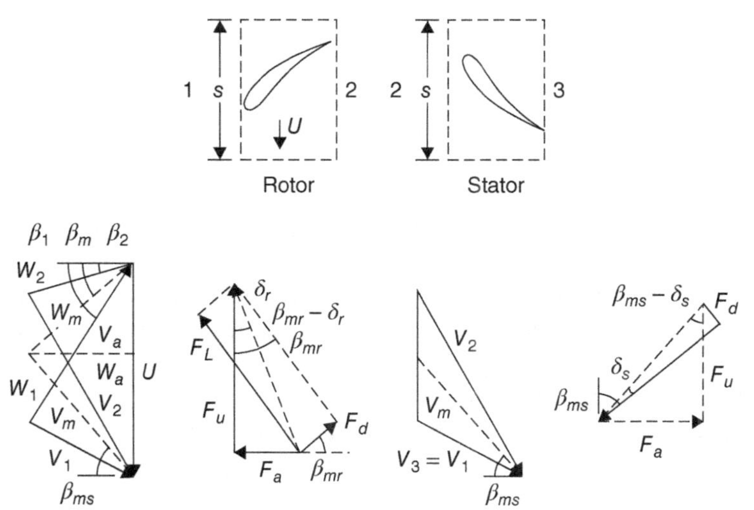

Figure 3.49 Velocity and force diagrams of the rotor and stator.

From the force diagram, we have $F_a = F_u \tan(\beta_m - \delta_r)$, and from the momentum equation in the tangential direction, we have

$$F_u = \rho V_a s(V_{u2} - V_{u1}) = \rho U^2 s\phi\Lambda, \tag{3.24}$$

where $\phi = V_a/U$ is the flow coefficient, $\Lambda = (V_{u2} - V_{u1})/U$ is the loading coefficient, and $\delta_r = \tan^{-1}(C_d/C_L) \approx C_d/C_L$. Substituting these expressions into Equation (3.23), we have

$$\begin{aligned}(p_2 - p_1)_r &= \left(\frac{F_u}{s}\right)\tan(\beta_m - \delta_r) = \rho U^2\phi\Lambda\left(\frac{\tan\beta_m - \tan\delta_r}{1 + \tan\beta_m \tan\delta_r}\right)\\ &= \rho U^2\phi\Lambda\left(\frac{R_r/\phi - \delta_r}{1 + R_r\delta_r/\phi}\right) = \rho U^2\phi\Lambda\left(\frac{R_r - \phi\delta_r}{\phi + R_r\delta_r}\right).\end{aligned} \tag{3.25}$$

Since the stator will deflect the flow direction from V_2 to V_3, which is equal to V_1, as shown on the velocity diagram, we have $\tan\beta_{ms} = V_{mu}/V_a$. Hence the pressure rise across the stator can be expressed as

$$\begin{aligned}(p_3 - p_2)_s &= \frac{F_a}{s} = \left(\frac{F_u}{s}\right)\tan(\beta_{ms} - \delta_s) = \rho V_a(V_{u2} - V_{u1})\tan(\beta_{ms} - \delta_s)\\ &= \rho U^2\phi\Lambda\left(\frac{\tan\beta_{ms} - \delta_s}{1 + \delta_s\tan\beta_{ms}}\right) = \rho U^2\phi\Lambda\left[\frac{(1 - R_r)/\phi - \delta_s}{1 + \delta_s(1 - R_r)/\phi}\right]\\ &= \rho U^2\phi\Lambda\left[\frac{1 - R_r - \phi\delta_s}{\phi + \delta_s(1 - R_r)}\right].\end{aligned} \tag{3.26}$$

Hence we have the pressure rise across the stage as

$$\begin{aligned}\Delta p &= (p_3 - p_1) = (p_2 - p_1)_r + (p_3 - p_2)_s\\ &= \rho U^2\phi\Lambda\left[\frac{R_r - \phi\delta_r}{\phi + R_r\delta_r} + \frac{1 - R_r - \phi\delta_s}{\phi + \delta_s(1 - R_r)}\right],\end{aligned} \tag{3.27}$$

where $\phi = V_a/U$, $\Lambda = \Delta V_u/U$, $R_r = W_{mu}/U$, and $\delta_{r/s} = \tan(C_d/C_L) \approx C_d/C_L$. For an ideal case without drag, that is, $\delta_r = \delta_s = 0$, the pressure rise becomes $\Delta p_i = \rho U^2\phi\Lambda[R_r/\phi + (1 - R_r)/\phi] = \rho U^2\Lambda$. Hence the hydraulic efficiency can be obtained as

$$\eta_h = \frac{\Delta p}{\Delta p_i} = \phi\left[\frac{R_r - \phi\delta_r}{\phi + \delta_r R_r} + \frac{1 - R_r - \phi\delta_s}{\phi + \delta_s(1 - R_r)}\right], \tag{3.28}$$

which is Equation (3.21).

3.2.4.1 Diffusion Casing

The purpose of the diffusion casing of an axial-flow pump or fan is to straighten the flow into the axial direction and convert part of the kinetic energy, leaving the impeller to pressure rise. This is accomplished with a set of stationary vanes (or stators) and a divergent flow area.

The diffusion vane curvature is selected so that the fluid enters the vanes with a minimum loss and leaves the casing axially. The fluid entering the vanes should have a flow angle greater than the vane angle, that is, $\beta_{f2} > \beta_b$, with angle of attack $\alpha = \beta_{f2} - \beta_b$ between 2° and 3°. The clearance d between the impeller trailing edge and diffusion vanes should be between 5 and 10% of the impeller diameter. This is to allow the flow adjustment under off-design conditions. A closer clearance is preferred if the machine is to be operated mainly around the design condition.

The solidity c/s at the mean radius should be around unity. The casing enclosing the diffusion vanes can also be divergent to slow down the axial-flow velocity and to match with the discharge dimension specified. However, the included angle of divergence should be less than 8° to prevent possible boundary layer separation.

Example 3.15 Design an axial-flow fan with the following specifications: $Q = 5\,\text{m}^3/\text{s}$, $\Delta p_s = 500\,\text{Pa}$, $N = 1500\,\text{rpm}$, assuming $\rho_a = 1.22\,\text{kg/m}^3$.

Solution

a) Convert $Q = 5\ \text{m}^3/\text{s} = 10{,}595$ cfm, $\rho_a = 0.0761\ \text{lb}_\text{m}/\text{ft}^3$, $\text{SP} = \Delta p/(\rho_w g) = 500/(1000 \times 9.8) = 0.051$ m $= 2.01$ in. wg.
b) Calculate the specific speed:

$$N_s = \frac{1500(10,575)^{0.5}}{(2.01)^{0.75}} = 91,376\ \text{rpm(cfm)}^{0.5}/(\text{in. H}_2\text{O})^{0.75}.$$

and $\omega = N\pi/30 = 1500\pi/30 = 157$ rad/s. So $\omega_s = 157(5)^{0.5}/(500/1.22)^{0.75} = 3.85$. From Figure 3.47b, we select $\nu = D_h/D_t = 0.5$, and the blade number $Z_b = 6\nu/(1-\nu) = 6$. From Figure 2.2, the specific diameter is obtained as $\Delta_s \approx 1.5$. Hence $D_t = 1.5 \times (5)^{0.5}/(500/1.22)^{0.25} = 0.74$ m and $D_h = \nu D_t = 0.37$ m. Also $A = \pi D_t^2(1-\nu^2)/4 = 0.322\ \text{m}^2$ and $D_m = \left[\left(D_t^2 + D_h^2\right)/2\right]^{0.5} = 0.585$ m, $V_a = Q/A = 15.5$ m/s, $U_m = \omega D_m/2 = 45.9$ m/s, and $\phi_m = V_a/U_m = 0.337$ can be obtained. Then from Figure 3.47c, we obtain $\phi_m(s/c) = 0.65$ or $s/c = 1.93$, where $s = \pi D_m/Z_b = 0.306$ m. Assuming $V_1 = V_3 = V_a = 15.5$ m/s, the total head can be calculated from $gH_t = \Delta p_s/\rho + V_a^2/2 = 530\ (\text{m/s})^2$, or TP $= 2.59$ in. wg.

c) Next try $\eta_h = 0.77$. Then we obtain

$$V_{u2} = \frac{gH_t}{\eta_h U_m} = \frac{530}{0.77 \times 45.9} = 15\ \text{m/s},$$

$$\beta_1 = \tan^{-1}\left(\frac{U_m}{V_a}\right) = \tan^{-1}\left(\frac{45.9}{15.5}\right) = 71.3°,$$

$$\beta_2 = \tan^{-1}\left(\frac{U_m - V_{u2}}{V_a}\right) = \tan^{-1}\left(\frac{45.9-15}{15.5}\right) = 63.4°,$$

$$\tan\beta_m = 0.5(\tan\beta_1 + \tan\beta_2) = 2.47, \quad \beta_m = 68°.$$

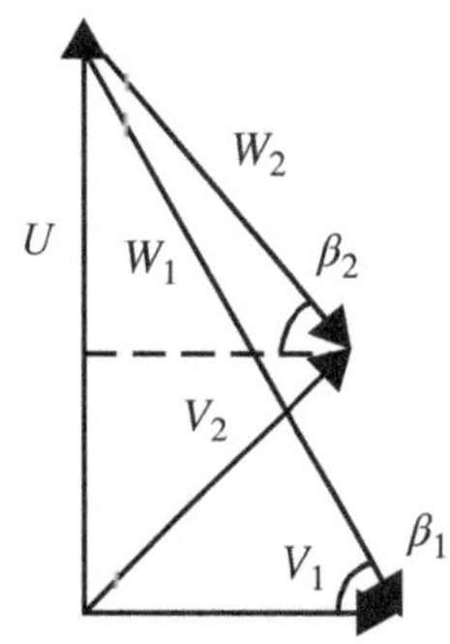

From Equation (3.20) we have $C_L = 2(s/c)(\tan\beta_1 - \tan\beta_2)\cos\beta_m = 2 \times 1.93(2.95 - 2.0) \times 0.374 = 1.37$. This is the cascade lift coefficient required. To use the isolated airfoil data, we obtain $K = 1.2$ from Figure 3.46 with $\gamma = 60°$. Hence $C_{Li} = 1.37/1.2 = 1.14$.

d) If NACA 4312 airfoil section is selected, at $\alpha = 12°$, $C_{Li} = 1.14$ and $C_L/C_d = L/D = 12$. Substituting the above data into Equation (3.28), where $R_r = W_{mu}/U_m = \phi\tan\beta_m = 0.337 \times 2.475 = 0.834$, $\delta_r \approx \delta_s = 0.08$, we have

$$\eta_h = 0.337\left(\frac{0.834 - 0.337 \times 0.08}{0.337 + 0.08 \times 0.834} + \frac{1 - 0.834 - 0.337 \times 0.08}{0.337 + 0.08(1 - 0.834)}\right)$$
$$= 0.337(1.99 + 0.397) = 0.80(\approx 0.77).$$

Also $\gamma = \beta_m - \alpha = 68 - 12 = 56°$, $c = s/1.93 = 0.16$ m.

e) Double check the data obtained with those given in Figure 3.38. It is shown that $\eta_s = 0.74$, $D_s = 0.35 = D_t(\text{SP})^{0.25}/(\text{CFM})^{0.5}$. Hence $D_t = 0.35(10{,}595)^{0.5}/(2.01)^{0.25} = 30.2$ in. $= 0.77$ m. It is close to what we have. However, some alternative designs may be performed with selection of a little higher hub–tip ratio ν and other available airfoil sections.

3.2.5 Three-Dimensional Flow Effects: Radial Equilibrium Requirement

The flow through a rotating impeller is in general a complicated three-dimensional process, even in an axial-flow machine. Some three-dimensional flow effects on total pressure loss have been mentioned in the previous section, such as the secondary flow, boundary layer on the hub and casing surfaces, and leakage through the tip clearance.

Another important three-dimensional flow effect is the tangential-flow velocity distribution along the blade radius from the hub to the tip. This can be controlled in the impeller design by selecting a proper blade section setting along the radius.

The general three-dimensional momentum equation of inviscid fluid (also called the Euler equation in fluid mechanics, not to be confused with the Euler equation for energy transfer in turbomachinery) in a radial direction expressed in a cylindrical coordinate system can be written as follows:

$$-\frac{1}{\rho}\frac{\partial p}{\partial r} = V_r\frac{\partial V_r}{\partial r} + \frac{V_u}{r}\frac{\partial V_r}{\partial \theta} + V_a\frac{\partial V_r}{\partial z} - \frac{V_u^2}{r} \tag{3.29}$$

Assuming axisymmetric flow and imposing a zero radial-flow velocity, it can be simplified to

$$\frac{1}{\rho}\frac{dp}{dr} = \frac{V_u^2}{r}. \tag{3.30}$$

This is called the radial equilibrium condition. It can also be derived from the force balance between the pressure gradient and the centrifugal force of a fluid element in the radial direction as shown.

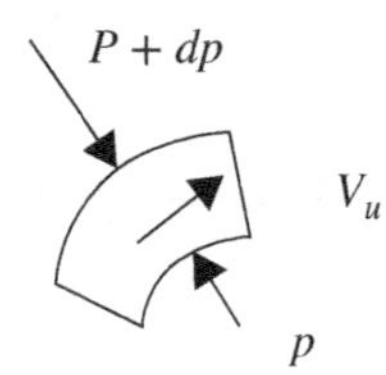

The momentum equation can also be written along a streamline (again neglecting the radial-flow velocity component) as

$$\frac{dp}{\rho} + V_u\ dV_u + V_a dV_a = 0. \tag{3.31}$$

Combining Equations (3.30) and (3.31), we obtain

$$\left(\frac{V_u}{r}\right)\ d(V_u r) + V_a dV_a = 0. \tag{3.32}$$

Several different tangential-flow velocity distributions along the radius can satisfy this momentum equation. They are discussed below.

a) Free-vortex flow, that is, $V_u r = \text{const.}$, with the axial-flow velocity set as a constant is the simplest type of velocity distribution. It also results in uniform loading (i.e., energy transfer) along the impeller radius from hub to tip since $\Delta E = UV_u = \omega r V_u = \text{const.}$ It is commonly adapted in axial-flow impeller design.
b) Forced-vortex flow can be generalized as $V_u r^n = \text{const.}$ For a free vortex, $n = 1$, and if $n = -1$, that is, $V_u/r = \text{const.}$, it is called forced-vortex or solid-body rotation because its tangential velocity distribution is the same as that of a rotating solid body. The corresponding pressure distribution can be derived from

$$\frac{dp}{dr} = \frac{\rho V_u^2}{r} = \frac{\rho(\omega r)^2}{r} = \rho\omega^2 r, \qquad p_t - p_h = \frac{\rho\omega^2\left(r_t^2 - r_h^2\right)}{2}. \tag{3.33}$$

No simple expression can be obtained for the axial-flow velocity distribution in this case. Also, the forced-vortex pattern must be set up artificially by the inlet guide vanes at the inlet of the rotor. And there is a limiting hub–tip ratio, since the axial-flow velocity decreases from the hub to the tip, with a limiting value of zero at the tip. It may be impossible to obtain radial equilibrium without V_a becoming negative, which is an impossible design condition.
c) The third type of flow pattern, useful in practice, is one in which the blade angle is constant, that is, an untwisted blade. In the case of fixed blades such as those in a nozzle or diffuser, by combining $V_a = V_u \cot\beta_b$ with

$$\frac{dp}{\rho} + V_u dV_u + V_a dV_a = 0, \quad \frac{dp}{\rho} = \left(\frac{V_u^2}{r}\right) dr, \tag{3.34}$$

we obtain $\left(V_u^2/r\right) dr + (1 + \cot^2\beta_b)V_u dV_u = 0$, or $V_u r^{\sin^2\beta_b} = \text{const.}$, also $V_a r^{\sin^2\beta_b} = \text{const.}$ Hence both V_u and V_a vary along the radius. When β_b is close to 90°, $\sin\beta_b \cong 1.0$, the tangential-flow velocity will be close to that of the free vortex. Because the untwisted blade means a saving in manufacturing, this flow pattern can be adapted in the design of low-cost products. However, with today's five-axis CNC machines, the twisted blades can be easily produced with some extra cost.

Example 3.16 The whirl velocity at the trailing edge of an axial-flow impeller as shown is constant with 50 m/s from hub to tip. What should the axial-flow velocity distribution be to satisfy the radial equilibrium condition? If the axial-flow velocity is 70 m/s at the hub, determine its magnitude at the tip.

Solution

From Equation (3.32), with V_u = const., we have $V_u^2/rdr + V_a dV_a = 0$, or $V_u^2 \ln r + V_a^2/2 = C$, or $V_u^2 \ln(r/r_h) = \left(V_{ah}^2 - V_a^2\right)/2$, or $V_a^2 = V_{ah}^2 - 2V_u^2 \ln(r/r_h) = 4900 - 2 \times 2500 \ln(r/0.2)$. Hence we have $V_a = [4900 - 5000 \ln(5r)]^{0.5}$ m/s, and at the tip, $V_{at} = 27.1$ m/s.

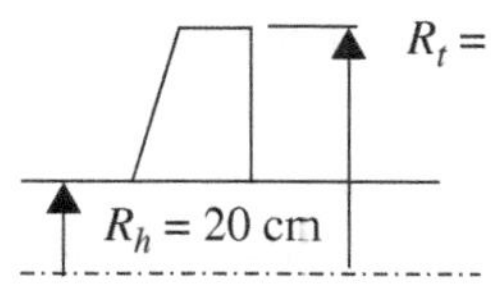

Example 3.17 Determine the relative flow angles at the hub and tip for the fan impeller of Example 3.15 using the free-vortex design.

Solution

Assume uniform axial flow at the inlet:

a) At the hub

$$V_a = 15.5 \text{ m/s}, \quad U_h = \frac{1}{2}\omega D_h = 157 \times 0.37/2 = 29.04 \text{ m/s},$$

$$V_{u2} = \frac{V_{mu2} D_m}{D_h} = \frac{15 \times 0.585}{0.37} = 23.7 \text{ m/s},$$

Hence

$$\beta_1 = \tan^{-1}\left(\frac{U_h}{V_a}\right) = \tan^{-1}\left(\frac{29.04}{15.5}\right) = 61.9°.$$

$$\beta_2 = \tan^{-1}\left(\frac{U_h - V_{u2}}{V_a}\right) = \tan^{-1}\left(\frac{29.04 - 23.7}{15.5}\right) = 18.9°.$$

b) At the tip

$$U_t = \frac{1}{2}(157 \times 0.74) = 58.09 \text{ m/s}, \quad V_{u2} = \frac{15 \times 0.585}{0.74} = 11.86 \text{ m/s}.$$

Hence

$$\beta_1 = \tan^{-1}\left(\frac{58.09}{15.5}\right) = 75.0°, \quad \beta_2 = \tan^{-1}\left(\frac{58.09 - 11.86}{15.5}\right) = 71.4°.$$

3.2.6 Fan Test

In addition to the axial-flow fan discussed in this chapter, centrifugal fans (or radial-flow fans) are commonly used for higher head and lower capacity. They will be discussed in the next chapter along with the centrifugal blower and compressor since they are geometrically similar and require thermodynamic consideration due to the compressibility effect. However, in the next three sections, dealing with fan tests, application, and selection, the centrifugal-type fan will be included.

The setup and procedure to test a fan by the manufacturers are specified and can be certified by the AMCA (Air Movement and Control Association). Ten different test setups are specified in AMCA 210. Three typical ones are shown in Figure 3.50.[24] The flow rate can be measured with either a pitot tube traverse or a nozzle; the former

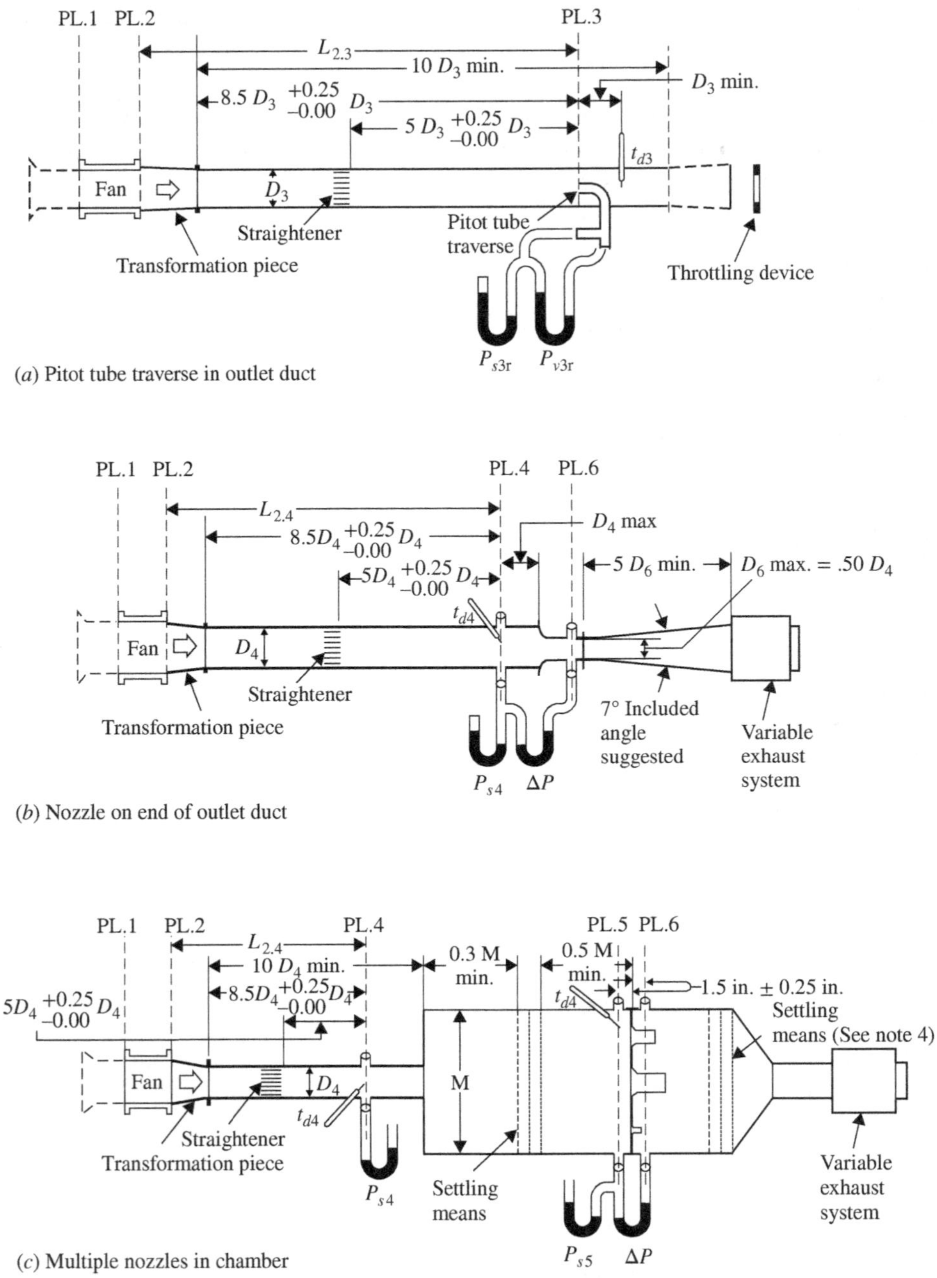

Figure 3.50 AMCA fan test setups. (Reprinted by permission from Laboratory Methods of Testing Fans for Rating, *AMCA Standard 210-99, and ASHRAR Standard 51-1999*, Air Movement and Control Association, Inc., Arlington Heights, IL, www.amca.org; © American Society of Heating and Air-Conditioning Engineers, Inc., Atlanta, GA, 2000, www.ashrae.org.)

is less expensive but more tedious to perform. The static pressure is measured with a pressure tap on the pipe wall. The fan can be installed either upstream or downstream. The dynamic pressure measured with the pitot tube, $p_{v3} = \rho_w g \Delta h_v$, can be used to calculate the flow velocity V_3 through $\rho_3 V_3^2/2 = p_{v3}$. Then the inlet flow rate Q_0 can be calculated from $\rho_0 Q_0 = \rho_3 V_3 A_3$, where A is the cross-sectional area. Also, the total pressure at the fan outlet, p_{t2}, is calculated as $p_{t2} = p_{s3} + p_{v3} + \Delta p_L$, where Δp_L is the loss due to the straightener and transformation piece. Or the static pressure at the fan outlet can be measured with a pressure tap and the dynamic pressure there can be obtained from $p_{v2} = p_{v3}(\rho_3/\rho_2)(A_3/A_2)^2$, where ρ can be calculated from temperature T and pressure p measured. The detailed formula and the test procedure are given in Ref. 24.

Example 3.18 A centrifugal fan is tested with the setup shown in Figure 3.50*a*. The airflow velocity at P_{L3} is measured with a pitot tube (p_{v3}), the static pressure at P_{L2} is measured with some pressure taps (p_{s2}), and the rotating speed N and shaft power H are measured with a dynamometer mounted on the shaft between motor and impeller, respectively. These data are listed below in the first four columns. The pipe inside diameter and fan discharge area are $d_i = 6.065$ in. and $A_2 = 5 \times 6.5 = 32.5$ in.^{2}, respectively. The lab barometric pressure and temperature are $p_a = 14.6$ psia and $T_a = 75°\text{F} = 535$ R. Calculate the relevant data and plot the fan performance curves of total head, shaft power, and total efficiency versus the flow rate at $N = 3500$ rpm.

Solution

Since the pressure changes are small compared with the barometric pressure, constant densities are assumed, that is, $\rho_3 \cong \rho_2 \cong \rho_a$, where $\rho_a = p_a/(RT_a) = 14.6 \times 144/(53.3 \times 535) = 0.0737\ \text{lb}_\text{m}/\text{ft}^3$. The areas A_2 and A_3 are calculated as $A_2 = 5 \times 6.5 = 32.5\ \text{in.}^2 = 0.225\ \text{ft}^2$ and $A_3 = \pi \times 6.065^2/4 = 28.9\ \text{in.}^2 = 0.2007\ \text{ft}^2$. From $\rho_3 V_3^2/2 = \rho_w g p_{v3}$, V_3 can be calculated as

$$V_3 = \sqrt{\frac{2\rho_w g p_{v3}}{\rho_3}} = \sqrt{\frac{2 \times 62.4 \times 32.2}{0.0737}}\sqrt{\frac{p_{v3}}{12}} = 67.4\sqrt{p_{v3}}\ \text{(ft/s)}.$$

Then the inlet flow rate can be calculated as $Q \cong Q_3 = V_3 A_3 = 0.2007 \times 60 \times V_3 = 12.04 \times V_3$ (cfm). Also the dynamic pressure at P_{L2} can be calculated as $p_{v2} = p_{v3}(A_3/A_2)^2(\rho_3/\rho_2) = (0.2/0.225)^2 p_{v3} = 0.79 p_{v3}$ and the total pressure as $p_{t2} = p_{s2} + p_{v2}$.

To correct these data to a fixed rotating speed of 3500 rpm, the fan laws can be used as $Q' = Q(3500/N)$, $P'_{t2} = p_{t2}(3500/N)^2$, and $H' = H(3500/N)^3$. Finally, the total efficiency can be calculated as

$$\eta_t = \frac{\rho_w g p'_{t2} Q'}{H'} = \frac{62.4 p'_{t2} Q'}{H' \times 550 \times 12 \times 60} = \frac{Q' p'_{t2}}{6346H'}.$$

In order to plot all three parameters on the same scale, $10p'_{t2}$, $10H'$, and η_t are plotted in percent.

p_{v3} ("wg)	p_{s2} ("wg)	N (rpm)	H (hp)	V_3 (fps)	Q (cfm)	pv_2	pt_2	Q'	$10pt_s'$	$10H'$	etat (%)
2.4	2.5	3450	1.1	104.4	1257.2	1.896	4.4	1275	45.24	11.5	79.17
2.1	4.2	3520	1.25	97.67	1176	1.659	5.86	1169	57.93	12.3	86.86
1.9	6	3500	1.49	92.9	1118.6	1.501	7.5	1119	75.01	14.9	88.74
1.5	6.8	3420	1.55	82.55	993.88	1.185	7.99	1017	83.63	16.6	80.68
1.2	7.6	3430	1.67	73.83	888.95	0.948	8.55	907	89	17.7	71.7
0.8	8.5	3500	1.72	60.28	725.82	0.632	9.13	726	91.32	17.2	60.73
0.4	9.5	3520	1.81	42.63	513.24	0.316	9.82	510	97.05	17.8	43.86

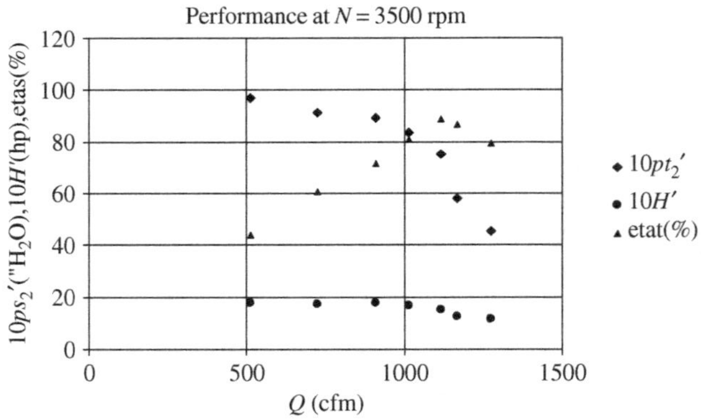

3.2.7 Fan Application

Fans are extensively used in HVAC systems, as shown in Figure 1.7. The fans are used in the air-cooled condenser and the supply/return air duct system. In a large system, they are used in a cooling tower, as shown in Figure 3.51.[25]

Fans are also used in building ventilation, such as building HVAC systems and the typical industrial and tunnel ventilation systems shown in Figure 3.52.[26]

The ventilation flow rate Q can be estimated from $Q = \text{CGR/MAC}$ (in cubic feet per minute), where CGR (parts per minute) is the contaminant generation rate and MAC (parts per cubic feet) is the maximum allowable concentration. Or from the specified air change per hour, N, Q can be calculated from $Q = NV/60$, where V (in cubic feet) is the total volume of space to be ventilated. The American Society of Heating and Air Conditioning Engineers (ASHRAE) specifies the following criteria:

Office buildings	$N = 4, \dots, 10$
Garages	$N = 4, \dots, 6$
Restaurants	$N = 8, \dots, 12$
Warehouse	$N = 1, \dots, 4$

Other applications of fans, blowers, or exhausters include material conveying, dust and fume removal, and handling of hot air or industrial gases.

The operating condition of a fan in a duct system is located at the intersection point of the fan performance curve and the system curve similar to that in liquid pumping systems, shown in Figure 3.27. Both parallel and series combinations of fans are also possible. The similarity laws (called fan laws when applied to fans) mentioned in Chapter 2 are also applicable to fan performance. They will be discussed in detail in Chapter 4.

In fan applications, noise level L_p is an important consideration. Noise can be generated from both mechanical and aerodynamic sources. It is a complicated and difficult topic. Detailed discussion on this topic is beyond the scope of this text. It will only be briefly mentioned here that noise level is specified in terms of decibels (dB), which is defined as $L_p = 20 \log(p/p_0) = 10 \log(w/w_0)$, where p and w are the rms (root-mean-square) magnitude of the noise pressure and power intensity and $p_0 = 2 \times 10^{-5}$ Pa, $w_0 = 10^{-12}$ W are the reference values.

The noise due to various sources can be added to or subtracted from each other. But they have to be done on the power values, not the decibel values directly. So the noise due to sources A and B can be added as $w_a + w_b = w_c$, or

$$10^{\text{dB}_a/10} + 10^{\text{dB}_b/10} = 10^{\text{dB}_c/10} \quad \text{or} \quad \text{dB}_c = 10\left[\log\left(10^{\text{dB}_a/10} + 10^{\text{dB}_b/10}\right)\right].$$

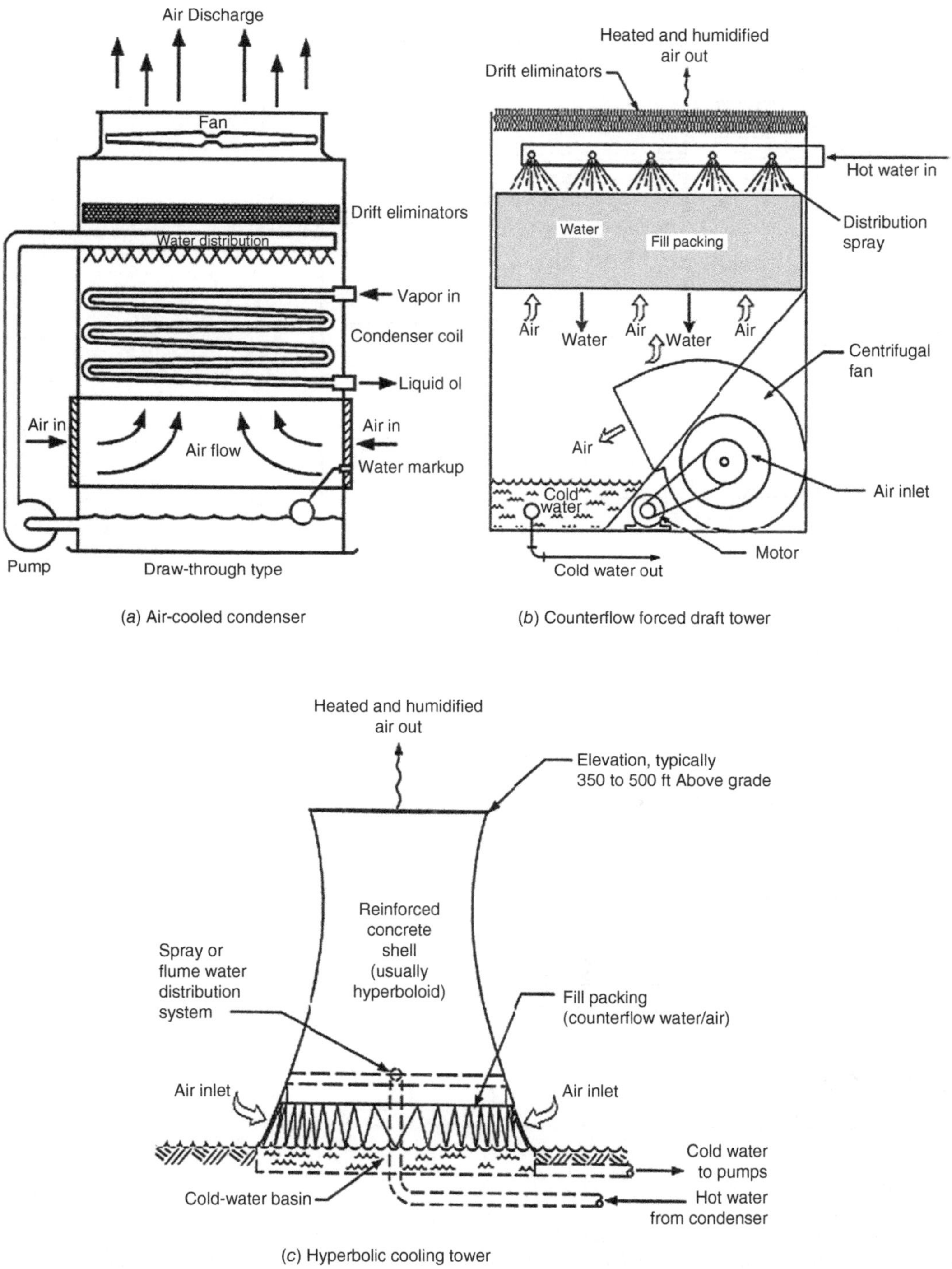

Figure 3.51 Fans used in air-cooled condenser and cooling tower. (Reprinted by permission from Parsons, R. A. (Ed.), *ASHRAE Handbook, HVAC Systems and Equipment*, © American Society of Heating, Refrigerating and Air-Conditioning Engineers, Inc., Atlanta, GA, 2004, www.ashrae.org.)

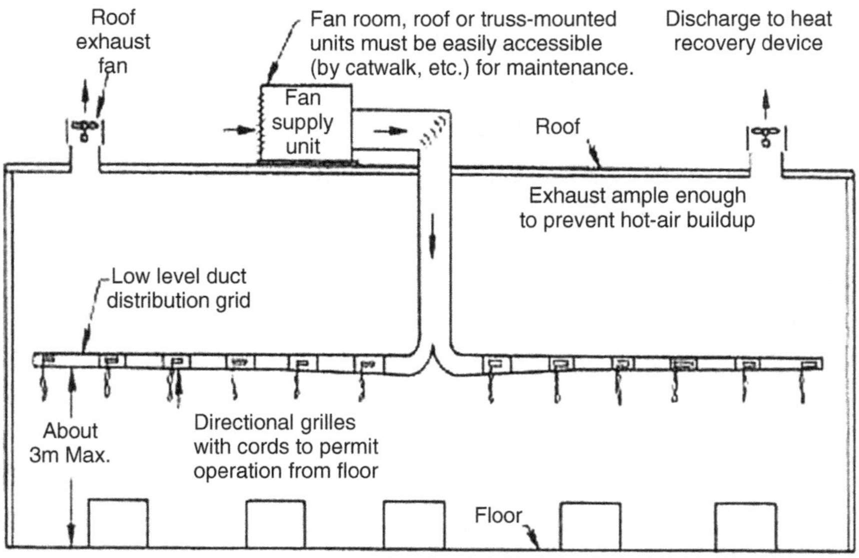

(*a*) Industrial ventilation

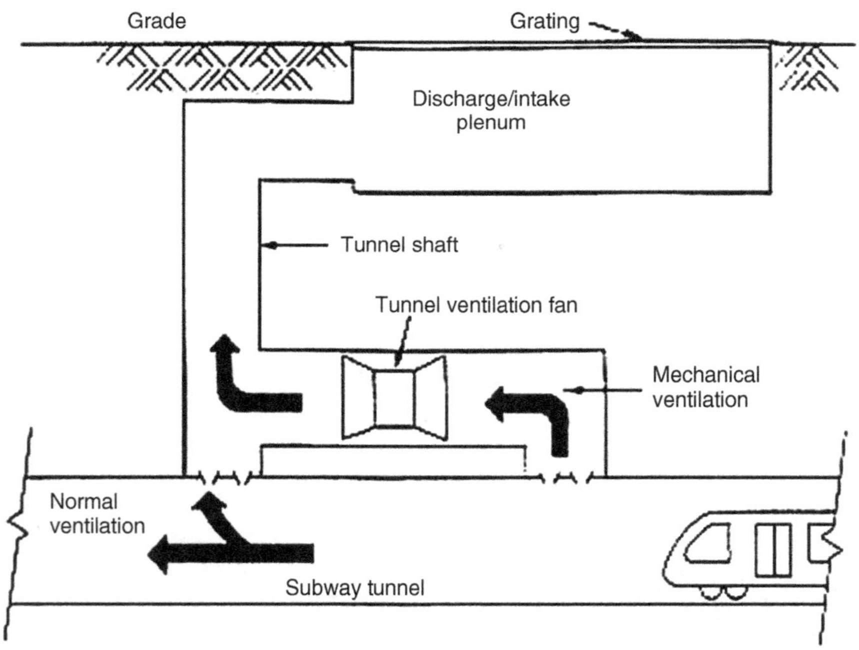

(*b*) Tunnel ventilation

Figure 3.52 Typical ventilation systems. (Reprinted by permission from Parsons, R. A. (Ed.), *ASHRAE Handbook, HVAC Applications*, © American Society of Heating, Refrigerating and Air-Conditioning Engineers, Inc., Atlanta, GA, 2003, www.ashrae.org.)

Example 3.19 If the background noise level is 65 dB and the total combined noise due to a fan and the background is measured as 70 dB, calculate the noise level due to the fan itself.

Solution

Since $dB_t = 10\left[\log\left(10^{dB_b}/10 + 10^{dB_f}/10\right)\right]$, we have

$$dB_f = 10\left[\log\left(10^{dB_t/10} - 10^{dB_b/10}\right)\right] = 10[\log(10^7 - 10^{6.5})]$$
$$= 10[\log(6.84 \times 10^6)] = 68.3.$$

3.2.8 Fans and Systems

The fan and air duct system in a typical central air-conditioning system is shown in Figure 3.53.[27] Similar to the pumping system discussed in Section 3.1, the operating condition of the fan in an air duct system is determined by the intersection point of the fan and system curves. Flow resistance in the air duct can be calculated with a procedure similar to that for the hydraulic system. But in the air duct system, the pressure change due to gravitation is negligible and the flow pattern adjustment takes a longer distance, as shown in Figure 3.54.[27, 28] Hence, in analyzing the air duct system, consideration should be given to the additional effects due to (a) fan–system interaction and (b) the interaction between the elements of the system itself.

3.2.8.1 Fan–System Interaction (System Effect Factor)

The system effect factor (SEF) is due to the difference in fan inlet and outlet conditions under laboratory testing and those when the fan is installed in the system. The effects are associated with (1) improper outlet connection, (2) nonuniform inlet flow, and (3) swirl at the fan inlet.

A set of system effect curves is published by the AMCA and reproduced in Figure 3.55.[28] Each line is for a particular configuration. For example, if a vane-axial fan is installed without an outlet duct, the line *U* should be used. If 12% of the effective duct is installed, line *V* should be used. To determine the SEF for a certain configuration, from the abscissa with appropriate airflow velocity, read up to the applicable curve, then read across to the ordinate to find the SEF at standard air density. To correct for a different density, multiply the standard SEF by the density ratio. A more detailed discussion of other configurations is given in Ref. 28.

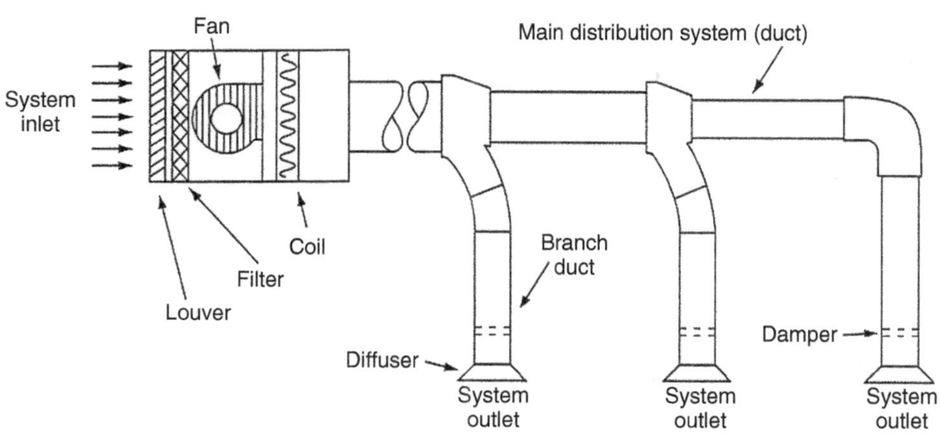

Figure 3.53 Air duct system for central air conditioning (Ref. 27). (Reprinted by permission from AMCA International Inc., *AMCA Publication 200-95, Air Systems*, Air Movement and Control Association, Inc., Arlington Heights, IL, 2007. www.amca.org.)

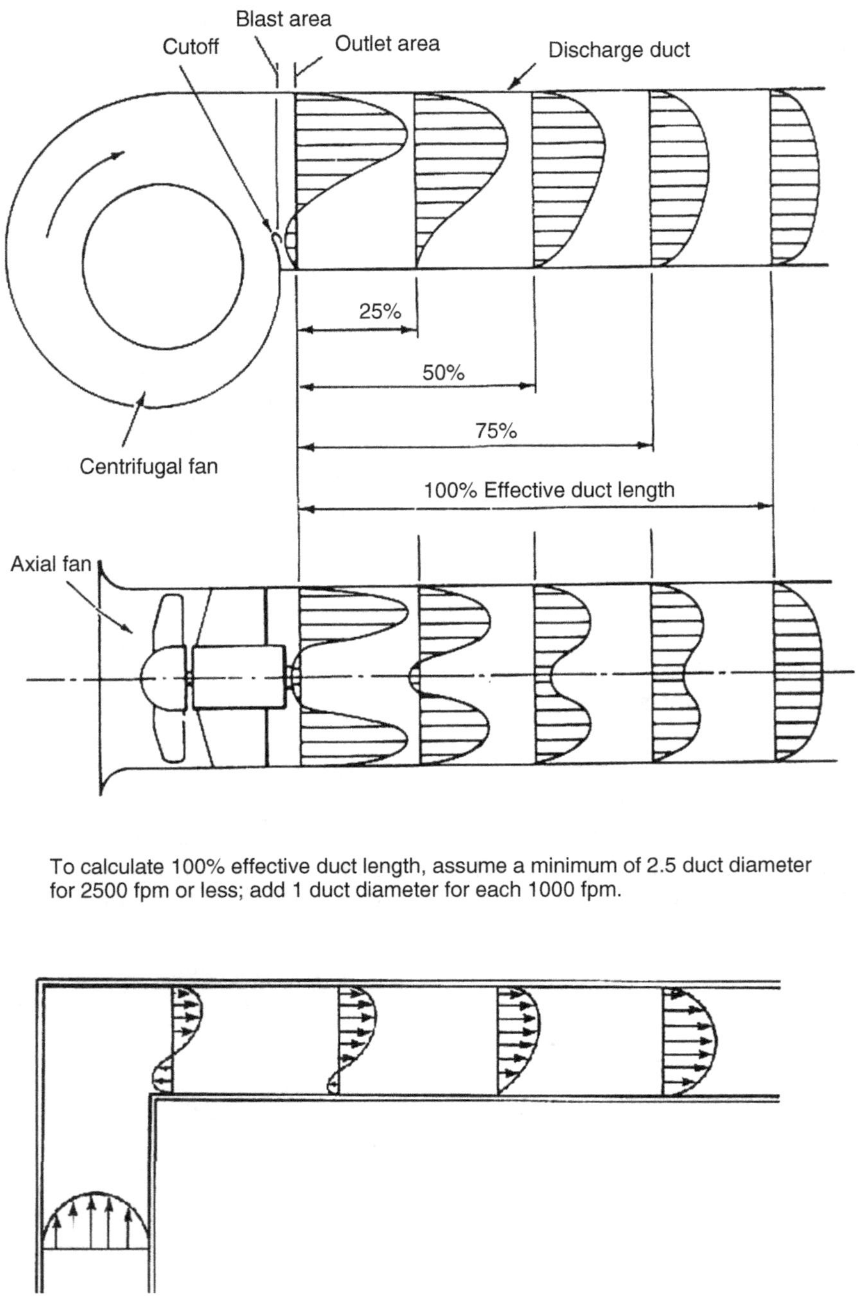

Figure 3.54 Air flow velocity profiles in the vicinity of fan and elbow element. (Reprinted by permission from AMCA International Inc., *AMCA Publication 200-95, Air Systems*, 2007 and *AMCA Publication 201-02, Fans and Systems*, Air Movement and Control Association, Inc., Arlington Heights, IL, 2002, www.amca.org.)

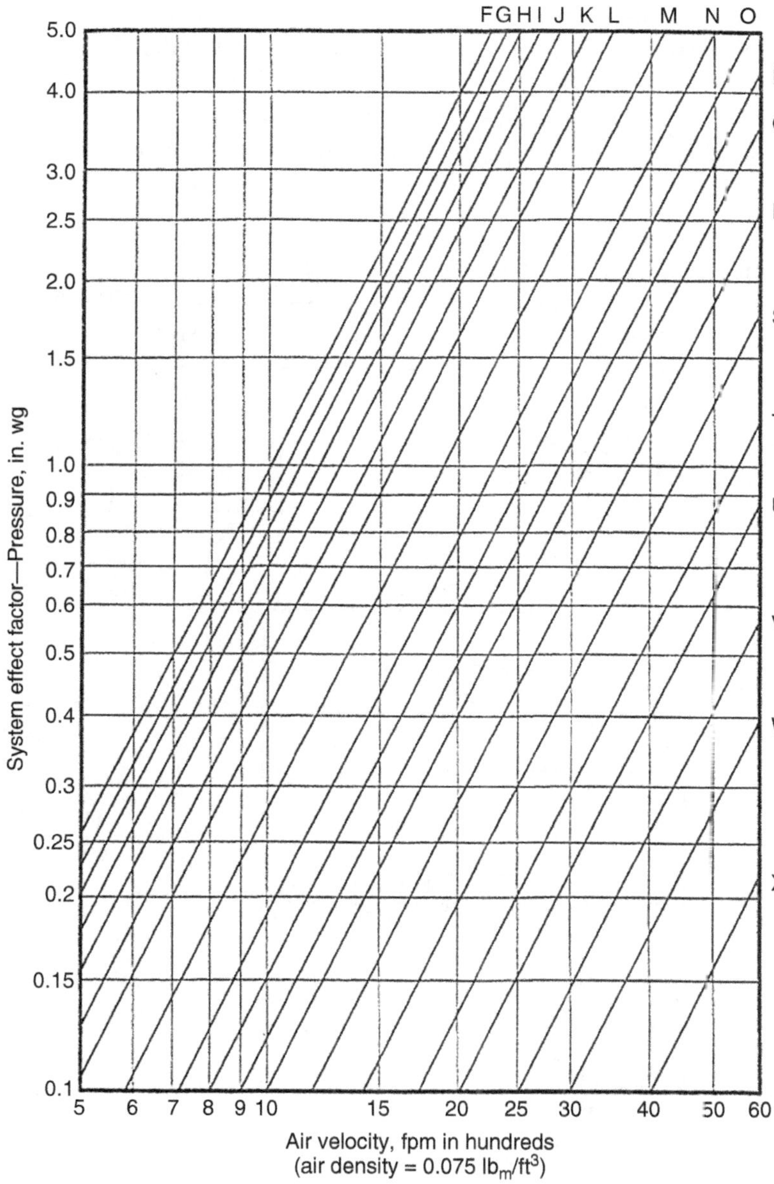

Figure 3.55 System effect cures. (Reprinted by permission from AMCA International Inc., *AMCA Publication 201-02, Fans and Systems*, Air Movement and Control Association, Inc., Arlington Heights, IL, 2002, www.amca.org.)

3.2.8.2 Element System Effects

Element system effects account for the additional losses when two or more elements are installed in close proximity, while the data published for an individual element are tested with an appropriate setup (normally measured at 10 diameter straight duct upstream and downstream of the element). Some examples of these effects are shown in Figure 3.53.[27] The dimensionless loss coefficient C is defined as $dp = C(\rho V^2/2)$. It is shown that in the case of two identical elbows when two elbows are located too closely, the total loss is more than the summation of two individual ones. In the case of two different elements, the system effect factor K defined as $C_c = K(C_a + C_b)$ approaches 1 when two elements are separated farther apart. A more detailed discussion is given in Ref. 27.

Example 3.20 Two mitered square elbows are installed at the discharge of a vane-axial fan as shown. If the airflow rate is 7000 cfm, determine the additional pressure loss due to the system effect factor (assume air density of 0.075 lb_m/ft^3).

Solution

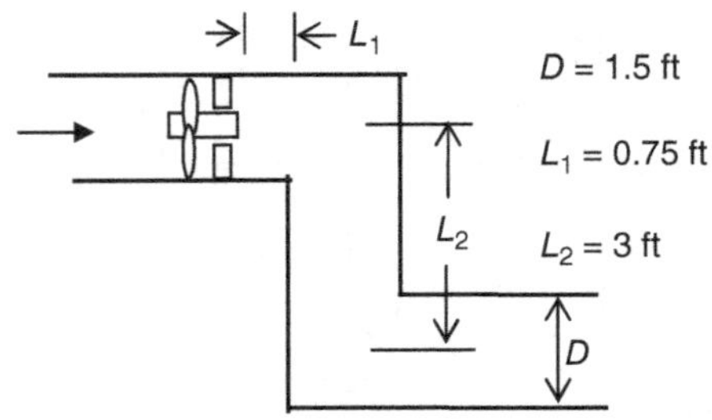

The average flow velocity V is determined as $V = Q/A = 7000/(1.5^2\pi/4) = 3961$ fpm ≈ 4000 fpm. So we have 100% effective discharge duct length $L_e = [2.5 + (4000 - 2500)/1000]D = 4D = 6$ ft and $L_1/L_e = 0.75/6 = 0.125$. So the line V in Figure 3.55 can be used to obtain $SEF_1 = 0.255$ in. wg. For the two elbows, we have $L_2/D = 3/1.5 = 2$ and $C_2 = 4.2$ from Figure 3.56*a*. Hence we obtain $SEF_2 = 4.2(\rho V^2/2) = 4.2[(0.075/32.2)(4000/60)^2/2](12/62.4) = 4.18$ in. wg.

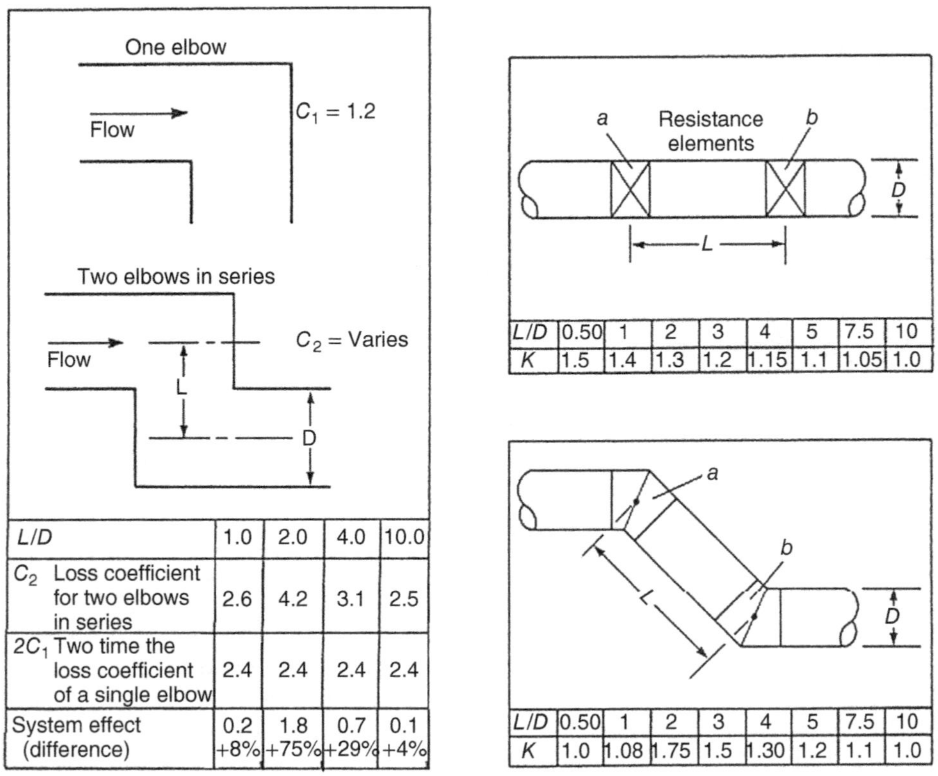

L/D	1.0	2.0	4.0	10.0
C_2 Loss coefficient for two elbows in series	2.6	4.2	3.1	2.5
$2C_1$ Two time the loss coefficient of a single elbow	2.4	2.4	2.4	2.4
System effect (difference)	0.2 +8%	1.8 +75%	0.7 +29%	0.1 +4%

L/D	0.50	1	2	3	4	5	7.5	10
K	1.5	1.4	1.3	1.2	1.15	1.1	1.05	1.0

L/D	0.50	1	2	3	4	5	7.5	10
K	1.0	1.08	1.75	1.5	1.30	1.2	1.1	1.0

(*a*) Two identical mitered square elbow elements (*b*) Two identical straight and turning elements

Figure 3.56 Estimated system effects. (Reprinted by permission from AMCA International Inc., *AMCA Publication 200–95, Air Systems*, Air Movement and Control Association, Inc., Arlington Heights, IL, 2007, www.amca.org.)

3.2.9 Fan Selection

Fans or blowers can normally be classified into the following eight types according to the aerodynamic aspect of the impeller and housing, as shown in Figure 3.57.[25] The performance curves of static pressure p_s, brake horsepower w_0, and efficiency η versus volume flow rate are also given:

a) A propeller fan with two to four blades and no housing, hence a high swirl flow and maximum efficiency that is reached near free delivery, is used for low-static-pressure and high-volume-flow-rate applications.
b) A tube-axial fan with a single propeller in a tube can handle static pressures up to 4 in. of water at flow rates of 500,000 cfm.
c) A vane-axial fan with discharge guide vanes to smooth gas flow for static pressures up to 9 in. of water and flow rates to 600,000 cfm is unstable at low flow rates, with peak horsepower at shutoff conditions due to the flow mismatch between rotor and stator.
d) A centrifugal tubular impeller housed in a tube, with flow discharged from the impeller having to turn 90° into the axial direction, has lower efficiency but higher static pressure than the axial fans for the same size and speed.
e) An airfoil section blade is for stable and efficient operation with less turbulence; it can be either radial or backward inclined.
f) A backward-inclined or curved blade is used for stable operation but is less efficient than the airfoil section blade, which is the most widely used centrifugal fan.
g) A radial blade, for easy fabrication but even less efficient, can generate static pressure up to 60 in. of water.
h) A forward-curved blade, for higher discharge heads with the same impeller diameter, is less efficient and stable with a positive pressure slope on the left of b.e.p.

Their corresponding N_s–D_s and static efficiency curves are also shown in Figure 3.38. This diagram will be used as the starting point for selecting a fan to meet the performance required.

Other factors that should affect the choice include noise level, initial operating cost, safety, reliability, installation, and maintenance. For the same application, a correctly sized axial fan will generally be smaller and lighter than the radial type, but it has to be operated at a much higher speed. This implies that the axial-flow fan will cost less to purchase and be less expensive to install but usually will cost more to operate because its higher speed incurs noise, vibration, and more mechanical problems. There are several different fan drive arrangements, as shown in Figure 3.58.[28] Their choice will depend on the space and the system configuration.

Standard fan performance curves are shown in Figure 3.59.[28] The static or total head, efficiencies, and shaft power are plotted versus flow rate at constant rotating speed, similar to the pump performance curves shown in Section 3.1. However, for convenience of selection, they can be plotted with the system curves for several different rotating speeds in log-log charts, as shown in Figure 3.60.[29] The dashed curves are constant brake horsepower contours. Since the system curves can be represented by $\Delta p = cQ^2$, where c is the proportional constant, they will be straight lines, such as the S–B–A–C–S line in the log-log chart. Also, from the similarity law, $\Delta p \sim N^2$, $Q \sim N$, the corresponding operation point(Δp, Q) of constant efficiency should be related as $\Delta p \sim Q^2$ when the rotating speed changes. Hence, the efficiency is constant along a given system line as indicated. In other words, the operating conditions of a fan running at different rotating speeds for several different systems can be obtained from this chart.

Other types of fan performance characteristics published by manufacturers can be either in chart or table form, as shown in Figures 3.61a and b are also included.

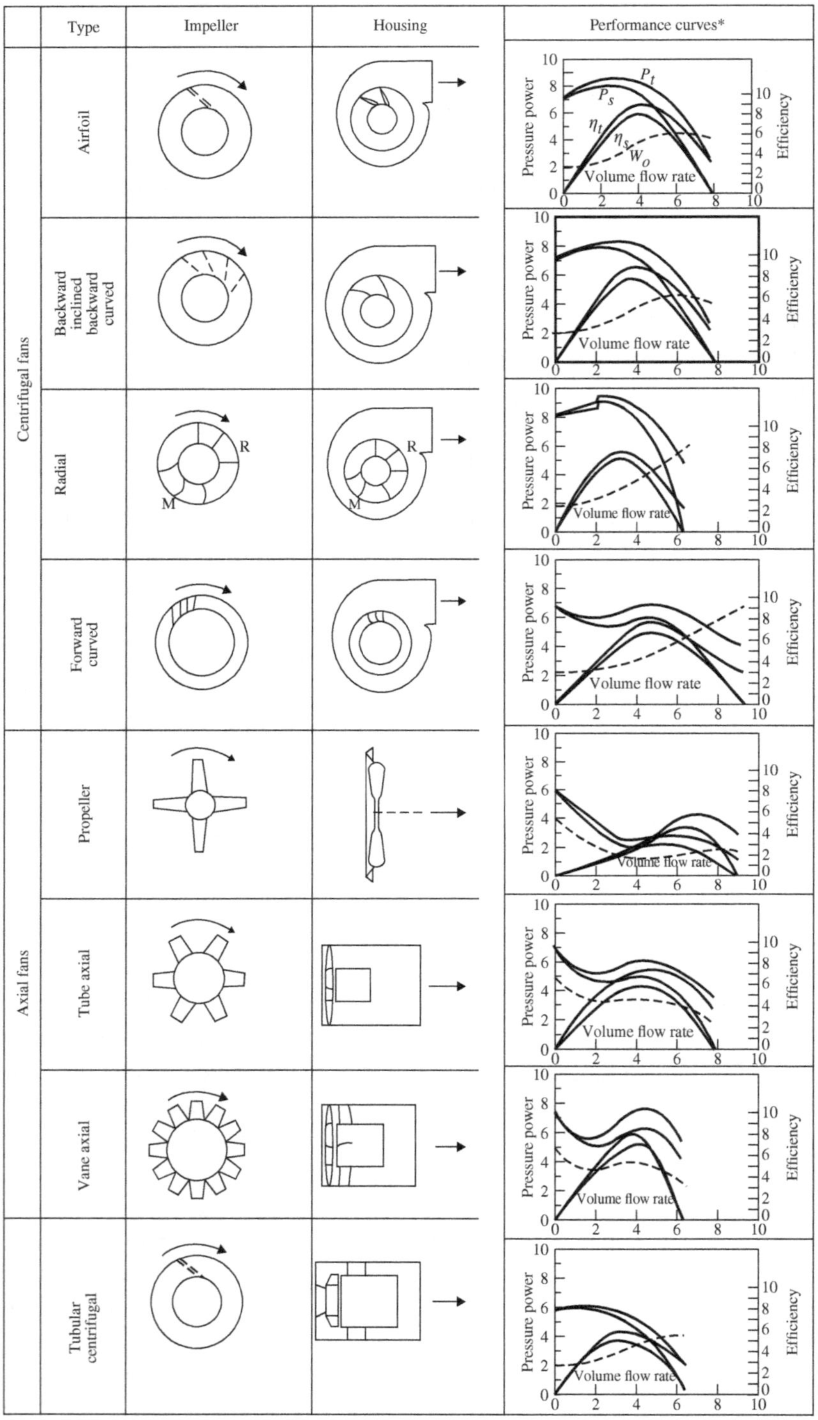

Figure 3.57 Classification of axial and centrifugal Fans. (Reprinted by permission from Parsons, R. A. (Ed.), *ASHRAE Handbook-HVAC System and Equipment*, © American Society of Heating, Refrigerating and Air-Conditioning Engineers, Inc., Atlanta, GA, 2004, www.ashrae.org.)

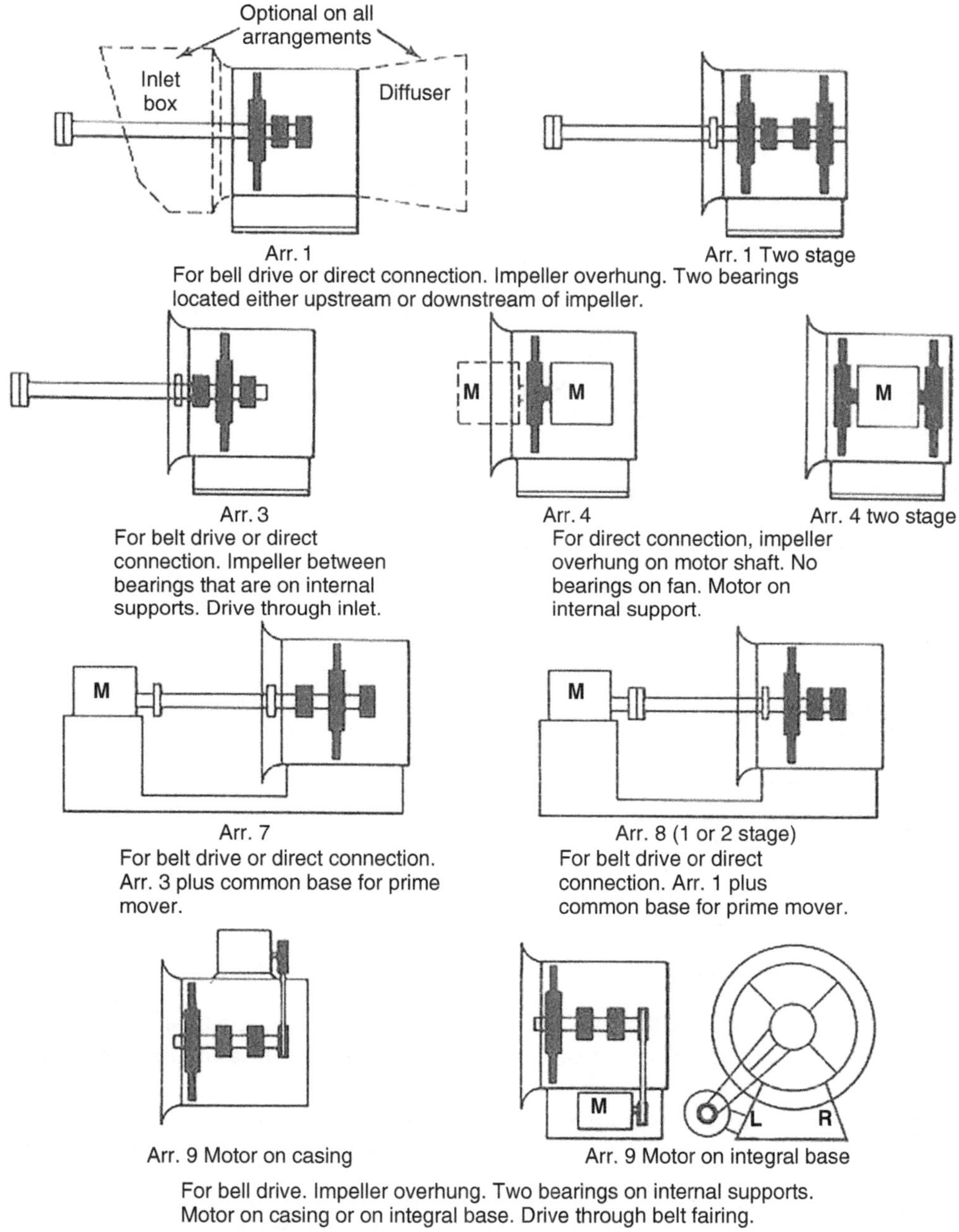

Figure 3.58 Drive arrangement for axial fans with or without a diffuser and inlet box. (Reprinted by permission from *AMCA Publication 201-02, Fans and Systems*, Air Movement and Control Association, Inc., Arlington Height, IL, 2002, www.amca.org.)

In Figure 3.61*a*,[30] for a given flow rate, the rotating speed is specified for a certain static pressure. This is convenient to use if the rotating speed can be easily adjusted. In Figure 3.61*b*,[31] the operating point indicated is read for the rotating speed of 1750 rpm, the flow rate of 24,000 cfm, and the static pressure of 3.60 in. wg. The horsepower of 20.5 hp is read from the top curves. If the fan selected does not fall under the conditions given in the table or curve, linear interpolation or extrapolation can be employed.

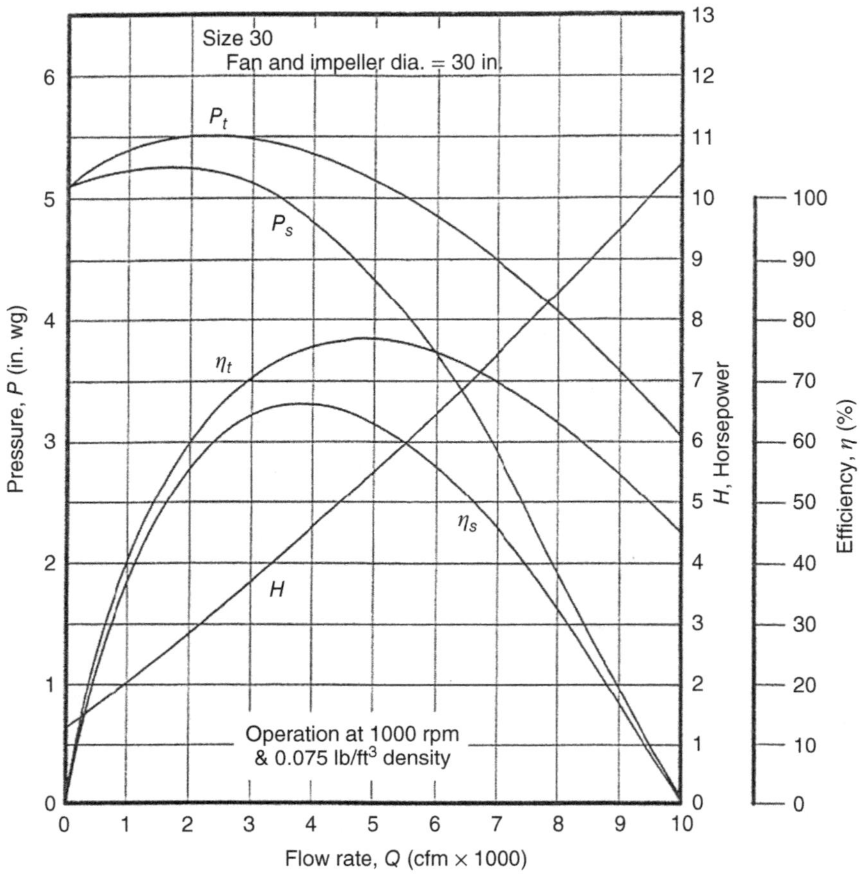

Figure 3.59 Standard fan performance curves. (Reprinted by permission from *AMCA Publication 201–02, Fans and Systems*, Air Movement and Control Association, Inc., Arlington Height, IL, 2002, www.amca.org.)

The steps for fan selection can be summarized as follows:

1) Study the system arrangement and determine the number of fans required, flow rate (Q), total or static pressure rise (Δp_t or Δp_s), and gas properties such as corrosivity, temperature, and density (ρ) at the inlet of each fan.
2) If the inlet gas density in the field is different from the standard condition, the equivalent pressure rise can be calculated from $\Delta p_0 = (\rho_0/\rho)\,\Delta p_t$, where ρ_0, Δp_0 are for the standard condition specified in the vendor's catalog and ρ, Δp_t are for the field operating condition.
3) Pick the rotating speed N based on the drive to be used and calculate the specific speed $N_s = NQ^{1/2}/\Delta p_0{}^{3/4}$.
4) Use the N_s–D_s diagram (Figure 3.38) to select the fan type and calculate the approximate size from $D_s = D\Delta p_0{}^{1/4}/Q^{1/2}$. Make sure the units are consistent with those specified.
5) Use a composite chart to select the appropriate model.
6) Use the appropriate rating charts or tables to select the right size and rotating speed based on the required operation condition (Q, Δp_0).
7) Read out the shaft power P_{so} or efficiency $\eta_{t,s}$ and noise power level, if available.
8) Calculate the expected shaft power in the field from $P_s = (\rho/\rho_0)P_{so}$ or $P_s = Q\Delta p_{t,s}/\eta_{t,s}$.
9) Select the proper electric motor and enclosure.
10) Select the proper bearing arrangement, inlet configuration, and construction material for the critical parts if hot, dirty, or corrosive gas is expected in operation or the floor space is limited.

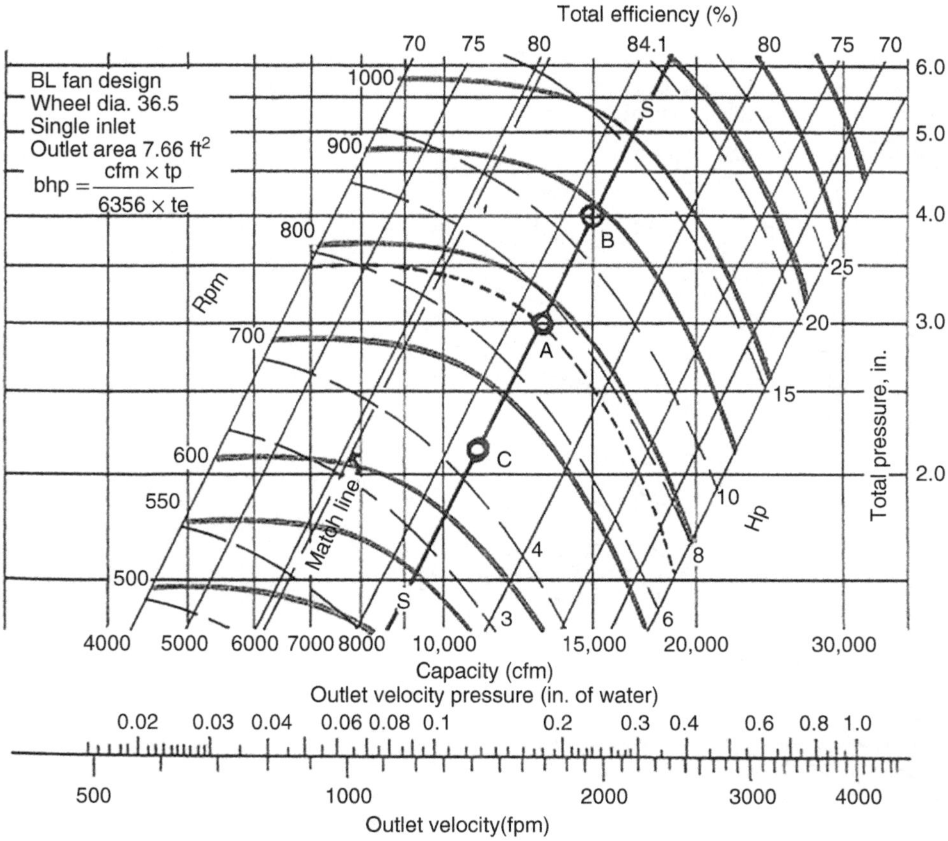

Figure 3.60 Combined fan and system performance curves. (Reprinted by permission from Parsons, R. A. (Ed.), *ASHRAE Journal*, Vol. 14, Part 1, No. 1, © American Society of Heating, Refrigerating and Air-Conditioning Engineers, Inc., Atlanta, GA, 1972, www.ashrae.org.)

		1/4 SP		3/8 SP		1/2 SP		5/8 SP		3/4 SP	
C.F.M.	VEL	RPM	BHP	RPM	BHP	RPM	BHP	RPM	BHP	RPM	BHP
6128	800	270	.38	303	.51	335	.65	365	.82	397	.99
6894	900	289	.46	317	.61	347	76	375	.92	402	1.10
7660	1000	309	.56	335	.72	360	.88	387	1.05	412	1.22
8426	1100	330	.67	354	.85	376	1.01	400	1.19	424	1.38
9192	1200	352	.80	374	1.00	395	1.18	415	1.36	438	1.57
9958	1300	374	.95	395	1.16	414	1.35	434	1.56	452	1.76
10724	1400	396	1.12	416	1.33	434	1.55	453	1.77	470	1.98
11490	1500	419	1.30	437	1.53	455	1.77	472	2.00	490	2.24

(Underlines indicate best efficiency conditions,
VEL: discharge velocity in ft/s)

(*a*) Centrifugal fan multirating table

Figure 3.61 Typical fan performance characteristics published by manufacturers. (Courtesy of Loren Cook Co., Springfield, MO.)

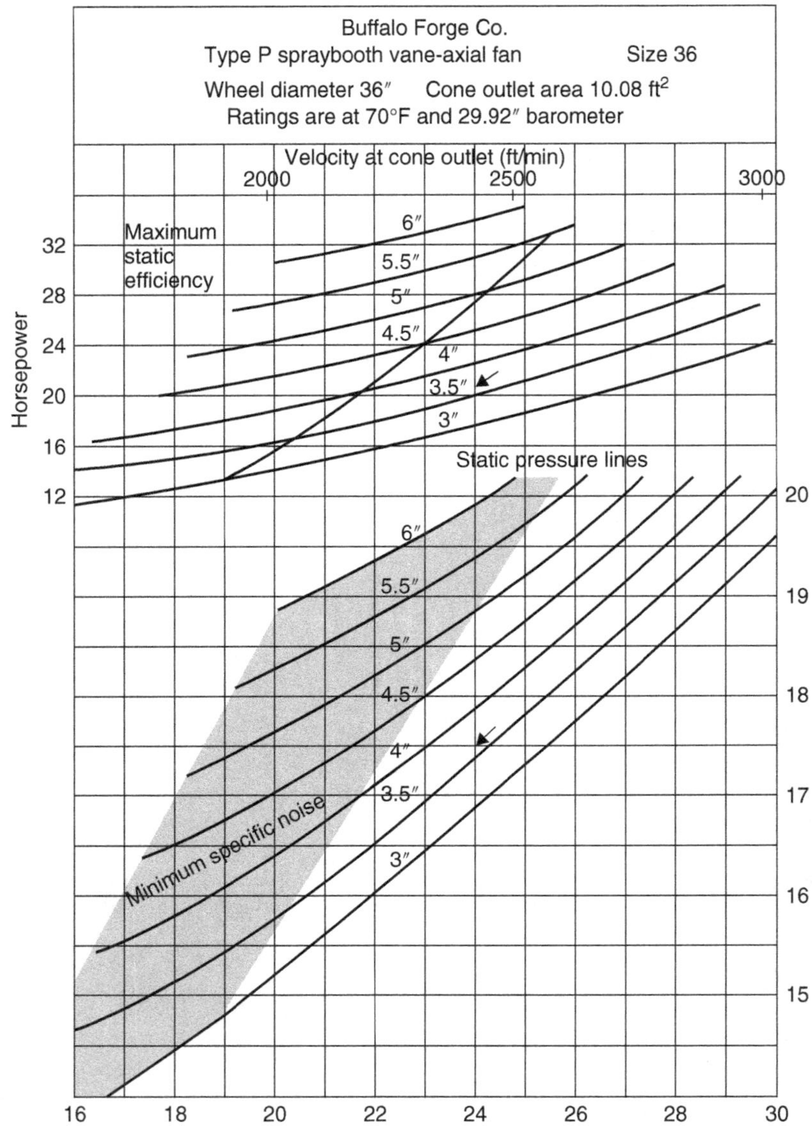

(*b*) Fan multirating chart

Figure 3.61 (*Continued*)

Example 3.21 A fan is required to exhaust 7000 cfm air of 85°F out of a building through an air duct. The building is located 5000 ft above sea level with a barometric pressure of 12.5 psia. The air duct has been tested to have 0.25 in. wg resistance at 5000 cfm. Select an appropriate fan from the tables in Figure 3.61*a*. Also, determine the required input shaft power (brake horsepower).

Solution

From $\Delta p \sim Q^2$, we have the duct resistance $\Delta p = 0.25 \times (7000/5000)^2 = 0.49$ in. wg at 7000 cfm. The air density in the building can be calculated from.

$$\rho = \frac{p}{RT} = \frac{12.5 \times 144}{53.3(85 + 460)} = 0.0619 \ \text{lb}_\text{m}/\text{ft}^3.$$

So the equivalent SP_e of the required fan at (STP) is $\text{SP}_e = 0.49 \times 0.075/0.0619 = 0.593$ in. wg.

From Figure 3.61a, at SP = 0.5, $Q_1 = 6894$, $N_1 = 347$, $\text{BHP}_1 = 0.76$ and $Q_2 = 7660$, $N_2 = 360$, $\text{BHP}_2 = 0.88$. Hence, for $Q = 7000$ cfm we have

$$N' = 347 + (360 - 347) \times \frac{7000 - 6894}{7660 - 6894} = 347 + 13 \times 0.138 = 348.8 \text{ rpm},$$

$$\text{BHP}' = 0.76 + (0.88 - 0.76) \times 0.138 = 0.776.$$

At SP = 5/8 = 0.625 in. wg, $Q_1 = 6894$, $N_1 = 375$, $\text{BHP}_1 = 0.92$ and $Q_2 = 7660$, $N_2 = 387$, $\text{BHP}_2 = 1.05$, so for $Q = 7000$ cfm, we have

$$N'' = 375 + (387 - 375) \times 0.138 = 376.6 \text{ rpm},$$

$$\text{BHP}'' = 0.92 + (1.05 - 0.92) \times 0.138 = 0.938 \text{ hp}.$$

Again interpolating for $\text{SP}_e = 0.593$, we have

$$N = 348.8 + (376.6 - 348.8) \times \frac{0.593 - 0.5}{0.625 - 0.5} = 348.8 + 27.8 \times 0.744 = 369 \text{ rpm},$$

$$\text{BHP} = 0.776 + (0.938 - 0.776) \times 0.744 = 0.896 \text{ hp}.$$

Correcting for the reduced air density, the actual BHP is $0.896 \times 0.0619/0.075 = 0.739$ hp.

Example 3.22 An air duct system is tested to have 0.5 in. wg total pressure loss at a flow rate of 5000 cfm. If the fan of Figure 3.60 running at 700 rpm is selected for this system, determine its operating condition, including flow rate, total pressure, and brake horsepower.

Solution

1) Convert the system specification to $Q' = 10{,}000$ cfm with the corresponding total pressure loss $\Delta p'_L = 0.5(10,000/5000)^2 = 2.0$ in. wg.
2) Draw a system line (which is a straight line with a slope of 2 on a log–log chart) passing this point and intercept with the fan curve of 700 rpm.
3) Read the operating condition as $Q = 11{,}000$ cfm, $\Delta p = 2.51$ in. wg, and $P_s = 5.6$ hp.

3.2.10 Propellers

The propellers used for low-speed aircraft and marine propulsions are similar to the axial-flow fan and pump in geometry and analysis. The conventional propeller has fewer blades (two to four) and a lower rotating speed. Its main objective is to produce thrust rather than pressure rise. However, the complete propeller power plant is too bulky for aircraft due to the need for a gearbox to reduce rotating speed. Hence, the turbofan gas turbine engine uses a lower-speed jet to produce thrust, replacing the conventional propeller for low-speed aircraft applications.

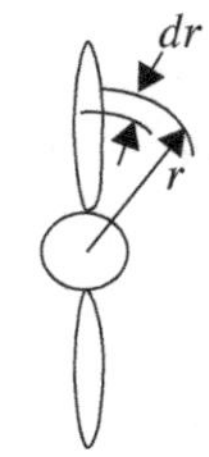

By using the blade element analysis discussed in Section 3.2.2, the torque and axial force acting on a spanwise length dr of the blade at radius r can be expressed as

$$\begin{aligned} d\tau &= rF_u dr = r(F_L \cos\beta_m + F_d \sin\beta_m)dr \\ &= \frac{1}{2} rc\rho W_m^2 (C_L \cos\beta_m + C_d \sin\beta_m)dr, \\ dF_a &= (F_L \sin\beta_m - F_d \cos\beta_m)dr \\ &= \frac{1}{2} c\rho W_m^2 (C_L \sin\beta_m - C_d \cos\beta_m)dr. \end{aligned} \tag{3.35}$$

With $W_m = W_a/\cos\beta_m \cong V_0/\cos\beta_m$, the total torque and thrust can be obtained as

$$\begin{aligned} \tau &= \frac{1}{2}\rho V_0^2 Z_b \int_{r_h}^{r_t} \frac{rc}{\cos^2\beta_m}(C_L \cos\beta_m + C_d \sin\beta_m)dr, \\ F_a &= \frac{1}{2}\rho V_0^2 Z_b \int_{r_h}^{r_t} \frac{c}{\cos^2\beta_m}(C_L \sin\beta_m - C_d \cos\beta_m)dr, \end{aligned} \tag{3.36}$$

where V_0 is forward speed, c is chord length, and Z_b is blade number. The propeller efficiency can be given as

$$\eta_p = \frac{\text{Thrust power output}}{\text{Shaft power input}} = \frac{F_a V_0}{\omega\tau} \tag{3.37}$$

From the dimensional analysis discussed in Chapter 2, the relevant nondimensional coefficients for a thrust-producing device are given by

$$\begin{aligned} &\text{Thrust coefficient: } C_f = \frac{F_a}{\rho\omega^2 D^4}, \\ &\text{Power coefficient: } C_p = \frac{P_s}{\rho\omega^3 D^5}. \end{aligned} \tag{3.38}$$

Hence $\eta_p = (C_f/C_p)(V_0/\omega D)$, where $V_0/(\omega D)$ is the advance ratio.

More detailed analysis can be found in Ref. 20.

References

1 Karassik, I. J., Krutzsch, W. C., Fraser, W. H., and Messina, J. P. (Ed.), *Pump Handbook*, McGraw-Hill, New York, 1976.

2 Stepanoff, A. J., *Centrifugal and Axial Flow Pumps*, John Wiley & Sons, New York, 1957.

3 Hydraulic Institute, *Hydraulic Institute Centrifugal Pumps Tests—2000*, Hydraulic Institute, Parsippany, NJ, 2015.

4 Goulds Pumps, Industrial Products Groups, *Goulds Pump Manual GPM5*, Goulds Pumps, ITT Corp., Seneca Falls, New York, 1988.

5 Fraser, W. H., "Recirculation in Centrifugal Pumps," *Proceeding of 10th Annual Turbomachinery Symposium*, pp. 95–100, Texas A&M University, College Station, TX, Dec. 1981.

6 Tuzon, J. J., "Inlet Recirculation in Centrifugal Pumps," *Symposium Proceeding of Performance Characteristics of Hydraulic Turbines and Pumps*, pp. 195–200, ASME Annual Meeting, Boston, MA, Dec. 1983.

7 Alpan, K. and Peng W. W., "Suction Reverse Flow in an Axial-Flow Pump," *J. Fluids Eng. ASME Trans.*, Vol. 113, No. 1, pp. 90–97, Mar. 1991.

8 Ingersoll Rand Pumps, *Ingersoll Rand's High Speed Pump: Your Logical Choice*, Ingersoll Rand Company, Phillipsburg, NJ, 1983.

9 Wislicenus, G. F., *Fluid Mechanics of Turbomachinery*, McGraw-Hill, New York, 1947.

10 Hydraulic Institute, *Hydraulic Institute Vertical Pump Tests—2000*, Hydraulic Institute, Parsippany, NJ, 2015.

11 Schivley, G. P., and Dussourd, J. L., "Analytical and Experimental Study of a Vortex Pump," *J. Basic Eng. ASME Trans.*, Vol. 92, No. 4, pp. 889–900, Dec. 1970.

12 Sheth, K. K., Morrison, G. L., and Peng, W. W., "Slip Factors of Centrifugal Slurry Pumps," *J. Fluids Eng. ASME Trans.*, Vol. 109, pp. 313–318, Sept. 1987.

13 Peng, W. W., "Nonequilibrium and Dissolved Gas Effects on Cavitation in Centrifugal Pumps," *Proceeding of Cavitation and Polyphase Flow Forum*, pp. 46–48, ASME Fluids Engineering Meeting, June 1983.

14 Church, A. H., *Centrifugal Pumps and Blowers*, Robert E. Krieger Publishing, Huntington, NY, 1972.

15 Hydraulic Institute, *Hydraulic Institute Engineering Data Book*, Hydraulic Institute, Parsippany, NJ, 1979.

16 American Petroleum Institute, *Centrifugal Pumps for Petroleum, Heavy Duty Chemical and Gas Industry Services, API Standard 610*, 8th ed, American Petroleum Institute, Washington, DC, 1995.

17 Neerken, R. F., "Selecting the Right Pump", in *Fluid Movers, Pumps, Compressor, Fans and Blowers*, J. Matley and The staff of Chemical Week Associates (Eds.), McGraw-Hill, New York, 1979, pp. 122–133.

18 Bhatia, M. V., and Cheremisinoff, P. N. (Eds.), *Air Movement and Vacuum Devices, Process Equipment Series*, Vol. 3, Technomic Publishing Company, Westport, CT, 1981.

19 Dixon, S. L., *Fluid Mechanics and Thermodynamics of Turbomachinery*, 5th ed., Pergamon, Oxford, 1998.

20 Lakshminarayana, B., *Fluid Dynamics and Heat Transfer of Turbomachinery*, John Wiley & Sons, New York, 1996.

21 Stepanoff, A. J., *Centrifugal and Axial Flow Pumps*, John Wiley & Sons, New York., 1957.

22 Wallis, R., *Axial Flow Fans and Ducts*, John Wiley & Sons, New York, 1983.

23 Hay, N., Metcalfe, R., and Reizes, J. A., "A Simple Method for the Selection of Axial Fan Blade Profiles," *Int. Mech. Eng.*, Vol. 192, No. 25, pp 269–275, 1978.

24 Laboratory Methods of Testing Fans for Rating, *AMCA Standard 210-99*, and *ASHRAR Standard 51-1999*, Air Movement and Control Association, Inc., Arlington Heights, IL, and American Society of Heating and Air-Conditioning Engineers, Atlanta, GA, 2000.

25 Parsons, R. A. (Ed.), *ASHRAE Handbook, HVAC Systems and Equipment*, American Society of Heating and Air-Conditioning Engineers, Atlanta, GA, 2004.

26 Parsons, R. A. (Ed.), *ASHRAE Handbook, HVAC Applications*, American Society of Heating and Air-Conditioning Engineers, Atlanta, GA, 2003.

27 AMCA International Inc., *AMCA Publication 200-95, Air Systems*, Air Movement and Control Association, Arlington Heights, IL, 2007.

28 AMCA International, Inc., *AMCA Publication 201-02, Fans and Systems*, Air Movement and Control Association, Arlington Heights, IL, 2002.

29 Parsons, R. A. (Ed.), *ASHRAE Journal*, Vol. 14, Part 1, No. 1, American Society of Heating and Air-Conditioning Engineers, Atlanta, GA, 1972.

30 Loren Cook Co., *Cook Centrifugal Plug Fan*, Loren Cook Co., Springfield, MO, 1989.

31 Jorgensen, R. (Ed.), *Fan Engineering*, 9th ed., Buffalo Forge Company, Buffalo, NY. 1999.

32 Peabody Floway, Inc., *Turbine Data Handbook*, 1st ed., Peabody Floway, Inc., Fresno, CA, 1987.

Problems

Problems from 3.1

3.1 Test results on a single-stage, single-suction mixed-flow water pump operating at 575 rpm are given as follows:

Flow rate (gpm)	0	1000	2000	3000	4000	5000	6000	7000	7600
Head (ft)	39.5	37.2	35.0	32.5	30	27	23.5	19.2	16
Efficiency (%)	0	27.6	50.0	65.6	75.5	84	88	86	81

a) Plot curves for these values and the brake horsepower curve.
b) On the same sheet, draw the brake horsepower curve when the liquid pumped has a specific gravity of 0.9 but otherwise is the same as water.

3.2 A centrifugal pump, whose performance curves are shown in Figure 3.7a is used to pump water of 150°F from a tank as shown. If an impeller trimmed to 8-in. is used, determine the flow rate and the motor power input at the best efficiency point and the minimum tank pressure required to avoid cavitation at the pump inlet (friction loss in the suction pipe is 0.8 ft).

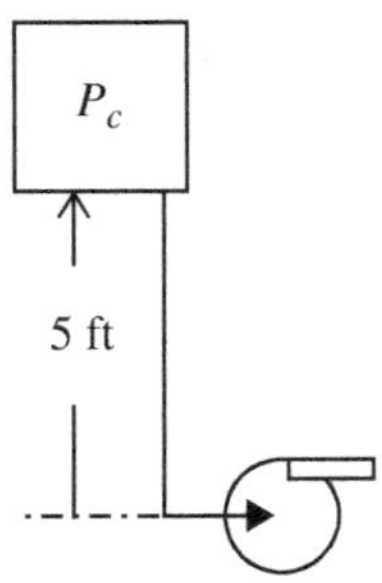

3.3 A centrifugal pump is required to develop a pressure rise of 30 psi when delivering 1000 gpm. Which motor should be used for maximum efficiency if three motors with 2000, 3600, and 4200 rpm are available? What are the approximate impeller diameter and the pump efficiency? Should a double-suction pump be considered?

3.4 A single-stage radial-type impeller pump has dimensions as shown. It develops a head of 100 ft at the design flow rate of 500 gpm when rotating at 1750 rpm. If the shaft power is measured to be 21 hp and the losses are estimated to be 1.4 hp for disk friction, 0.5 hp for bearing and seals, 20 gpm for leakage, determine (a) inlet blade angle β_{b1}, (b) slip factor μ_s, (c) hydraulic efficiency η_h, (d) volumetric efficiency η_v, and (e) mechanical efficiency η_m.

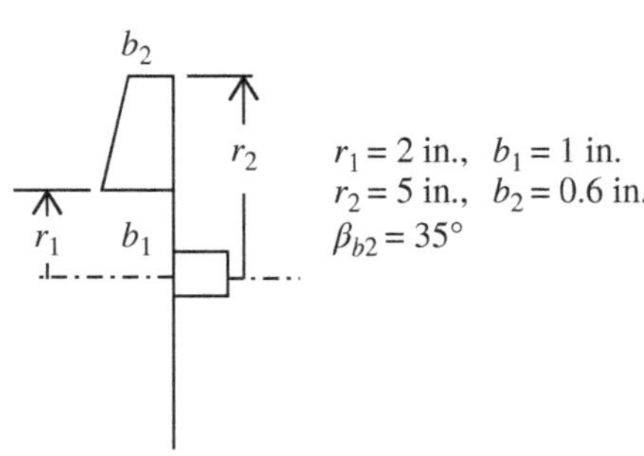

3.5 Calculate the principal dimensions at the inlet and outlet of a centrifugal water pump impeller, which develops a 100-ft head at 800 gpm when operating at 1200 rpm under design conditions.

$r_1 = ?$	$b_1 = ?$	$\beta_1 = ?$
$r_2 = ?$	$b_2 = ?$	$\beta_2 = ?$

3.6 Check the manufacturer's claim that a centrifugal pump with a rotor diameter of 12 in. can supply water at a rate of 4500 gpm and produce a head of 30 ft at a rotating speed of 980 rpm. What should be the more likely size or performance?

3.7 A six-stage vertical turbine pump has a design flow rate of 1500 gpm and a total discharge head of 700 psi when rotating at 1750 rpm. Determine its specific speed. Then, estimate its efficiency and approximate impeller diameter.

3.8 Design the impeller vane profile for the pump in problem 3.5. Use a linear distribution for the blade angle, varying from inlet to outlet.

3.9 A plant is located at an altitude of 2500 ft above sea level (atmospheric pressure 27.5 in. Hg). The open feedwater heater pump system is shown with the suction line equivalent length of 25 ft, pipe inside diameter of 3 in., and friction coefficient of 0.02. If the water flow rate is 180 gpm and temperature is 190°F, what is the available NPSH under this condition?

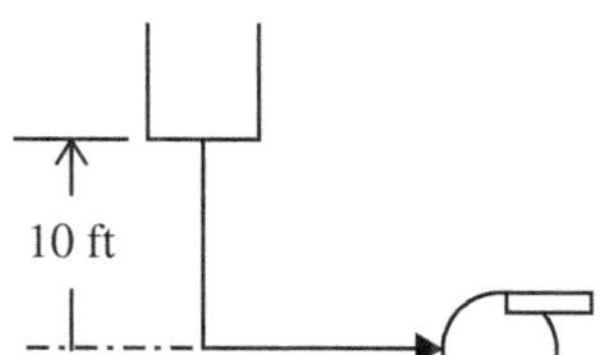

3.10 Plot the performance curves for the pump with a 7-in. outside diameter (OD) impeller shown in Figure 3.7*a*, in Figure 3.6*b* for head, shaft power, and efficiency versus flow rate.

3.11 A centrifugal pump is tested with water of 70°F in the setup shown. The suction pipe inside diameter is 4 in., and the discharge pipe inside diameter is 3 in. The impeller rotating speed is 1200 rpm, and the following data are taken when cavitation starts: shaft torque 12 ft-lb$_f$, flow rate 180 gpm, $H_1 = 2$ ft of water, $H_2 = 25$ ft of water. Determine the pump's overall efficiency and the required NPSH.

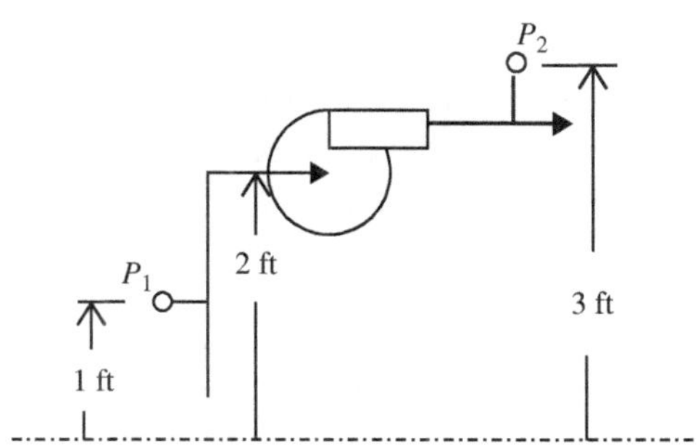

3.12 Two identical pumps with impeller trimmed to 7-in. with the performance curves shown in Figure 3.7*a* are installed in parallel and in series as shown in Figure 3.29. Plot the combined performance curves of head versus capacity.

3.13 Oil with a specific gravity of 0.75 is pumped by a centrifugal pump. When the flow rate is 375 gpm, the pump requires 19.3 hp input, and its efficiency is 81.2%. Determine the pressure rise produced by the pump. Express the result as (a) feet of water and (b) feet of oil.

3.14 Dimensions of a centrifugal pump impeller are given as $r_1 = 400$ mm, $b_1 = 120$ mm, $\beta_{b1} = 40°$, $r_2 = 1200$ mm, $b_2 = 80$ mm, $\beta_{b2} = 15°$. If the pump is driven at 575 rpm and pumping water, calculate the theoretical head, torque, and power input at the design point and plot them from shutoff to the maximum volumetric flow rate.

3.15 Design a single volute for the impeller of problem 3.5.

Problems from 3.2

3.16 If an axial-flow blade with a radius ratio of 1.65 has outlet angles from root to tip such that the whirl velocity is 300 ft/s at all radii, determine the axial velocity at the tip for the condition of radial equilibrium if the axial velocity at the root is 500 ft/s.

3.17 The axial flow fan impeller shown is rotating at 2500 rpm. The design flow rate is 2400 cfm, the total head is 1.2 in. of water, and hydraulic efficiency along the mean radius is 85%. If a free-vortex flow distribution is designed, determine the relative flow angles at the hub, mean, and tip radii for both the inlet and outlet of the impeller and draw the corresponding velocity diagrams.

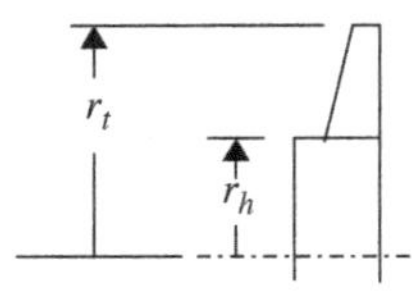

$r_t = 7$ in. $r_h = 3$ in.

3.18 An axial-flow fan rotor with the dimensions shown rotates at 1800 rpm. Draw the velocity diagrams at the inlet and outlet's mean radii at the design condition. Estimate the lift coefficient, the design flow rate, and the ideal static pressure rise across the rotor at the mean radius. Neglect the flow interference between the blades.

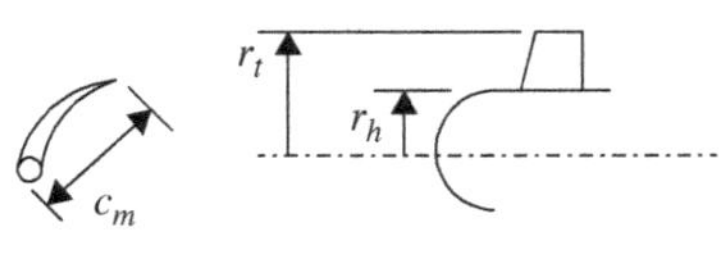

$r_t = 6.0$ in. $r_h = 4.6$ in.
$\beta_{b1} = 35$ deg. $\beta_{b2} = 15$ deg.
$Z_b = 6$ $c_m = 4.5$ in.

3.19 An axial-flow pump impeller consists of five blades with chord length $c = 3$ in., hub radius $r_h = 4.5$ in., and tip radius $r_t = 6.5$ in. If the airfoil data of Figure 3.40*a* corrected with the cascade coefficient given in Figure 3.46 is used and the impeller is rotating at 1800 rpm, determine the static pressure across the impeller and the rated flow rate. Assume the stagger angle, γ is set at 30°.

3.20 Design a vane-axial fan using the specifications in problem 3.17, except for the hydraulic efficiency. Namely, $N = 2500$ rpm, $Q = 2400$ cfm, and SP = 1.2 in. wg. (*Hint*: Use Figures 3.38, 3.40, 3.47, and 3.48.)

3.21 A fan is needed to deliver air of 5000 cfm for a static pressure of 3 in. of water. If motors of 1800, 2400, and 3600 rpm are available, what type of fan should be selected for maximum efficiency? What are the approximate diameter and efficiency of the selected fan?

3.22 (a) Use the fan similarity laws to convert the multi-rating table of Figure 3.61*a* into the standard fan curves as shown in Figure 3.59 for $N = 400$ rpm (including static pressure, total pressure, brake horsepower, static efficiency, and total efficiency). (b) Plot the fan data of Figure 3.61*b* into the standard fan curves for $N = 1800$ rpm.

3.23 An axial-flow fan with a 6-hp motor has a total efficiency of 80%. The fan delivers 10,000 cfm of air through a duct of 3 ft inside diameter. Calculate the static pressure rise associated with this fan.

3.24 Select a vane-axial fan to supply air for an underground mine shaft with a cross-section of 8 ft by 7 ft and 1 mile long. The inside surface roughness is estimated at 0.015, and the required airflow rate is 33,000 cfm. Determine the optimal rotating speed, its size, and required power.

3.25 A vane-axial fan is to be designed for 4500 cfm with a tip diameter of 14 in. and a hub-tip ratio of 0.35. Estimate the desired rotating speed, static pressure, and required shaft power.

4

Compressors

4.1 Centrifugal Fans, Blowers, and Compressors

4.1.1 Introduction

Centrifugal compressors have great advantages for low flow compression both in efficiency and size. There are wide applications in jet engine compressors, oil and gas compression, turbochargers, air-conditioning, and industrial compressors. In most cases, gas enters a centrifugal compressor axially and then turns in the radial direction out from the impeller with the action of the centrifugal force. The gas is then directed to a radial annular vaned or vaneless diffuser. Finally, it moves into a volute or collector to deliver the compressed gas to the next stage or send it to the next components.[1–10] Unlike a fan or a blower,[11] the work input for a centrifugal compressor is almost independent of the nature of the flow. A centrifugal compressor can be designed with a much higher De Haller number (dH = velocity ratio of exit and inlet of the fluid) than an axial compressor can achieve. Therefore, a centrifugal compressor can have a much higher stage pressure ratio than an axial one. Furthermore, centrifugal compressors have wide applications for a mass flow rate of less than 10 kg/s.[1–10, 12, 13]

Recently, multidisciplinary optimization has drawn more attention, but it is still challenging due to the complications of the engineering system and its multidisciplinary nature. He and Wang[14, 15] developed a process for an adjoined approach to concurrent blading aerodynamic and aero-elastic design optimizations at earlier stages of a design process. Blade flutter stability and forced response were obtained from this work. Ghisu et al.[16, 17] recently developed a system for the integrated design optimization of gas turbine engines: postponing the setting of the interface constraints to a point where more information is available facilitates better exploration of the available design space and better exploitation of the trade-offs between different disciplines and modules. Alsalihi et al.[18, 19] developed a multidisciplinary optimization system and used it to design a small radial compressor impeller. The method only discussed the aerodynamic performance and stress interactions. However, the impeller reliabilities relied on the stress and vibrations, that is, low cycle fatigue (LCF) and high cycle fatigue (HCF).

Turbomachine pumping gases can generally be classified into three types: (a) fan, (b) blower, and (c) compressor. The dividing lines are somewhat vague. Fans are machines with discharge pressures of less than 2 psig (55.4 in. wg.) and relatively high flow rates. Blowers develop pressures from 2 psig to about 35 psig; machines that discharge pressures above 35 psig fall into the compressors category. The classification can also be based on their construction, as illustrated in Figure 4.1. The terminology of these machine components is illustrated in Figures 4.2,[1] 4.3,[2] and 4.4.[2]

Similar to the centrifugal pump, a rotating impeller transfers mechanical energy from the drive to the fluid in the form of velocity and pressure/temperature rise. The stationary diffuser is employed downstream of the impeller to convert some kinetic energy into more pressure/temperature increase. Variation in gas conditions within a typical centrifugal compressor is illustrated in Figure 4.5.[2]

Fundamentals of Turbomachinery: Theory and Applications, Second Edition. Ryoichi S. Amano and William W. Feng.

Companion website: www.wiley.com/go/amano/turbomachinery2e

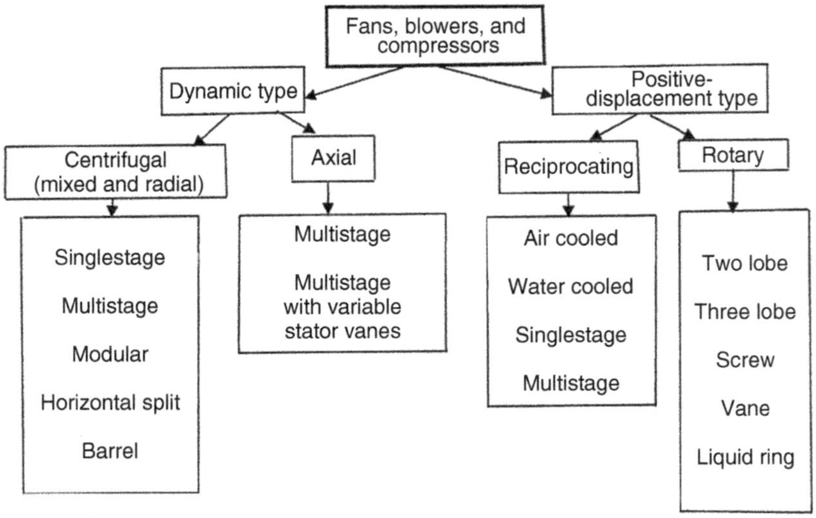

Figure 4.1 Classification of fans, blowers, and compressors.

The basic theory and performance of an axial-flow fan have been discussed in Chapter 2; the multistage axial-flow compressor will be discussed in Section 4.2. Mixed- and radial-flow machines, called centrifugal machines, including fan, blower, and compressor, will be discussed in this chapter since they are similar in construction and performance analysis theory.

4.1.2 Performance Parameters and Characteristics

Typical performance curves of a fan, blower, or compressor are similar to those of a centrifugal pump, as shown in Figure 4.6*a*, except that the low flow rate is normally limited by surging, a phenomenon similar to suction flow recirculation in a liquid pump. Surging in a gas-pumping device causes more severe vibration than that in a liquid pump. At a high flow rate, operation is limited by choking (called stonewalling in some literature) associated with shock waves. This is similar to cavitation in a liquid pump.

Static or total pressure Δp expressed in inches or millimeters of water is used for fans, and head of air, H, and pressure ratio p_2/p_1 are employed in blowers and compressors. The flow rate or capacity is given in volumetric flow rate at the inlet, Q_1, or in mass flow rate m. Typical performance curves or tables of centrifugal fans are similar to those for axial-flow fans, as shown in Chapter 2. A typical set of performance curves published by the manufacturer is shown in Figure 4.6*b*, adapted from Ref. 3. Constant horsepower contours are also plotted. For a given flow rate in cfm (cubic feet per minute) and rotating speed, the static pressure can be read from the solid line, and the horsepower is interpolated with the constant dashed contours.

An example of centrifugal fan performance characteristics in table form is shown in Table 4.1.[4] In each row for a specified flow rate, the rotating speed and brake horsepower are listed at different static pressures.

Similarity laws governing the volumetric flow rate, head, and shaft power also apply to fans, blowers, and compressors. They are sometimes called fan laws and are listed here again: $Q \sim ND^3$, $H \sim N^2D^2$, $P_S \sim N^3D^5$.

The specific speed N_s and specific diameter D_s are also important parameters in the preliminary design and selection of fans, blowers, and compressors. They are defined in dimensional forms as follows:

For fans,

$$N_s = \frac{NQ_1^{0.5}}{(sp)^{0.75}}, \qquad D_s = \frac{D(sp)^{0.25}}{Q_1^{0.5}}$$

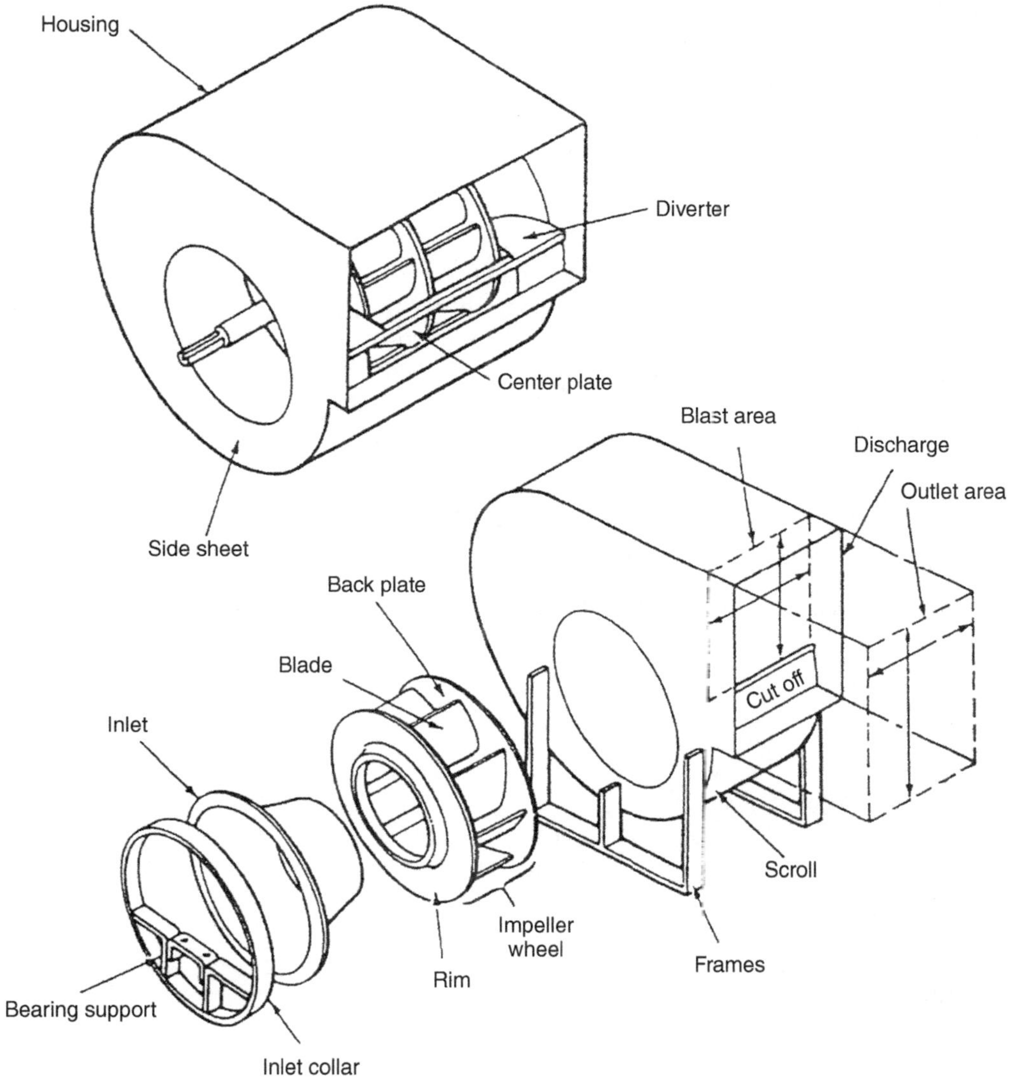

Figure 4.2 Terminology for centrifugal fan and blower components. (Reproduced from AMCA publication 201–02, Fans and Systems Air Movement and Control Association Inc. Arlington Heights, IL 2002 and ASHRAE Handbook-HVAC System and Equipment, 2004/Air Movement and Control Association, Inc.)

and for blowers or compressors,

$$N_s = \frac{NQ_1^{0.5}}{H_{\mathrm{ad}}^{0.75}}, \qquad D_s = \frac{DH_{\mathrm{ad}}^{0.25}}{Q_1^{0.5}},$$

where Q_1 = inlet flow rate (in cfm) for a fan/blower and V_1 (in cfs) for a compressor, N = rotating speed (in rpm), sp. = static pressure (in inches or millimeters of water gauge), H_{ad} = adiabatic head (in ft-lb$_f$/lb$_m$) is energy transfer or enthalpy increase per unit mass of fluid, and D = diameter of impeller (in inches for a fan or feet for a blower and compressor).

The relationship between these two parameters and efficiency was developed by O. E. Balje and is shown in Figure 4.7,[5,11] for compressors. For fans, it was modified by the fan industry and is shown in Figure 3.38. These

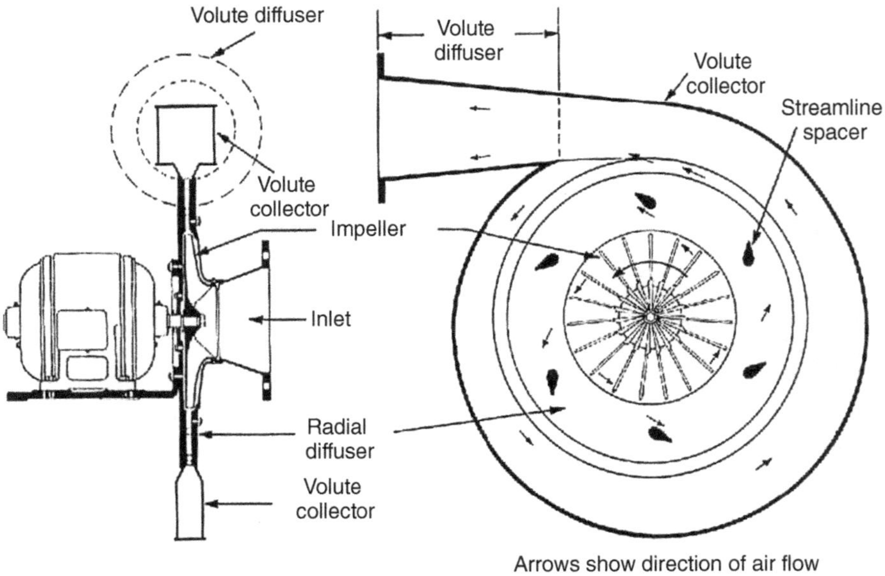

Figure 4.3 Typical single-stage centrifugal compressor. (Reproduced from Gibbs, 1971/Ingersoll-Rand.)

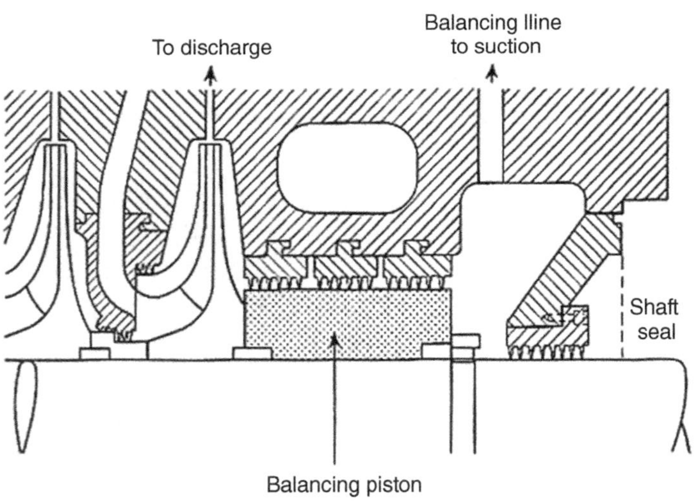

Figure 4.4 Labyrinth seals and balancing drum for multistage compressor. (Reproduced from Gibbs, 1971/Ingersoll-Rand.)

diagrams represent typical optimum design performance and can be used for the preliminary design and selection of machines with specified performance requirements. The Cordier diagram relating the nondimensional specific speed and specific diameter, shown in Figure 2.2, is also applicable to fans.

A thermodynamic process should be considered when pumping compressible gas. The head interpreted as the energy transfer per unit mass (ft-lb_f/lb_m) is the total enthalpy increase per unit mass Δh_0. From the compression process shown in the Carnot cycle as shown below, neglecting the flow velocity difference at the inlet and outlet, the ideal head is

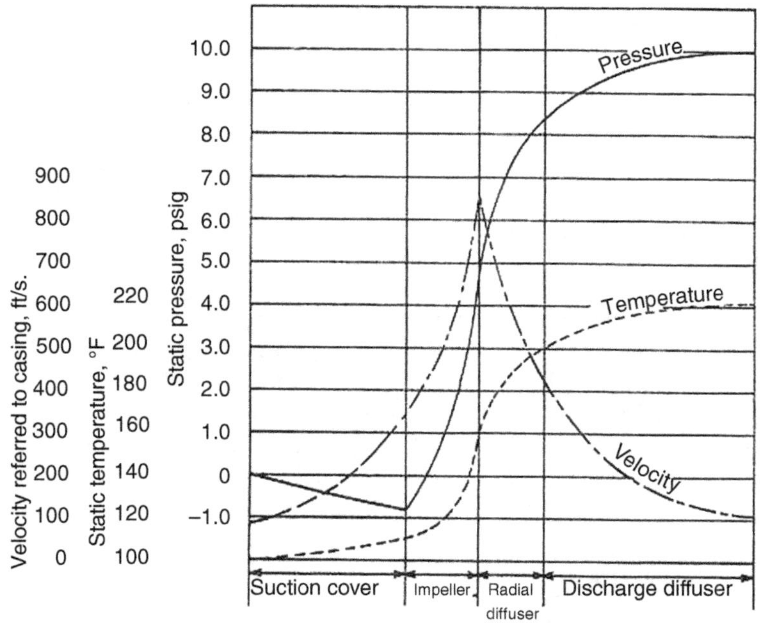

Figure 4.5 Variation of gas conditions within a typical centrifugal compressor. (Reproduced from Gibbs, 1971/ Ingersoll-Rand.)

Carnot cycle.

$$H_i = h_{s2} - h_1 = C_p T_1 \left(\frac{T_{s2}}{T_1} - 1\right) = C_p T_1 \left[\left(\frac{p_2}{p_1}\right)^{(k-1)/k} - 1\right]$$
$$= \frac{kRT_1}{k-1}\left[\left(\frac{p_2}{p_1}\right)^{(k-1)/k} - 1\right], \tag{4.1}$$

since we have $k = C_p/C_v$ and $R = C_p - C_v$. The actual head, including the frictional effect, assuming a polytropic process, can be derived as follows. From $pv^n = C'$ or $v = C/p^{1/n}$ and the energy equation, we have $\delta q = de + p\,dv = dh - v\,dp$. Hence $dh = v\,dp = C\,dp/p^{1/n}$ for the adiabatic but frictional process. Then the following integration can be carried out:

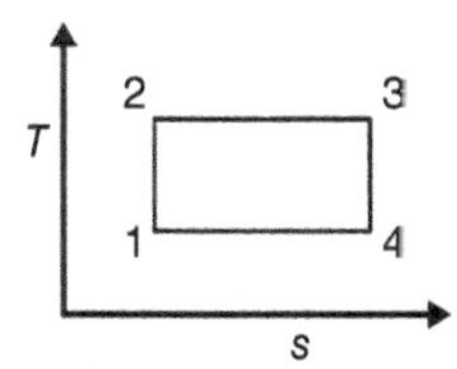

Carnot cycle.

$$H_{ad} \equiv h_2 - h_1 = \int_1^2 dh = \int_1^2 v\,dp = C\int_1^2 \frac{dp}{p^{1/n}} = C\left.\frac{p^{(-1/n+1)}}{-1/n+1}\right|_1^2$$
$$= \frac{Cn}{n-1}\left[p_2^{(n-1)/n} - p_1^{(n-1)/n}\right] = \frac{Cn}{n-1}p_1^{(n-1)/n}\left[\left(\frac{p_2}{p_1}\right)^{(n-1)/n} - 1\right]$$
$$= \frac{n}{n-1}p_1^{1/n}v_1 p_1^{(n-1)/n}\left[\left(\frac{p_2}{p_1}\right)^{(n-1)/n} - 1\right]$$
$$= \frac{n}{n-1}p_1 v_1\left[\left(\frac{p_2}{p_1}\right)^{(n-1)/n} - 1\right] = \frac{n}{n-1}RT_1\left[\left(\frac{p_2}{p_1}\right)^{(n-1)/n} - 1\right]. \tag{4.2}$$

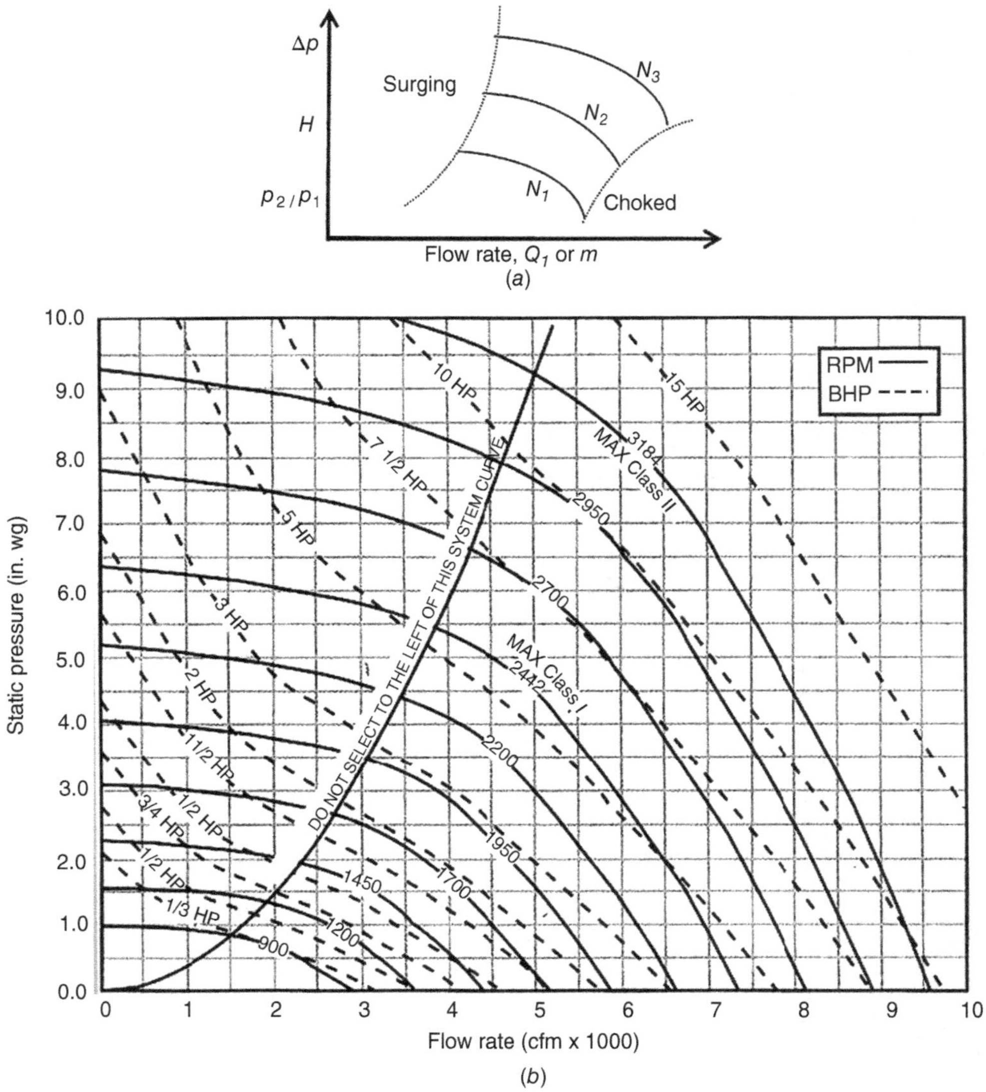

Figure 4.6 (*a*) Typical performance curves for a fan, blower, or compressor. (*b*) Performance curves of a centrifugal fan published by the manufacturer. (Reproduced from Technical Data Tubular Centrifugal, Industrial Air Products, Inc., Philips, WI/Air Products, Inc.)

Hence, the adiabatic efficiency can be expressed in the form

$$\eta_{\text{ad}} = \frac{H_i}{H_{\text{ad}}} = \frac{k(n-1)}{n(k-1)}\left(\frac{(p_2/p_1)^{(k-1)/k} - 1}{(p_2/p_1)^{(n-1)/n} - 1}\right), \tag{4.3}$$

where n is the polytropic exponent if the actual process is related to pv^n = const.

If we use the Euler equation as shown in Equation (2.36), with $U_1V_{u1} = 0$ for the design condition and include the slip factor μ_s at the impeller outlet as defined for centrifugal pumps earlier, we obtain

$$H_{\text{ad}} = \Delta E_p = h_2 - h_1 = \mu_s U_2 V_{u2} = \mu_s U_2\left(U_2 - \frac{Q_2}{A_2 \tan \beta_{b2}}\right).$$

Table 4.1 Sample Performance Table for a Centrifugal Fan.

		1/4sp		3/8sp		1/2sp		5/8sp		3/4sp		7/8sp		1 sp		1–1/4sp		1–1/2sp	
CFM	VEL	RPM	BHP	RPM	BHP	RPM	BHP	RPM	BHP	RPM	BHP	RPM	BHP	RPM	BHP	RPM	BHP	RPM	BHP
7448	800	245	0.46	275	0.62	304	0.80	331	0.99	360	1.20								
8379	900	262	0.55	287	0.73	314	0.91	340	1.12	365	1.34	389	1.57						
9310	1000	280	0.68	303	0.87	327	1.07	361	1.27	374	1.49	396	1.73	418	2.00				
10 241	1100	300	0.82	321	1.03	341	1.23	363	1.45	385	1.68	438	1.92	427	2.19	466	2.75		
11 172	1200	319	0.97	339	1.21	358	1.43	376	1.66	397	1.90	417	2.14	436	2.40	474	2.97	510	3.59
12 103	1300	339	1.15	358	1.40	376	1.65	393	1.89	410	2.15	429	2.40	448	2.67	483	3.22	518	3.87
13 034	1400	359	1.36	377	1.62	394	1.89	411	2.16	427	2.42	442	2.69	460	2.97	494	3.54	527	4.17
13 965	1500	380	1.58	396	1.86	413	2.16	428	2.43	444	2.72	459	3.01	473	3.30	506	3.89	537	4.51
14 896	1600	401	1.84	416	2.13	432	2.44	447	2.76	462	3.08	478	3.39	490	3.66	518	4.27	549	4.92
15 827	1700	423	2.13	436	2.43	451	2.75	486	3.09	479	3.41	493	3.72	507	4.05	632	4.69	551	5.37
16 758	1800	444	2.45	467	2.77	471	3.10	485	3.45	496	3.79	511	4.14	524	4.47	549	5.15	574	5.86
17 688	1900	468	2.80	479	3.16	491	3.49	505	3.86	517	4.20	530	4.59	542	4.96	557	5.67	589	6.38
18 620	2000	488	3.20	600	3.65	511	3.90	524	4.28	537	4.67	549	5.08	550	5.43	584	6.19	608	6.92
20 492	2200	532	4.01	543	4.47	554	4.87	584	5.26	576	5.68	587	6.08	599	6.54	620	7.39	641	8.18
22 344	2400	576	5.14	587	6.58	597	6.00	606	6.41	616	6.86	627	7.31	637	7.75	658	8.70	677	9.63
24 205	2600	621	6.39	631	6.85	640	7.30	649	7.76	658	8.23	657	8.71	677	9.19	695	10.18	715	11.18
26 068	2800	666	7.83	675	8.31	684	8.81	692	9.28	701	9.20	709	10.30	717	10.60	738	11.87	753	12.90
27 930	3000	711	9.45	719	9.97	728	10.53	736	11.05	744	11.58	752	12.12	759	12.63	775	13.72	792	14.85
29 792	3200	766	11.34	764	11.89	772	12.46	789	13.03	787	13.56	795	14.18	802	14.71	816	15.85	832	17.04
31 654	3400	801	13.44	809	14.05	816	14.81	824	15.24	831	15.82	838	16.41	845	17.02	859	16.25	872	19.45

Note: INTAKE AREA = 9.72 ft^2. WHEEL DIAMETER = $40\frac{1}{4}$

TIP SPEED FPM = 10.537 × RPM MAXIMUM BHP = $31.14\left[\frac{\text{RPM}}{1000}\right]^3$.

BHP does not include drive loss; underlines indicate maximum static efficiencies, VEL is inlet flow velocity in ft./min, and blank spaces are surging conditions.
Source: Reproduced from Cook Centrifugal Plug Fan, Loren Cook Co., Springfield, MO., 1989.

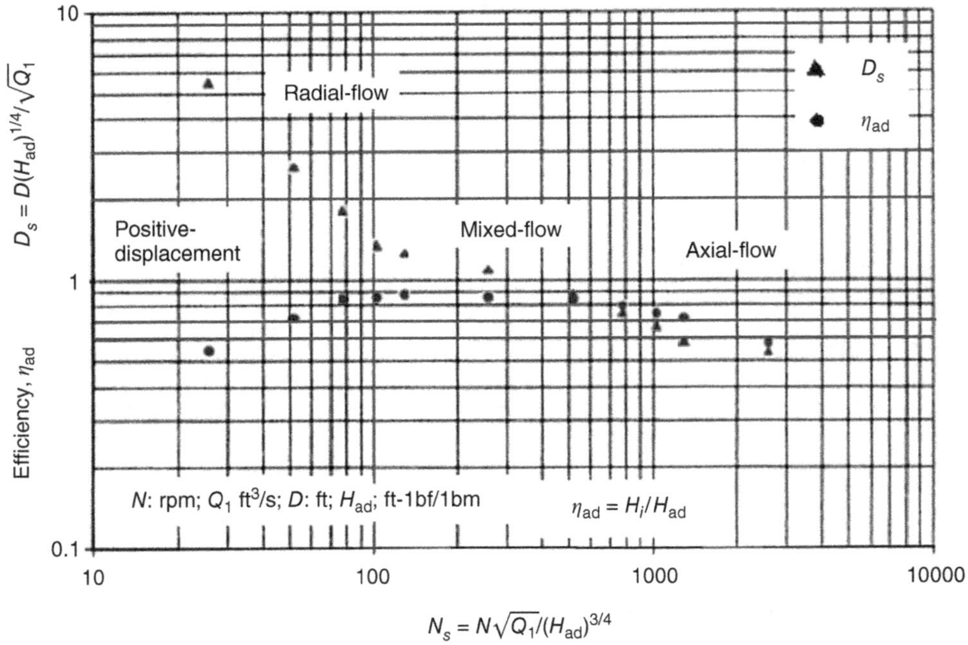

Figure 4.7 D_s and adiabatic efficiency versus N_s for compressors (Adapted from [5]).

The pressure ratio can then be related to the flow rate through the expression.

$$\eta_{ad} = \frac{C_p T_1 \left[(p_2/p_1)^{(k-1)/k} - 1\right]}{\mu_s U_2[U_2 - Q_2/(A_2 \tan \beta_{b2})]},$$

or

$$\frac{p_2}{p_1} = \left[1 + \frac{\eta_{ad}\mu_s}{C_p T_1} U_2 \left(U_2 - \frac{Q_2}{(A_2 \tan \beta_{b2}}\right)\right]^{k/(k-1)}. \tag{4.3A}$$

Since Q_2 varies with respect to the discharge pressure for a given mass flow rate, the performance curve is normally plotted in terms of the pressure ratio P_2/P_1 versus the mass flow rate m or the volumetric flow rate Q_1 at the inlet conditions of pressure and temperature, as shown in Figure 4.6*a*.

Example 4.1 Convert two sets of data with a constant flow rate in Table 4.1 into the performance curves of constant rotating speed at 300 rpm as those shown in Figure 4.6.

Solution

Pick the data rows for the flow rates of 10,241 and 13,965 cfm and list them in the first four columns of the table below. Convert the flow rate Q, static pressure SP, and brake horsepower BHP values for various rpm into 300 rpm based on the fans' laws. That is,

$$Q_1 = Q\left(\frac{300}{N}\right), \quad SP_1 = SP\left(\frac{300}{N}\right)^2, \quad BHP_1 = BHP\left(\frac{300}{N}\right)^3$$

and calculate the static efficiency using $\eta_s = Q\,SP/(6346\,BHP)$. The results are plotted in the chart, which shows a peak efficiency of 73.7% at 7195 cfm.

Q (cfm)	SP (in.wg)	N (rpm)	BHP	Q_1 (cfm)	SP_1 (in. wg)	BHP_1	η
10241	0.25	300	0.82	10241	0.25	0.82	0.492
10241	0.375	321	1.03	9600.94	0.32959	0.84869	0.58754
10241	0.5	341	1.23	9009.68	0.38699	0.83754	0.65601
10241	0.625	363	1.45	8463.64	0.42688	0.81849	0.69559
10241	0.75	385	1.68	7980	0.45539	0.79486	0.72043
10241	1	427	2.19	7195.08	0.49361	0.75949	0.73688
10241	1.25	466	2.75	6592.92	0.51806	0.73373	0.73353
13965	0.25	380	1.58	11025	0.15582	0.77745	0.3482
13965	0.375	396	1.86	10526.4	0.21306	0.79658	0.44367
13965	0.5	413	2.16	10144.1	0.26382	0.82788	0.5094
13965	0.625	428	2.43	9788.55	0.30707	0.83683	0.566
13965	0.75	444	2.72	9435.81	0.3424	0.83904	0.60678
13965	1	473	3.3	8857.29	0.40227	0.84197	0.66685
13965	1.25	506	3.89	8279.64	0.43939	0.8107	0.70713
13965	1.5	537	4.51	7801.68	0.46815	0.78635	0.73191

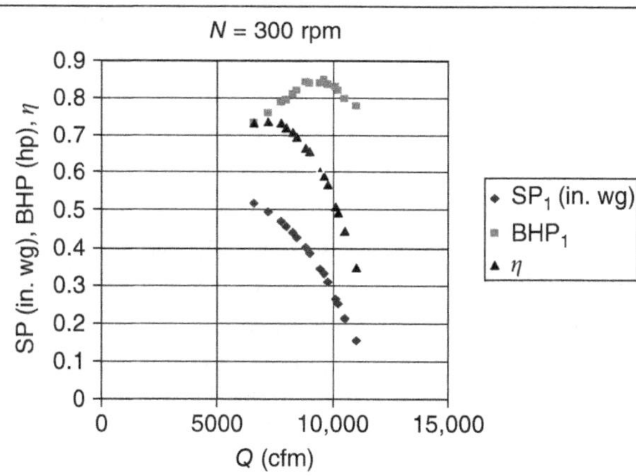

4.1.3 Change of Performance

Performance curves published by manufacturers are normally based on laboratory tests at a standard rotating speed and standard inlet conditions. However, some applications in the field require changes in the rotating speed or the inlet condition to meet the optimal system conditions. Occasionally, slight trimming of the impeller is required to reduce the shaft power. The change in machine performance due to these changes can be estimated according to the similarity laws discussed earlier and will be studied in more detail in this section.

4.1.3.1 Change of Rotating Speed and Diameter

Similarity laws for the pump derived from dimensional analysis also apply to the fan, blower, and compressor. Hence, for the corresponding points of one machine or two similar machines operating at two different conditions,

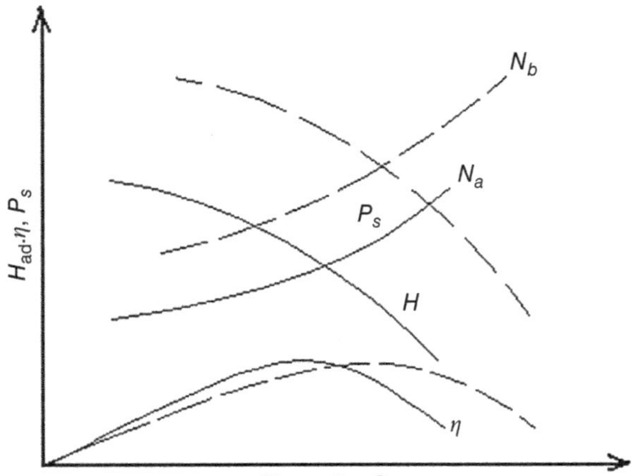

Figure 4.8 Change of centrifugal blower and compressor performance due to rotating speed change.

a and b, the parameters can be related as

$$\frac{Q_b}{Q_a} = \left(\frac{N_b}{N_a}\right)\left(\frac{D_b}{D_a}\right)^3, \quad \frac{H_b}{H_a} = \left(\frac{N_b}{N_a}\right)^2\left(\frac{D_b}{D_a}\right)^2,$$

$$\frac{P_{s,b}}{P_{s,a}} = \left(\frac{N_b}{N_a}\right)^3\left(\frac{D_b}{D_a}\right)^5, \quad \eta_b = \eta_a,$$

where the head can be either the ideal head or the actual head; since the efficiencies are the same at these two points, the effect of the change in rotating speed is shown in Figure 4.8. The head ratio rule can also be expressed in terms of the pressure ratio,

$$\frac{(p_2/p_1)_b^{(k-1)/k} - 1}{(p_2/p_1)_a^{(k-1)/k} - 1} = \left(\frac{N_b}{N_a}\right)^2\left(\frac{D_b}{D_a}\right)^2,$$

since the pressure ratio is the parameter normally used for compressors.

4.1.3.2 Inlet Condition Changes with Constant Rotating Speed

Sometimes, the machine inlet conditions of a system in the field differ from laboratory standard conditions. The change in performance characteristics is mainly related to the change in the fluid density at the inlet, ρ_1. Since the rotating speed is unchanged, curves of head and efficiency versus volumetric flow rate are unchanged. The shaft power and the mass flow rate will be changed according to

$$P_{s,b} = P_{s,a}\left(\frac{\rho_{1b}}{\rho_{1a}}\right), \quad m_b = m_a\left(\frac{\rho_{1b}}{\rho_{1a}}\right).$$

The change in fluid density at the inlet can be due to either of two cases:

1) Inlet Throttling (with constant stagnation temperature, but drop in stagnation pressure p_{01}). A throttling process occurs through a valve with significant friction loss so that part of the flow kinetic energy is converted to heat (or thermal energy). Since no heat or work is added to or removed from the fluid, the stagnation enthalpy is unchanged (changes in static enthalpy and temperature are also negligible). However, the stagnation and static pressure decrease with an increase in entropy, as shown in Figure 4.9. From $p = \rho RT$, ρ will decrease accordingly, and it results in a change of mass flow rate and shaft power such that

$$m_b = m_a\left(\frac{\rho_{1b}}{\rho_{1a}}\right) = m_a\left(\frac{p_{1b}}{p_{1a}}\right) \quad P_{s,b} = P_{s,a}\left(\frac{p_{1b}}{p_{1a}}\right).$$

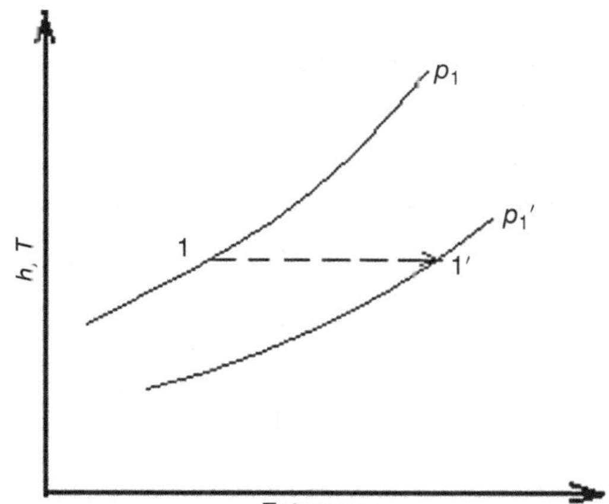

Figure 4.9 Thermodynamic diagram for a throttling process.

However, the pressure ratio is unchanged since the head and inlet temperatures are constant. Similarly, the efficiency at the corresponding volumetric flow rate is unchanged.

2) Inlet Temperature Change (pressure and rotating speed are unchanged). With $p = \rho RT$, a change in the inlet temperature will cause the density to change inversely if pressure is kept constant. Therefore, the mass flow rate and shaft power will change accordingly such that $m_b = m_a(T_{1a}/T_{1b})$ and $P_{s,b} = P_{s,a}(T_{1a}/T_{1b})$. The head and efficiency remain constant. But the pressure ratio will change, since T_1 changes, in the expression $H_i = [kRT_1/(k-1)]\,[(p_2/p_1)^{(k-1)/k} - 1]$. It can be related as

$$\left(\frac{p_2}{p_1}\right)_b^{(k-1)/k} - 1 = \left(\frac{T_{1a}}{T_{1b}}\right)\left[\left(\frac{p_2}{p_1}\right)_a^{(k-1)/k} - 1\right].$$

The inlet volumetric flow rates Q_a and Q_b mentioned are measured. If the flow rates at the new and original conditions expressed in terms of the standard condition, Q'_a, Q'_b, are needed, they have to be further related as $\rho_o Q'_a = \rho_{1a} Q_a$, $\rho_o Q'_b = \rho_{1b} Q_b$, where ρ_o is the density of the gas at the standard condition.

4.1.3.3 Summary

The relationships for correcting the performance curves due to changes of N, D, p_1, and T_1 may be summarized as follows:

$$\text{Volumetric flow rate:}\quad Q_b = Q_a\left(\frac{N_b}{N_a}\right)\left(\frac{D_b}{D_a}\right)^3, \tag{4.4}$$

$$\text{Mass flow rate:}\quad m_b = m_a\left(\frac{N_b}{N_a}\right)\left(\frac{D_b}{D_a}\right)^3\left(\frac{p_{1b}}{p_{1a}}\right)\left(\frac{T_{1a}}{T_{1b}}\right), \tag{4.5}$$

$$\text{Head:}\quad H_b = H_a\left(\frac{N_b}{N_a}\right)^2\left(\frac{D_b}{D_a}\right)^2, \tag{4.6}$$

$$\text{Pressure ratio:}\quad \left(\frac{p_2}{p_1}\right)_b^{(k-1)/k} - 1 = \left[\left(\frac{p_2}{p_1}\right)_a^{(k-1)/k} - 1\right]\left(\frac{N_b}{N_a}\right)^2\left(\frac{D_b}{D_a}\right)^2\left(\frac{T_{1a}}{T_{1b}}\right), \tag{4.7}$$

$$\text{Efficiency:}\quad \eta_b = \eta_a, \tag{4.8}$$

$$\text{Shaft power:}\quad P_{s,b} = P_{s,a}\left(\frac{N_b}{N_a}\right)^3\left(\frac{D_b}{D_a}\right)^5\left(\frac{p_{1b}}{p_{1a}}\right)\left(\frac{T_{1a}}{T_{1b}}\right). \tag{4.9}$$

Example 4.2 A catalog shows that a centrifugal blower operating in the ambient condition of 58 °F and 14.7 psia at 2400 rpm will deliver 1800 cfm at 8.5 psig with an efficiency of 70%. It is to be operated in the field at 3600 rpm with the ambient condition ranging from 50° to 95 °F and a constant pressure of 14.7 psia. Determine the expected flow rate, range of discharge pressure, and required shaft power. Also determine the flow rates if it is measured in the standard condition.

Solution

Assume as given $N_a = 2400$ rpm, $T_{1a} = 68\,°\text{F} = 528$ R, $p_1 = 14.7$ psia, $p_{2a} = 8.5$ psig $= 23.2$ psia, $Q_a = 1800$ cfm, $\eta = 0.70$, $N_b = 3600$ rpm, and $T_{1b} = 50$ to $95\,°$F. To find Q_b, p_{2b}, P_{sb}, and Q'_b. From

$$Q_b = Q_a\left(\frac{N_b}{N_a}\right) = 1800\left(\frac{3600}{2400}\right) = 2700\ \text{cfm},$$

$$\left(\frac{p_2}{p_1}\right)_b^{(k-1)/k} - 1 = \left(\frac{T_{1a}}{T_{1b}}\right)\left[\left(\frac{p_2}{p_1}\right)_a^{(k-1)/k} - 1\right]\left(\frac{N_b}{N_a}\right)^2$$

$$= \left(\frac{T_{1a}}{T_{1b}}\right)\left[\left(\frac{23.2}{14.7}\right)^{0.2857} - 1\right]\left(\frac{3600}{2400}\right)^2 = 0.313\left(\frac{T_{1a}}{T_{1b}}\right),$$

we obtain $p_{2b} = 14.7[1 + 0.313\,(T_{1a}/T_{1b})]^{3.5} = 36.6$ to 39.3 psia. Also,

$$P_{sb} = \frac{\rho_{1b}Q_bH_i}{\eta} = \frac{(p_{1b}/RT_{1b})Q_bC_pT_{1b}\left[(p_2/p_1)_b^{(k-1)/k} - 1\right]}{\eta}$$

$$= \frac{[k/(k-1)]p_{1b}Q_b \times 0.313(T_{1a}/T_{1b})}{0.7}$$

$$= \frac{[3.5 \times 14.7 \times 144 \times 2700/(550 \times 60)] \times 0.313(T_{1a}/T_{1b})}{0.7}$$

$$= 271\left(\frac{T_{1a}}{T_{1b}}\right) = 257.8\ \text{to}\ 280.6\ \text{hp}.$$

From $\rho_o Q' = \rho_1 Q$, we have the flow rate measured at the standard condition, $Q'_b = (\rho_{1b}/\rho_o)Q_b = (T_o/T_{1b})Q_b = 2795$ to 2568 cfm.

Example 4.3 An airplane compressor rotating at 25,000 rpm with an efficiency of 70% and discharge pressure of 34.5 in. Hg is required to supply 180 lb_m/min of air at high altitudes, where the ambient air is −10 °F and 13.7 in. Hg. If it is tested at sea level with the ambient air at 75 °F and 30 in. Hg and a rotating speed of 12,000 rpm, determine the inlet volumetric flow rate, discharge pressure, and required shaft power.

Solution

With

$$\rho_a = \frac{p}{RT} = \frac{13.7 \times 0.491 \times 144}{53.33 \times (-10 + 460)} = 0.0403\ \text{lb}_\text{m}/\text{ft}^3,$$

$$Q_b = Q\left(\frac{N_b}{N_a}\right) = \left(\frac{180}{0.0403}\right)\left(\frac{12{,}000}{25{,}000}\right) = 2144\ \text{cfm},$$

$$H_b = H_a\left(\frac{N_b}{N_a}\right)^2,$$

we obtain

$$\left(\frac{p_2}{p_1}\right)_b^{(k-1)/k} = \left[\left(\frac{p_2}{p_1}\right)_a^{(k-1)/k} - 1\right]\left(\frac{N_b}{N_a}\right)^2\left(\frac{T_{1a}}{T_{1b}}\right) + 1$$

$$= \left[\left(\frac{34.5}{13.7}\right)^{0.2857} - 1\right]\left(\frac{12{,}000}{25{,}000}\right)^2\left(\frac{450}{535}\right) + 1 = 1.0585$$

So

$$p_{2b} = (1.0585)^{3.5}(p_{1b}) = 1.22 \times 30 = 36.6 \text{ in.Hg},$$

$$H_{ib} = \frac{kRT_{1b}}{k-1}\left[\left(\frac{p_2}{p_1}\right)_b^{k-1/k} - 1\right]$$

$$= \left(\frac{1.4 \times 53.33 \times 535}{0.4}\right)\left[\left(\frac{36.6}{30}\right)^{0.2857} - 1\right] = 5830 \text{ ft} - \text{lb}_f/\text{lb}_m,$$

$$m_b = \rho_b Q_b = \left(\frac{30 \times 0.491 \times 144}{53.33 \times 535}\right) \times 2144 = 159.4 \text{ lb}_m/min.$$

Thus

$$P_{sb} = \frac{m_b H_{ib}}{\eta} = \frac{159.4 \times 5830}{60 \times 550 \times 0.70} = 40.2 \text{ hp}.$$

Example 4.4 A blower is designed to compress 1000 cfm of carbon dioxide at 90 °F and 17.7 to 19.7 psia when operating at 4000 rpm. It is to be tested with air at 14.7 psia and 70 °F and driven by a motor running at 3550 rpm. Determine the flow and discharge pressure that the machine should deliver at the design point on the test to be acceptable.

Solution

With $Q_a = Q_c(N_a/N_c) = 1000 \times (3550/4000) = 887.5$ cfm, $H = C_p T_1[(p_2/p_1)^{(k-1)/k} - 1]$ and $H_a/H_c = (N_a/N_c)^2$, where $k_a = 1.40$, $C_{pa} = 0.24$ Btu/(lb_m °F), $T_{1a} = 530$ R, $(k_a - 1)/k_a = 0.2857$ and $k_c = 1.30$, $C_{pc} = 0.20$ Btu/(lb_m °F), $T_{1c} = 550$ R, $(k_c - 1)/k_c = 0.2307$, we have

$$\frac{C_{pa}T_{1a}\left[(p_2/p_1)^{(k-1)/k} - 1\right]_a}{C_{pc}T_{1c}\left[(p_2/p_1)^{(k-1)/k} - 1\right]_c} = \left(\frac{3550}{4000}\right)^2 = 0.787,$$

or

$$\frac{0.24 \times 530\left[(p_2/p_1)_a^{0.2857} - 1\right]}{0.20 \times 550[(19.7/17.7)^{0.2307} - 1]} = 0.787.$$

Hence

$$\left(\frac{p_2}{p_1}\right)_a^{0.2875} = 1 + \frac{0.787 \times 0.025}{1.156} = 1.017, \quad p_{2a} = 15.6 \text{ psia}.$$

4.1.4 Polytropic Efficiency

The adiabatic efficiency for compressors defined in Equation (4.3) and illustrated in Figure 4.7 does not truly reflect the characteristics of the machine itself. Because the constant pressure lines in an h–s diagram are not parallel, the same machine tested with a different inlet temperature/enthalpy (even at the same pressure) may result in different adiabatic efficiency. Another compressor efficiency, called polytropic efficiency, is defined to reflect its true characteristics. It is defined through a tiny amount of compression such that $\eta_p = dh_i/dh$. The polytropic efficiency is normally greater than the adiabatic efficiency, depending on the degree of divergence of the constant pressure lines in the h–s diagram. Their relationship can be derived in analytical form for compressing an ideal gas.

From $T\,ds = dh - dp/\rho$, we have $dh_i = dp/\rho$. Hence, the polytropic efficiency can be expressed as

$$\eta_p = \frac{dp/\rho}{C_p\,dT} = \frac{RT}{C_p p}\frac{dp}{dT}.$$

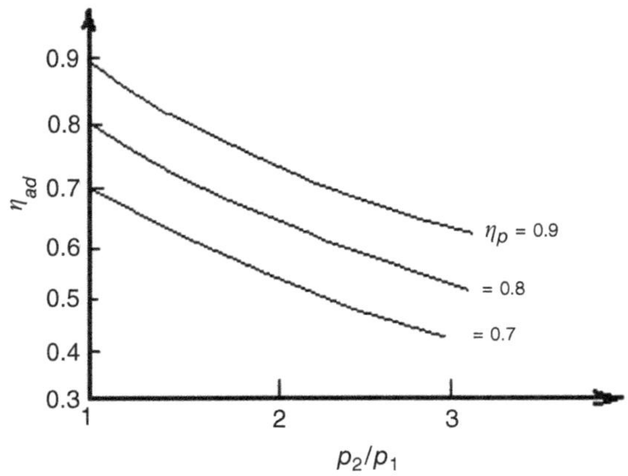

Figure 4.10 Relationship between adiabatic and polytropic efficiencies.

Since η_p reflects the true characteristics of the flow through the entire compressor, it is reasonable to assume η_p to be constant if the flow pattern does not change through the whole passage. The above equation can be rearranged and integrated from the compressor inlet to the outlet. That is,

$$\int_{T_1}^{T_2} \frac{dT}{T} = \frac{R}{C_p \eta_p} \int_{p1}^{p_2} \frac{dp}{p},$$

or it can be expressed in the form

$$\frac{T_2}{T_1} = \left(\frac{p_2}{p_1}\right)^{R/(C_p\eta_p)} = (p_2/p_1)^{(k-1)/(k\eta_p)}.$$

The adiabatic efficiency can, therefore, be expressed as

$$\eta_{ad} = \frac{\Delta h_s}{\Delta h} = \frac{T_{s2} - T_1}{T_2 - T_1} = \frac{T_{s2}/T_1 - 1}{T_2/T_1 - 1} = \frac{(p_2/p_1)^{(k-1)/k} - 1}{(p_2/p_1)^{(k-1)/(k\eta_p)} - 1} \tag{4.10}$$

This relationship can be plotted as shown in Figure 4.10. It is shown that η_{ad} is always lower than η_p, as mentioned earlier. They become equal at the limit of $p_2/p_1 = 1$.

When the compressed fluid is steam or other nonideal gases, the relationship can be illustrated as follows. From Figure 4.11, the polytropic efficiency is given as

$$\eta_p = \frac{\Delta h_i}{\Delta h} = \frac{h_{sx} - h_1}{h_x - h_1} = \frac{h_{sy} - h_x}{h_y - h_x} = \cdots,$$

or

$$\eta_p(h_x - h_1) = h_{sx} - h_1, \quad \eta_p(h_y - h_x) = h_{sy} - h_x, \cdots.$$

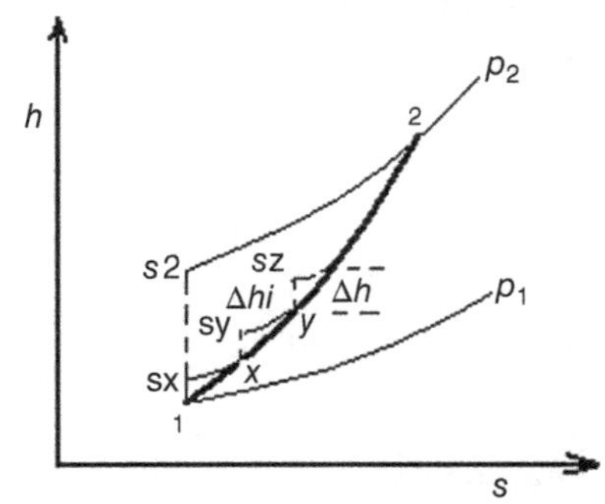

Figure 4.11 h–s Diagram to define polytropic efficiency.

Adding up these expressions, we obtain $\eta_p(h_2 - h_1) = (h_{sx} - h_1) + \cdots$, or

$$\eta_p = \frac{(h_{sx} - h_1) + (h_{sy} - h_x) + (h_{sz} - h_y) + \cdots}{h_2 - h_1}. \tag{4.11}$$

Comparing this with the adiabatic efficiency $\eta_{ad} = \dfrac{(h_{s2} - h_1)}{(h_2 - h_1)}$, it can be concluded that $\eta_{ad} < \eta_p$ since the constant-pressure lines on the h–s diagram are divergent in the direction of increasing entropy, as shown. Comparing Equation (4.10) with Equation (4.3), the polytropic efficiency can also be expressed in terms of the specific heat ratio k and the polytropic exponent n in the form $\eta_p \cong [(k-1)/k]/[(n-1)/n]$, where n can be determined from test or empirically.

Example 4.5 Two compressors are compared for bidding. Compressor A is submitted with a total pressure ratio of 4.5 and adiabatic efficiency of 85%. Compressor B is submitted with a discharge static pressure of 90 psia, temperature of 480 °F, and velocity of 500 ft./s as tested at the ambient condition of 14.7 psia and 60 °F. Which compressor is actually more efficient?

Solution

From po2b = 90 + 0.5 × 0.00237 × 5002/144 = 94.1 psia and To2/To1 = (480 + 460)/(60 + 460) = (94.1/14.7)0.2857/ηp = 1.807, or (0.2857/ηp) ln (94.1/14.7) = ln(1.807), we obtain the polytropic efficiency for compressor B, ηpb = 0.896.

From Equation (4.10)

$$0.85 = \frac{(4.5)^{0.2857} - 1}{(4.5)^{0.2857/\eta_{pa}} - 1} = \frac{0.5368}{(4.5)^{0.2857/\eta_{pa}} - 1},$$

we have the polytropic efficiency for compressor A, $\eta_{pa} = 0.878$. Hence, compressor B is more efficient.

4.1.5 Preliminary Design of Centrifugal Fans

The centrifugal fans illustrated in Figures 3.38 and 3.57 can be classified into radial, backward-curved, and forward-curved bladed impellers with corresponding blade exit angles of 90°, less than 90°, and greater than 90° (concerning the peripheral direction). Their ideal head-flow rate curves are the same as those for the pumps shown in Figure 3.6*a*. Since the fans are normally designed to deliver gas for low head with a high flow rate, compared with blowers and compressors, centrifugal fans are characterized by a high inlet–outlet radii ratio and a large constant blade width.

The radial-bladed fan is simple and less expensive to construct, with about 6–10 blades. But its efficiency is lower. The backward-curved bladed fan has higher efficiency, with about 9–16 blades. However, the size is larger for the same head requirement. To improve efficiency further, airfoil section blades can be used.

Forward-curved blades, extensively used in the multi-vane or Sirocco fan, are shown in Figure 4.12. Its efficiency is lower than that of the backward-curved bladed fan, and there is a dip in the head at low flow rates, as shown in Figure 3.57. However, it has a smaller size and lower rotating speed (hence lower noise level) for the same performance requirement at the design point. The ideal curves in Figure 4.12 can explain this. U_2 is directly proportional to the square root of the ideal shutoff head from the Euler equation. The dip in head at the lower flow rate for the forward-curved bladed fan is due to the positive *H*–*Q* curve slope shown.

The circular arc blade is commonly used for its simplicity and lower construction costs. From the geometry shown in Figure 4.12*b*, with

$$R_o^2 = R_b^2 + r_2^2 - 2R_b r_2 \cos(\pi - \beta_2) = R_b^2 + r_1^2 - 2R_b r_1 \cos(\pi - \beta_1),$$

we have

$$R_b = \frac{r_2^2 - r_1^2}{2[r_2 \cos(\pi - \beta_2) - r_1 \cos(\pi - \beta_1)]}, \quad R_o = \left[R_b^2 + r_2^2 - 2R_b r_2 \cos(\pi - \beta_2)\right]^{0.5}.$$

Typical ranges of the parameters are $r_1/r_2 = 0.80$–0.95, blade width $b_1 = b_2 = 0.80$–1.2 of r_2, and blade number $Z_b = 24$–64.

4.1.5.1 Inlet Configuration

The radial-flow impeller is easy to construct, but the flow at the inlet must make a sharp 90° turn to enter the blades. In the fan design, the main concern is to prevent or minimize the flow separation there. Several suggestions have been made. Some suggest flow acceleration, and hence, the inlet dimensions can be related as

$$\zeta = \frac{A_e}{A_1} = \frac{\pi r_1^2 (1 - v^2)}{2\pi r_1 b_1}$$

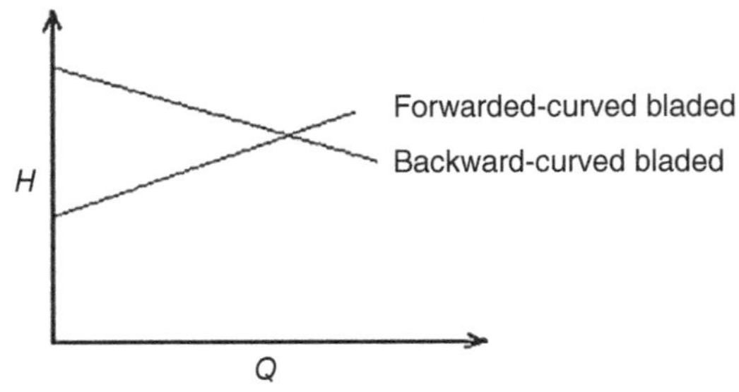

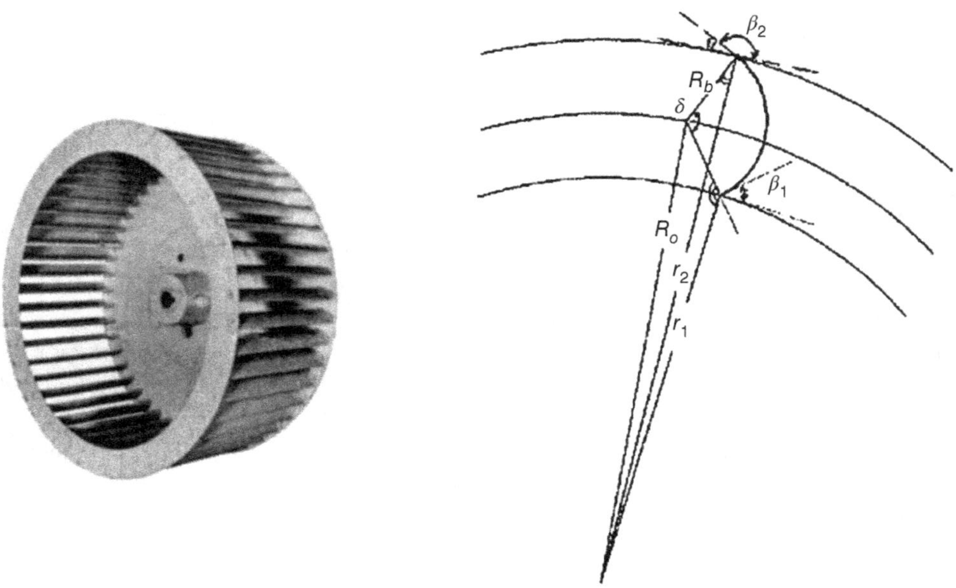

(*a*) (*b*)

Figure 4.12 Multivane impeller with forward-curved blades. (*a*) Photograph of the forward-curved bladed impeller. (*b*) Circular arc blade layout. (Courtesy of Fan Services, Inc., West Chicago, IL.)

or

$$b_1 = \frac{r_1(1 - \upsilon^2)}{2\zeta},$$

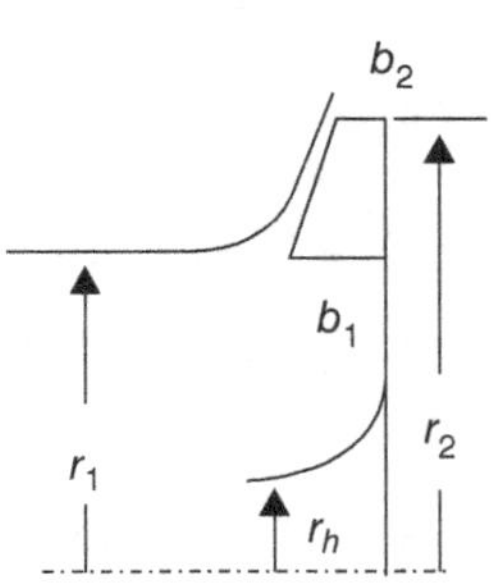

where $\upsilon = r_h/r_1$ and $\varsigma = 1.05$–1.2. Another suggestion is to keep the relative flow velocity at a minimum at the leading edge. From

$$W_1^2 = U_1^2 + V_{r1}^2, \quad V_{r1} = \frac{Q_1}{2\pi r_1 b_1} = \frac{Q_1\zeta}{\pi r_1^2(1 - \upsilon^2)},$$

we have

$$W_1^2 = \left[\frac{Q_1\zeta}{\pi r_1^2(1 - \upsilon^2)}\right]^2 + \omega^2 r_1^2.$$

To minimize W_1, we set

$$\frac{d\left(W_1^2\right)}{dr_1} = \frac{-4\zeta^2 Q_1^2}{\pi^2 r_1^5(1 - \upsilon^2)^2} + 2\omega^2 r_1 = 0.$$

Or we have

$$r_{\text{opt1}} = \left[\frac{\sqrt{2}\zeta Q_1}{\pi\omega(1-\upsilon^2)}\right]^{1/3},$$

$$\tan\beta_{f1} = \frac{V_{r1}}{U_1} = \frac{Q_1\zeta}{\pi r_1^2(1-\upsilon^2)r_1\omega} = \frac{1}{\sqrt{2}}.$$

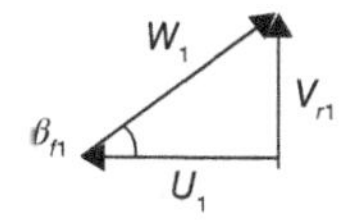

That is, $\beta_{f1} = 35.2°$.

4.1.5.2 Outlet Configuration

The outlet dimensions can be determined similarly to those for the centrifugal pump discussed in Section 3.1. Since the pressure rise across the fan is small, the density change is negligible, so the Euler equation is expressed as

$$\frac{\Delta p_t}{\rho} = \eta_{\text{imp}}\mu_s U_2\left(U_2 - \frac{Q}{A_2\tan\beta_{b2}}\right)$$

can be employed.

The outlet radius r_2 and the efficiency η_{imp} can be estimated from the specific speed N_s given in Figure 3.38. The slip coefficient μ_s can be estimated with the empirical formula available. The following is suggested by Eck[12]:

$$\mu_s = \left(1 + \frac{a + b\beta_{b2}^{\circ}/90}{Z_b\left[1-(r_1/r_2)^2\right]}\right)^{-1},$$

where $a = 0.9$, $b = 3.7$ for $20° < \beta_{b2} < 45°$ and $a = 1.5$, $b = 1.1$ for $45° < \beta_{b2} < 170°$. It is slightly different from the Stodola formula mentioned in Section 3.1. However, the qualitative effects due to blade number and blade angle in these two formulas are similar. The number of blades can be determined with $Z_b = 8.5\frac{\sin\beta_{b2}}{(1-r_1/r_2)}$.

With these parameters and a selected β_{b2}, the Euler equation can be used to determine the outlet area A_2 and, hence, the blade width b_2. Or b_2 can be selected first (e.g., $b_2 = b_1$); then β_{b2} is calculated, with some iterations. Some designers use solidity, $\sigma = Z_b L_b/(2\pi r_2)$, to judge the adequacy of the blade number. Here L_b is the blade length from the inlet to the outlet. The designers claim that σ should be between 2 and 3.

4.1.5.3 Blade Profile

The complete blade profile can be determined with the fixed inlet and outlet configurations. It can be determined with the procedure for the centrifugal pump impeller described in Section 3.1, that is, to specify the blade angle distribution from inlet to outlet. However, for the fan, the blades are normally fabricated with sheet metal forged into the desired shape rather than by machining or casting. Some simple shapes are employed. The simplest one is a straight blade, shown in Figure 4.13*a*. The blade angles are related to the law of sines as

$$\frac{r_1}{\sin(\pi/2-\beta_{b2})} = \frac{r_2}{\sin(\pi/2+\beta_{b1})}$$

or $r_2\cos\beta_{b2} = r_1\cos\beta_{b1}$. For a circular arc blade, the geometry is similar to that in Figure 4.12*b*. It is shown in Figure 4.13*b* again for the backward blade. So if one of these simple blade profiles is selected, β_{b2} has to be related to these equations. Then A_2 or b_2 can be determined from the Euler equation.

It is also instructive to mention the profile of an inactive blade here. From the Euler equation, $\Delta E = U_2V_{u2} - U_1V_{u1}$, there will be no energy transfer between the impeller blades and the fluid if the blade profile is such that $rV_u = r_1V_{u1} = C_1$. And from the continuity equation, we have $rV_r = r_1V_{r1} = C_2$ if the blade width is constant. Hence we have

$$\tan\beta_b = \frac{dr}{r\,d\theta} = \frac{V_r}{U-V_u} = \frac{C_2/r}{r\omega - C_1/r}$$

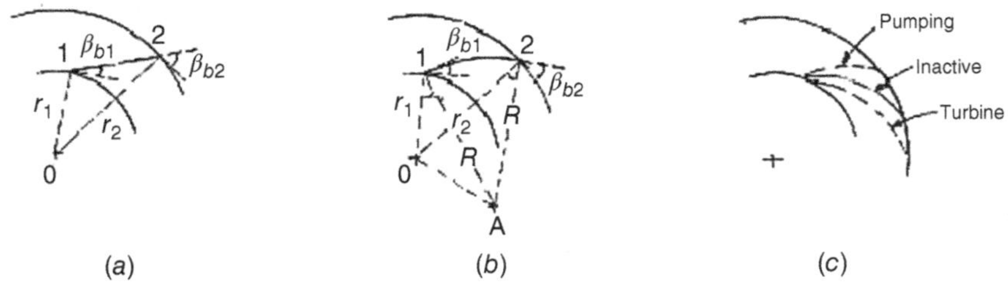

Figure 4.13 Radial-flow impeller blade profiles. (*a*) Straight blade. (*b*) Circular arc blade. (*c*) Profile of inactive blade.

or

$$d\theta = \frac{dr}{r}\frac{(r\omega - C_1/r)}{C_2/r} = \frac{1}{C_2}\left(r\omega - \frac{C_1}{r}\right)dr$$

or

$$\theta = \frac{1}{C_2}\left[\frac{\omega}{2}(r^2 - r_1^2) - C_1 \ln\left(\frac{r}{r_1}\right)\right].$$

If there is no Prewhirl, $C_1 = 0$, and we have

$$\theta = \frac{\omega}{2V_{r1}r_1}r_1^2\left[\left(\frac{r}{r_1}\right)^2 - 1\right] = \frac{U_1}{2V_{r1}}\left[\left(\frac{r}{r_1}\right)^2 - 1\right]$$
$$= \frac{1}{2\tan\beta_{f1}}\left[\left(\frac{r}{r_1}\right)^2 - 1\right]. \tag{4.12}$$

The profile is plotted as shown in Figure 4.13*c*. Beyond this, the machine becomes a turbine. This, of course, is an inefficient turbine because it is radial outflow.

4.1.5.4 Cross-Flow Fan

A special type of centrifugal fan, called a crossflow fan, should be mentioned briefly. It is characterized by its compactness for a high flow rate despite low-pressure rise and efficiency. Its application is found in electronic cooling, portable heaters, air curtains, and so on.

A typical arrangement is shown in Figure 4.14, along with velocity diagrams at four points. The impeller is similar to the multi-vane forward-curved bladed fan. But the flow passes through the impeller twice. It is basically a two-stage fan with one impeller. Flow inside the impeller is close to free-vortex flow, and hence we have $V_2 = V_3$, $\alpha_2 = \alpha_3$, and $V_4 > V_1$, as illustrated in the velocity diagrams. The desired flow rate can be achieved with any finite length of the impeller, and a higher-pressure rise can also be achieved with a multistage arrangement mounted on one shaft, as shown in Figure 4.15. A more detailed discussion of this type of fan can be found in Refs. 5, 12

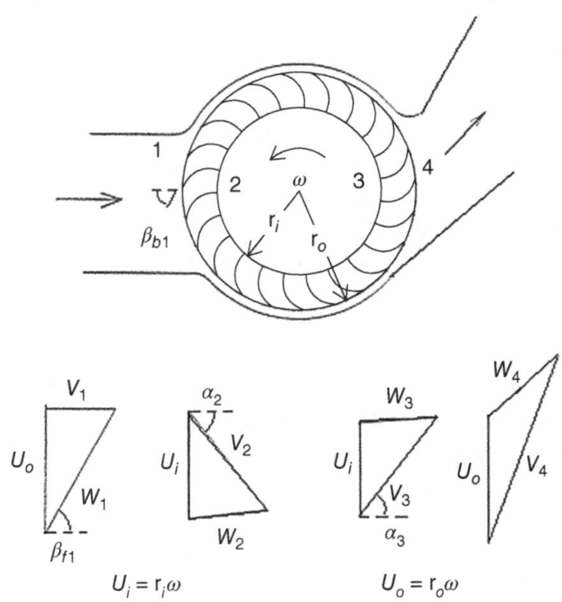

Figure 4.14 Cross-flow fan.

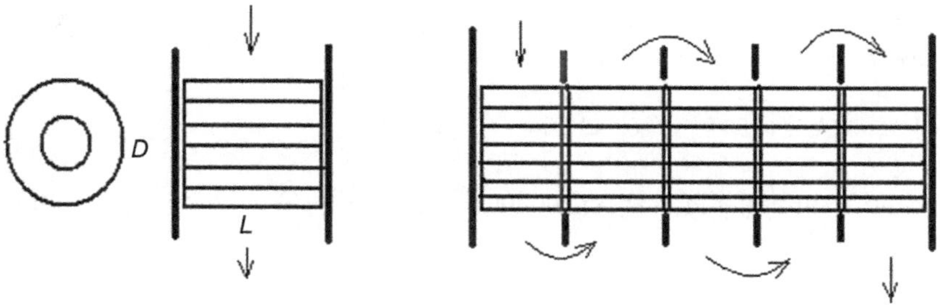

Figure 4.15 Cross-flow fans of wide impeller for higher flow rate and multistage for higher pressure rise.

Example 4.6 Design an impeller for a centrifugal fan rotating at 1800 rpm to deliver air of 14,000 cfm with a discharge static pressure of 5 in. wg.

Solution

The specifications are $N = 1800$ rpm $= 188.5$ rad/s, $Q = 14{,}000$ cfm $= 233.3$ cfs, and sp. = 5 in. wg. The specific speed is calculated as

$$N_s = \frac{N\sqrt{Q}}{(\text{sp})^{0.75}} = \frac{1800\sqrt{14,000}}{5^{0.75}} = 63,700\frac{\text{rpm}\sqrt{\text{cfm}}}{(\text{in.H}_2\text{O})^{0.75}}.$$

From Figure 3.38, selecting the airfoil blades, we have $\eta_s = 0.80$, and

$$D_s = \frac{d_2(\text{sp})^{0.25}}{Q^{0.5}} = 0.33.$$

Hence

$$P_s = \frac{Q \cdot \text{sp}}{\eta_s \times 6356} = \frac{14,000 \times 5}{0.8 \times 6356} = 13.7 \text{ hp},$$

$$d_2 = \frac{0.33 \times (14,000)^{0.5}}{5^{0.25}} = 26 \text{ in.},$$

and the shaft torque $\tau = \frac{13.7\times550}{(1800\pi/30)} = 40$ lb$_f$-ft. The shaft diameter can be determined as

$$d_s = \left(\frac{16\tau}{\pi s_s}\right)^{1/3} = \left(\frac{16 \times 40 \times 12}{\pi \times 10,000}\right)^{1/3} = 0.63 \text{ in.},$$

where $s_s = 10{,}000$ psi is the assumed maximum allowable shear stress of the shaft material. Hence d_s is set as 0.75 in. and $d_h = 1.0$ in. or $r_h = 0.5$ in. Using the minimum W_1 criteria, we have $\beta_{f1} = 35.2°$ and

$$r_1 = \left[\frac{\sqrt{2}\zeta Q}{\pi\omega(1-\upsilon^2)}\right]^{1/3} = 0.824 \text{ ft} = 9.88 \text{ in.} \cong 10 \text{ in.},$$

where $\upsilon = 0.05$ and $\zeta = 1$. Also, we have $b_1 = r_1(1-\upsilon^2)(2\zeta) = 5''$ and $\beta_{b1} = \beta_{f1} + i = 37°$. To determine the outlet dimensions, first we have $U_2 = \omega\, r_2 = 188.5 \times 13/12 = 204.2$ ft./s. For a straight blade, we will have $\beta'_{b2} = 52°$, so try $\beta_{b2} = 45°$, then

$$Z_b = 8.5\frac{\sin\,\beta_{b2}}{1 - r_1/r_2} = 26, \quad \mu_s = \left(1 + \frac{0.9 + 3.7\beta_{b2}/90}{Z_b\left[1-(r_1/r_2)^2\right]}\right)^{-1} = 0.794.$$

Since the static pressure is specified but the total pressure is needed in the Euler equation, the discharge flow velocity is needed.

Setting $b_2 = b_1 = 5$ in., we have $A_2 = 2\pi r_2 b_2 = 2\pi \times 13 \times 5/144 = 2.83\ \text{ft}^2$. The fan discharge area A_3 is set as $A_3 = 1.5 \times A_2 = 4.25\ \text{ft}^2$. Hence we have $V_3 = Q/A_3 = 54.8$ ft./s and

$$\text{tp} = \text{sp} + \frac{1}{2}\rho V_3^2 = \frac{62.4 \times 5}{12} + \frac{1}{2} \times 0.00237 \times 54.8^2 = 26 + 3.56 = 29.5\ \text{lb}_\text{f}/\text{ft}^2.$$

Also from $\eta_t = \eta_s(\text{tp/sp}) = \eta_{\text{imp}}\ \eta_v\ \eta_m$ and assuming $\eta_v\ \eta_m = 0.95$, we have $\eta_{\text{imp}} = \frac{0.8 \times 29.5/26}{0.95} = 0.955$. So from the Euler equation, we have

$$\frac{29.5}{0.00237} = 0.955 \times 0.794 \times 204.2 \left(204.2 - \frac{233.3}{2.83 \tan \beta_{b2}}\right) = 154.8 \left(204.2 - \frac{82.4}{\tan \beta_{b2}}\right),$$

or $\beta_{b2} = 33.6°$. Since Z_b and μ_s are related to β_{b2}, we must repeat the calculations. The new values $Z_b = 20$, $\mu_s = 0.781$, and $\beta_{b2} = 33.9°$ are obtained. Hence the final values are $Z_b = 20$ and $\beta_{b2} = 33.9°$.

It will be interesting to check the blade angle for the inactive blade at this point. From Equation (4.12) we have

$$\frac{d\theta}{dr} = \frac{1}{2 \tan \beta_{f1}} \left[2\left(\frac{r}{r_1}\right)\frac{1}{r_1}\right].$$

Hence

$$\tan \beta'_{b2} = \left[\frac{dr}{r\, d\theta}\right]_{r_2} = 2\frac{\tan \beta_{f1}}{r_2}\frac{r_1^2}{2r_2} = \tan \beta_{f1}\left(\frac{r_1}{r_2}\right)^2 = \tan 35.2°\left(\frac{10}{13}\right)^2 = 0.417,$$

and $\beta'_{b2} = 22.6°$, which is smaller than the β_{b2} we have as expected from Figure 4.13c.

4.1.6 Preliminary Design of Centrifugal Compressors

Comparing with the fan/blower, the centrifugal compressor rotates at a higher speed and delivers higher discharge pressure expressed in terms of total pressures ratio p_{03}/p_{01}. Since a significant change in gas density is involved in the compression, a thermodynamic process should be considered.

A typical high-performance centrifugal compressor with a Francis-type impeller, inducer section, and splitters is shown in Figure 4.16. The frontal, meridional, and isometric views with the T–s diagram of the compression process are shown. It is noticed that across the impeller energy is transferred to the gas, so we have $T_{02} > T_{01}$. But across the diffuser there is no energy being added to the gas, only some kinetic energy is converted to pressure rise, and hence we have $(T_{02} - T_2) > (T_{03} - T_3)$.

Given the specifications, including mass flow rate m, inlet total pressure and temperature, p_{01}, T_{01}, total pressure ratio p_{03}/p_{01}, and rotating speed N, the specific speed N_s defined earlier can be calculated. Then Figure 4.7 can be used to ascertain that a centrifugal-type compressor is appropriate. Also, the approximate efficiency and impeller diameter can be estimated.

4.1.6.1 Impeller Inlet Configuration

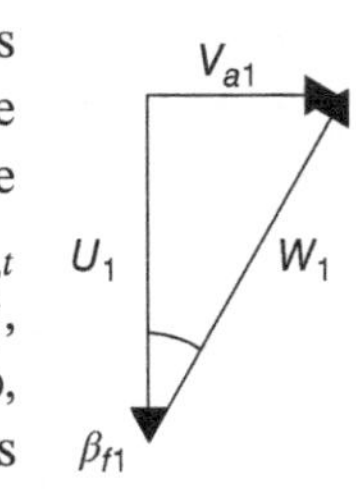

The inlet configuration is primarily determined by the relative Mach number, defined as $M_{r1,t} = W_{1,t}/a_1$, where $W_{1,t}$ is the relative flow velocity at the impeller inlet tip, and a_1 is the acoustic speed. Assuming uniform inlet flow velocity V_{a1}, from the velocity diagram, we have $M_1 = M_{r1,t} \sin \beta_{f1,t}$, where $\beta_{f1,t}$ is the relative flow angle at the inlet tip. Hence, with $M_{r1,t}$ and $\beta_{f1,t}$ selected, the static temperature at the inlet, T_1, can be calculated from $T_{01}/T_1 = 1 + (k-1)/2M_1^2$, and $a_1 = (kRT_1)^{0.5}$. Then W_1, V_1, and $U_{1,t}$ can be determined, and hence the tip radius r_{1t}. Also, from the mass equation $m = \rho_1 A_1\ V_{a1}$ and $\rho_1/\rho_{01} = (T_1/T_{01})^{1/(k-1)}$, we have the hub radius $r_{1h} = \left[r_{1t}^2 - m/(\rho_1 V_1 \pi)\right]^{1/2}$.

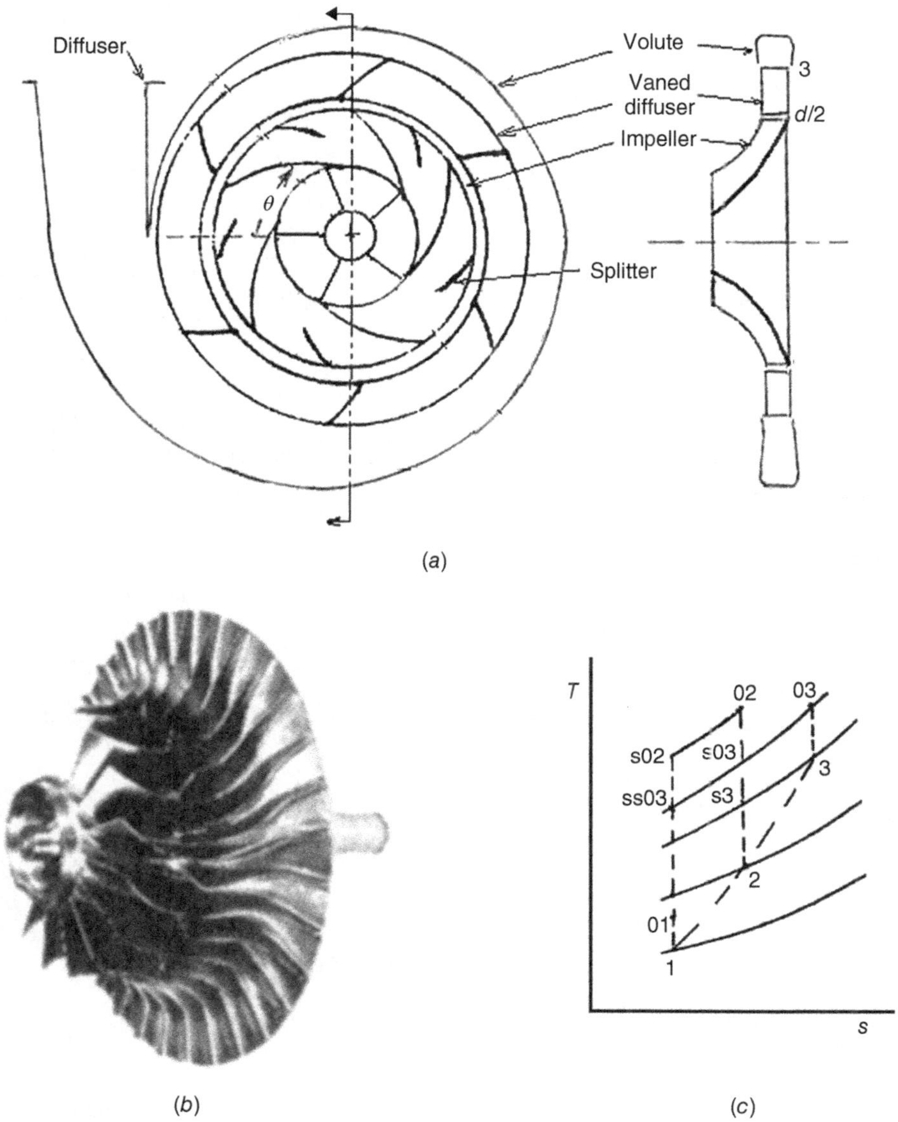

Figure 4.16 Centrifugal compressor with Francis-type impeller. (*a*) Schematic for a typical high-performance centrifugal compressor, (*b*) Francis-type impeller, and (*c*) T-S diagram of the compression process. [(*b*) Reproduced with permission from Morris, R. E. & Kenny, D. P., "High Pressure Ratio Centrifugal Compressors for Small Turbine Engines," in Advanced Centrifugal Compressor, ASME Gas Turbine Division, 1971, © the American Society of Mechanical Engineers.]

The process can be summarized in the following steps:

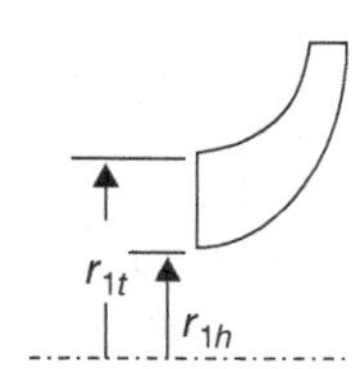

1) Select $M_{r1,t} = 0.7$ to 0.8 and $\beta_{f1,t} = 20$ to 30°.
2) Calculate M_1, T_1, and a_1, then ρ_1.
3) Calculate $W_{1,t} = M_{r1,t}\, a_1$, $V_{a1} = M_1 a_1$, and $U_{1,t} = V_{a1}/\tan\beta_{f1,t}$ then r_{1t}.
4) Calculate r_{1h} from $m = \rho_1 V_1 \pi\left(r_{1t}^2 - r_{1h}^2\right)$, then U_{1h} and $\beta_{f1,h}$.
5) Select $\beta_{b1} = \beta_{f1} + 2$ to 3° from hub to tip.

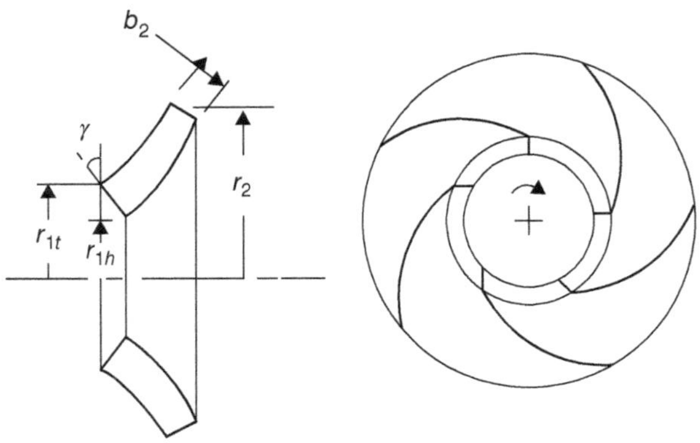

Figure 4.17 Inlet and outlet dimensions of a mixed-flow impeller.

For a mixed-flow impeller without an inducer, as shown in Figure 4.17, we have $A_1 = \pi(r_{1h} + r_{1t})b_1$, and the inclination angle γ_1 has to be specified such that $\cos\gamma_1 = (r_{1t} - r_{1h})/b_1$. The minimum shaft size is determined from the stress consideration, which is similar to that discussed in Chapter 3.

4.1.6.2 Impeller Discharge Configuration

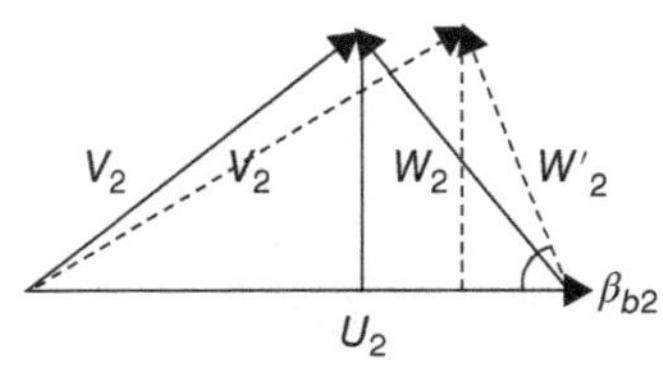

The discharge configuration is determined by the pressure ratio requirement at the design flow rate. From the Euler equation (neglecting inlet prewhirl, that is, $V_{u1} = 0$) with the energy equation and the definition of impeller efficiency, we have $-w/m = U_2V_{u2} = \mu_s U_2 V'_{u2} = h_{o2} - h_{o1}$ and $\eta_{imp} = (h_{so2} - h_{o1})/(h_{o2} - h_{o1})$, where μ_s is the slip factor at the impeller discharge. Also, we have $V'_{u2} = U_2 - W_{u2} = U_2 - Q_2/(A_2 \tan\beta_{b2})$, and

$$h_{so2} - h_{o1} = C_p(T_{so2} - T_{o1}) = C_pT_{o1}\left(\frac{T_{so2}}{T_{o1}} - 1\right)$$
$$= C_pT_{o1}\left[\left(\frac{p_{o2}}{p_{o1}}\right)^{(k-1)/k} - 1\right].$$

When these equations are combined, the pressure ratio can be expressed in the form

$$\frac{p_{o2}}{p_{o1}} = \left(1 + \frac{\eta_{imp}\mu_s U_2[U_2 - Q_2/(A_2\tan\beta_{b2})]}{C_pT_{o1}}\right)^{k/(k-1)}. \tag{4.13}$$

With a selected blade angle β_{b2} and η_{imp}, μ_s calculated from the empirical formula or chart, such as Figure 4.7 and the Stanitz formula, the discharge configuration can be determined in terms of r_2 and the blade width b_2. The blade profile from inlet to outlet is constructed using a procedure similar to that for the mixed-flow pump impeller, discussed in Chapter 3. The final design can be checked with the solidity and diffusion ratio requirements before the detailed design and analysis are conducted with computational fluid dynamics (CFD) software.

The procedure can be summarized in the following steps: (For convenience of reading, the velocity diagrams at the impeller inlet and outlet are depicted in Figure 4.18 again)

1) The given specifications are T_{o1}, p_{o1}, m, p_{o3}, N.
2) Calculate the ideal head, which is energy transferred to the gas per unit mass, $H_{ad} \equiv H_i = C_pT_{o1}[(p_{o3}/p_{o1})^{k-1/k} - 1]$ and inlet volumetric flow rate $Q_1 = m/\rho_1$.
3) Calculate the dimensional specific speed $N_s = N\sqrt{Q_1}/H_{ad}^{3/4}$, with the units given in Figure 4.7, based on radial-flow for $N_s = 30 \sim 80$, mixed flow for $N_s = 80 \sim 1000$, and axial flow for $N_s > 1000$ to ensure that the centrifugal-type impeller is appropriate as well as to determine the specific diameter D_s and overall efficiency η_c.

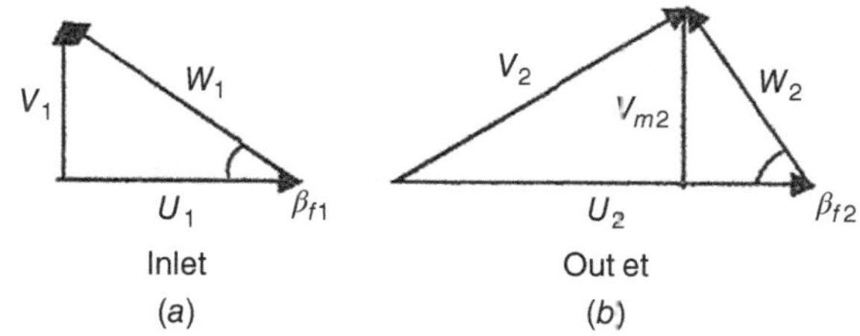

Figure 4.18 Velocity diagrams at the inlet and outlet of a mixed-flow impeller.

4) Calculate the outlet diameter D_2 from D_s, then the peripheral velocity U_2 from $U_2 = r_2\omega$.
5) Use the isentropic equation $T_{sso3}/T_{o1} = (p_{o3}/p_{o1})^{k-1/k}$ and the definition of efficiency, $\eta_c = (T_{sso3} - T_{o1})/(T_{o3} - T_{o1})$, to determine T_{o2}, which is equal to T_{o3}.
6) The mechanical efficiency η_m to account for power losses due to the bearing, seal, and disk friction is defined as $\eta_m = U_2V_{u2}[C_p(T_{o2} - T_{o1})]$, where the numerator is the impeller power input, and the denominator is the total power input, including those due to friction, assuming no heat transfer; typical mechanical efficiency is around 90 ~ 95%, so V_{u2} can be determined.
7) The flow coefficient $\phi = V_{m2}/U_2$ is selected around 0.2 ~ 0.4, so V_{m2} and $W_2 = \left[V^2_{r/m2} + (U_2 - V_{u2})^2\right]^{0.5}$ can be calculated.
8) At this point, the diffusion ratio $D_f = W_1/W_2$ can be checked. It must be less than 1.9 to ensure no boundary layer separation on the blade surface; otherwise, a different flow coefficient should be selected at step 7, or step 4 needs to be repeated to select a different outlet diameter.
9) Estimate the impeller efficiency η_{imp} from the loss ratio $\chi = (1 - \eta_{imp})/(1 - \eta_c)$ by setting χ to be 0.6–0.7.
10) Calculate p_{o2} from the isentropic equation and the definition of impeller efficiency [Equation (4.13)].
11) Calculate V_2 from $V_2{}^2 = V_{u2}{}^2 + V_{r/m2}{}^2$ and T_2 from $C_p(T_{o2} - T_2) = V_2{}^2/2$.
12) Calculate p_2 and ρ_2 from $p_2/p_{o2} = (T_2/T_{o2})^{k/k-1}$ and $\rho_2 = p_2/RT_2$.
13) The slip coefficient formula by Stanitz, which is adequate for 45° < β_{b2} < 90°, can be rearranged as follows for convenience: $\mu_s = V_{u2}/V'_{u2} = 1\text{–}0.63\pi/[Z_b(1 - \phi_2 \cot \beta_{b2})] = 1 - (0.63\pi/Z_b)(U_2/V'_{u2})$, or $V'_{u2} = V_{u2} + 0.63\pi U_2/Z_b$, so V'_{u2} can be calculated by selecting the blade number Z_b, which can be 10–20 (the formulas for μ_s by Stodola and Eck are preferred by some designers).
14) Calculate the outlet blade height b_2 and blade angle β_{b2} from $\rho_2 V_{r/m2}\pi D_2 b_2 \varepsilon_2 = m$ and $\tan \beta_{b2} = V_{r/m2}/(U_2 - V'_{u2})$, where $\varepsilon_2 = 1 - (Z_b t/\sin \beta_{b2})/(2\pi r_2)$ is the flow area contraction factor due to the blade thickness t at the outlet. Some iterations are needed between steps 13 and 14.
15) Construct the blade profile from inlet to outlet.

4.1.6.3 Diffuser

The diffuser downstream of the impeller is employed to convert the excessive kinetic energy of gas into increased static pressure rise. In the vaned diffuser, the divergence angle of the vanes is normally limited to less than 12° to avoid possible boundary layer separation. The diffuser inlet configuration should match that of the impeller discharge.

The clearance between the impeller trailing edge and the vaned diffuser leading edge should be around 10–20% of the impeller radius to provide flow adjustment. This clearance can be considered as a vaneless diffuser. The flow there can be analyzed with the conservation equations of mass and angular momentum. Hence, we have $r_d V_{ud} = r_2 V_{u2}$ and $\rho_d r_d V_{md} = \rho_2 r_2 V_{m2}$, where the subscript m designates the meridional component, and subscript d designates the vaned diffuser leading edge. Incorporating the equations $C_p(T_{o2} - T_d) = (V_{md}{}^2 + V_{ud}{}^2)/2$, $p_d/p_{o2} = (T_d/T_{o2})^{k/k-1}$, and $p_d = \rho_d RT_d$, the parameters ρ_d, p_d, T_d, V_{ud}, and V_{md} can be solved iteratively. Hence the blade angle and throat passage area can be determined, that is, $\beta_d = \beta_{fd} = \tan^{-1}(V_{md}/V_{ud})$, $A_d = \varepsilon_d(2\pi r_d/Z_d) \sin \beta_d$. A contraction factor ε_d can be included.

The diffuser blade number Z_d should not have a common multiple with the impeller blade number to avoid vibration and resonance. The diffuser outlet dimension can be determined with a specified outlet flow velocity and the compressor discharge condition p_{o3} and T_{o3}. Adjustable diffuser vanes can be used to accommodate the variable flow rate and maintain stable operating conditions.

4.1.6.4 Volute

Downstream of the vaned diffuser, a volute is used to collect the gas and discharge it to the outlet, similar to that discussed in Chapter 3 for the centrifugal pump. This arrangement is especially common for industrial compressor applications. The volute cross-sectional area increases from the tongue, with $\theta = 0, \ldots, 2\pi$ linearly as $A_\theta = Q_3\theta r_c/(2\pi K)$, where $K = r_c V_u = r_3 V_{u3}$, and r_c is the centroid radius of the cross-sectional area. So as $\theta/2\pi$ increases from zero to one, A_θ increases from zero to the exit area $A_3 = Q_3/V_3$.

In some designs without a vaned diffuser, the entire diffusion process will take place in the volute and the discharge diffuser, resulting in a less efficient machine. However, the possible surging due to the flow mismatch between the impeller and vaned diffuser under the off-design condition can be avoided.

Another correlation of compressor efficiency with respect to the specific speed was published and is reproduced in Figure 4.19 for reference. These data are obtained from tests of impellers with a diffusion factor $W_1/W_2 = 1.4$, hub–tip diameters ratio $D_{h1}/D_2 = 0.3$, and solidity $\pi D_2/(2Z_b L) = 0.25$ where Z_b is the blade number and L is the blade length. Hence, their application ranges are limited. The nondimensional specific speed is defined in terms of the average volumetric flow rate, such as $\omega_s = \omega[(Q_1 + Q_2)/2]^{0.5}/H^{0.75}$, where ω is in radians per second, Q is in cubic feet per second, H is the polytropic head or actual total enthalpy change per unit gas mass in (lb$_f$-ft/slug) across the impeller, and the efficiency is the polytropic efficiency of the impeller. The effects of other parameters are also shown. The optimum specific speed increases as the outlet blade angle with respect to the radial direction

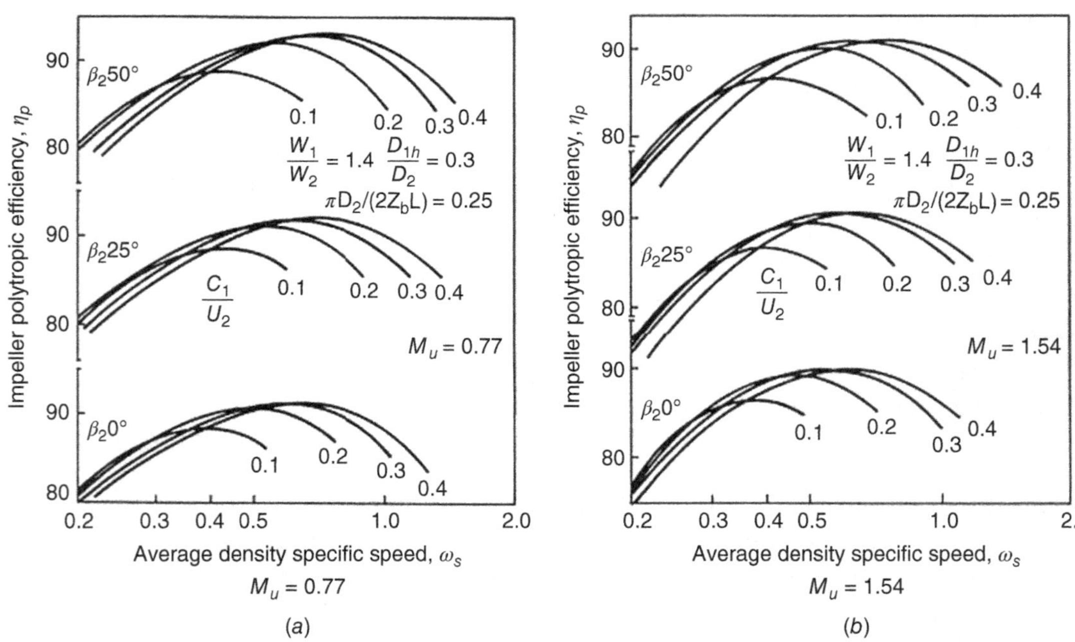

Figure 4.19 Specific speed chart of polytropic efficiency. (Reproduced from Specific speed and efficiency of centrifugal impellers; Performance prediction of centrifugal pumps, Proceedings of the Twenty-fifth Annual International Gas Turbine Conference and Exhibit and Twenty-second Annual Fluids Engineering Conference, New Orleans, LA, March 9–13, 1980/American Society of Mechanical Engineers (ASME).)

β_2 increases from zero to 50° for a given Mach number $M_u = U_1/a_{01}$. Also, at a given specific speed and Mach number, the efficiency is a function of the inlet flow coefficient $\phi = C_1/U_2$.

The centrifugal compressor design is a complicated topic, and a more detailed discussion is beyond the scope of this text. Interested readers can consult Refs. 14, 16, 18, 20.

Example 4.7 Design a centrifugal air compressor impeller with the following specifications: $N = 15{,}500$ rpm, $\dot{m} = 17\ \text{lb}_\text{m}/\text{s}$, $T_{01} = 530$ R, $p_{01} = 14.7$ psia, and $p_{03}/p_{01} = 2.50$.

Solution

a) Inlet Configuration Selecting $M_{r1,t} = 0.75$ and $\beta_{f1} = 25°$, we have

$$M_{1t} = 0.75\ \sin 25° = 0.317, \qquad T_1 = \frac{T_{01}}{1+(k-1)M_{1t}^2/2} = \frac{530}{1+0.2\times 0.317^2} = 519.6\ \text{R},$$

$$a_1 = (1.4\times 53.33\times 32.2\times 519.6)^{0.5} = 1117.6\ \text{ft/s},$$

$$V_{1t} = 0.317\times 1117.6 = 354.3\ \text{ft/s},$$

$$W_{1t} = 0.75\times 1117.6 = 838.2\ \text{ft/s},$$

$$U_{1t} = (838.2^2 - 354.3^2)^{0.5} = 759.6\ \text{ft/s}.$$

So we have $r_{1t} = U_{1t}/\omega = 759.6/1623 = 0.468$ ft. $= 5.6$ in. Also from $p_1 = p_{01}(T_1/T_{01})^{k/(k-1)} = 14.7(519.6/530)^{3.5} =$ 13.7 psia and $\rho_1 = p_1/(RT_1) = 13.7\times 144/(53.33\times 519.6) = 0.0713\ \text{lb}_\text{m}/\text{ft}^3$, we have $A_1 = \dot{m}/(\rho_1 V_1) = 17/(0.0713 \times 354.3) = 0.673\ \text{ft}^2 = 96.9\ \text{in.}^2$, assuming uniform inlet flow. From $\pi(r_{1t}{}^2 - r_{1h}{}^2) = A_1$, we have $r_{1h} = (r_{1t}{}^2 - A_1/\pi)^{0.5} = 0.72$ in., $U_{1h} = 97.0$ ft./s, and $\beta_{f1,h} = \tan^{-1}(V_{1h}/U_{1h}) = \tan^{-1}(354.3/97.0) = 74.7°$.

b) Outlet Configuration From

$$H_{ad} = C_p T_{01}\left[\left(\frac{p_{o2}}{p_{o1}}\right)^{(k-1)/k} - 1\right] = 0.24\times 778\times 530[(2.5)^{0.2857} - 1]$$

$$= 29614\ \text{ft-lb}_\text{f}/\text{lb}_\text{m},$$

$$Q_1 = \frac{\dot{m}}{\rho_1} = \frac{17}{0.0713} = 238.4\ \text{ft}^3/\text{s},$$

we obtain the specific speed

$$N_s = \frac{N\sqrt{Q_1}}{(H_{ad})^{0.75}} = \frac{15,500\sqrt{238.4}}{(29,614)^{0.75}} = 106\ \text{rpm}\sqrt{\text{ft}^3/\text{s}/(\text{ft}-\text{lb}_\text{f}/\text{lb}_\text{m})^{0.75}}.$$

From Figure 4.7 we have the estimated $\eta_c = 0.87$ and $D_s = D_2(H_{ad})/\sqrt{Q_1} = 1.50$. Hence $D_2 = 1.5\sqrt{238.4}/(29,614)^{0.25} = 1.765' = 21.2''$ or $r_2 = 10.6$ in. and $U_2 = r_2\omega = 1433.7$ ft./s. Referring to Figure 4.16c, we have $T_{sso3} = T_{01}(p_{03}/p_{01})^{(k-1)/k} = 530(2.5)^{0.2857} = 688.6$ R, and from $\eta_c = (T_{sso3} - T_{01})/(T_{03} - T_{01})$, we have $T_{02} = T_{03} = (1/0.87)\,(688.6\text{–}530) + 530 = 712.3\ R$. Also from $\eta_m = U_2V_{u2}/[C_p(T_{02} - T_{01})]$, with the estimated $\eta_m = 0.95$, we have $V_{u2} = 0.95\times 0.24\times 778\times 32.2(712.3\text{–}530)/1433.7 = 726.3$ ft./s. Selecting the flow coefficient $\phi = V_{m2}/U_2 = 0.30$ results in $V_{m2} = 0.3\times 1433.7 = 430.1$ ft./s, $W_2 = [V_{m2}{}^2 + (U_2 - V_{u2})^2]^{0.5} = 827.9$ ft./s, and $V_2 = [V_{u2}{}^2 + V_{m2}{}^2]^{0.5} = 844.1$ ft./s. Hence we have the diffusion factor $D_f = W_{1t}/W_2 = 838.2/827.9 = 1.012$ (<1.9), which is O.K. The impeller efficiency can be estimated from the losses fraction $\chi = (1 - \eta_{imp})/(1 - \eta_c) \cong 0.6$ or $\eta_{imp} = 1\text{–}0.6(1 - \eta_c) = 1\text{–}0.6(1\text{–}0.87) = 0.922$. Hence from $\eta_{imp} = (T_{so2} - T_{01})/(T_{02} - T_{01})$, we have $T_{so2} = T_{01} + \eta_{imp}(T_{02} - T_{01}) = 698$ R and $p_{02} = p_{01}(T_{so2}/T_{01})^{k/(k-1)} = 14.7(698.1/530)^{3.5} = 38.5$ psia. Then from the energy equation we have $T_2 = T_{02} - V_2{}^2/2C_p = 653.0$ R. Hence $p_2 = p_{02}(T_2/T_{02})^{k/(k-1)} = 28.4$ psia and $\rho_2 = p_2/(RT_2) = 0.117\ \text{lb}_\text{m}/\text{ft}^3$. Selecting $Z_b = 16$ and using the Stanitz formula for the slip coefficient, we have $V'_{u2} = V_{u2} + 0.63\pi U_2/Z_b = 726.3 + 0.63\pi \times 1433.7/16 = 903.6$ ft./s and $\tan\beta_{b2} = V_{m2}/(U_2$

$-V'_{u2}) = 430.1/(1433.7–903.6) = 0.81$ or $\beta_{b2} = 39°$. With blade thickness $t = 0.15$ in., the contraction factor is

$$\varepsilon_2 = 1 - \frac{Z_b t/\sin\ \beta_{b2}}{\pi D_2} = 1 - \frac{16 \times 0.15/\sin\ 39°}{\pi \times 21.3} = 0.94.$$

Hence from the mass equation $\rho_2 V_{m2} \pi D_2 b_2 \varepsilon_2 = m$, we have $b_2 = 17/(0.117 \times 430.1\ \pi \times 1.765 \times 0.94) = 0.0648' = 0.78$ in.

4.1.7 Application of Centrifugal Compressors

In gas turbine applications, centrifugal compressors are primarily used for small engines of less than 10 MW with a pressure ratio of less than 4. For applications in other industries, the types of gas handled vary widely, from regular air to different kinds of pure vapor or mixture. Some are employed in high-power applications. One example is shown in Figure 4.20 for repressuring service in crude oil production. The principal fields and general applications involved are as follows:

Industry	Service or Process
Chemical or petrochemical	Oxidation, gas recovery, refrigeration, air supply, and liquefaction
Iron and steel	Combustion, oxidation, and exhausting
Mining and metallurgy	Power tool and machinery, copper, and nickel purification
Natural gas	Production, processing, and distribution
Refrigeration	Industrial and commercial air conditioning
Utilities	Soot blowing, cyclone furnaces, and combustion
Miscellaneous	Sewage treatment, food processing, paper making, glassmaking, wind tunnel, pharmaceutical, etc.

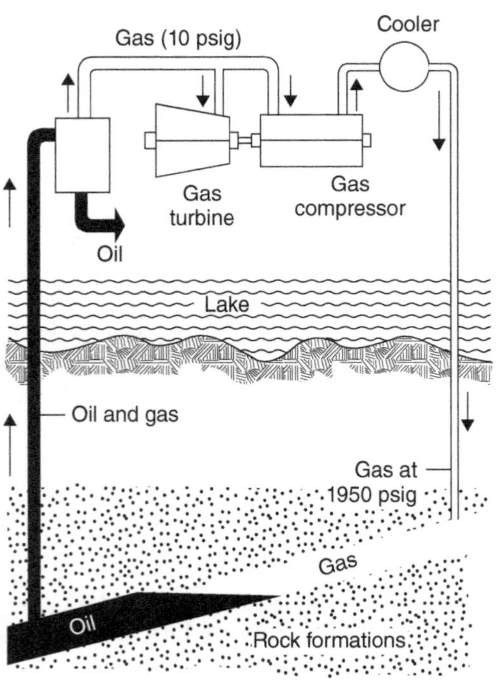

Figure 4.20 Compressor used on repressuring service in-crude oil production. (Reproduced from Gibbs, 1971/Ingersoll-Rand.)

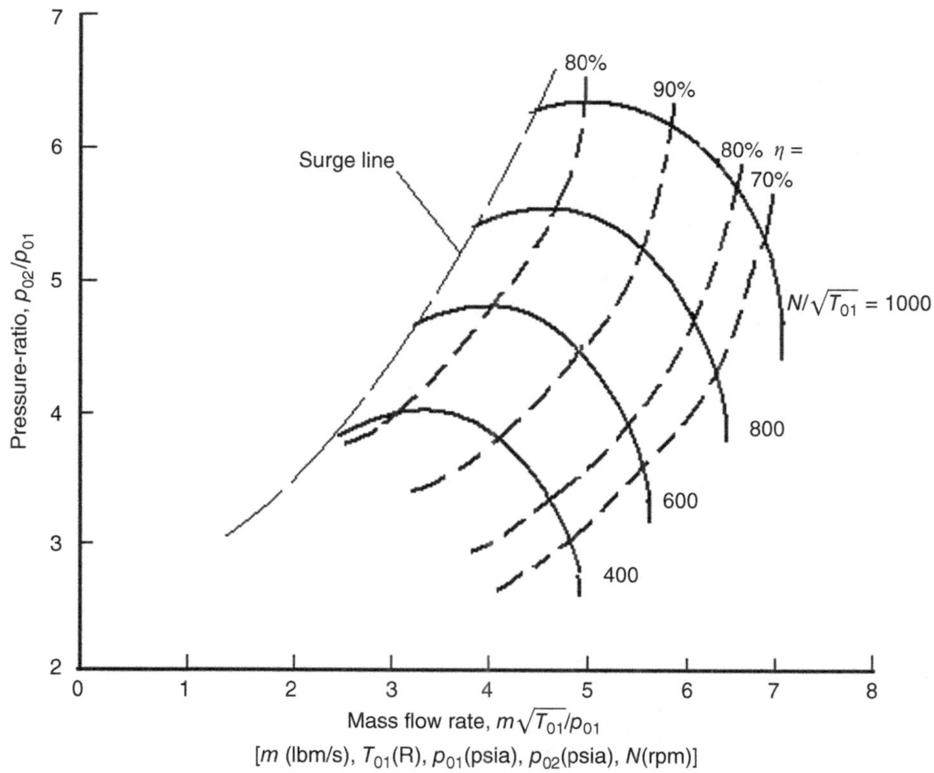

Figure 4.21 Typical centrifugal compressor curves in nondimensional parameters.

For convenience of application, centrifugal compressor performance curves are normally given in terms of nondimensional parameters, as discussed in Chapter 2. Relevant parameters obtained from the dimensional analysis are p_{02}/p_{01}, T_{02}/T_{01} (or η_{ad}), $\dot{m}\sqrt{T_{01}}/p_{01}$, and $N/\sqrt{T_{01}}$. Here p_{02} and T_{02} are the compressor discharge pressure and temperature. A set of typical characteristic curves are plotted in Figure 4.21. The surge line at low flow and fast pressure drop due to choking at high flow rate are obvious.

Due to the narrow range of stable operating conditions, a variable-speed drive is always desirable and employed whenever applicable. But other types of flow control are also possible:

1) suction throttling with butterfly valve or variable inlet guide vanes,
2) recirculation to suction, if very low capacity is needed, even for zero flow conditions, and
3) throttling at discharge or employing variable diffuser vanes.

A combination of the various methods can also be employed to achieve optimal operation. The centrifugal compressor with an inlet guide vane and its performance curves with various vane settings are shown in Figure 4.22. The guide vanes can be adjusted by pneumatic, electrical, or hydraulic means. Typical performance curves and constant efficiency contours are shown for five different vane positions. The volumetric flow rate and head are plotted in terms of the percentage of the design condition.

In general, the surge line can be extended to a lower flow rate with the inlet guide vanes than with the variable-speed approach. However, the efficiency of the former method is slightly poor in most cases.

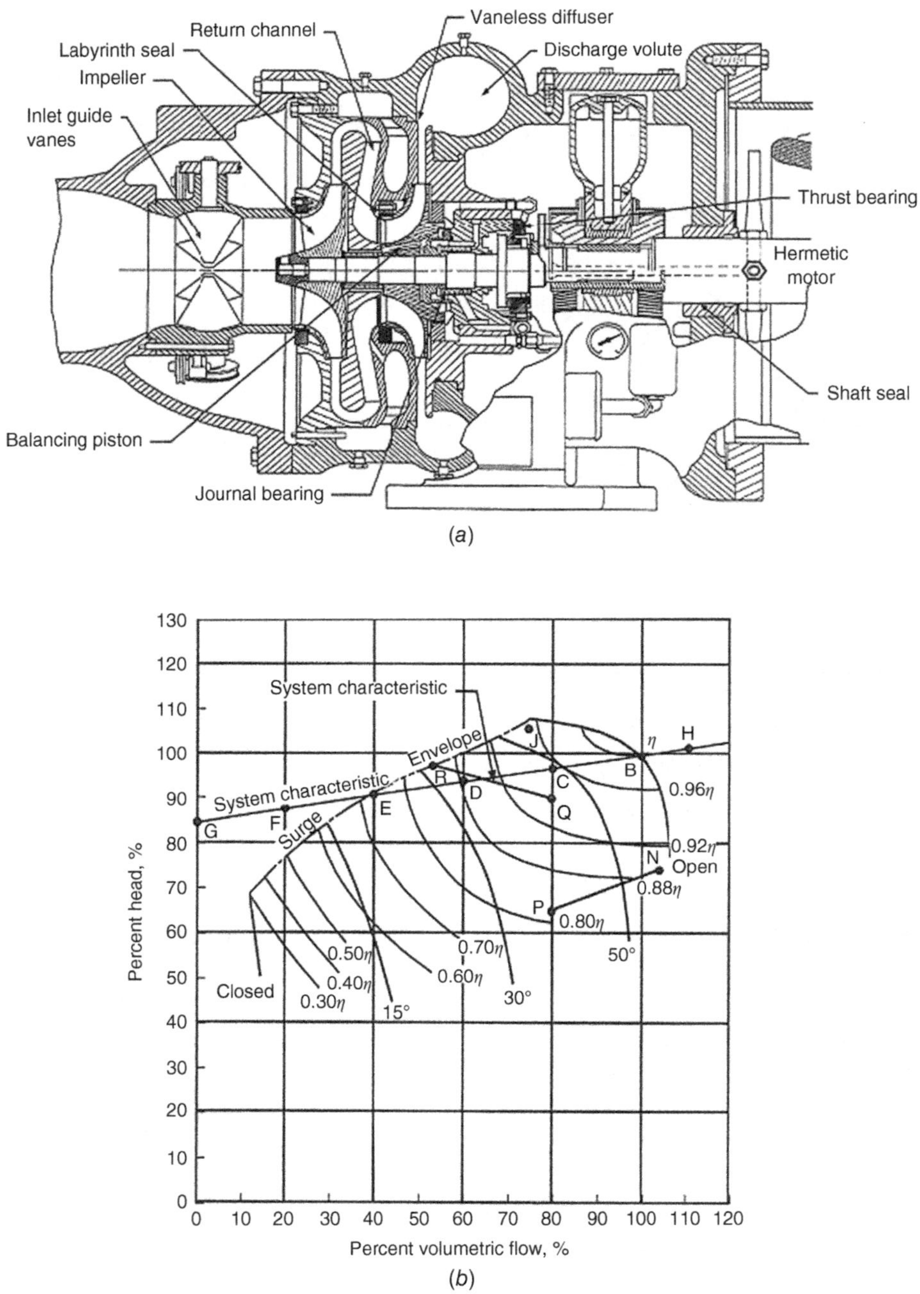

Figure 4.22 Centrifugal compressor with inlet guide vanes. (*a*) Two-stage compressor with inlet guide vanes. (*b*) Typical compressor performance with various inlet guide vane settings. (Reproduced from ASHRAE Handbook, HVAC Systems and Equipmen, 2004/ASHRAE.)

4.1.7.1 Intercooling

For high-pressure-ratio compression, a multistage compressor with intercooling is normally employed. As shown in Figure 4.23, if the compression is divided into two stages, the gas is compressed from condition 1 to 2, then cooled to 2′ before being compressed to 3′. The discharged gas pressure at 3′ is the same as the single-stage compression pressure at 3, but the temperature is lower. The input power per unit mass flow rate required is $P_s' = C_p\left[(T_3' - T_2') + (T_2 - T_1)\right]$, which is less than the single-stage value of $P_s = C_p(T_3 - T_1)$. Since $T_3' - T_2'$ is less than $T_3 - T_2$ because the constant pressure lines on a T–s diagram diverge. It can be shown in Equation (4.2) that the adiabatic head H_{ad} is proportional to the inlet temperature of each stage. Hence, the efficiency of the multistage compressor with intercooling is higher than that of the single stage. Another advantage of intercooling is that a lower maximum gas temperature implies that less expensive material can be used for the critical part of the flow passage. Also, low-temperature, high-pressure gas is needed for some applications.

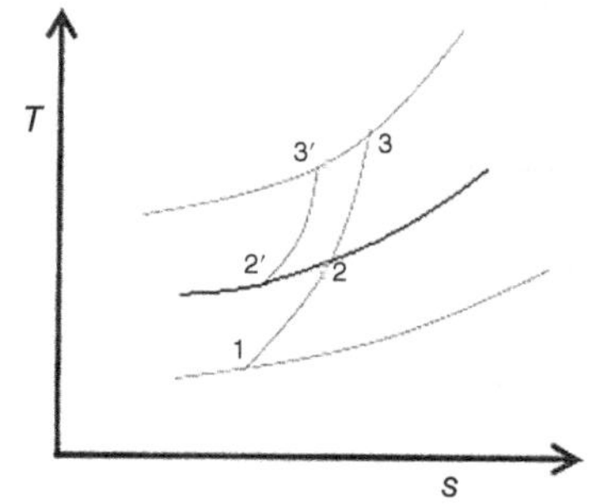

Figure 4.23 T–s diagram for compressor with intercooling.

Three different types of intercooling arrangements are shown in Figure 4.24. An external cooler can be inserted between any two stages after a few stages, wherever is needed. A water-cooled diaphragm is less effective, and problems associated with scaling may result. Spray injection can be applied with liquids compatible with the compressed gas and construction materials, but the injection amount is not easy to control.

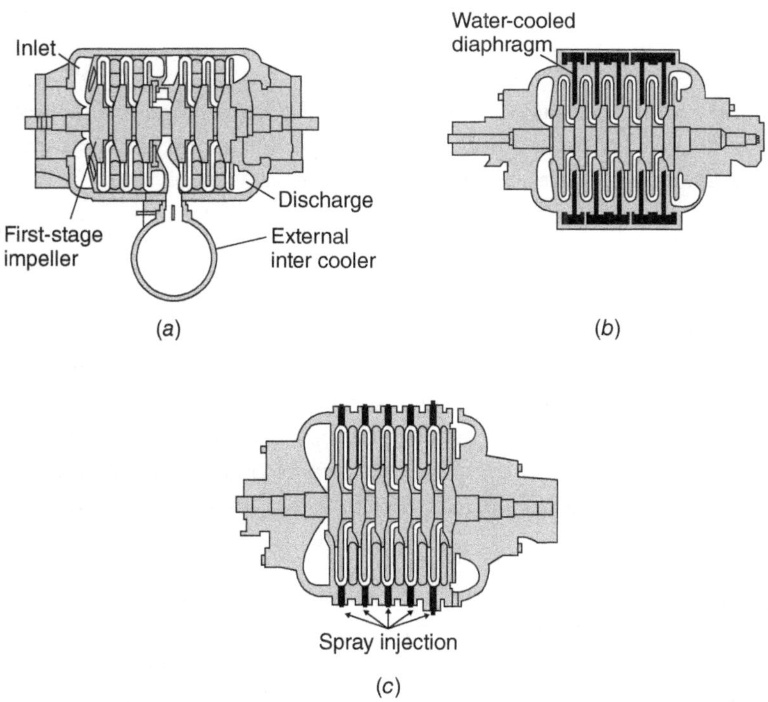

Figure 4.24 Multistage compressors with intercoolers. (*a*) External intercooling. (*b*) Intercooling with water-cooled diaphragm. (*c*) Intercooling with liquid spray injection. (Reproduced from Gibbs, 1971/Ingersoll-Rand.)

Table 4.2 Comparison of Compressor Types.

	Residential and Automobile	Commercial	Industrial
Capacity	0.1	10	100 (kW)
Ranges (1 ref. ton = 3.51 kW)	0.1	10	100 (tons)
Rotary			
Scroll			
Screw			
Reciprocating			
Centrifugal	Advantage	Disadvantage	
Rotary	Inexpensive	Noisy	
Screw	Simple construction, no valve, quiet	Expensive to make	
Reciprocating	Inexpensive to make	Noisy, low efficiency	
Scroll	Simple construction, no valve, quiet, high efficiency	High precision required, expensive to make	
Centrifugal	Simple construction, high efficiency at large capacity	Expensive to make, high precision and balancing required	

4.1.7.2 Application in Refrigeration Systems

In refrigeration and air-conditioning systems, the centrifugal compressor is primarily used in large industrial units. The positive-displacement types dominate the small- to medium-power units. A comparison of the capacity ranges of applications and the advantages and disadvantages of different types of compressors is given in Table 4.2.

4.1.8 Selection of Compressors

Selecting a compressor can be a straightforward job or a tedious, complicated task, depending on the application. For application in a gas turbine, a detailed discussion will be given in Chapter 5. The discussion here is limited to application in ground-based industrial systems. The basic guidelines and two examples will be included. Although a machine designed from scratch can result in a most efficient unit, the off-shelf unit, if available, will provide a more economic choice. A building-block approach using existing components is a compromise for large units.

Some factors influencing the choices among positive-displacement, centrifugal, or axial-flow type compressors suggested by Ref. 2 are as follows:

1) Flow Capacity: Inlet volumetric flow rate is a major consideration in the selection of compressor type. Typically, for flow rates higher than 1000 cfm, a dynamic compressor, either centrifugal or axial type, will be chosen. Otherwise, a positive-displacement type will be preferred, as shown in Figures 4.25 and 4.26.
2) Driver: Generally speaking, the choice of driver is dictated by the available source of power; the compressor that best fits the driver should receive first consideration. In other words, a centrifugal compressor always receives the first consideration if the driver must be a turbine. Otherwise, a positive displacement machine should be considered if the driver is an electric motor and the specific speed is appropriate.
3) Head: If a dynamic compressor is selected, the discharge pressure or pressure ratio requirement can be converted into a head requirement in ft-lb_f/lb_m. With the number of stages selected, the specific speed

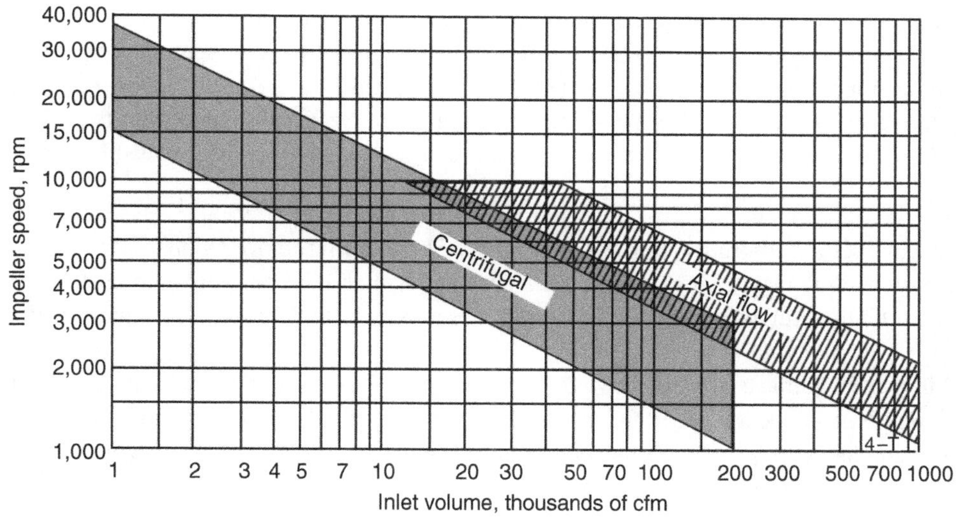

Figure 4.25 Range of speed versus inlet capacity for near-optimum efficiency for dynamic compressor. (Reproduced from Gibbs, 1971/Ingersoll-Rand.)

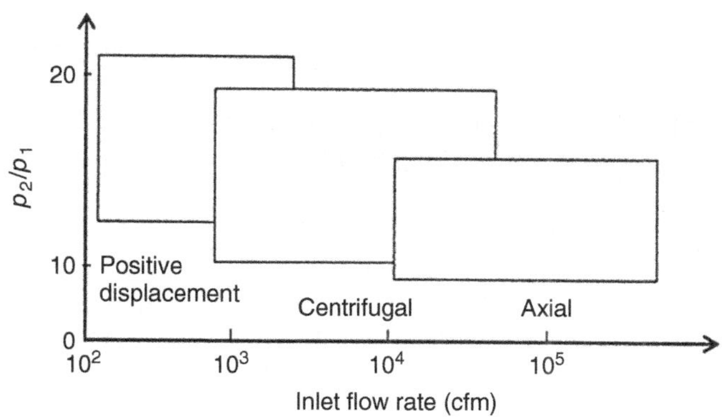

Figure 4.26 General guides for selecting industrial compressor (single or multistage). (Adapted from[11])

$N_s = N\sqrt{Q}/(H_{ad})^{0.75}$ (based on one stage) can be calculated. Then Figure 4.7 can be used to check its size and expected efficiency. Iteration can be performed to reach an optimum selection.

4) Floor Space: Floor space and its shape will, sometimes, influence the selection. The opportunity for selection will be broadened if the exact dimensions of the space available are given to all potential vendors.
5) Foundation Needs: Foundation requirements for rotating machinery are often less than that for an equivalent reciprocating machine. The latter has more vibration problems.
6) Capacity Control: If the variation range in gas demand is wide, the positive displacement type will be favored in most cases.
7) Oil-Free Air: Any requirement for oil-free air rules out the vane-type rotary. Helical-lobe and dynamic compressors require no oil.
8) For air power application, the load factor, defined as the ratio of total actual output over the rated full-load output of the machine, should be between 50 and 80%. This will provide more uniform pressure, even during peak demand and a cool-off period, especially for air-cooled units.

9) Gas Characteristics: For petrochemical process applications, the gas properties are important in compressor selection. A low gas inlet density (i.e., low-molecular-weight gas) will affect the centrifugal more than the positive displacement machine. It may require multistage for the centrifugal.
10) If the required brake horsepower is very high, it is likely that an axial-flow type will be considered first. In the other case, if the required pressure rise is very high, then the first choice will likely be the positive displacement type.

Sometimes, the gas may have certain limiting conditions that must be kept during compression, especially the temperature. Then, interstage cooling may become necessary. The centrifugal or axial type will be preferred for gas at very high or very low temperatures because the lubricant is required in the flow passage region in the reciprocator.

Considering the above items, economic evaluations of initial and operating costs can be performed on several choices. The brief discussion given in Section 3.1.10 for pump selection also applies here. Then a final decision can be made. Two simple examples below illustrate some basic calculations for a typical compressor selection.

Example 4.8 Air at ambient conditions of 14.7 psia and 70 °F has to be compressed to 50 psia at a rate of 100 $1b_m$/s. If a steam turbine running at 4800 rpm is available, select the most efficient type of compressor (single or multistage) for this job. Also, estimate its diameter and the number of stages.

Solution

Using Figure 4.7, the specific speed $N_s = NV_1^{0.5}/H_{ad}^{0.75}$ should be calculated based on the inlet volumetric flow rate V_1 and the adiabatic head per stage H_{ad}.

From $\rho_1 = p_1/(RT_1) = 14.7 \times 144/(53.3 \times 530) = 0.075\ \text{lb}_m/\text{ft}^3$, we have $V_1 = (100/0.075) = 1333\ \text{ft}^3/\text{s}$. Also from Equation (A.3), we have total adiabatic head $\sum H_{ad} = C_p T_1[(p_e/p_i)^{(k-1)/k} - 1] = 0.24 \times 778 \times 530[(50/14.7)^{0.2857} - 1]$ $= 41{,}436\ \text{lb}_f\text{-ft/lb}_m$.

The specific speeds for different numbers of stages are calculated as follows:

$$n = 1: \quad N_s = \frac{4800(1333)^{0.5}}{(41,436)^{0.75}} = 60,$$

$$n = 3: \quad N_s = \frac{4800(1333)^{0.5}}{(13,812)^{0.75}} = 137,$$

$$n = 4: \quad N_s = \frac{4800(1333)^{0.5}}{(10359)^{0.75}} = 170,$$

$$n = 6: \quad N_s = \frac{4800(1333)^{0.5}}{(6906)^{0.75}} = 231,$$

$$n = 8: \quad N_s = \frac{4800(1333)^{0.5}}{(5179)^{0.75}} = 287.$$

Reading the curves in Figure 4.7, the best efficient point is around $N_s = 280$ with $\eta \approx 0.90$ and $D_s = D(H_{ad})^{0.25}/V_1^{0.5}$ ≈ 0.7. Hence, 8-stage mixed-flow impellers are selected. The impeller diameter can be calculated as $D = 0.7 \times (1333)^{0.5}/(41{,}436/8)^{0.25} = 3.0$ ft., and the total required shaft power $P_s = m \sum H_{ad}/\eta = 100 \times 41{,}436/(0.9 \times 550) = 8371$ hp.

Example 4.9 An air compressor is needed to deliver an airflow of 1500 $1b_m$/min from 65 psia and 120 °F to 250 psia. The polytropic efficiency is estimated to be 75%. Select the appropriate type of compressor with its rotating speed. If multistage types of 2, 4, and 6 are considered, estimate their respective efficiencies and required horsepower.

Solution

From the ideal gas equation, the inlet air-specific volume can be calculated as

$$v_1 = \frac{RT_1}{144p_1} = \frac{53.3 \times 580}{65 \times 144} = 3.303\ \text{ft}^3/\text{lb}_\text{m},$$

and $Q_1 = mv_1 = 4954$ cfm. The overall adiabatic head can be calculated from

$$H_{o/a} = RT_1\left(\frac{n}{n-1}\right)\left[\left(\frac{p_2}{p_1}\right)^{(n-1)/n} - 1\right] = 53.3 \times 580 \times 2.625\left[\left(\frac{250}{65}\right)^{1/2.625} - 1\right]$$

$$= 54,417\ \text{ft-lb}_\text{f}/\text{lb}_\text{m},$$

where $n/(n-1) = \eta_p k/(k-1) = 0.75 \times 1.4/(1.4-1) = 2.625$.

From Figure 4.25, a centrifugal compressor with speed $N = 10,000$ rpm is appropriate for the present application.

To use Figure 4.7, the specific speed can be calculated from $N_s = NV_1^{0.5}/H_{ad}^{0.75}$, where $V_1 = 4954/60 = 82.6$ cfs and $H_{ad} = H_{o/a}/n_s$ is the head for each stage. Selecting the number of stages to be 2, 4, and 6, the head for each stage and specific speed can be calculated, then the total-to-total adiabatic efficiencies can be read. The required shaft horsepower can be calculated with the volumetric and mechanical efficiencies assumed to be 0.98 and 0.95, respectively. That is,

$$P_s = \frac{mH_{o/a}}{33,000 \times \eta_{ad}\eta_v\eta_m}.$$

The results are as follows:

Stage Number	H_{ad} (ft-lb$_f$/lb$_m$)	N_s	η_{ad}	P_s (hp)
2	27,208	43	0.72	3690
4	13,604	72	0.83	3201
6	9069	98	0.87	3054

Example 4.10 Select a centrifugal fan from Table 4.1 for a duct system requiring 12,000 cfm at a static pressure of 1.25 in. wg. (a) Determine its rotating speed, (b) plot the fan performance curves at this speed, and (c) plot the system curve.

Solution

From Table 4.1, at 1.25 SP, the rotating speeds for $Q_1 = 11,172$ cfm and $Q_2 = 12,103$ cfm are $N_1 = 474$ rpm and $N_2 = 483$ rpm, respectively. Hence the rotating speed for the selected fan can be determined by interpolation as

$$N = 474 + \frac{(483-474)(12,000-11,172)}{12,103-11,172} = 474 + 9 \times 0.889 = 482\ \text{rpm}.$$

Select a few data points around 482 rpm from Table 4.1 as

Q (cfm)	14,896	12,103	11,172	11,172	10,241	7448
N (rpm)	490	448	436	474	466	360
SP (in. wg)	1.0	1.0	1.0	1.25	1.25	0.75
BHP (hp)	3.66	2.67	2.40	2.97	2.75	1.2

Convert them into the conditions of 482 rpm according to the similarity laws, resulting in

Q (cfm)	14,653	13,021	12,350	11,360	10,593	9972
SP (in. wg.)	0.967	1.16	1.22	1.29	1.34	1.34
BHP (hp)	3.5	3.20	3.24	3.12	3.04	2.88

The system curve can be calculated from $H \sim Q^2$ as

Q (cfm)	10,000	11,000	12,000	13,000	14,000
H (in. wg.)	0.87	1.05	1.25	1.47	1.70

They are plotted as shown.

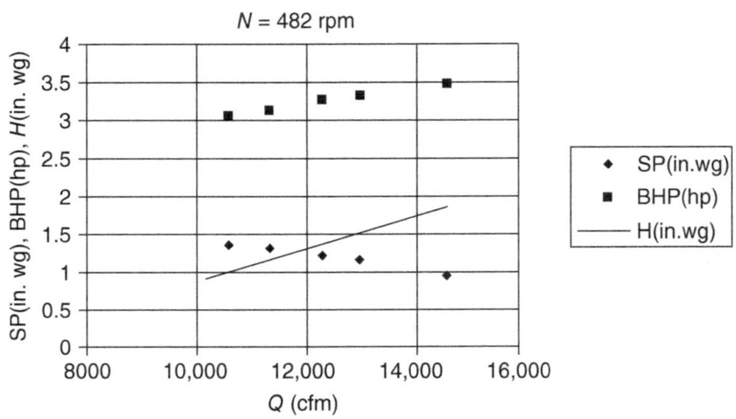

4.1.9 Design of Centrifugal Compressor

4.1.9.1 Impeller Design

Developments in computer technologies and advances in turbomachinery technology have enhanced the optimization of centrifugal compressor design. Turbomachine design normally starts with a mean-line program at each operating point on a map; the through-flow calculation is performed, and the impeller, diffuser, and volute are designed. It is also important to optimize an off-design performance.

Compressor design employs a current design process, including a viscous process aerodynamic design and structure optimization for achieving efficiency and stability targets. The compressor developed in this study consists of three major parts: an impeller, a low-solidity diffuser, and a volute. Good surge margins were achieved without using a variable geometry for a steady-state operation.

There are three design processes for impeller blades and diffuser vanes: (i) an inverse, (ii) a direct, and (iii) a hybrid direct-inverse.[20–23] The inverse design is faster because it is based on flow analysis to produce the blade or airfoil according to a pressure profile. However, the location of the profile control points in the blades' profiles is difficult to control, and the number of arcs of the blade profiles is generally large. This fact makes using the same curve definition routine for design and manufacturing difficult. Another difficulty of this method is that the inverse impeller may not meet the structure requirements. At the same time, direct design may require more time in aerodynamic analyses, but one can generate more accurate designs to meet the manufacturing requirements. Direct design can also easily incorporate structure analyses into the design process. The hybrid direct-inverse design method is still not mature enough to be used in impeller design.

A combined turbomachinery viscous aerodynamic optimal method[20–23] is one of the most effective methods with structure analyses to fully optimize a centrifugal compressor design.

Optimization procedures are widely used in centrifugal compressor developments to enable high efficiency and wide operating ranges of compressors. It is important to understand how to set the parameters and which parameter needs to be optimized. This paper demonstrates a case study on the importance of mechanical and structure combination optimization and discusses compressor performance's meridional impacts. Recently, the high-pressure ratio and the efficiency of centrifugal compressors require impeller designs to pay attention to multidisciplinary optimization. High efficiency and a wide operation range are always the design goal. The geometry of the blades and the meridional profile are important contributors to compressor performance and structure reliability. Therefore, it is critical to consider all aspects of compressor optimization. Currently used centrifugal compressors aim to design a compressor with the same back-disc diameter, the best efficiency, and the highest-pressure ratio.

The detailed design cycle based on direct design starts with a parametric optimization involving the vector diagram employing a mean-line study, as shown in Figure 4.27. After the impeller blade and diffuser are designed, analytical codes are used to calculate the compressor performance and optimize the two-dimensional and three-dimensional design.[21, 23] At the same time, the wheel stress, vibration, HCF, and LCF life are predicted to

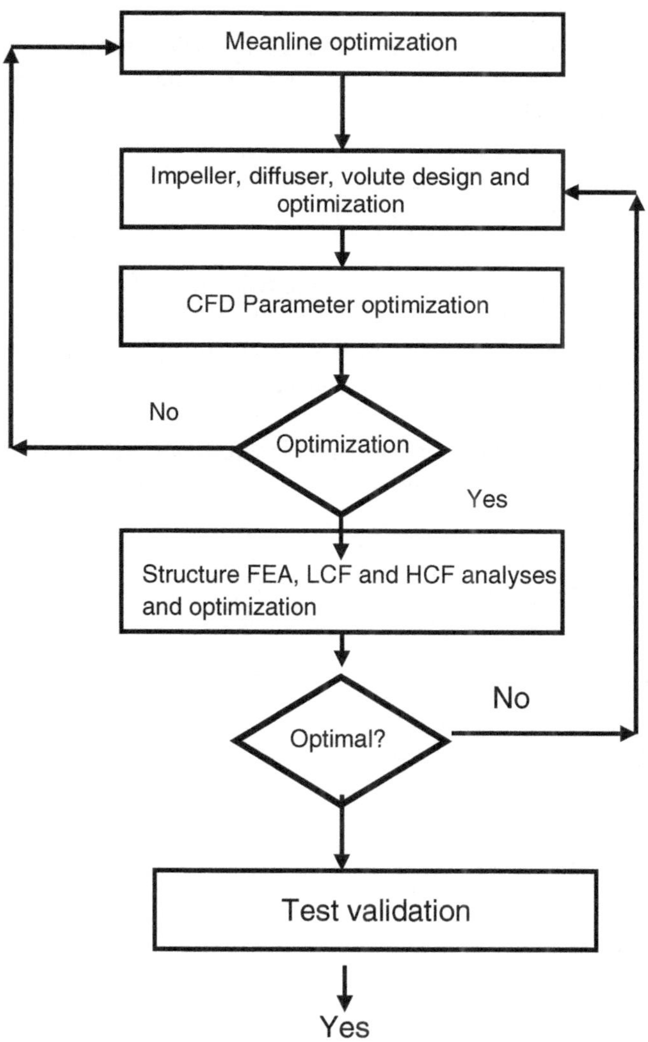

Figure 4.27 Centrifugal design process.

optimize the structure design. In this way, the final design is optimized for both aerodynamics and structure. The volute design is based on the conservation of momentum and the conservation of mass in this study.[24, 25]

This section focuses on the impeller design to demonstrate the importance of aerodynamics and structure integration. Optimization techniques are used in different steps for different components of the aerodynamic design.[18, 20, 26–29] The overall machine optimization was performed using mean-line optimization.

4.1.9.2 Impeller Design Process

A centrifugal compressor stage consists of three main components: a rotating impeller, diffusers, and a volute, as shown in Figure 4.28. The impeller design is critical regarding the aerodynamic efficiency and the operating range of the compressor stage. Impeller aerodynamics also affects the performance of the diffuser and the volute. It is very important to achieve an optimized configuration of the impeller. The traditional impeller design was performed using a one-dimensional analysis; then, the blade and end wall geometry were selected according to the velocity triangles so that the new impeller is usually a scale of an existing one. This design method could not provide the prediction of the impeller flows to improve the design; therefore, the flow separation in the impeller is a consequence of the stipulation of the impeller geometry in most cases. In this design, a high impeller backward exit angle was used to improve efficiency by reducing the impeller loading and improving the surge margin by providing a steeper energy addition characteristic. The high impeller exit blade angles required a coordinated aeromechanical design effort.

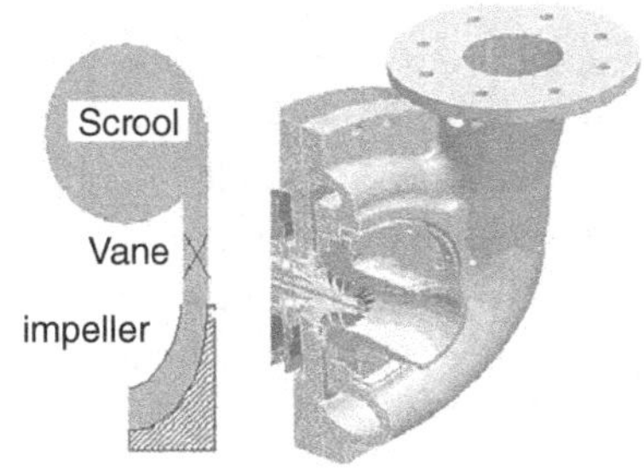

Figure 4.28 Cutaway view.

The final impeller was designed using a fast three-dimensional viscous code,[2, 30] as shown in Figure 4.29.

4.1.9.3 Diffuser Design

A vaned diffuser has become popular because of its high efficiency and a similar operating range for the pressure ratio over 4.0 compared with a vaneless diffuser. However, the problem with a conventional high-solidity, wedge-vaned diffuser design compressor is that the operating range is often relatively small compared with a vaneless diffuser for a pressure ratio lower than 3.0.[2, 7] Here, a compressor was designed for a wide pressure ratio

Figure 4.29 Calculation meshes. Overview of 360-degree mesh. (Xu et al., (2012)/John Wiley & Sons/CC BY 3.0.)

of 1.1 to 5.0, requiring wide operating ranges. For this reason, a low-solidity vaned diffuser was used.[2, 5–7, 10, 31] Diffuser research showed that a low-solidity diffuser could provide higher efficiency and a reasonable operating flow range. In the present study, a diffuser was designed to apply to this high-pressure ratio machine to meet the higher efficiency and wide flow operating range requirements. The diffuser was generated by optimizing sections and stacking up using three-dimensional analyses.[21, 23]

4.1.9.4 Closing Remarks

An aerodynamic and structure integration process was used to design a single-stage centrifugal compressor. The importance of aerodynamic and structure integration was emphasized. The selection of the optimization parameters is also critical to produce a better compressor design. A case study for different lean and meridional profiles was reported in detail. The final design was built and tested; test results showed fair agreement with the analysis. Our results have proven that the design system developed in this study is feasible; the design process thus provides useful guidance for future centrifugal compressor development.

This section provides a study of the meridional features of impeller design. It demonstrates that a proper design for the inlet and the exit of the meridional plans can improve the compressor's overall efficiency and reduce the sensitivity of the tip clearance. This study also indicates that optimizing the impeller meridional shape can improve the compressor stage efficiency and the pressure ratio without changing the other components. The study also indicated that selecting the optimization parameters is a critical step for a successful design during compressor optimization.

4.1.10 Impeller Design for Centrifugal Compressors

4.1.10.1 Importance of Impeller Design

An impeller is one of the key components of industrial centrifugal compressors and turbochargers. Aerodynamic and structural designs of the impeller are critical to the success of all the compressor stages. The requirements for efficiency and operating range of industrial centrifugal compressors and turbochargers have increased dramatically compared with the past. The efficiency of the newly developed low-pressure ratio centrifugal compressor has reached the possible level of the machine. However, the efficiency level of intermediate and high-pressure ratio machines still has a gap between the current state-of-the-art and possible levels. The challenge for centrifugal compressor design is to keep the efficiency level at state-of-the-art and increase the compressor operating range. Increasing the compressor operating range without sacrificing peak efficiency is difficult to achieve. Product globalization requires one product design that can be used in all locations. Due to technology differences in some countries, electricity frequency variations could be 10%. Turbocharger compressors work at different rotational speeds for the majority of the time. The compressor impeller rotating speeds change in a certain range. The variation in the impeller rotating speed makes the design of the impeller structure more challenging.

Centrifugal compressors are widely applied in the industrial, aerospace, and automobile industries. In a centrifugal compressor, the flow enters the compressor axially and then turns in the radial direction from the impeller. The flow then enters a radial annular vaned or vaneless diffuser. The flow exit from the diffuser needs a volute or collector to deliver the flow to the next stage or send it to the next components.[7–10] The basic components of the radial compressor are shown in Figure 4.29. Unlike an axial compressor or fan,[13, 32] the work input for the centrifugal compressor is almost independent of the nature of the flow. A centrifugal compressor can be designed with a much higher De Haller number than an axial compressor can achieve. So, a centrifugal compressor can have a much higher stage pressure ratio than an axial compressor. Centrifugal compressors have wide applications for a mass flow rate of less than 10 kg/s.[7, 13]

The small flow centrifugal compressors have wide applications in turboshaft aircraft engines, turbochargers, and industry. However, being in a low-power class, these compressors need to be inexpensive to manufacture and operate, requiring the compressor to be a simple design with fewer parts and a relatively large tolerance. Moreover, the five-axis machine is now a common tool for impeller machining, but most other parts should be fabricated using other machining types to reduce manufacturing costs.

The Turbomachinery industry is increasingly interested in optimizing procedures that enhance compressor efficiency and widen operating ranges. During compressor development, designers always compromise between a peak attainable efficiency and an overall operating range.[33] Compressor design normally starts with a meanline program at each operating point on a map, then throughflow calculation is performed, and finally, the impeller, diffuser, and volute are designed. This study extended a recently developed turbomachinery viscous optimal method[34, 35] for axial machines to a centrifugal compressor design.[36] The main focus of this study lies in the development of a small flow compressor where the flow coefficient $\phi = Q/N/D3 = 0.145$. Q, N, and D are the volume metric flow (m^3/s), rotational speed (rps), and impeller tip diameter (m). The design pressure ratio and the flow rate are 3.65 0.75 kg/s, respectively, at the design condition. The total static efficiency must be higher than 84% and the stability operating range (SB) greater than 38%. The compressor design employs a global optimization viscous process for achieving efficiency and stability targets.[35] Good surge margins were achieved without using a variable geometry for a stable operation. Special attention has been paid to the tip clearance profile to permit large clearance without lost efficiency too much during the impeller design. The compressor developed in this study consists of three major parts: an impeller, low solidity diffuser, and a volute. In this study, particular attention was paid mainly to impeller design and analyses.

A variable-speed centrifugal compressor impeller with a compound lean blade was developed.[36, 37] The newly developed impeller defines a blade pressure side that extends from the leading edge to the trailing edge, is convex near the leading edge, and is flat near the trailing edge. The impeller leaned opposite to rotation at the inlet portion and leaned back at the exit. The clearance between the impeller and casing is nonuniform, defined by a high-order equation. The flow and structure analyses were performed. Some basic performance tests and structure modal tests were performed. The analyses agreed with the test results.[38]

4.1.10.2 Impeller Design

We consider two impeller designs with the same diffuser and scroll stage. The performer test and data acquisition procedures were based on the ASME performance test procedure.[39] The test compressor consists of an inlet pipe, impeller, vaneless diffuser, low solidity vaned diffuser, vaneless diffuser, and volute, as shown in Figure 4.28. The test circuit is an open loop testing system, as shown in Figure 4.30. The flow rate is regulated by a butterfly valve

Figure 4.30 Test rig of a single-stage compressor. (Xu et al., (2005)/John Wiley & Sons/CC BY 3.0.)

Table 4.3 Main Design Information of the Test Compressor.

Shaft Speed $N = 60\,000$ rpm	Mass Flow rate $m = 0.77$ kg/s
Impeller outlet radius $R_2 = 82.90$ mm	Impeller outlet width $b_2 = 7.85$ mm
Out blade angle $\beta_2 = 55°$	Design pressure ratio $\pi = 3.60$
Impeller blade number $Z_i = 19$	Numbers of diffuser vanes $Z_d = 9$
Diffuser inlet radius $R_3 = 88.70$ mm	Volute inlet radius $R_5 = 170.00$ mm

located at the discharge pipe of the compressor. The main dimensions of the test compressor are listed in Table 4.3. Xu and Amano[36] provide more details for the test compressor.

The compressor developed by this research has a wide operating range 40% with high efficiency in all operating conditions. When the compressor operated near the choke, computations overpredicted the performance compared with tests. For full 3D design, calculation shows a few performance advantages, better efficiency, and head at the same flow conditions, compared with traditional impeller design. However, the tests near the design flow show that there are no significant benefits for both efficiency and head coefficient near the design point. The test showed a few benefits near choke and surge operations. To reduce the test tolerances, a few repeated tests were performed, and similar results were obtained. Further inspecting the impellers, it was found that the impeller surface finish is about 6.3 μm near the hub and about 3.2 μm on the blades. Because the calculations assumed that all walls on the impeller are smooth walls, the friction loss estimations may be smaller than the test impeller. When a compressor is operated at a near-design condition, both the full 3D impeller and the traditional impeller have no massive separation. The friction loss for the full 3D impeller is larger than that of the traditional blade due to the bowed and lean features, increasing the wet area. Further examination of the impeller calculations at the design point for compounded lean 3D impeller and conventional impeller showed that the total efficiency for 3D impeller and traditional impeller was 93.33% and 92.99%, respectively. It can be seen that the efficiency gain for 3D design is about 1/3%. This performance gain in the impeller becomes less significant for the whole stage. If the test part's surface finish is not very high, the performance effect of a 3D design will be washed out by frictional loss.

The tip relative Mach number contours near the tip of the impellers for both fully 3D and traditionally designed impellers are shown in Figure 4.31, as the traditional design leading edge area had a relatively larger Mach number compared with the full 3D design impeller. This may indicate that the full 3D design had less portion of the flow passing through the blade tip range. It is favorable for impeller efficiency if it shows less flow in the tip-high loss range. This is one of the reasons why the full 3D design shows better efficiency.

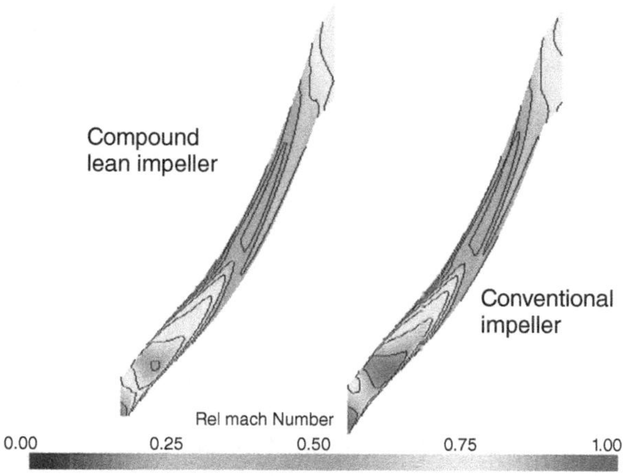

Figure 4.31 Tip relative Mach number distribution near shroud.

The relative Mach number contours near the suction surface, pressure surface, and mid-span plans are shown in Figures 4.32–4.34. The compounded lean blade has a relatively larger Mach number area than the conventional blade near the suction side of the blade in the middle of the flow path. At a nearby inlet area, the full 3D design depicts a smaller high Mach number area than the traditional impeller design. On the pressure side, the low Mach number region of the full 3D impeller is smaller than the conventional design. Mid-span Mach number distributions also show that a full 3D design provides more uniform Mach number distributions. The uniform Mach number distributions would benefit the compressor's performance.

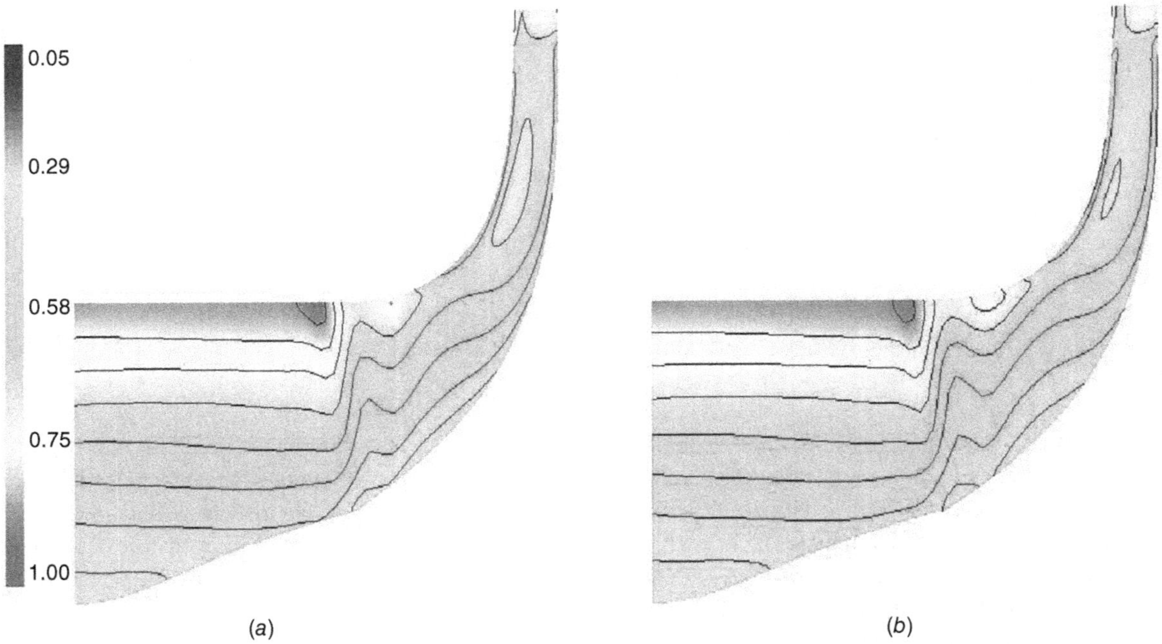

Figure 4.32 Relative Mach number contour near the suction side of the blade. (*a*) Full 3-dimensional blade and (*b*) traditional 3-dimensional blade.

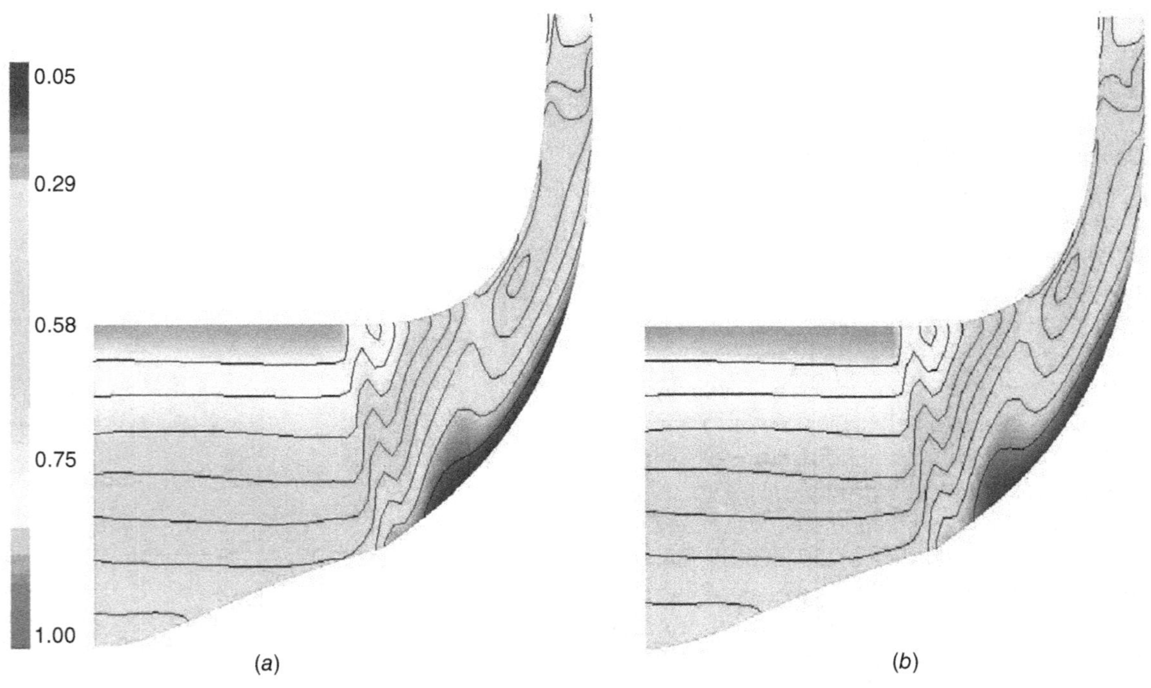

Figure 4.33 Relative Mach number contour near the pressure side of the blade. (*a*) Full 3-dimensional blade and (*b*) traditional 3-dimensional blade.

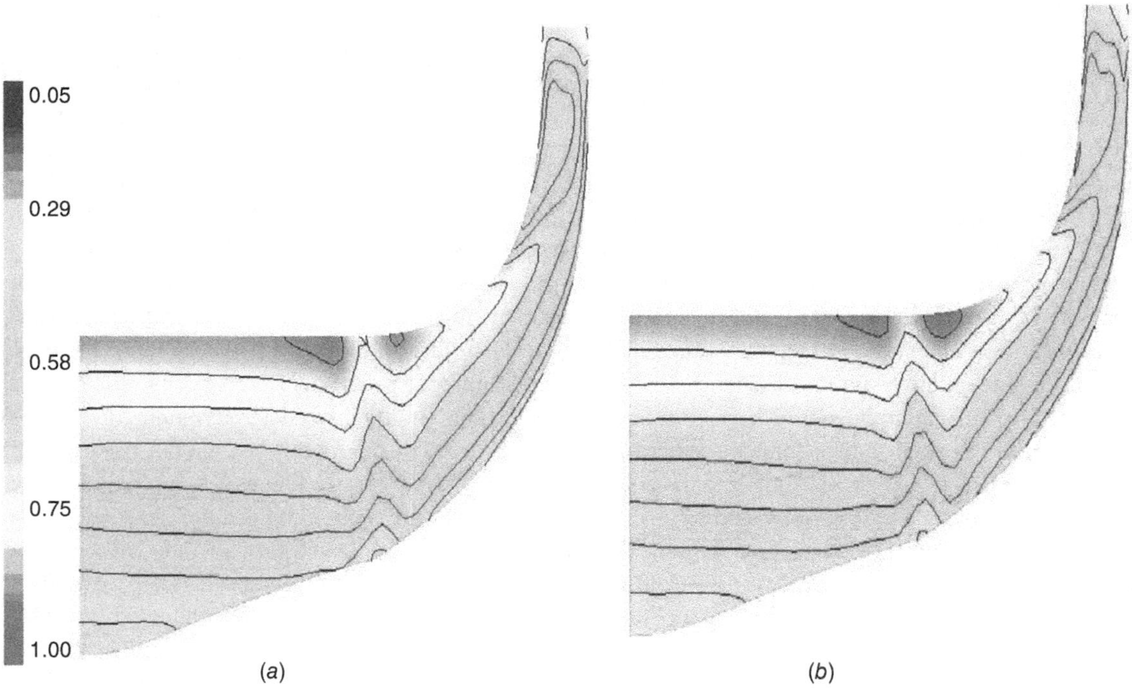

Figure 4.34 Relative Mach number contour near mid-plan between the blades. (*a*) Full 3-dimensional blade and (*b*) traditional 3-dimensional blade.

4.1.10.3 Concluding Remarks

The test results show that the full 3D impeller did not show any significant performance benefits near the design point due to increased friction losses on the surface of the bowed blades, whose surface area is greater than that of the conventional impeller. In cases of near surge and choke, full 3D impeller separation is reduced in the bowed blade impeller to increase the stagnation pressure ratio and efficiency.

4.1.11 Tong Volute Design for a Centrifugal Compressor

4.1.11.1 Importance of Tong Volume Design

The volute and diffuser interactions at design and off-design conditions were found to be much smaller for the large cut-back volute in comparison with the reported cases in the literature. This subsection presents that the large cut-back tongue volute design benefits the compressor performance and reduces the impeller exit static pressure nonuniformity.

Early studies[40–44] revealed that nonuniform pressure distributions are very strong inside the volute of the centrifugal compressors. Studies showed that the pressure circumferential distortions would cause nonuniformity of the upstream flows. Pressure distortions create an unbalanced aeromechanical force in a centrifugal compressor impeller. Thus, it is important to consider static pressure distortions both from the aerodynamic and rotor-dynamic design standpoints. An innovative volute concept was proposed and patented.[45] One of the research results demonstrated that the large cut-back tongue can reduce the circumferential distortions of the static pressure.

During the last couple of decades, extensive studies of the flow interactions between volute and impeller have been performed. Sideris and Van den Braembussche[2, 40] measured the velocity and pressure fields in the vaneless diffuser. They found that the flow distortion exists even at the impeller exit. Flathers and Bache[41] conducted a numerical investigation of the centrifugal compressor impeller due to downstream static pressure distortions. Flathers and Bache[42] estimated the influence of the volute on the pressure field surrounding an impeller. Hagelstein[5, 43] performed an experimental investigation of the circumferential pressure distortion in centrifugal compressor stages. Sorokes and Koch[33] studied the static pressure distortions caused by the discharge volute in a re-injection centrifugal compressor. Their results showed that the nonuniformity in the magnitude of the pressure was substantially different from the diffuser to the impeller exit. Gu et al.[12, 44] performed a simulation study on a single-stage centrifugal compressor. A steady flow was obtained by circumferentially averaging the exit fluxes of the impeller. The recirculation and twin vortex structure were found to explain the volute loss increase at lower and higher mass flows. Most recently, Meakhail and Park[14, 46] reported the study of velocity measurement between the impeller and vaned-diffuser and a numerical simulation for an entire machine (impeller, vaned-diffuser, and volute) of a single-stage centrifugal fan by employing a commercial code. The simulation results show that the separated flow region near the diffuser hub extends to the volute. However, quantitative agreement between numerical study and experiment remained rather poor.

The nonuniformity of the static pressure produced by a volute creates an unbalanced aerodynamic force in the centrifugal impeller rotor group. This force may cause unacceptable levels of subsynchronous radial vibrations of the rotor and excessive bearing loads. It is important to reduce or eliminate the static pressure distortion at the exit of the impeller. Most of the previous studies[2–4, 11, 12, 14, 33, 40–42, 44, 46] were focused on the pressure field, flow field measurements and most of them were numerical studies. The studies indicated that a volute causes static pressure distortion in the impeller exit for vaneless diffuser centrifugal compressors. Early studies were focused on the low-pressure ratio compressor. For reducing the static pressure distortions, Hagelstein[5, 43] conducted extensive studies by using a throttling ring and a vaned diffuser. However, attempts to eliminate the circumferential variation of the pressure field were not successful. Early studies showed that substantial difference in the magnitude of the pressure dip (the sudden pressure drop) at the impeller and diffuser exits. The nonuniformity of the static pressure at the impeller or diffuser exits is machine dependent. There is very limited literature reported on how to reduce static pressure distortions. In this research, a high-pressure ratio compressor was developed in order to study the reduction of the impeller exit static pressure distortions. A volute with a large cut-back tongue was designed and tested. The study showed that the pressure distortions at the impeller exit were eliminated by using a large cut-back tongue. The comparisons of the test and experimental study demonstrated good agreement with the simulation results.

4.1.11.2 Volute Design

A centrifugal compressor consists of three main components: a rotating impeller, vaned or vaneless diffusers, and a volute or collector (see Figure 4.35). The volute of a centrifugal compressor is the last component in the compression process. The fluid leaves the impeller with a logarithmic spiral shape due to the rotation of the impeller. A volute has two functions: one is to collect and transport the fluid into the downstream system, and the other to

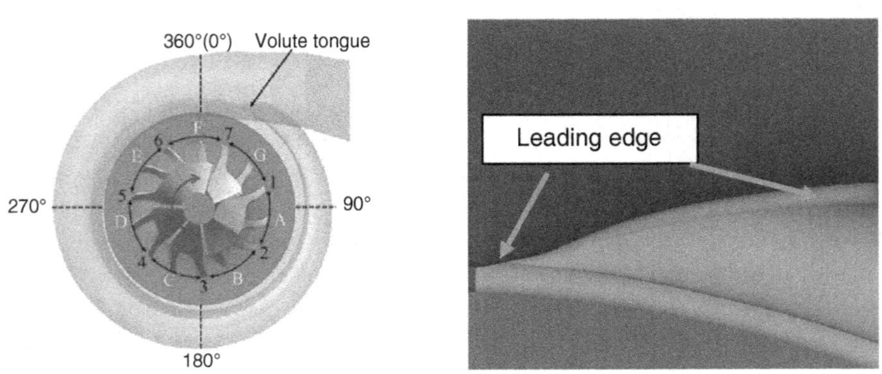

Figure 4.35 Centrifugal compressor and volute tongue detail.

raise the static pressure by converting kinetic energy into potential energy. Typically, most of the volute designs are based on two-dimensional analyses. However, a volute design is very complicated due to the strong swirl flow exiting from the impeller, which makes the flow strongly three-dimensional.[16]

The volute design needs to be considered in detail to meet the overall compressor efficiency and operating range goal. The shape of the cross-section was designed mainly from manufacturing and secondary flow structure considerations. It was found that there is only a small difference for all the asymmetric volute sections at reasonable volute floor diameters. The volute radial position was chosen by applying an angular momentum conservation law. Tongue geometry design is very important for both aerodynamic performance and structure design. Recent studies[18, 20, 47, 48] report the investigation of the effects of the tongue shape on the head and noise in a centrifugal pump. It was found that the head of the pump could be increased at low flow by retracting and rounding the tongue.

Volute design in this study embraces two basic considerations. First, angular momentum conservation was used. Second, the pressure and velocity are independent of angular angles during the first design phase. The tongue leakage was assumed during the beginning of the design process. The velocity enters the volute at an angle of about 65° from the tangential line at design flow, thus the dominant part of the total kinetic energy available is associated with the velocity in the radial direction. Based on the three-dimensional analysis at the design flow rate, the volute losses and the pressure recovery were estimated. Some final modifications were made after performing three-dimensional analyses. The exit cone was designed based on the CFD analyses. A convergent and divergent type of pipe was used to have the best static pressure recover capacity and the shortest length. The leading edge of the tongue is an intersection region between the scroll and the exit cone of the volute. The leading edge of the tongue is analogous to an airfoil, a well-rounded leading edge is insensitive to incidence but has large blockage. Increased efficiency has been the emphasis of compressor design. However, a wide operating range has become important in an age of increasing productivity. A "multitasking" compressor has various economic benefits. A volute with a rounded tongue has better off-design performance compared with a sharp tongue. The manufacturing process is also straightforward for a large, rounded tongue. The rounded tongue, however, has a larger blockage than the sharp tongue, which may reduce the design point efficiency. With an understanding of the influence of the tongue, a compressor with higher efficiency and a wider stable operating region can be designed. For obtaining a wide operating range of the compressor, a large radial rounded tongue can be used. The tongue radial is about 8% of the volute inlet width for this volute design.

Primary studies showed that the location of the tongue influences the static pressure distortion of the diffuser exit. This study showed that the cut-back tongue reduces static pressure distortions. Different tongue location and geometries were analyzed. The final volute tongue was chosen as a large cut back rounded tongues (the cut-back angle is about 27°), which allows about 25% of flow recirculation at a design flow condition. The details of the compressor stage and tongue structure are shown in Figure 4.35.

4.1.11.3 Volute Testing Results

The test compressor consists of an inlet pipe, impeller, vaneless diffuser, low solidity vaned diffuser, vaneless diffuser, and volute, as shown in Figure 4.35. The test compressor is integrated into an open-loop testing system. The airflow rate is regulated by a butterfly valve located at the discharge pipe. The details on the test compressor geometry are presented in Table 4.3. The test and data acquisition procedure follows the ASME performance test procedure.[39] The test uncertainty for the mass flow rate is less than 2% at 95% confidence level. The head and efficiency uncertainties were kept under 2.2% and 2.5%, respectively, with the same confidence level based on the system uncertainty analyses.[49] The performance characteristics were tested through five total temperature and total pressure measurements at the volute inlet and discharge of the volute exit. For determining the static pressure ratio, five pressure transducer taps were mounted on the circumferentially distributed on the wall at discharge of volute. The mass flow was measured by using an ASME[39] nozzle located at the end of the discharge pipe. Three static pressure taps were located in the scroll at 240° and 360° for CFD calibrations. There are five static pressure

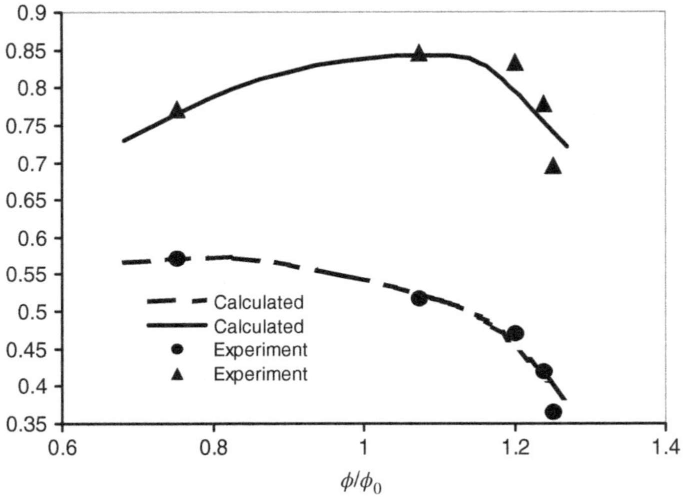

Figure 4.36 Comparison of the performance curves.

taps mounted on the vaned diffuser inlet, exit, and volute inlet location on the hub wall in order to study the static pressure distortions. The tip clearance of the impeller was measured with an Alumina pin. The design running tip clearances at the impeller exit is about 4.6% of the impeller exit width. Performance test results are shown in Figure 4.36. It is shown that the compressor has a wide operating flow range (RG = 40%) with high efficiency in the operating range. The large cut-back tongue provided a wide operating range without suffering a dramatically large loss in compressor efficiency.

Figure 4.37 shows the wall pressure at $\theta = 240°$ and $\theta = 360°$ locations for both calculations and experiments. The calculation results are in good agreement with the experiments. Further, the experimental results are slightly

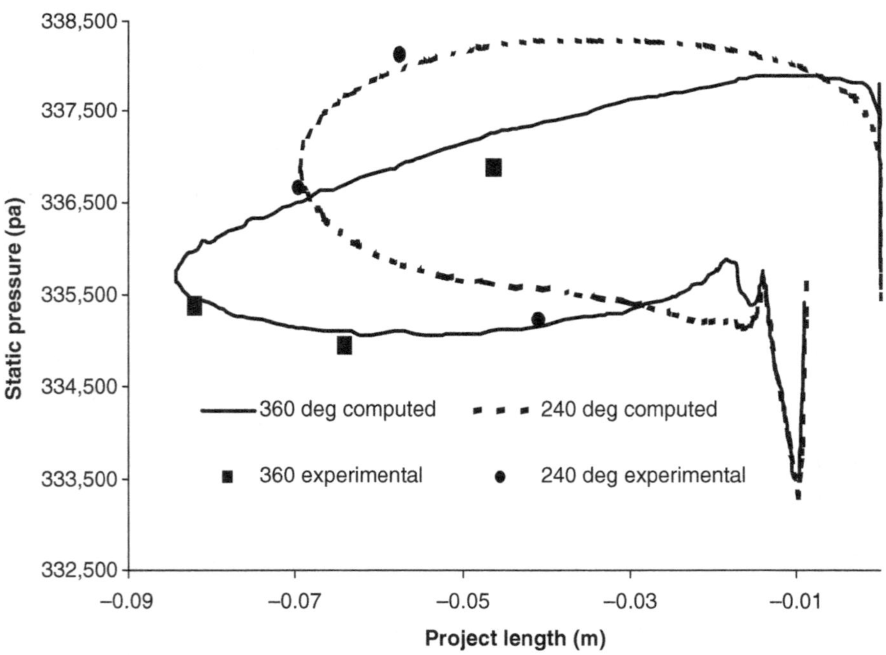

Figure 4.37 Static pressure distributions along the wall of volute cross-sections ($\phi/\phi_0 = 1.08$).

lower than the computations at the scroll exit $\theta = 360°$. This phenomenon is mainly attributed to the difference in boundary layer development between the prediction and the experiment, which causes different blockages in both the simulations and measurements.

The comparisons between computed and measured static pressure distributions at the volute inlet; diffuser exit and impeller exit are shown in Figures 4.38–4.40.

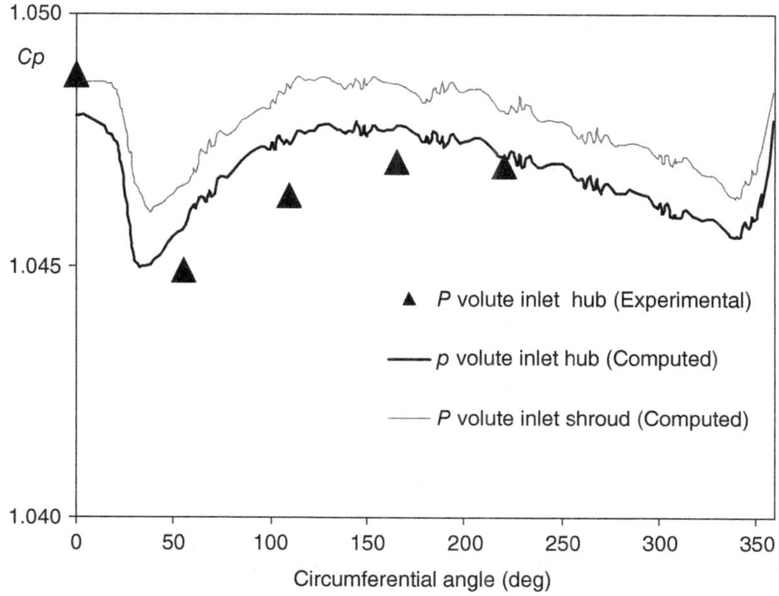

Figure 4.38 Static pressure coefficient at the volute inlet on the hub wall ($\phi/\phi_0 = 1.08$).

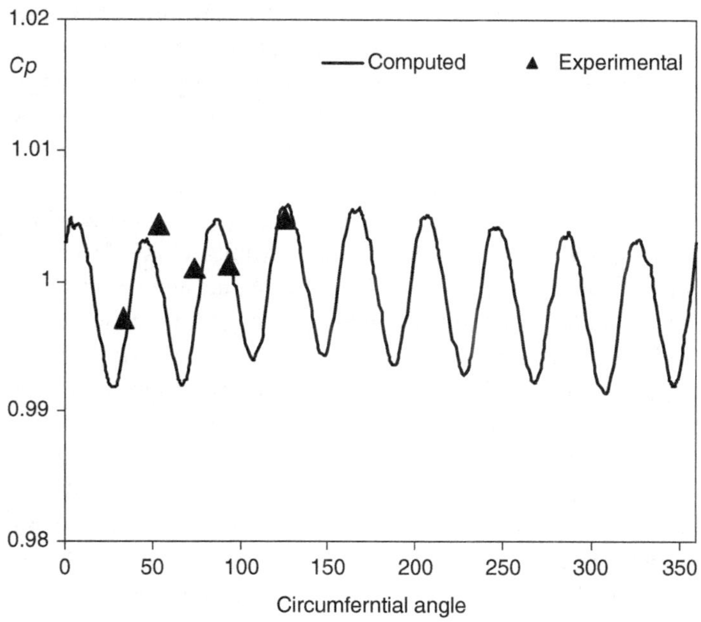

Figure 4.39 Static pressure coefficients at diffuser exit on the hub wall ($\phi/\phi_0 = 1.08$).

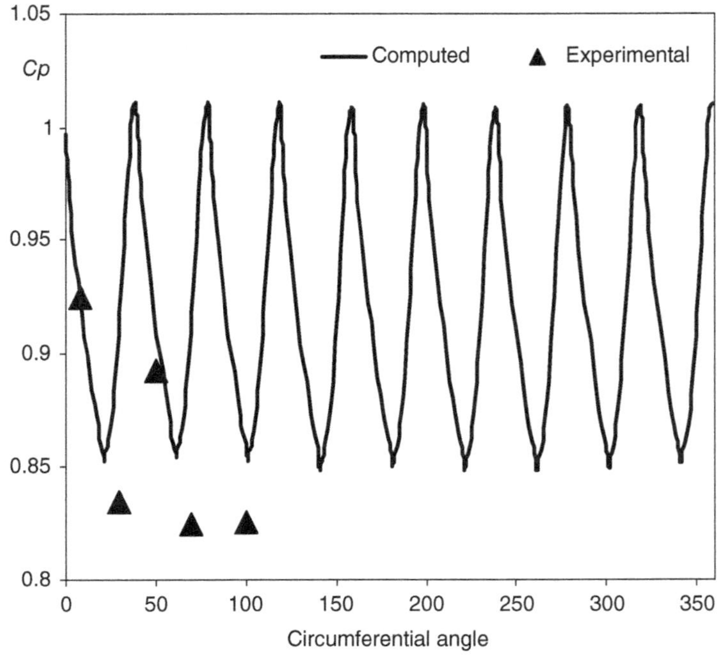

Figure 4.40 Static pressure coefficients at impeller exit at the hub wall ($\phi/\phi_0 = 1.08$).

The static pressure is measured at five locations. Multiple samples were taken over a period of time and algebraic average values were plotted. The leading edge of the tongue starts from 0° location. The results show that the static pressure is not circumferentially uniform at the inlet of the volute. The minimum static pressure is near the volute tongue. The general trends of the static pressure fluctuation were similar to the early studies.[2–5, 11, 40–43] However, there is a substantial difference in the fluctuation magnitude of the volute inlet static pressure in the current large cut-back tongue. The early studies showed that the static pressure fluctuations were from 10% to within a few percentages of the mean level at the design flow condition, whereas this study shows that a large cut-back tongue only produces 0.25% static pressure magnitude change related to the mean level. This value is much smaller compared with the early studies. For example, Hagelstein et al.[5, 43] reported about 10% of the mean level of pressure, and Sorokes and Koch[11, 33] reported about 3% of the mean level of static pressure variations. It is shown that the current design has over 10 times smaller static pressure fluctuation than those reported earlier. This study further indicates that the location of the minimum static pressure depends on the volute tongue structure. For example, Hagelstein et al.[5, 43] showed the minima located about 90° away from the tongue, and Sorokes and Koch[11, 33] showed that the minima is located about 50° to 60° away from tongue depending on the flow rate of the compressor. However, this study shows that the large cut back rounded tongue causes minima static pressure near 36° locations at near design flow. The static pressure fluctuation at the diffuser exit was observed mainly from the influence of the diffuser vanes as shown in Figure 4.39. The magnitude of fluctuation caused by the volute is even smaller than the static pressure fluctuation induced by the diffuser vanes. The unsymmetrical static pressure was eliminated at the impeller exit as shown in Figure 4.40. The experimental study also showed that the unsymmetrical aerodynamic force was very small. The static pressure fluctuations were mainly caused by the flow passing through nine diffuser vanes. Figure 4.5 also shows the static pressure variation on the shroud side of the volute inlet. It is shown that the pressure variation is in the same trend as the hub side. The minimum pressure location also occurred at the same circumferential location as the hub side. However, the static pressure level of the shroud side is higher than the hub side. This is due to the nonsymmetrical blockage of the volute scroll.

4.1.12 Conclusions

A new large cut back volute design is presented to eliminate the unsymmetrical pressure distributions at the impeller exit. This investigation confirmed that the volute tongue caused the static pressure distortion at the volute inlet and this nonuniformity pressure would also impact the upstream flows. The large cut-back tongue volute not only provides a good operating range but also eliminates the pressure asymmetric character at the impeller exit. The study demonstrated that the large cut-back tongue caused the level of static pressure distortion to be much smaller than the conventional tongue. It is also important to point out that the large cut-back tongue volute produces good features for the manufacturing process, which is easier for casting.

4.2 Axial-Flow Compressors

4.2.1 Introduction

It is generally more difficult to design a compressor than a turbine. This is because the flow in a compressor is decelerating/diffusion, and hence, it is more likely to suffer from boundary layer separation, surging, and stall, especially when the pressure increases too fast in the direction of flow. Also, for a mixed- or radial-flow compressor, a high-pressure ratio requires an impeller of an excessively large radius. Hence, a multistage axial-flow compressor design is the logical choice for the high-pressure-ratio requirement. This type of compressor is extensively used in gas turbine engines, especially for aircraft propulsion. Its advantages compared with the centrifugal type are summarized as follows:

(a) It has a smaller frontal area for a given high flow rate.
(b) A more compact multistage arrangement can achieve a higher-pressure ratio.
(c) Overall weight is lighter compared with the centrifugal compressor for the same performance.
(d) It operates at a higher speed due to its smaller radius and is more efficient in general.

The disadvantages of the axial-flow compressor are its steeper head-capacity curve and, hence, a narrow operating range. However, the adjustable stator blades can be used to extend the range somewhat. The typical performance characteristics of these two types of compressors are compared in Figure 4.41.[1] The effect of stator blade setting on the axial compressor is also shown. It is shown that the trend of the brake horsepower curves is opposite.

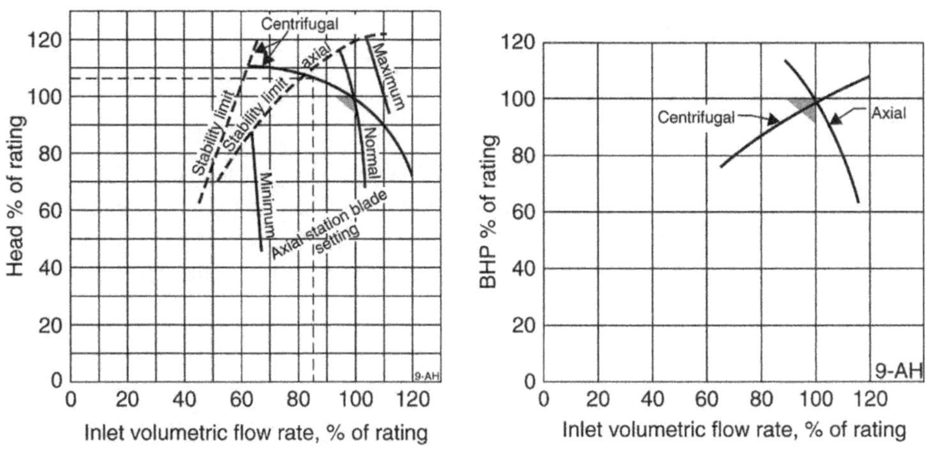

Figure 4.41 Comparison of axial and centrifugal compressors performance characteristics. (Reproduced from Gibbs, 1971/Ingersoll-Rand.)

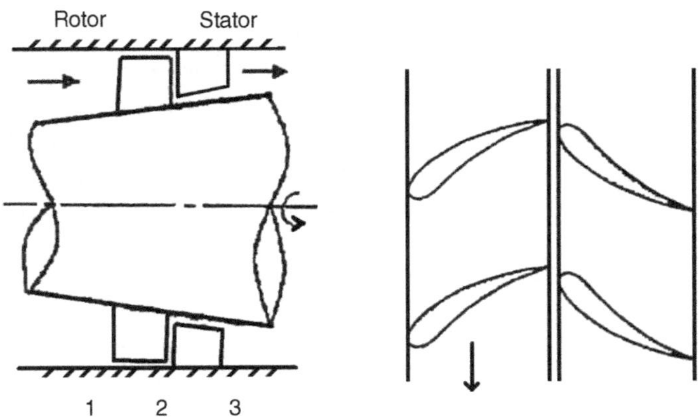

Figure 4.42 Axial-flow compressor stage.

Each axial-flow compressor stage consists of a rotating impeller (rotor) and a stationary diffuser (stator), as shown in Figure 4.42. Mechanical energy is converted into total enthalpy (thermal/pressure and kinetic energy) of flow in the rotor. At the same time, the stator is used to convert the excessive kinetic energy into thermal/pressure energy (or static enthalpy) and redirect the flow into the next-stage rotor. For convenience of analysis, the rotor and stator blades are developed onto a plane, also shown in Figure 4.43. The flow can be treated as two-dimensional in the preliminary discussion, especially if the blade height is small compared with the mean radius.

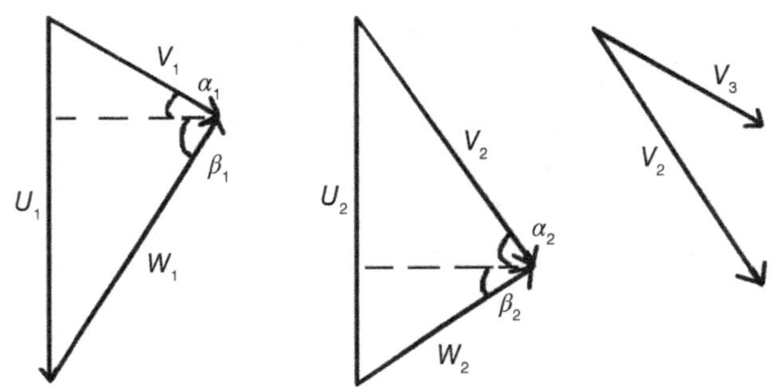

Figure 4.43 Velocity diagram across rotor and stator.

There are some similarities between the design of an axial-flow compressor and that of an axial-flow pump or fan, as discussed in Section 3.2, except that the compressibility effects and thermodynamic processes must be included for the compressor. However, only the basic principles will be covered in this chapter. More detailed treatment can be found in texts on gas turbines, such as Refs. 2–4.

4.2.2 Basic Theory

The velocity diagrams for stations 1, 2, and 3, as indicated in Figure 4.43, where $V_3 = V_1$ and $U_1 = U_2 = U$ for the normal compressor stages. The corresponding flow angles are also shown. Across the rotor, the relative flow velocity is guided by the blades, while across the stator, the absolute flow velocity is guided by the blades. The energy transfer per unit mass of fluid across each stage, expressed in terms of the total enthalpy change, can be related to these velocities according to the Euler equation, that is,

$$\begin{aligned} h_{o2} - h_{o1} &= U_2 V_{u2} - U_1 V_{u1} = U(V_{u2} - V_{u1}) \\ &= U[(U - V_a \tan\beta_2) - (U - V_a \tan\beta_1)] = UV_a(\tan\beta_1 - \tan\beta_2). \end{aligned} \tag{4.14}$$

Expressed in terms of the nondimensional parameters, with the stage loading factor and flow coefficient defined as $\psi = (h_{o2} - h_{o1})/U^2$ and $\varphi = V_a/U$, respectively, the Euler equation can therefore be written as $\psi = \varphi\,(\tan\beta_1 - \tan\beta_2)$. In actual situations, the energy transfer is reduced due to slip, blockage, and flow nonuniformity at the rotor discharge (similar to the slip coefficient μ_s, discussed earlier for centrifugal pumps and compressors), and it is accounted for with a coefficient called *work done factor*, λ, such that

$$\psi_a = \lambda\psi = \lambda\varphi(\tan\beta_1 - \tan\beta_2), \tag{4.15}$$

where ψ_a is the actual value and ψ is the one based on the flow parameters calculated with the mean flow velocities at the mean radius. Lambda, λ, is always less than unity and can be approximated as $\lambda = 0.85 + 0.15$ exp.$[-(n_s - 1)/2.73]$, suggested by Howell,[50] where n_s is the stage number. In other words, $\lambda = 1$ in the first stage, where the annulus wall boundary layer is thin. It decreases progressively in the downstream stages to the value of approximately 0.85. The flow friction effect converts the kinetic energy and static pressure into thermal energy (heat). Hence, Equation (4.15) is valid as long as the adiabatic process is assumed.

Another important parameter in compressor design is the *degree of reaction*, discussed in Section 2.2.6. It is defined as the ratio of static enthalpy change across the rotor to that across the whole stage, $R = (h_2 - h_1)/(h_3 - h_1)$, since the pressure change is related to the enthalpy change. Too much enthalpy increase in either the rotor or stator tends to cause a boundary layer separation in that element. Hence, an appropriate distribution of the enthalpy increase through the stage is important.

From the Euler equation and the definition of total enthalpy, we have

$$h_{o2} - h_{o1} = h_2 - h_1 + \frac{1}{2}\left(V_2^2 - V_1^2\right) = U(V_{u2} - V_{u1}), \tag{4.16}$$

With $V_{a1} = V_{a2}$ and $U_2 = U_1 = U$, we can rewrite it into

$$\begin{aligned} h_2 - h_1 &= \frac{1}{2}(V_{u2} - V_{u1})(2U - V_{u2} - V_{u1}) \\ &= \frac{1}{2}(V_{u2} - V_{u1})(W_{u2} + W_{u1}) \qquad \text{(a)} \\ &= \frac{1}{2}(V_{u2} - V_{u1})(U + W_{u2} - V_{u1}) \qquad \text{(b)} \\ &= (V_{u2} - V_{u1})\left[U - \frac{1}{2}(V_{u2} + V_{u1})\right] \qquad \text{(c).} \end{aligned} \tag{4.17}$$

Also, since $V_3 = V_1$, we have

$$h_3 - h_1 = h_{o3} - h_{o1} = h_{o2} - h_{o1} = U(V_{u2} - V_{u1}) \qquad \text{(d).}$$

So,

$$R = \left(\frac{1}{2U}\right)(W_{u2} + W_{u1}) = \left(\frac{\varphi}{2}\right)(\tan\beta_1 + \tan\beta_2) \tag{4.18a}$$

from (a) and (d), or,

$$\begin{aligned} R &= \frac{h_2 - h_1}{h_3 - h_1} = \frac{1}{2} + \frac{V_a[(W_{u2}/V_a) - (V_{u1}/V_a)]}{2U} \\ &= \frac{1}{2} + \left(\frac{\varphi}{2}\right)(\tan\beta_2 - \tan\alpha_1) \end{aligned} \tag{4.18b}$$

from (b) and (d), or

$$R = 1 - \left(\frac{\varphi}{2}\right)(\tan\alpha_2 + \tan\alpha_1) \tag{4.18c}$$

from (c) and (d).

For a 50% reaction stage, $R = 0.5$, we have $\alpha_1 = \beta_2$ from Equation (4.18b). The velocity diagrams are symmetrical, and the likelihood of boundary layer separation is equally minimized in both the rotor and stator. If $R > 0.5$ or $R < 0.5$, the boundary layer separation will be likely in either the rotor or the stator, and hence the overall stage efficiency is lower. It was mentioned earlier (in Section 2.2.6) that for incompressible and reversible flow, the reaction can be expressed in terms of the pressures such that $R = (p_2 - p_1)/(p_3 - p_1)$, since $\Delta h = \Delta p/\rho$. However, this parameter is not so commonly used in literature on pumps and fans.

The other two cases, rarely used in compressor designs, are (a) $R = 0$, and hence $\beta_1 = -\beta_2$, and all static enthalpy or pressure increase occurs in the stator, while the rotor is used to increase the kinetic energy of the fluid; and (b) $R = 1$, hence $\alpha_1 = -\alpha_2$, and all static enthalpy or pressure rise occurs in the rotor, the stator is used to change flow direction only. The velocity diagrams for these three cases are shown in Figure 4.44.

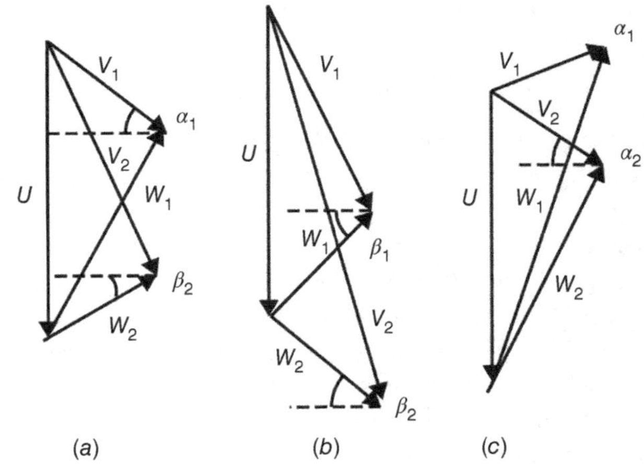

Figure 4.44 Velocity diagrams for different degrees of reaction. (*a*) $R = 0.5$, (*b*) $R = 0$, and (*c*) $R = 1$.

Thermodynamic processes through a compressor stage can be depicted in a Mollier diagram, shown in Figure 4.45. It is shown that $h_{01} = h_1 + V_1^2/2$ from the definition of total enthalpy and $h_{02} = h_{03}$. Since there is no energy transfer across the stator, only energy conversion occurs there. It is shown that $h_3 > h_2$ and $V_2 > V_3$.

Across the rotor, it was shown from the discussion in Section 2.6 that

$$h_1 + \frac{1}{2}W_1^2 = h_2 + \frac{1}{2}W_2^2, \tag{4.19}$$

since $\Delta E_p = h_{02} - h_{01}$ and $U_1 = U_2$.

The adiabatic and polytropic efficiencies that account for the irreversibility or friction effects in the flow process are the same as those defined for centrifugal blowers and compressors discussed in the previous chapters. They can also be related to the total pressure ratio across each stage for an ideal gas in the expression

$$\frac{p_{o3}}{p_{o1}} = \left(1 + \frac{\Delta T_o \eta_{ad}}{T_{o1}}\right)^{k/(k-1)} = \left(1 + \frac{\Delta T_o}{T_{o1}}\right)^{\eta_p k/(k-1)}, \tag{4.20}$$

where ΔT_o is the total temperature change across each stage. In general, η_{ad} is less than η_p. For small stages, the efficiencies are very close to each other, and hence η_p is also called small-stage efficiency, or simply stage efficiency. In the multistage axial-flow compressor, we have $\eta_s \cong \eta_p$, and η_{ad} is the adiabatic efficiency of the overall compressor.

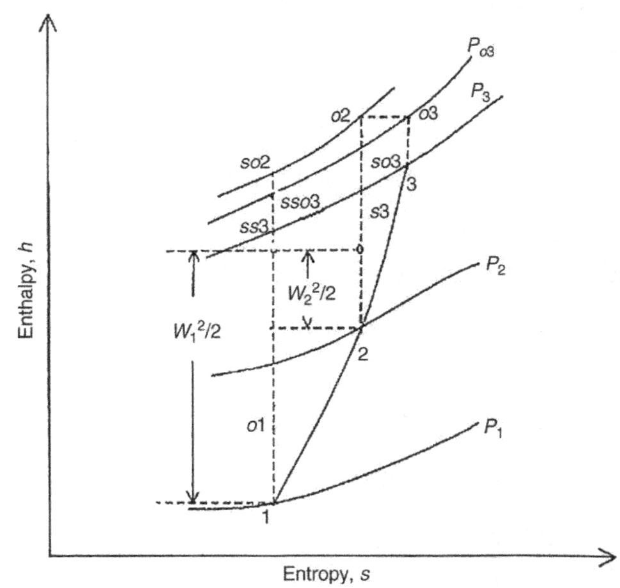

Figure 4.45 Mollier diagram for an axial-flow compressor stage.

From cascade test data, it is shown that the nominal deflection angle ε^*, which is the design condition, is a function of the pitch–chord ratio s/c and exit flow angle α_2, that is, $\varepsilon^* = f\ (s/c, \alpha_2)$. Typical data are plotted as shown in Figure 4.46.[51] These curves can also be approximated by $\tan \alpha_1 - \tan \alpha_2 = 1.55/[1 + 1.5(s/c)]$ in the range $0 \le \alpha_2 \le 40°$. In the design process, if β_1 and β_2 (which are α_1 and α_2 in the cascade data) are given or determined, s/c can be determined from these curves. Also, from the specified aspect ratio b/c and the calculated blade height $b = r_t - r_h$, the chord c can be determined. Hence s and then the blade number can be calculated as $Z_b = 2\pi r_m/s$.

Figure 4.46 Flow deflection through airfoil cascade.

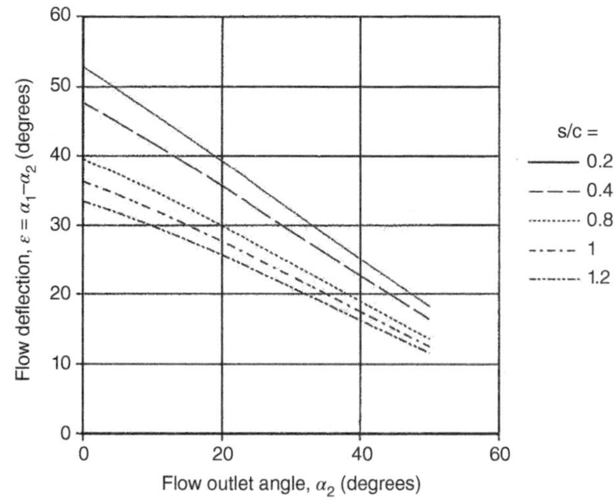

In general, the blade numbers of the rotor and stator should not be common multiples to avoid resonance. However, from a maintenance consideration, it is preferred to have an even number of rotor blades for balancing when a damaged blade has to be replaced.

In the design process, the incidence angle is chosen more or less arbitrarily (normally less than 2°), but the deviation angle δ is not under the designer's control. From tests,[51] it is found that δ can be related as $\delta \equiv \beta_2 - \beta_{b2} = m\theta(s/c)^{0.5}$, where $m = 0.23(2a/c)^2 + 0.1(\beta_2/50)$, $a \equiv x_{c,max}$ is the distance of the maximum camber from the leading edge, β_2 is in degrees, and $\theta \equiv \beta_{b1} - \beta_{b2}$ is the camber angle, also in degrees.

The three-dimensional flow effect due to the radial equilibrium requirement in axial-flow machines was discussed in Section 3.2. Only the free-vortex design will be elaborated here. If the blade hub-tip ratio is high ($r_h/r_t > 0.8$), which is typical in the downstream stages, the two-dimensional approach is a good approximation. However, in the upstream stages, with $r_h/r_t \sim 0.4$, the variation of whirl velocity from hub to tip is significant in counterbalancing the pressure variation in the radial direction. Hence, the blade profile will be twisted from the hub to tip.

Under the free-vortex condition, we have V_a = const. and $V_u r$ = const. In other words, if all three conditions Δh_o = const., V_a = const., and $V_u r$ = const. along the radial direction is imposed, the radial equilibrium condition will be satisfied. Under these conditions, from Equation (4.18c) the degree of reaction R for different radii can be expressed as

$$R = 1 - \frac{V_a}{2U}(\tan\alpha_2 + \tan\alpha_1) = 1 - \frac{1}{2U}(V_{u2} + V_{u1}) = 1 - \frac{V_{u2}r + V_{u1}r}{2\omega r^2}$$
$$= 1 - \frac{k_1}{r^2} = 1 - \frac{k_2}{\nu^2} = 1 - \frac{1 - R_m}{\nu^2} \tag{4.21}$$

where k_1, k_2 are constants, $\nu = r/r_m$, and R_m is R at the mean radius r_m. Since a negative R means decreasing static enthalpy in the rotor, which should be avoided, the minimum radius is obtained by setting $R = 0$, or $\nu_{min} = (1 - R_m)^{0.5} = 0.707$, with $R_m = 0.5$.

Example 4.11 An axial-flow compressor consisting of seven reaction stages is used to compress air from 14.7 psia, 70 °F to a pressure ratio of 8. Each stage requires the same amount of power with a stage efficiency of 87% while rotating at 12000 rpm. Determine the flow angles at the inlet and outlet of the first-stage rotor if its mean radius $r_m = (r_t + r_h)/2$ and blade height are 9 and 3 in., respectively, and the flow rate at the inlet is 450 ft^3/s. Also, the total power required and the compressor efficiency should be determined if the mechanical efficiency is 95%.

Solution

From $(p_{oe}/p_{oi}) = [(T_{oi} + n_s\,\Delta T_o)/T_{oi}]^{k\eta p/(k-1)}$ or $8 = (1 + 7 \times \Delta T_o/530)^{3.5\times 0.87}$, we obtain the temperature rise per stage $\Delta T_o = 74.2\,°\text{F}$. Also, from

$$V_{a1} \approx \frac{Q}{2\pi r_{m1} b} = 450 \times 144/(2\pi \times 9 \times 3) = 382\ \text{fps},$$

$$U = \frac{N\pi \times 9}{30 \times 12} = \frac{12,000\pi \times 9}{30 \times 12} = 942.5\ \text{fps},$$

$$\Delta h_o = UV_a(\tan\beta_1 - \tan\beta_2), \qquad R = \left(\frac{1}{2}\phi\right)(\tan\beta_1 + \tan\beta_2),$$

we have

$$\tan\beta_1 - \tan\beta_2 = \frac{0.24 \times 778 \times 74.2 \times 32.2}{942.5 \times 382} = 1.24,$$

$$\tan\beta_1 + \tan\beta_2 = \frac{0.5 \times 2 \times 942.5}{382} = 2.467.$$

So $\tan\beta_1 = 1.853$, $\beta_1 = 61.6°$, and $\tan\beta_2 = 0.613$, $\beta_2 = 31.5° = \alpha_1$. The total power required is

$$P_s = \frac{\dot{m} C_p n_s\,\Delta T_o}{\eta_m} = \frac{0.075 \times 450 \times 0.24 \times 778 \times 7 \times 74.2}{550 \times 0.95} = 6264\ \text{hp}.$$

The adiabatic efficiency is given as

$$\eta_{ad} = \frac{(p_{oe}/p_{oi})^{k-1/k} - 1}{T_{oe}/T_{oi} - 1} = \frac{0.811}{0.979} = 0.827.$$

Hence, $\eta_c = \eta_m \eta_{ad} = 0.95 \times 0.827 = 0.786 = 78.6\%$.

Example 4.12 An axial-flow compressor stage with hub and tip radii $r_h = 15$ in. and $r_t = 24$ in. is designed with the following data: $N = 6000$ rpm, $\Delta T_{os} = 35$ R, $V_a = 450$ ft./s, $\lambda = 0.92$, and $R_m = 0.5$.

a) Determine the flow angles at the hub, as well as the mean and tip radii of both the inlet and outlet of the rotor.
b) What are the degrees of reaction at both the hub and tip?

Solution

At the mean radius, we have

$$r_m = \left(\frac{24^2 + 15^2}{2}\right)^{0.5} = 20\ \text{in.}, \qquad U_m = \frac{N\pi r_m}{30} = \frac{6000\pi \times 20}{12 \times 30} = 1047.2\ \text{ft/s}.$$

a) From Equation (4.15), we have $0.24 \times 778 \times 32.2 \times 35 = 0.92 \times 1047.2 \times 450(\tan\beta_{m1} - \tan\beta_{m2})$, or

$$\tan\beta_{m1} - \tan\beta_{m2} = 0.485. \qquad \text{(E4.12.1)}$$

From Equation (4.18b), for $R_m = 0.5$ we have

$$\alpha_{m1} = \beta_{m2}. \qquad \text{(E4.12.2)}$$

From the velocity diagrams in Figure 4.44, we have

$$\tan\alpha_{m1} + \tan\beta_{m1} = \frac{U_m}{V_a} = 2.325, \qquad \text{(E4.12.3)}$$

$$\tan\alpha_{m2} + \tan\beta_{m2} = \frac{U_m}{V_a} = 2.325. \qquad \text{(E4.12.4)}$$

Substituting (E4.12.2) into (E4.12.1), we have

$$\tan\beta_{m1} - \tan\alpha_{m1} = 0.485. \qquad \text{(E4.12.5)}$$

Combining (E4.12.5) and (E4.12.3), we have $\tan\alpha_{m1} = 0.92$, or $\alpha_{m1} = 42.6°$. From Equations (E4.12.4) and (E4.12.2), we have $\tan\alpha_{m2} = 1.405$, or $\alpha_{m2} = 54.5°$. To determine the flow angles at the hub and tip, we use the free-vortex condition of $V_u r = \text{const.}$, or $r_h \tan\alpha_h = r_t \tan\alpha_t = r_m \tan\alpha_m$. Hence the flow angles can be determined as

$$\tan\alpha_{h1} = \frac{0.92\times 20}{15} = 1.226, \quad \text{or} \quad \alpha_{h1} = 50.8°,$$

$$\tan\alpha_{t1} = \frac{0.92\times 20}{24} = 0.766, \quad \text{or} \quad \alpha_{t1} = 37.5°,$$

$$\tan\alpha_{h2} = \frac{1.405\times 20}{15} = 1.873, \quad \text{or} \quad \alpha_{h2} = 61.9°,$$

$$\tan\alpha_{t2} = \frac{1.405\times 20}{24} = 1.171, \quad \text{or} \quad \alpha_{t2} = 49.5°.$$

b) The degree of reaction at the hub and tip can be determined from $R = 1 - (1 - R_m)/v^2$, so we have

$$R_h = 1 - \frac{1-0.5}{(15/20)^2} = 0.11, \quad R_t = 1 - \frac{1-0.5}{(24/20)^2} = 0.65.$$

Example 4.13 Continue the last example. If the aspect ratio of the rotor blades is set to 3, estimate the required number of blades, chord length, and blade angles at the mean radius.

Solution

The relative flow angles have to be determined first. From Equations (E4.12.3) and (E4.12.4) or for $R = 0.5$, we have $\beta_{m1} = \alpha_{m2} = 54.5°$, $\tan\beta_{m1} = 1.405$, and $\beta_{m2} = \alpha_{m1} = 42.6°$. Assume $\tan\beta_{m2} = 0.92$. Hence from Figure 4.46 or $\tan\beta_{m1} - \tan\beta_{m2} = 1.55/[1 + 1.5(s/c)]$, we have $s/c = [1.55/(1.405–0.92) - 1]/1.5 = 1.47$. Also, from $b/c = 3$, we have $c = (r_t - r_h)/3 = (24–15)/3 = 3$ in., $s = 1.47 \times 3 = 4.4$ in., and $Z_b = 2\pi r_m/s = 2\pi \times 20/4.4 = 28$. The blade angles can be estimated from $\beta_{b1} = \beta_{m1} - i$ and $\beta_{b2} = \beta_{m2} - \delta$. With $i \approx 3°$ and $\delta = m\theta(s/c)^{0.5}$, where $m = 0.23(2a/c)^2 + 0.1(\beta_{m2}/50)$, $\theta = \beta_{b1} - \beta_{b2} \approx \beta_{m1} - \beta_{m2} = 54.5–42.6 = 11.9°$, and $a/c = 0.5$ for a circular arc blade, we have $m = 0.23 + 0.1 \times 42.6/50 = 0.315$, or $\delta = 0.315 \times 11.9(1.47)^{0.5} = 4.5°$. Hence $\beta_{b1} = 54.5–3 = 51.5°$ and $\beta_{b2} = 42.6–4.5 = 38.1°$.

4.2.3 Preliminary Design of Compressor Stage

The preliminary design procedure will be illustrated with the theory covered in the previous sections. For convenience of reading, the relevant equations and design criteria are listed here again:

$$T_o = T + \frac{V^2}{2C_p}, \quad p = \rho RT, \quad \frac{p}{p_o} = \left(\frac{\rho}{\rho_o}\right)^k = \left(\frac{T}{T_o}\right)^{k/(k-1)} \tag{4.22}$$

$$\dot{m} = \rho V_a A = \rho V_a \pi r_t^2 \left[1 - \left(\frac{r_h}{r_t}\right)^2\right]$$

$$C_p \Delta T_{os} = \lambda U V_a(\tan\beta_1 - \tan\beta_2), \tag{4.23}$$

or

$$\psi = \lambda\varphi(\tan\beta_1 - \tan\beta_2),$$

$$\frac{p_{o3}}{p_{o1}} = \left(1 + \frac{\eta_s \Delta T_{os}}{T_{o1}}\right)^{k/(k-1)},$$

$$R = \frac{V_a}{2U}(\tan\beta_1 + \tan\beta_2).$$

The free-vortex condition is

$$R = 1 - \frac{1 - R_m}{\nu^2} \tag{4.24}$$

$$\nu \dot{m}_{\min} = (1 - R)^2 \tag{4.25}$$

The flow deviation at the trailing edge is given as $\delta \equiv \beta_2 - \beta_{b2} = \dot{m}\theta\,(s/c)^{0.5}$, where $m = 0.23(2a/c)^2 + 0.1(\beta_2/50)$.

The **design criteria** are as follows: $U_t \sim 1100$ ft./s, $V_a = 500$ to 650 ft./s, $M_{r1t} < 1.0$, $r_h/r_t = 0.4$ to 0.6, $1 > W_2/W_1 > 0.7$, $\Delta T_{os} = (T_{oe} - T_{oi})/n_s = 20$ to 60 °F, $\nu > \nu_{min}$.

The **design specifications are** (a) overall pressure ratio p_{oe}/p_{oi} and p_{oi}, T_{oi} and (b) mass flow rate m. The subscripts i and e designate the overall inlet and outlet of the compressor. Meanwhile, subscripts 1, 2, and 3 designate the inlet/outlet of the rotor and stator in each stage.

The **design steps** are as follows:

1) Choice of Rotating Speed, N, and r_t: With the selected $V_1 = V_a = 500$ ft./s and $r_h/r_t = 0.4$, then r_t, r_h, ρ_1, p_1, and T_1 can be solved from Equations (4.22) and (4.23). And the mean radius, defined as $r_m = [(r_t^2 + r_h^2)/2]^{0.5}$ or $r_m = (r_t + r_h)/2$ in some of the literature, can be calculated.
   With the selected $U_t = 1100$ ft./s, then N and U_m can be obtained from $N = 30 \times U_t/(\pi r_t)$ and $U_m = \pi N r_m/30$. Assuming a constant mean radius, the following calculations are performed at the mean radius, until step 4, such that $\beta_1 = \tan^{-1}(U_m/V_a)$, $W_1 = V_a/\cos\beta_1$. With selected $W_2/W_1 = 0.72$ greater than 0.7, W_2 and β_2 can be calculated from $\beta_2 = \cos^{-1}(V_a/W_2)$. The temperature rise per stage, ΔT_{os}, can then be calculated from Equation (4.15), with $\lambda \leq 1$ assumed for the first stage. The result is checked with the design criteria. Also, the relative flow Mach number Mrt1 = Wr1/(kRT1)0.5 can be calculated to ensure that it is less than or around unity. If R_m or ΔT_{os} is selected, β_2 can be calculated from Equations (4.15) or (4.18a), and then W_2/W_1 can be checked.
2) Determination of Number of Stages, n_s: With an estimated η_p, the overall temperature ratio T_{oe}/T_{oi} can be calculated from $T_{oe}/T_{oi} = (p_{oe}/p_{oi})^{(k-1)/k\eta p}$. Then the number of stages required can be determined from $n_s = (T_{oe} - T_{oi})/\Delta T_{os}$.
3) Stage-by-Stage Design at Mean Radius: With the selected R_m, ΔT_{os}, and estimated λ values, β_1 and β_2 can be calculated from Equations (4.15) and (4.18a). However, the result should be checked with the de Haller number $W_2/W_1 = \cos\beta_1/\cos\beta_2$, which should be greater than 0.7; otherwise, different values of R_m and ΔT_{os} should be used. The work done factor λ starts with a value close to but less than 1.0 and decreases to around 0.85 as the boundary layer grows downstream of the stages, and R_m starts with a value of around 0.8 and decreases to 0.5. However, the selected value should be checked with Equation (4.25) such that $r_h > r_{min}$. That means R cannot become negative at any radius. The absolute flow angles α_1 and α_2 can be determined from β's with the velocity diagrams shown in Figure 4.43. Or if $R = 0.5$, we have $\alpha_1 = \beta_2$, $\alpha_2 = \beta_1$. And p_{03}, T_{03} for each stage can be calculated from Equation (4.20) and $T_{03} = T_{02} = T_{01} + \Delta T_{os}$. They are the p_{01} and T_{01} for the next stage, and α_3 is equal to the α_1 of the next stage.
   The design with a constant mean radius will result in a nonconstant outer radius, and the design with a constant outer radius will result in an increasing mean radius since the flow passage area decreases with increasing pressure downstream of the stages. In the latter case, we will have $U_2 > U_1$ at the mean radius, and hence higher $C_p\,\Delta T_{os} = U_2 V_{u2} - U_1 V_{u1}$ for each stage, and it may require less number of stages to achieve the total required pressure rise.
4) Number of Blades: The number of blades in each row can be determined with the aid of Figure 4.46, which is discussed in the last section. With the aspect ratio b/c being set (primarily from the stress consideration), the chord c and pitch s can be determined.
5) Variation of Angles from Hub to Tip: For the preliminary design, calculations are made only at three radii: r_h, r_m, and r_t. At each stage, with r_m specified, r_h and r_t can be determined from Equations (4.22) and (4.23) to accommodate the required mass flow rate. For the free-vortex design, we have V_a = const. and V_u varies with

respect to the radius as $V_u = V_{um}r_m/r$, and hence α and β at hub and tip can be determined from $\tan\alpha = V_u/V_a$, $\tan\beta = (U - V_u)/V_a$. Again, it is necessary to make sure that R is positive at all radii with Equation (4.24). Detailed numerical calculations for a sample case will be given in the following examples.

6) Construction of Blade Shape: Constructing the blade shape means determining the blade angles β_{b1}, β_{b2} and the stagger angle γ with the pitch s and chord c determined. For a circular arc camber, we have $a/c = 0.5$. Hence the deviation equation becomes $\delta \equiv \beta_2 - \beta_{b2} = [0.23 + 0.1(\beta_2/50)](\beta_{b1} - \beta_{b2})(s/c)^{1/2}$. Since β_{b1} can be determined from $\beta_{b1} = \beta_1 - i$ with a selected incidence angle i, β_{b2} can be determined from the equation directly. And the stagger angle can be calculated from $\gamma = \beta_{b1} - \theta/2$, where $\theta = \beta_{b1} - \beta_{b2}$, as shown in Figure 3.39. At the other radii, the pitch s can be calculated from $s = 2\pi r/Z_b$. Then, the corresponding chord c can be determined from Figure 4.32 for the desired flow angles. The blade can be constructed using a similar procedure.

Example 4.14 A multistage axial compressor is required to compress air from the ambient pressure of 14.7 psia and temperature of 70 °F to a pressure ratio of 4.5. Each stage has a 50% degree of reaction, mean blade speed of 920 ft./s, flow coefficient of 0.5, stage loading factor of 0.35, stage efficiency of 90%, and work done factor of 0.92. Determine the number of stages, the flow angles at both inlet and outlet and the overall adiabatic efficiency.

Solution

From $\psi = \lambda\,\varphi(\tan\beta_1 - \tan\beta_2)$ and $R = \frac{1}{2}\varphi(\tan\beta_1 + \tan\beta_2)$, we obtain $\tan\beta_1 - \tan\beta_2 = 0.35/(0.92\times0.5) = 0.761$ and $\tan\beta_1 + \tan\beta_2 = 2\times0.5/0.5 = 2$. Hence $\tan\beta_1 = 1.38$, $\beta_1 = 54.1°$ and $\tan\beta_2 = 0.619$, $\beta_2 = 31.8°$. For each stage, we have $\psi = C_p\,\Delta T_{os}/U_m^2$. Hence $\Delta T_{os} = 0.35\times920^2/6012 = 49.3$ R, where $C_p = 0.24\times778\times32.2 = 6012$ ft-lb$_f$/slug. For the overall compressor, from Equation (4.20), we have $P_{oe}/P_{oi} = (1 + \Delta T_{oe}/T_{oi})^{k\eta p/(k-1)}$. Hence $\Delta T_{oe} = 530[(4.5)^{0.2857/0.9} - 1] = 324$ R. The number of stages can be obtained as $n_s = \Delta T_{oe}/\Delta T_{os} = 324/49.3 = 6.58 \approx 7$. Hence the actual values are $\Delta T_{oe} = 7\times49.3 = 345.1$ R, $p_{oe}/p_{oI} = (1 + 345.1/530)^{0.9\times3.5} = 4.85$. The adiabatic efficiency is $\eta_{ad} = [(p_{oe}/p_{oi})^{(k-1)/k} - 1]/[(T_{oe}/T_{oi}) - 1] = [4.85^{0.2857} - 1]/[345.1/530] = 0.875 = 87.5\%$.

Example 4.15 Use the result of Example 4.14 to design a multistage axial compressor rotating at 15,000 rpm for a mass flow rate of 65 lb$_m$/s and pressure ratio of 4.5. The ambient condition is 70 °F and 14.7 psia. The degrees of reaction, work done factor, and energy transfer for each stage can vary, although all stages' efficiencies are 90%. Keeping the mean radii constant, determine the hub and tip radii and inlet/outlet flow angles for each stage.

Solution

Assume a constant axial flow velocity $V_a = 500$ ft./s, no prewhirl, and hence $V_1 = V_a$ for the first stage, and free-vortex conditions for all stages. The calculations can be performed as follows:

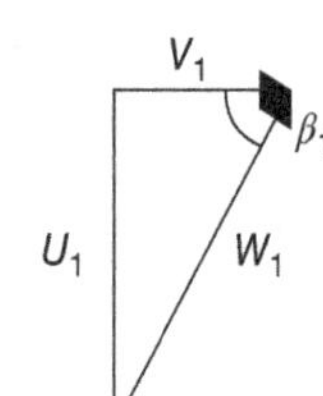

1. First Stage

$$T_1 = 530 - \frac{500^2}{2\times6012} = 509.2\ \text{R},$$

$$p_1 = 14.7\left(\frac{509.2}{530}\right)^{3.5} = 12.78\ \text{psia},$$

$$\rho_1 = \frac{12.78\times144}{53.33\times509.2} = 0.0677\ \text{lb}_m/\text{ft}^3.$$

With a selected $\nu = r_h/r_t = 0.45$, we have $A = m/(\rho_1 V_a) = 1.92\ \text{ft}^2 = 276.5\ \text{in.}^2 = \pi r_t^2\,(1 - \nu^2)$. Hence $r_t = 10.5$ in., $r_h = 4.7$ in., $r_m = 7.6$ in. and $U_t = r_t\omega = (10.5/12)\times1570.8 = 1374.4$ ft./s, $U_h = r_h\omega = 615.2$ ft./s, $U_m = r_m\omega = 996.6$ ft./s. Without inlet whirl flow, we have $W_t = [U_t^2 + V_a^2]^{0.5} = 1463$ ft./s, $W_h = 792.8$ ft./s, and $W_m = 1115.0$ ft./s. Also, $\tan\beta_{1m} = U_m/V_a = 1.99$, $\beta_{1m} = 63.3°$. With $a_1 = (kRT_1)^{0.5} = 1309$ ft./s, the relative Mach numbers will be $M_{rt} = W_t/a_1 = 1.12$, $M_{rh} = 0.60$, and $M_{rm} = 0.85$. Hence, the relative flow at the leading edge is transonic, and a supersonic blade may be needed. From the last example, to achieve a pressure ratio of 4.5, we have $\Delta T_{os} = 324/7 = 46.3$ R, so $\Delta T_{os} = 45$ R is selected for the first stage. From $C_p\,\Delta T_{os} =$

$\lambda U_m V_a(\tan\beta_{1m} - \tan\beta_{2m})$, with $\lambda = 0.95$, we obtain $\tan\beta_{2m} = 1.42°$ and $\beta_{2m} = 54.8°$. Checking the degree of reaction, from Equation (4.18a)

$$R = \frac{V_a}{2U_m}(\tan\beta_1 + \tan\beta_2) = 0.855,$$

we have $v_{\min} = (1 - R)^{0.5} = 0.38$, which is less than the selected value and is acceptable. Also, checking the de Haller number $W_2/W_1 = \cos\beta_1/\cos\beta_2 = 0.78 > 0.70$, it is also acceptable.

Before proceeding to the next stages, the following parameters have to be specified. The summation of ΔT_{os} should add up to 324°. So the following arrangement is chosen:

Stage number	1	2	3	4	5	6	7
ΔT_{os} (R)	45	46	47	47	47	47	45
Λ	0.95	0.93	0.90	0.89	0.88	0.86	0.85
R	0.855	0.8	0.7	0.63	0.60	0.55	0.50

2. Second Stage

$$T_{01} = 530 + 45 = 575\text{R} \qquad p_{01} = 14.7\left(1 + \frac{0.9 \times 45}{530}\right)^{3.5} = 19.0\ \text{psia}.$$

With the specified ΔT_{os}, λ, and R, Equations (4.15) and (4.18a) can be solved for $\tan\beta_1$ and $\tan\beta_2$. Hence $\beta_1 = 62.1°$ and $\beta_2 = 52.4°$, where $W_2/W_1 = 0.76 > 0.70$. Also, we have

$$V_1 = \left[V_a^{\,2} + (U_m - V_a\ \tan\beta_1)^2\right]^{0.5} = 502.6\ \text{ft/s},$$

$$T_1 = T_{01} - \frac{V_1^{\,2}}{2C_p} = 554\ \text{R}, \qquad p_1 = p_{01}\left(\frac{T_1}{T_{01}}\right)^{3.5} = 16.7\ \text{psia},$$

$$\rho_1 = \frac{p_1}{RT_1} = 0.0813\ \text{lb}_\text{m}/\text{ft}^3, \qquad A = \frac{m}{\rho_1 V_a} = 230\ \text{in.}^2 = 2\pi r_m b,$$

where $b = r_t - r_h$ and r_m is the same as that for the first stage. Hence, we have $r_t = 10.0$ in., $r_h = 5.2$ in., and $v = 0.52$, which is greater than v_{min} (0.447).

3. Third to Seventh Stages: The calculation procedure for these stages will be similar to the second stage and hence will not be repeated here. The results are as follows:

Stage number	1	2	3	4	5	6	7
T_{01} (R)	530	575	621	668	715	762	809
p_{01} (psia)	14.7	19.0	24.3	30.5	37.9	46.3	55.9
r_t (in.)	10.5	10.0	9.6	9.3	9.1	8.9	8.8
r_h (in.)	4.7	5.2	5.6	5.9	6.1	6.3	6.5
r_m (in.)	7.6	7.6	7.6	7.6	7.6	7.6	7.6
β_1 (deg)	63.3	62.1	59.7	57.5	56.5	54.7	52.4
β_2 (deg)	54.9	52.3	47.2	43.2	41.4	38.0	34.8

The final discharge stagnation pressure can also be checked with the flow across the last stage such that $p_{oe} = p_{03} = p_{01}(1 + \eta_s\ \Delta T_{os}/T_{01})^{k/(k-1)} = 55.9(1 + 0.9 \times 45/809)^{3.5} = 66.3$ psia, which checks with $p_{oe}/p_{oi} = 66.3/14.7 = 4.51$.

4.2.4 Determination of Stage Efficiency

After the blade design is completed, the stage efficiency η_s (or polytropic efficiency η_p) assumed in the design procedure has to be checked with those calculated from the test data of the cascade or airfoil chosen. To do this, we need the relationships between cascade/airfoil parameters and those of the compressor flow parameters.

There are two issues to be considered: (a) The tested cascade/airfoil is stationary while the rotor is rotating. So, there will be additional losses in the rotor flow, that is, the secondary flow losses associated with the vortex and blade tip clearance and the annulus drag. They are given in terms of drag coefficients. (b) The stage efficiency η_s is defined in terms of energy transfer, while the loss in the cascade/airfoil test, called blade efficiency η_b, is defined in terms of the pressure losses due to friction. They are not identical.

The first issue was discussed in Section 3.2.4 for the axial-flow pump and fan. It is also applicable to axial-flow compressors. But the second issue needs more elaboration. Since the flow through the pump and fan can be treated as an incompressible fluid, the blade efficiency obtained from cascade/airfoil data is equivalent to the hydraulic efficiency for pumps and fans, as shown in Equation (3.21).

To relate the stage efficiency with blade row efficiency for the compressor, some approximations are needed. Referring to Figure 4.45, from the definition of stage efficiency

$$\eta_s = \frac{T_{sso3} - T_{o1}}{T_{o3} - T_{o1}} \cong \frac{T_{ss3} - T_1}{T_3 - T_1} = \frac{T_1(T_{ss3}/T_1 - 1)}{\Delta T_s},$$

we have

$$\frac{p_3}{p_1} = \left(1 + \frac{\eta_s \Delta T_s}{T_1}\right)^{k/(k-1)}.$$

Also we have

$$\frac{p'_3}{p_1} = \left(1 + \frac{\Delta T_s}{T_1}\right)^{k/(k-1)},$$

where p'_3 is the ideal pressure corresponding to T_3 along the isentropic process upward from point 1.

Hence, the blade row efficiency can be expressed as

$$\eta_b = \frac{p_3/p_1 - 1}{p'_3/p_1 - 1} = \frac{\left(1 + \frac{\eta_s \Delta T_s}{T_1}\right)^{k/(k-1)} - 1}{\left(1 + \frac{\Delta T_s}{T_1}\right)^{k/(k-1)} - 1} \cong \frac{\frac{k}{k-1}\frac{\eta_s \Delta T_s}{T_1}\left[1 + \frac{\eta_s \Delta T_s}{2(k-1)T_1}\right]}{\frac{k}{k-1}\frac{\Delta T_s}{T_1}\left[1 + \frac{\Delta T_s}{2(k-1)T_1}\right]}$$

$$\cong \eta_s \left[1 - (1 - \eta_s)\frac{\Delta T_s}{2(k-1)T_1}\right].$$

Further simplification can be made if needed. For $k = 1.40$ and $\eta_s \cong 0.8$ we will have $\eta_b \cong \eta_s\,(1–0.25\,\Delta T_s/T_1)$. Some sample calculations for $k = 1.40$, $\Delta T_s = 20$ R, and $T_1 = 530$ R can be obtained as $\eta_s = 0.6, 0.7, 0.8, 0.9$ and $\eta_b = 0.589, 0.690, 0.792, 0.896$. Or we have η_s being always slightly greater than η_b.

The above approximations are made, based on the Taylor expansion for a small value of ε such that $(1 + \varepsilon)^x = 1 + x\varepsilon + x(x-1)\,\varepsilon^2/2 + \ldots \cong 1 + x\varepsilon\,[1 + (x-1)\varepsilon/2] \cong 1 + x\varepsilon$.

4.2.5 Effects of Compressibility and Shock Waves

The effects of Mach number on losses in a typical subsonic cascade are shown qualitatively in Figure 4.47 (comparable to Figure 3.46). At a low Mach number, the range of incidences for which losses are acceptable is wide. As the Mach number increases, this range decreases, and the overall loss level increases substantially.

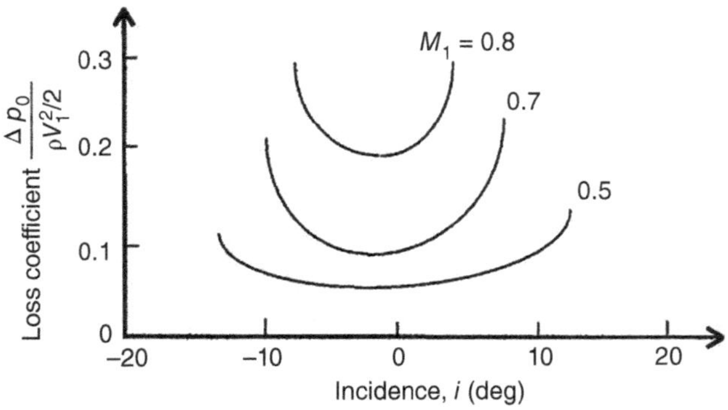

Figure 4.47 Qualitative effect of Mach number on losses in cascade.

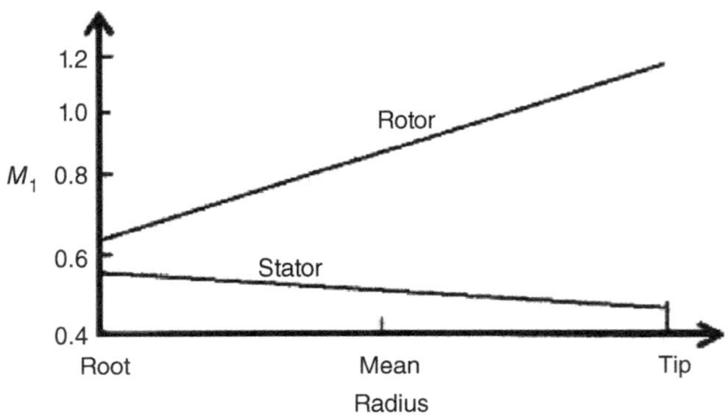

Figure 4.48 Variation of entry Mach number along blade radius.

In a normal design, the maximum relative Mach number at the rotor inlet occurs at the tip due to the maximum tip velocity. But the absolute maximum Mach number at the stator inlet occurs at the root. Because the whirl velocity is maximum at the root, there is constant work input along the radial direction. A typical distribution is shown in Figure 4.48.

For the Mach number those changes from subsonic at the root to supersonic at the tip, called transonic flow, the pressure loss through a blade row can be analyzed with the model shown in Figure 4.49. Supersonic flow is accelerated from A to B through the Prandtl–Meyer expansion with a turning angle of $\Delta\theta$ along the suction side surface. A shock wave emanates from C due to the concave surface of the pressure side of the next blade encountered. It is approximately normal to the mean streamline and produces a pressure jump. But the loss through the normal shock wave is the pressure loss.

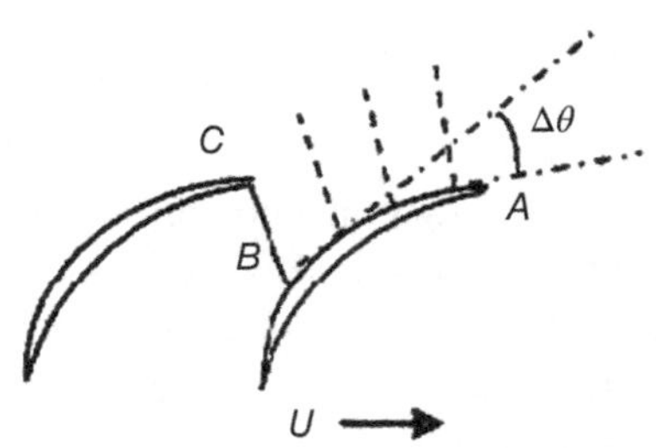

Figure 4.49 Shock loss model.

Additional loss can be due to shock wave–boundary layer interaction and recirculation flow. Other problems that the shock wave may cause are secondary flow due to the different flow patterns at the different radii and the flow may be choked, unable to respond to the downstream condition changes.

For the high-Mach-number supersonic flow at the inlet, a supersonic diffuser with a decreasing flow area or a slightly concave suction side can be used to bring the flow to a transonic speed ($M \sim 1.2$) before entering the blade row.

Some basic topics on compressible flow and shock waves are given in Appendix A. But a detailed discussion of their effects on compressor performance is beyond the scope of this text.

4.2.6 Axial-Flow Compressor Stage Performance

The performance of a compressor stage can be expressed in terms of the nondimensional parameters loading factor ψ versus flow coefficient φ. Equation (4.15), can be rearranged into

$$\psi = \lambda\varphi(\tan\beta_1 - \tan\beta_2) = \lambda\frac{V_a}{U}\left(\frac{U - V_{u1}}{V_a} - \tan\beta_2\right)$$

$$= \lambda\left(1 - \frac{V_{ul}}{U} - \frac{V_a}{U}\tan\beta_2\right) = \lambda[1 - \varphi(\tan\alpha_1 + \tan\beta_2)] = \lambda(1 - \varphi k), \tag{4.26}$$

where $\tan\alpha_1 + \tan\beta_2 = k$ is constant when the flow rate changes for the ideal case, since α_1 is fixed by the inlet guide vane or the previous stage stator and β_2 is relatively fixed by the rotor outlet angle.

As mentioned earlier, the work done factor λ is used to account for the reduction of energy transfer due to the flow slip, nonuniformity, and blockage by the boundary layer at the rotor outlet. For the ideal case, $\lambda = 1.0$, while in the actual case, λ decreases when the flow rate decreases or increases from the design condition. The pressure ratio across a single stage as shown in Equation (4.20) can be replaced by the pressure coefficient, defined as

$$\overline{C}_p = \frac{\Delta p_{os}}{\rho_{01}U^2},$$

by rewriting it as

$$\frac{p_{03}}{p_{01}} = \frac{p_{01+\Delta p_{os}}}{p_{01}} = 1 + \frac{\Delta p_{os}}{p_{01}} = \left(1 + \frac{\eta_s\Delta T_{os}}{T_{01}}\right)^{k/(k-1)} \cong 1 + \frac{k}{k-1}\frac{\eta_s\Delta T_{os}}{T_{01}}$$

$$= 1 + \frac{C_p}{R}\frac{\eta_s\Delta T_{os}}{T_{01}},$$

where $\Delta T_{os}/T_{01} \ll 1.0$. Hence

$$\frac{\Delta p_{os}}{p_{01}} = \frac{C_p}{R}\frac{\eta_s\Delta T_{os}}{T_{01}} \quad \text{or} \quad \overline{C}_p = \frac{\Delta p_{os}}{\rho_{01}U^2} = \eta_s\frac{C_p\Delta T_{os}}{U^2} = \eta_s\psi.$$

Hence the stage performance characteristics can be represented by ψ and η_s versus φ, as shown in Figure 4.50. And the pressure rise can be obtained from these two curves.

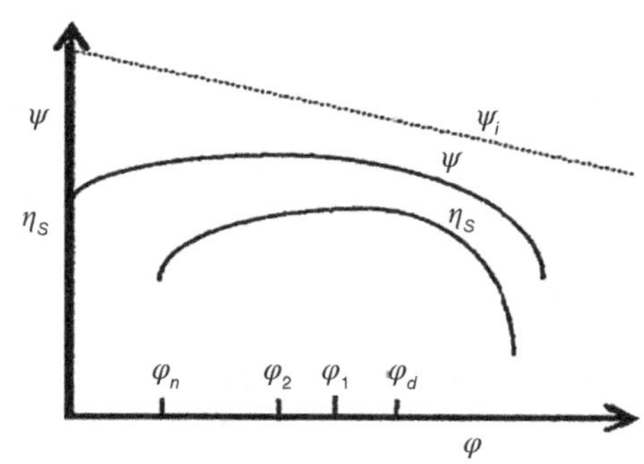

Figure 4.50 Compressor stage performance characteristics in nondimensional parameters.

As shown in Figure 4.50, the performance deteriorates rapidly on both ends of the design flow rate. At the low flow, surging occurs due to the flow mismatch; while at the high flow, choking occurs due to the shock waves associated with supersonic flow. In a multistage compressor, surging and choking are more likely to occur in the downstream stages, while it is free in the upstream stages. This is due to a phenomenon called stage stacking. Starting from the design flow rate, if the compressor is subjected to a decrease in the mass flow rate, say, the first-stage flow decreases from φ_d to φ_1 as shown above. It will result in an increase in pressure and density at its discharge. Both factors combined (decrease of mass flow rate and increase of density) cause a further decrease of axial-flow velocity at the inlet of the next stage, φ_2, from $\dot{m}_{\cdot 1} = \dot{m}_{\cdot 2} = \rho_2 V_{a2} A_2$. This effect is propagated downstream. Eventually, some stage will stall at φ_n. A similar process occurs when the mass flow rate increases from the design condition and causes some stage downstream to choke.

A typical performance chart of an axial-flow compressor can be plotted in terms of the nondimensional parameters as that for the centrifugal compressor shown in Figure 4.19. Or it can be plotted for different rotating speeds as shown in Figure 4.51.[4]

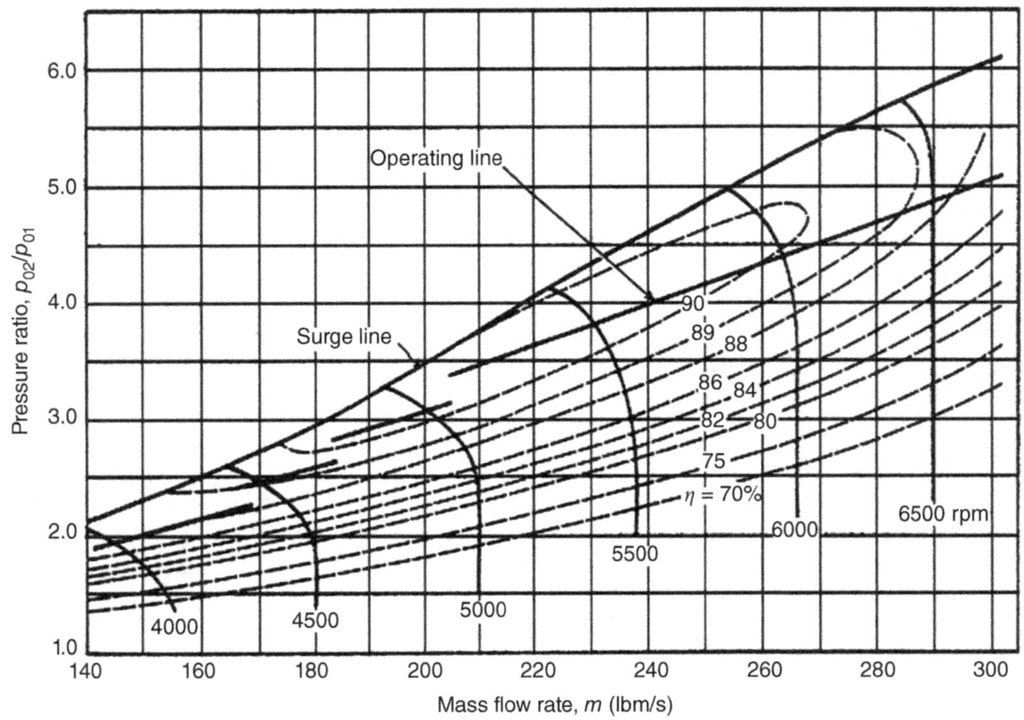

Figure 4.51 Typical axial-flow compressor performance curves.

The constant efficiency curves are also plotted. It is shown that for a constant operating speed the flow rate is limited to an even more narrow range by surging and choking.

However, the axial-flow compressor is normally operated as a variable-speed machine. In designing a gas turbine power plant, the compressor must be matched with the turbine and other components to assure that the operating line is sufficiently far away from the potential surging and choking, even if its efficiency has to be sacrificed slightly. This will be discussed in more detail in the next chapter when dealing with a procedure matching compressor and turbine.

Example 4.16 An axial-flow compressor stage operating at design condition has the following parameters: $\psi = 0.40$, $R = 0.65$ at $\varphi = 0.55$. If the work done factor $\lambda = 0.9$ and the exit flow angles from both rotor and stator are unchanged when the flow rate is reduced by 10%, determine the loading coefficient and reaction at this low flow condition. Also draw the velocity diagrams at the rotor inlet and outlet for both conditions.

Solution

Since the exit flow angle of the stator is the absolute flow angle at the inlet of the next stage, we have constant α_1 and β_2 as shown in the velocity diagrams for both conditions.

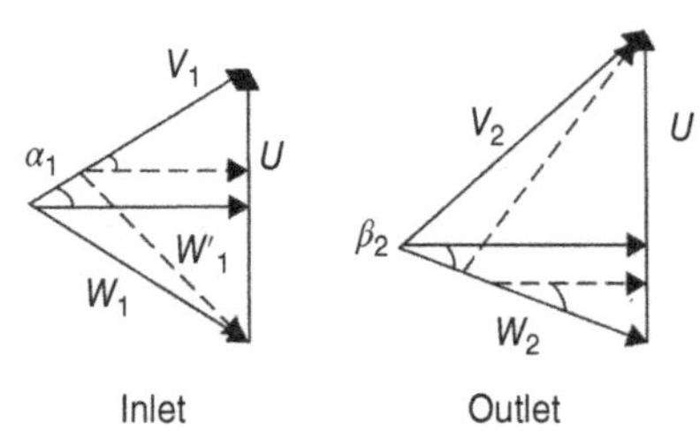

From Equations (4.15) and (4.18b), $\psi = \lambda\ [1 - \varphi(\tan\alpha_1 + \tan\beta_2)]$, $R = 1/2 + (\varphi/2)(\tan\beta_2 - \tan\alpha_1)$, we have $\tan\alpha_1 + \tan\beta_2 = (1 - \psi/\lambda)/\varphi = 1.01$ and $\tan\alpha_1 - \tan\beta_2 = (1 - 2R)/\varphi = -0.545$, or $\psi = \lambda(1\text{–}1.01\varphi)$ and $R = 0.5 + 0.272\varphi$. Hence at $\varphi' = 0.495$, assuming constant λ, we have

$$\psi' = \lambda[1 - \varphi'(\tan\alpha_1 + \tan\beta_2)] = 0.9[1 - 0.495 \times 1.01] = 0.45,$$

$$R' = \frac{1}{2} + \left(\frac{1}{2}\varphi'\right)(\tan\beta_2 - \tan\alpha_1) = 0.5 + 0.272\varphi' = 0.635.$$

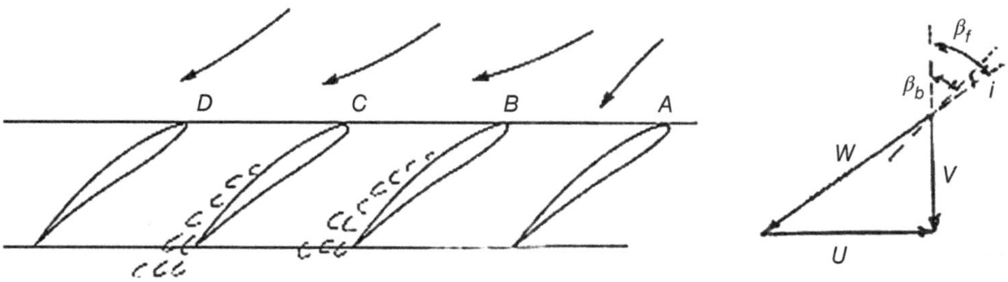

Figure 4.52 Rotating stall in axial-flow compressor.

4.2.7 Surge and Stall in Compressor and Remedies

Surge and stall are phenomena that occur when the flow angle of attack at the leading edge of the blades reaches a value such that boundary layer separation occurs. It is aerodynamically similar to the stall of an aircraft wing. At the stall, a wing loses much of its lift, while a compressor suffers noise, vibration, flow fluctuation, and a sharp loss in pressure rise, similar to the flow recirculation encountered in pumps. The onset of this phenomenon is difficult to predict analytically. The surge lines on the performance maps are typically determined from laboratory tests.

Surge and stall can occur in either the rotor or stator blades or both. Typically, they start from the stator in a low-pressure-ratio compressor and from the rotor in a high-pressure-ratio compressor. The stall can start from one blade and propagate around the annulus, causing a high-frequency noise and vibration. It is called a "rotating stall." This phenomenon can be explained intuitively as shown in Figure 4.52. For example, at a certain low flow rate, the stall (or boundary layer separation called a patch) starts from a blade B with a certain incidence i, as shown in the velocity diagram. This patch will cause a partial obstruction to the flow and deflect the flow on both sides of this blade. The incidence of the flow on the right of the stall cell is reduced, with an increased axial-flow velocity. But the incidence of flow on the left is increased, with a decreased axial-flow velocity. As these blades are already close to stalling, the net effect is for the stall patch to move to the left (blade C). And the process is repeated to the next left blade. The vibration frequency caused by the rotating stall is related to the rotor rotating speed and blade number. Hence it may be close to the natural frequency of the rotor and cause a vibration resonance and severe damage to the compressor system.

Stalling and choking can occur at either the front stages or the rear stages, depending on the change of flow rate or change of rotating speed from the design condition. This can be explained by the negative slope performance curve of a compressor stage.

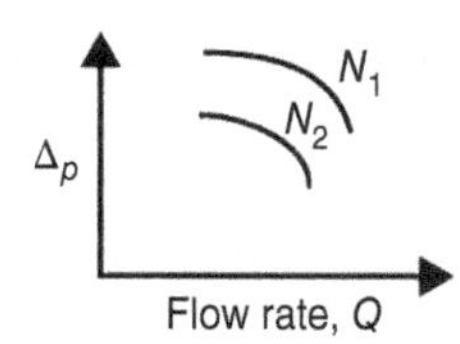

A sudden decrease in rotating speed, as shown in Figure 4.53*a*, will cause a decrease in pressure/density and an increase of axial-flow velocity progressively flow rate, Q into the rear stages, and hence choking will start at a certain rear stage. Also, the mass flow rate generally falls off more rapidly than the rotating speed, so the first stage may stall. Since the relative flow angle decreases from the design value, as indicated by dashed lines. Reversibly, an increase in rotating speed will cause an increase in pressure/density and a decrease in axial-flow velocity, so the stall will occur at the rear stages and the first stage may be choked.

Under a constant rotating speed, if the flow rate is reduced, the axial-flow velocity decreases progressively to the rear stages, as shown in Figure 4.53*b*, and a stall may occur as explained in the last section.

4.2.7.1 Twin-Spool Compressor

To avoid the rear-stage choking, air can be bled off at the intermediate stage, but it is a wasteful operation. Another approach is to use a twin-spool arrangement. The compressor is split into two sections, the LP (low-pressure) section in the front and the HP (high-pressure) section in the rear, which are driven by two separate turbines at different speeds. Two shafts are mechanically independent. Under the off-design condition, when the mass flow

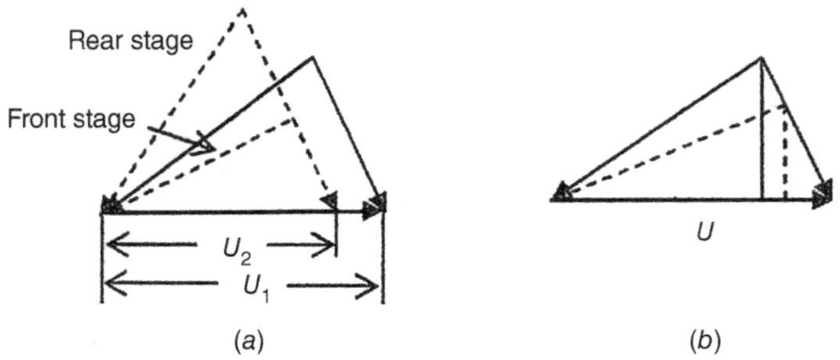

Figure 4.53 Velocity diagrams for changing operation. (*a*) Rotating speed decreases and (*b*) mass flow rate decreases.

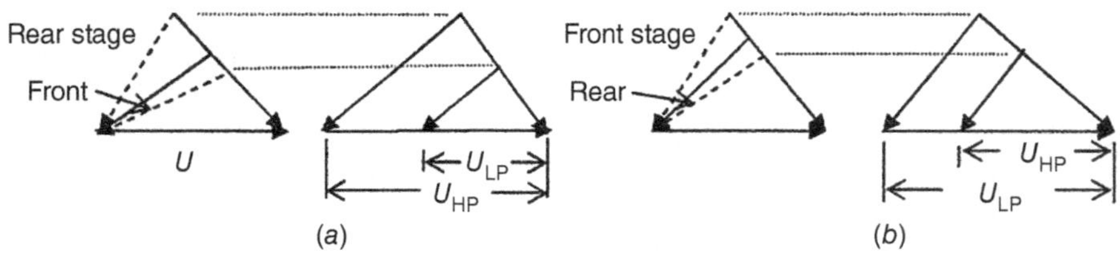

Figure 4.54 Changes of velocity diagram for twin-spool compressor. (*a*) Mass flow rate decreases and (*b*) mass flow rate increases.

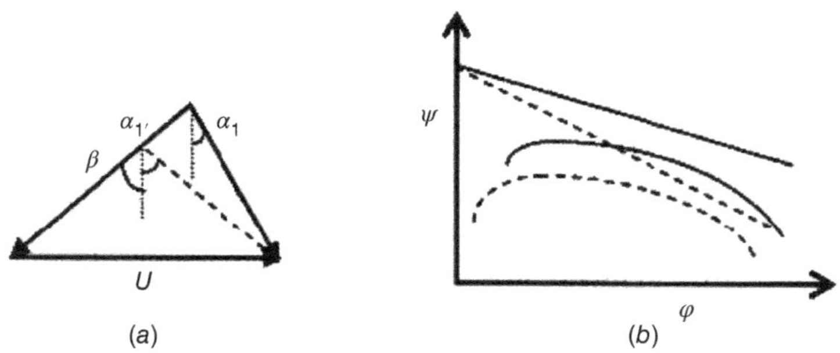

Figure 4.55 Effects of stator blade angle change on compressor performance.

rate has to be reduced, the LP section will be running at a lower speed, with the HP section running at a higher speed, as shown in Figure 4.54*a*. So, the incidence angles of all stages can be maintained relatively close to the design angles. When the mass flow rate must be increased, the LP section will be running at a higher speed than the HP section, as shown in Figure 4.54*b*.

4.2.7.2 Variable-Geometry Compressor

An alternative approach to improve the off-design performance is to use several rows of variable stators and IGVs (inlet guide vanes) at the front of the compressor stages to change the flow angle α_1, which is the α_3 of the upstream stage. When the flow rate decreases, α_1 increases to α_1' as, shown in Figure 4.55*a*, so β_1 can be maintained constant. A typical stage characteristic is shifted as shown, where the straight lines are for the ideal case $\psi = 1 - k\phi$ as mentioned earlier. And the surge point is moved to the left, as shown in Figure 4.55*b*.

References

1 AMCA Publication 201-02. *Fans and Systems, Air Movement and Control Association*, Arlington Heights, IL, 2002.
2 Gibbs, C. W. (Ed.), *Compressed Air and Gas Data*, 2, Ingersoll-Rand Co. Phillisburg, N.J., 1971.
3 IAP, Inc., *Technical Data Tubular Centrifugal*, Industrial Air Products, Philips, WI, 1993.
4 Loren Cook, inc., *Cook Centrifugal Plug Fan*, Loren Cook Co., Springfield, MO, 1989.
5 Balje, O. E., *Turbomachines*, John Wiley & Sons, New York, 1981.
6 Boyce, M.P, 2003, "*Centrifugal Compressor: A Basic Guide*," PennWell Corporation, Tulsa, OK, USA.
7 Japikse, D., 1996, "*Centrifugal Compressor Design and Performance*," Concepts ETI, VT, USA.
8 Zangeneh, M., Vogt, D. and Roduner, C., "Improving a Vaned Diffuser for a Given Centrifugal Impeller by 3D Inverse Design," Presented at the ASME TURBO EXPO 2002, Amsterdam, Netherlands, June 3–6, 2002. ASME paper no. GT-2002-30621, 2002.
9 Amineni, N. K., and Engeda, A., "Pressure Recovery in Low Solidity Vaned Diffusers for Centrifugal Compressors," 97-GT-472, 1997.
10 Senoo, Y., Hayami, H. and Ueki, H., "Low Solidity Tandem Cascade Diffusers for Wide Flow Range Centrifugal Blowers," 83-GT-3, 1983.
11 Boyce, M. P., How to Achieve Online Availability of Centrifugal Compressors, in *Fluid Movers–Pumps, Compressors, Fans and Blowers*, Matley, J. et al. (Eds.) Chemical Week Associates, New York, 1979.
12 Eck, B., *Fans*, Pergamon, Oxford, 1973.
13 Kenny, D. P., "The History and Future of the Centrifugal Compressor in Aviation Gas Turbine,' 1st Cliff Garrett Turbomachinery Award Lecture, Society of Automotive Engineers, SAE/SP-804/602, 1984.
14 Morris, R. E. and Kenney, D. P., High Pressure Ratio Centrifugal Compressors for Small Gas Turbine Engines, in *Advanced Centrifugal Compressor*, Dean, R. C. Jr., (Ed.), Gas Turbine Division, American Society of Mechanical Engineers, New York, 1971.
15 He, L. and Wang, D.X., 2011, "Concurrent Blade Aerodynamic-Aero-Elastic Design Optimization Using Adjoint Method," *Journal of Turbomachinery*, Vol. 133, No.1, pp. 110211–1102110.
16 Rodger, C., "Specific Speed and Efficiency of Centrifugal Impellers," in Proceedings of Symposium on Performance Prediction of Centrifugal Pumps and Compressor, Gopalakrishnan, S. & etc. (Eds.) American Society of Mechanical Engineers, New Orleans, LA, Mar. 1980.
17 Ghisu, T., Parks, G. T., Jarret, J. P. and Clarkson, P.J., 2011, "An Integrated System for the Aerodynamic Design of Compression Systems—Part 1: Development," *Journal of Turbomachinery*, Vol. 133, No.1, pp. 110121–110128.
18 Wilson, D. G., *The Design of High-Efficiency Turbomachinery and Gas Turbines*, MIT Press, Cambridge, MA, 1984.
19 Verstraete, V. Alsalihi, Z., Van den Braembussche, R.A., 2010, "Multidisciplinary Optimization of a Radial Compressor for Microgas Turbine Applications," *Journal of Turbomachinery*, Vol. 132, No.1, pp. 310041–310047.
20 Saravanamuttoo, H., Roger, C., and Cohe, H., *Gas Turbine Theory*, 5, Prentice-Hall, New York, 2001.
21 ASHRAE Handbook, *HVAC Systems and Equipment*, American Society of Heating and Air-Conditioning Engineers, Atlanta, GA, 2004.
22 Luo, J., Xiong, J., Liu, F. and McBean, I., 2011, "Three-Dimensional Aerodynamic Design Optimization of a Turbine Blade by Using an Adjoint Method," *Journal of Turbomachinery*, Vol. 133, No.1, pp.110211-1102110-110261-1102611.
23 Xu, C. and Amano, R.S., 2009, "On the Development of Turbomachine Blade Aerodynamic Design System," *International Journal for Computational Methods in Engineering Science & Mechanics* Vol. 10, Issue 3, pp. 186–196.

24 Morris, R. E., and Kenny, D. P., High Pressure Ratio Centrifugal Compressor for Small Gas Turbine Engines, Advanced Centrifugal Compressors, Dean, R. C., Jr., (Ed.) *Gas Turbine Division*, American Society of Mechanical Engineers, New York, 1971.
25 Weber C. R., and Koronowski, M. E., "Meanline Performance Prediction of Volutes in Centrifugal Compressors," ASME 31st Gas Turbine Conference and Exhibit, Dusseldorf, Germany, 86-GT-216, 1987.
26 Kerrebrock, J.L, *Aircraft Engines and Gas Turbines*, MIT Press, Cambridge, MA, 1992.
27 Arora, J.S., 1998, *Introduction to Optimum Design*, NY, MCGraw-Hill.
28 D. Bonaiuti, A. Arnone, M. Ermini, and L. Baldassarre, n.d. "Analysis and Optimization of Transonic Centrifugal Compressor Impellers Using the Design of Experiments Technique," GT-2002-30619.
29 Harry, M. J., 1997, "*The Nature of Six Sigma Quality*," Motorola Univ. Press., Shaumburg, Illinois.
30 Dong, R., Chu, S., and Katz, J., 1997, "Effect of Modification to Tongue and impeller Geometry on Unsteady Flow, Pressure Fluctuations and Noise in a Centrifugal Pump," *ASME J. Turbomach.*, 119, PP. 506–515.
31 Xu, C. and Valentine D, "Diffuser for a Centrifugal Compressor," U.S. Patent # 7581925, 1997.
32 Xu, C, Amano, R. S. and Lee, E. K., 2004, "Investigation of an Axial Fan—Blade Stress and Vibration due to Aerodynamic Pressure Field and Centrifugal Effects," *JSME International Journal, series b*, vol. 47, No. 1, pp. 75–90.
33 Sorokes, J.M., Range Versus Efficiency – A Dilemma for Compressor Designers and Users, IMECE2003–44221, Proceedings of PID. Industrial and Pipeline Compressors Sessions, ASME - IMECE 2003, November 16–21, 2003 Washington, DC.
34 Xu, C. and Amano R.S., "On the Development of Turbine Blade Aerodynamic Design System," ASME IGTI Turbo & Expo, 2001-GT-0443, 2001.
35 Xu, C and Amano, R. S., "Turbomachinery Blade Aerodynamic Design and Optimization," ASME paper, Paper No. GT-2002-30541, 2002.
36 Xu, C, & Amano, RS. "A Study of a Single Stage Centrifugal Compressor." *Proceedings of the ASME 2006 Power Conference. ASME 2006 Power Conference.* Atlanta, Georgia, USA. May 2–4, 2006. pp. 273-280. ASME. https://doi.org/10.1115/POWER2006-88023.
37 Xu, C., "Impeller for Centrifugal Compressor", U. S. Patent pending, 20070059179, 2007.
38 Xu, C. and Amano R.S., Computational Analysis of Scroll Tongue Shape to Compressor Performance by Using Different Turbulence Models, ASME Paper, Paper No. GT-2007-28224.
39 ASME. (1997). *Performance Test Code on Compressors and Exhausters (PTC 10-1997)* New York, NY: American Society of Mechanical Engineers.
40 Sideris M. and Van den Braembussche, R., 1987, "Influence of a Circumferential Exit Pressure Distortion on the Flow in an Impeller and Vaneless Diffuser." *ASME J. of Turbomach.*, Vol. 109, No 1, pp. 48–54.
41 Flathers, MB, & Bache, GE. "Aerodynamically Induced Radial Forces in a Centrifugal Gas Compressor: Part 2 — Computational Investigation." *Proceedings of the ASME 1996 International Gas Turbine and Aeroengine Congress and Exhibition. Volume 1: Turbomachinery.* Birmingham, UK. June 10–13, 1996. V001T01A086. ASME. https://doi.org/10.1115/96-GT-352.
42 Flathers, M. B., and Bache, G. E., "Aerodynamically Induced Radial Forces in a Centrifugal Gas Compressor—Part 2: Computational Investigation," ASME Paper, 96-GT-352, 1996.
43 Hagelstein, D., Van den Braembussche, R., Keiper, R., and Rautenberg, M., 1997, "Experimental Investigation of the Circumferential Pressure Distortion in Centrifugal Compressor Stages," *ASME paper* No. 97-GT-50.
44 Gu, F., Engeda, A., Cave, M. and Liberti, J., 2001, "A Numerical Investigation on the Volute/Diffuser Interaction due to the Axial Distortion at the Impeller Exit", *ASME, Journal of Fluid Eng.* Vol. 123, pp.475–483.
45 Xu, C., "Volute for a Centrifugal Compressor", U.S. Patent No. 7604457, 2009.
46 Meakhail, T. and Park, S. O., 2005, "A Study of Impeller-Diffuser-Volute Interaction in a Centrifugal Fan," *ASME J. of Turbomach.*, Vol. 127, No 1, pp. 84–90.

47 Chu, S., Dong, R., Katz, J., "Relationship Between Unsteady Flow, Pressure Fluctuations, and Noise in a Centrifugal Pump-Part B: Effects of Blade-Tongue Interactions", *ASME Journal of Fluids Engineering*, Vol. 117, pp.30–35, 1995.

48 Irabu, K., Tamazato, E and Teruya, I., "Velocity Measurement of Flow a rounded a Volute Tongue of the Vaneless Radial Diffuser," Proc. 3rd Asia Symposium on Visualization, Japan, 1994.

49 Kline, S. J., "The Purpose of Uncertainty Analysis," *ASME J. Fluids Eng.*, 107, pp.153–160, 1985.

50 Howell, A. R., and Bonham R. P., Overall and Stage Characteristics of Axial-Flow Compressors, *Proc. Inst. Mech. Eng.*, vol. 1 163, 1950.

51 Howell, A. R., Fluid Dynamics of Axial Compressor, *Proc. Inst. Mech Eng.*, vol. 153, 1945. pp. 441.

Problems

Problems from 4.1

4.1 A centrifugal blower operating at 15,000 rpm compresses air from 68 °F and 14.7 psia to 10 psig at 1350 cfm and requires 80 hp. (a) Determine the efficiency of the blower at this point. (b) What are the speed, flow rate, and shaft power if it is desired to increase the discharge pressure to 12 psig. (c) If the suction is throttled to 12 psia, what is the discharge pressure and flow rate (referred to 68 °F and 14.7 psia)?

4.2 If the above blower is operated in winter with an inlet air temperature of 40 °F, determine the discharge pressure, flow rate (referred to 68 °F and 14.7 psia), and shaft power.

4.3 If the outside diameter D of the above blower impeller is trimmed by 5%, what is the probable discharge pressure and brake horsepower operating at the conditions specified in problem 4.1?

4.4 A radial vane compressor with dimensions as shown is operated at 12,000 rpm, with air entering the inlet at 75 kPa, 290 K, and 10 m/s under design conditions. If the slip factor at the discharge is 0.85, adiabatic efficiency is 0.82, and mechanical efficiency is 0.98, determine the discharge volumetric flow rate, static pressure, and brake horsepower. Neglect the blockage effects due to blade and boundary layer thickness.

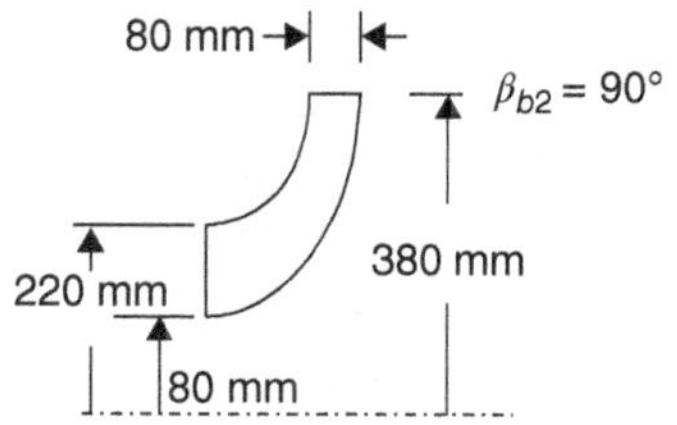

4.5 A centrifugal compressor rotating at 35,000 rpm is tested in the open atmosphere of 101 kPa and 15 °C. The mass flow rate of 0.7 kg/s, discharge stagnation pressure of 300 kPa, and temperature of 150 °C are measured. If the mechanical efficiency associated with the bearing and seal friction losses is estimated to be 95% (but leakage is neglected), determine the overall compressor efficiency and the brake horsepower, also determine the expected discharge pressure and brake horsepower if the rotating speed is reduced to 30,000 rpm.

4.6 Design a centrifugal compressor to draw 7 lb_m/s of air from an ambient atmosphere at 14.7 psia, 75 °F, and to discharge at a total pressure of 60 psia, when it runs at 21,000 rpm.

4.7 A blower is designed to draw in 1000 cfm of carbon dioxide at 90 °F and 3 psig and compress it to 5 psig when rotating at 4000 rpm. It is to be tested with air at 14.7 psia and 70 °F at 3550 rpm. Determine the flow rate and discharge pressure which the machine should deliver at the design point on test to be acceptable.

4.8 A blower is designed to compress 5000 cfm of 60 °F air from 14.7 psia to 4 psig. If the speed remains unchanged, but the inlet temperature is changed to 80 °F, what suction pressure can be maintained if the discharge pressure is to be 3 psig?

4.9 A compressor initially tested under the atmospheric pressure of 100 kPa and 15 °C results in a discharge pressure of 320 kPa and air mass flow rate of 18 kg/s. After the impeller diameter had been trimmed slightly, a second test was performed under the atmospheric pressure of 101 kPa and −2 °C results in a discharge pressure of 340 kPa and a mass flow rate of 19 kg/s. What are the rotating speed ratio and the trimmed diameter ratio?

4.10 A central air-conditioning duct system requires a fan that will deliver 12,100 cfm of air at a total pressure of 1.4 in. of H_2O. If the fan of Table 4.1 is selected, determine its speed, shaft power, and total efficiency.

4.11 Plot the performance curves at constant rpm for the fan selected in the previous problem, including the system curves in a form similar to Figure 3.60.

4.12 Plot the above fan performance curves at 400 and 600 rpm and the system curve in a form similar to Figure 3.59.

4.13 The test results of a centrifugal compressor are given as follows:

$$N = 46,000 \text{ rpm}, \quad m = 1.32 \text{ lb}_m/\text{s}, \quad p_{01} = 14.4 \text{ psia}, \quad T_{01} = 518 \text{ R},$$
$$p_2 = 27.8 \text{ psia}, \quad p_{03} = 43 \text{ psia}, \quad T_{03} = 772 \text{ R}.$$

If the impeller outlet dimensions are $D_2 = 6.5$ in., $b_2 = 0.4$ in., $\beta_{b2} = 90°$, $Z_b = 17$, determine (a) the compressor adiabatic efficiency, (b) stagnation pressure at the impeller outlet, and (c) the impeller adiabatic efficiency.

4.14 Considering the compressor of the above problem, if the relative Mach number at the impeller inlet tip should be 0.80 and the relative flow angle is 35° at the design flow rate, what are the impeller inlet dimensions?

4.15 Eight-stage centrifugal air compressor with an impeller diameter of 50 cm., outlet blade angle $\beta_2 = 65°$, rotating at 8500 rpm, delivers a mass flow rate of 10 kg/s and discharges total pressure at 900 kPa, total temperature at 600 K under the design condition. If $P_{01} = 100$ kPa, $T_{01} = 300$ K, and the slip coefficient, volumetric and mechanical efficiencies are estimated to be 0.89, 95%, and 97%, respectively. Determine: (a) compressor overall efficiency, (b) impeller discharge blade height, and (c) shaft power required.

Problems from 4.2

4.16 Each stage of an axial-flow compressor with 0.5 reaction has the same mean blade speed and the same flow outlet angle of 30° with respect to the axial direction. The mean flow coefficient is constant for all stages at 0.5. At the entry to the first stage, the stagnation temperature is 70 °F, the stagnation pressure is 14.7 psia, the static pressure is 12.5 psia, and the flow area is 3.5 ft^2. Determine the axial-flow velocity and the mass flow rate as well as the shaft power needed to drive the compressor if there are six stages and the mechanical efficiency is 95%.

4.17 An axial-flow compressor operating at design condition has a reaction of 0.6, a flow coefficient of 0.5, and a stage loading defined as $\Delta h_o/U^2$ of 0.35. If the flow exit angles for each blade row may be assumed to remain unchanged when the mass flow rate is reduced as shown in the velocity diagrams, determine the reaction and the stage loading factor of the stage when the airflow is reduced by 10% at constant blade speed.

4.18 A 16-stage axial-flow compressor is to have a total pressure ratio of 6.3. Tests have shown that a stage efficiency of 90% can be obtained for each of the first 6 stages and 87% for each of the remaining 10 stages. Determine the compressor's overall efficiency assuming constant work is done in each stage. For a mass flow rate of 80 lb_m/s, determine the power required by the compressor. Assume an inlet total temperature of 75 °F.

4.19 Determine the airflow angles at the mean radius and the blade length for the first stage of an axial-flow multistage compressor. The performance conditions are $N = 9000$ rpm, $\Delta T_o = 36$ °F/stage, $m = 44$ lbm/s, $V_a = 492$ ft./s, $p_{01} = 14.7$ psia, and $T_{01} = 518\ R$, and at the mean radius, $U = 590$ ft./s, $\lambda = 0.96$, and reaction $R = 0.5$.

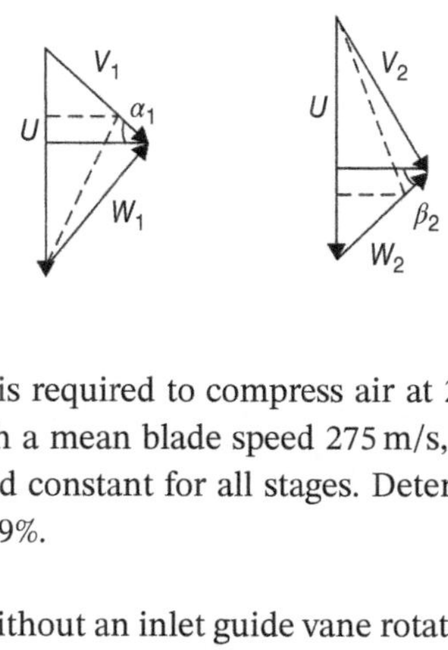

4.20 A multistage axial-flow compressor is required to compress air at 20 °C through a pressure ratio of 5:1. Each stage is to be 50% reaction with a mean blade speed 275 m/s, flow coefficient 0.5, and actual stage loading factor 0.3 assumed to be held constant for all stages. Determine the flow angles and number of stages required if stage efficiency is 89%.

4.21 A five-stage axial-flow compressor without an inlet guide vane rotating at 12,000 rpm takes ambient air of 101 kPa, 20 °C, at 150 m/s. The compressor stages are tapered such that the mean radius of each stage is the same at 20 cm and the axial-flow velocity ΔT_{os} and pressure ratio are constant. Also the flow is turned by 20° through the rotor of each stage, and the stage efficiency is estimated to be 92%. If the first-stage blade height is 5 cm, determine (a) the total shaft power required, (b) the overall pressure ratio, and (c) the blade height at the last stage.

4.22 A multistage axial-flow compressor draws 3.0 kg/s of air from the ambient atmosphere at 101 kPa, 20 °C and produces a total pressure ratio of 5. The total temperature rise per stage is to be 23 °C when the rotating speed is 20,000 rpm. Select a constant mean radius of 15 cm and flow coefficient of 0.45 and the polytropic efficiency is estimated to be 87%. Calculate (a) number of stages required, (b) first-stage rotor blade height, and (c) last-stage rotor blade height.

4.23 Determine the relative airflow angles at the hub and tip of the first and last stages of the compressor in Example 4.15. Also check their degrees of reaction there.

4.24 A 50% reaction compressor stage is designed for a stagnation temperature rise of 25 K with mean blade velocity of 220 m/s, axial flow velocity of 150 m/s, and estimated work done factor of 0.92. Determine the airflow angles at the mean radius.

4.25 A nine-stage axial-flow compressor rotating at 7000 rpm is used to compress the ambient air of 14.7 psia and 519 R. The first-stage rotor radii are $r_t = 12$ in., $r_h = 7$ in. Every stage has the same mean radius with the rotor blade angle changing by 12°, axial flow velocity of 500 ft./s, and produces the same amount of power. Assuming the ideal case without friction losses and neglecting incidence at the inlet and flow deviation at the outlet, determine

a) all blade angles at mean radii for both rotor and stator,
b) the total pressure ratio of each stage and the complete compressor, and
c) total shaft power required.

4.26 Solve problem 4.25 if an IGV is used to provide the inlet flow angle $\alpha_1 = 25°$ and the degree of reaction is 50% at the mean radius for each stage. All the other given data are the same (except $\beta_{b1} - \beta_{b2} \neq 12°$).

5

Gas Turbines

5.1 Introduction

Gas turbines widely used in aircraft engines and power generations are primarily axial-flow types except for the low-power units. From the second law of thermodynamics, the thermal efficiency of a heat engine is strongly affected by the inlet temperature of gases going to the turbine. The higher the inlet temperature of the turbine, the higher the efficiency is. Today, gas turbines with advanced materials and blade-cooling systems can operate with inlet temperatures up to about 1700 °C (3000 °F) for aircraft engines and to 1500 °C (2700 °F) for industrial units. However, the pressure is lower, typically from 1.4 to 4 MPa (200–600 psia). Hence, gas turbines are relatively lightweight, especially for the open-cycle systems of aircraft engines.

Due to its multiple-fuel application, gas turbine has become more popular in the power and process industries in the past few decades. It can run on natural gas, diesel fuel, methane, crude, gasified coal, and waste. In addition to increasing the turbine inlet temperature to improve the overall efficiency and power output of the engine, the basic gas turbine system can be modified by

a) Adding intercooler between compressor stages,
b) Reheating between turbine stages, and
c) Using turbine exhaust gas to preheat the compressed gas before entering the combustor or a combination of these modifications.

To satisfy some application requirements, the gas turbine can be designed as a closed system or be combined with a steam cycle. In a closed system, some special gas is used as a working fluid medium, and heat is transferred to and rejected out of the working fluid with closed-type heat exchangers, similar to the steam turbine cycle. In the combined gas and steam turbine cycle, the high-temperature exhaust gas from the gas turbine generates steam in a steam generator to drive the steam turbine. Sometimes, the generated steam is used for space heating or other processes. Then, the system is called cogeneration or combined heat and power (CHP).

Various types of gas turbine cycles are discussed in the Appendix, along with some topics on thermodynamics and compressible flow. Four of them are shown in Figure 5.1.[1] The turbofan engine consists of a single-stage fan upstream to produce a stream of air bypassing the compressor, combustor, and turbine. This cold air joins the core stream of hot air at the nozzle inlet, resulting in some additional thrust. In a ramjet, the compression is accomplished through the shock waves ahead of the engine at supersonic speed; hence, the compressor and turbine are unnecessary.

Each stage of the turbine consists of a stator and a rotor. The stator in the front serves as a nozzle to convert the gas's thermal/pressure energy (enthalpy) from fuel combustion into kinetic energy and direct it into the rotor. Through the rotor, kinetic or thermal energy is transferred to mechanical energy to drive the compressor, fan, and other accessories. The extra thermal energy is converted into a high-speed air jet for thrust. For power gas turbines, all useful output power is the excessive shaft power of the turbine rotor available to drive the external

Fundamentals of Turbomachinery: Theory and Applications, Second Edition. Ryoichi S. Amano and William W. Peng.

Companion website: www.wiley.com/go/amano/turbomachinery2e

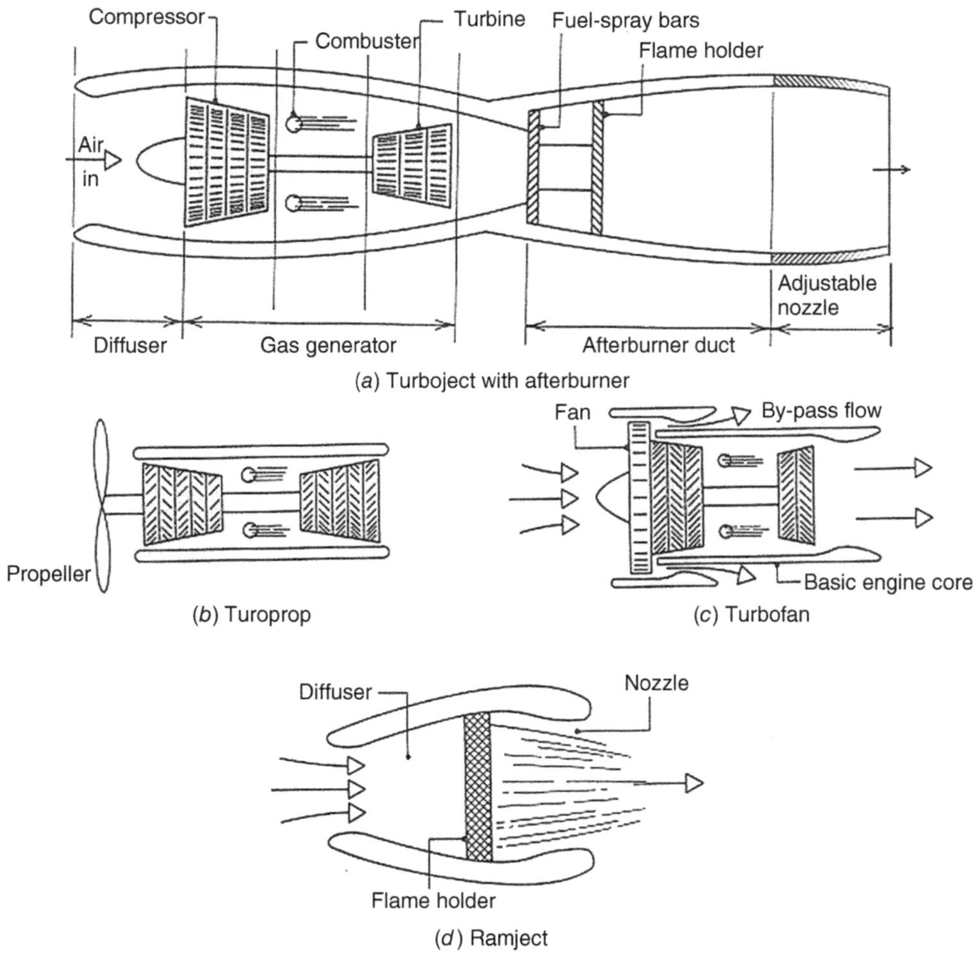

Figure 5.1 Aircraft engines. (*Source:* Reproduced from Moran and Shapiro.[1])

machine. The turbine can be divided into two parts. The upstream drives the compressor, and the downstream delivers output power. They are uncoupled mechanically.

The degree of reaction is also defined for the turbine. Depending on whether there is static enthalpy drop or not through the rotor, the turbines can be classified as impulse and reaction turbines. They are shown in Figure 5.2. In impulse turbines, the rotor passage area is constant, and hence, the relative flow velocity has constant magnitude and only changes direction from inlet to outlet. However, the absolute flow velocity decreases through the rotor, and the kinetic energy is converted into mechanical energy.

The passage area decreases in the reaction turbine rotor; hence, the relative flow velocity increases from the inlet to the outlet. In other words, both thermal energy and kinetic energy are converted into mechanical energy through the rotor. The next few sections will cover a more detailed discussion of the degree of reaction and other parameters.

5.2 Radial-Inflow Gas Turbine

A gas turbine can also use the radial- or mixed-flow impeller. It is primarily employed in the lower power units (up to a few hundred horsepower). The impeller is similar to that of a centrifugal pump or compressor, except

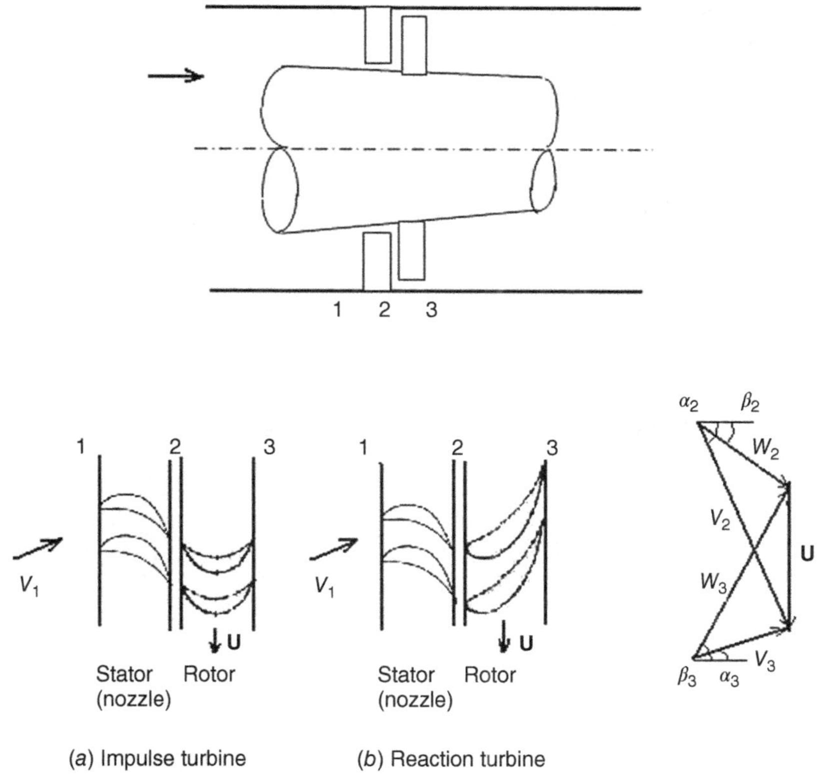

Figure 5.2 Axial-flow turbine stages.

the direction of the flow is just opposite. This can be understood from the energy transfer equations discussed in Chapter 2. Eq. (2.37), under design condition, becomes $\Delta E_t = U_2 V_{u2}$, where $U_3 V_{u3} = 0$. The subscripts 2 and 3 designate the impeller inlet and outlet, respectively. Since U_2 is proportional to inlet diameter D_2, to maximize the energy transfer ΔE_t, the flow inlet should be at the impeller outer diameter. Since the outlet area at the inner diameter is always smaller than the inlet, W_3 is greater than W_2. Hence, the radial-inflow turbine normally has a degree of reaction greater than zero. Also, to maximize the V_{u2} value, a nozzle close to the tangential direction at the inlet is always employed. Sometimes, a diffuser is added downstream of the impeller to bring the pressure up and match the ambient pressure. Thus, the kinetic energy loss from discharge is minimized. The nozzle/impeller/diffuser arrangement shown in Figure 5.3 is typical for a radial-inflow turbine.

Two extreme cases of impeller type are shown in Figure 5.4. Case (*a*), called a cantilever type, has short, pure radial blades. The inlet and outlet radii ratio is close to 1, so the airfoils can approximate the blades. Case (*b*), called a centripetal type, has blades extended from the pure radial to the pure axial direction. Sometimes, the exit region is called an exducer, similar to the inducer at the inlet of a centrifugal pump or compressor.

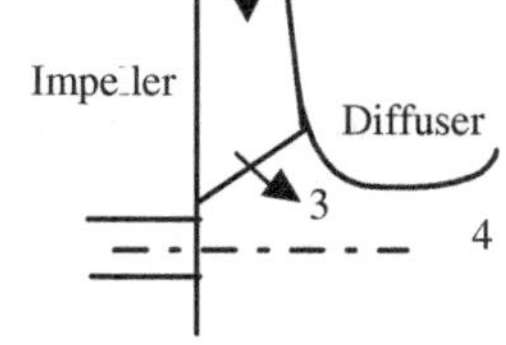

Figure 5.3 Typical radial-inflow turbine.

Compared with the axial-flow type, the radial-inflow turbine has the advantages of ruggedness, simplicity (hence lower cost), and higher efficiency in general. The disadvantage is its limit to a single stage (hence lower total pressure ratio and low power). The multistage arrangement is mechanically complicated, except for the radial-outflow Ljungstrom turbine, which will be discussed later for the steam turbine.

Typical applications of radial-inflow turbines are found in auxiliary power units, turbochargers for internal combustion engines, stand-alone ground vehicle power plants, low-power units of air motors driven with compressed air, and expanders for energy recovery in large refrigeration and cryogenic systems. A sample arrangement of the

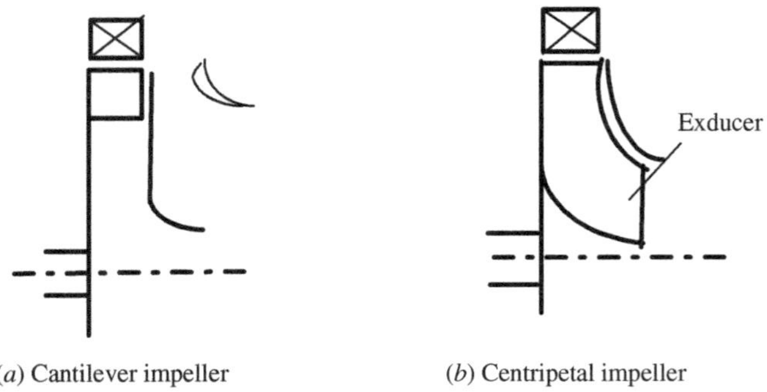

Figure 5.4 Types of radial-inflow turbine impellers.

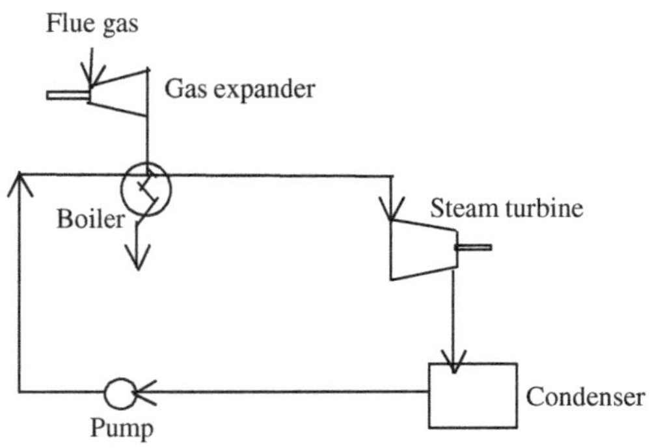

Figure 5.5 Energy recovery system in process plant.

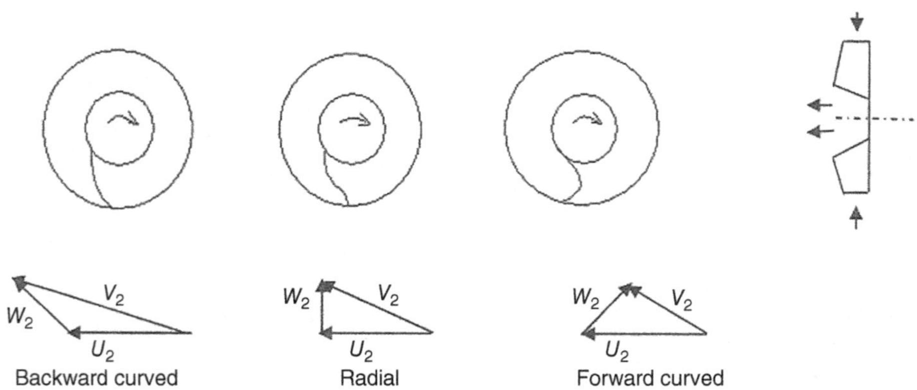

Figure 5.6 Blade profiles and velocity diagrams for radial-inflow turbine.

expander/steam turbine to recover energy in a process plant is shown in Figure 5.5. Some energy in the flue gas is extracted.

The performance characteristics of a radial-inflow turbine are mainly related to its inlet configuration. The blade profile at the inlet can be backward-curved, radial, or forward-curved. The schematic and flow velocity diagrams are shown in Figure 5.6.

The velocity diagrams show that the backward-curved impeller has the highest energy transfer per unit mass of fluid (ΔE_t) for a given size and rotating speed (or U_2). In other words, its size or rotating speed can be lower to transfer a fixed amount of energy. However, the radial blade is sometimes preferred due to its better structure integrity.

The nozzle at the inlet can be either a vaned type or a volute (this is basically a vaneless nozzle). They are similar to the diffusers in a centrifugal compressor. Due to flow mismatch, a vaned nozzle is more efficient at the design point but less efficient under the off-design condition. The volute type is better adapted for a wider range of operations but is less efficient in design conditions.

5.3 Thermodynamic Processes in Radial-Inflow Gas Turbine

As mentioned, the radial-inflow turbine is more compact and can be integrated with the compressor similar to the one shown in Figure 5.7.[5] The schematic and velocity diagrams at both the inlet and outlet of the rotor for the design condition are shown in Figure 5.8 for the case of pure radial blades at the inlet. A ring of nozzles is designed to produce a high tangential gas flow velocity. It flows inward and leaves the rotor with a negligible whirl velocity. A diffuser is used to reduce the velocity and increase the exhaust pressure to match the ambient condition. Under the design condition, the relative flow velocity at the inlet is pure radial, and the absolute flow velocity at the outlet is pure axial. Hence, the specific work output is expressed as $P_s/m = C_p(T_{01} - T_{03}) = U_2 V_{u2} = U_2^2$.

Thermodynamic processes through a radial-inflow turbine are similar to those for the axial-flow turbine discussed earlier. They are depicted in a T–s diagram, as shown in Figure 5.9, where p_a is the ambient pressure. Hence, the diffuser outlet pressure is $p_4 = p_a$. The maximum possible specific work output will be in the ideal process, without any loss and zero outlet velocity $P'_{s/m} = C_p(T_{01} - T_{4'}) = V_0^2/2$. Or in terms of the turbine pressure

Figure 5.7 Integrated radial-inflow turbine and compressor impellers. (Courtesy of Garrett Turbine Engine Co., Honeywell Aerospace, Phoenix, AZ.)

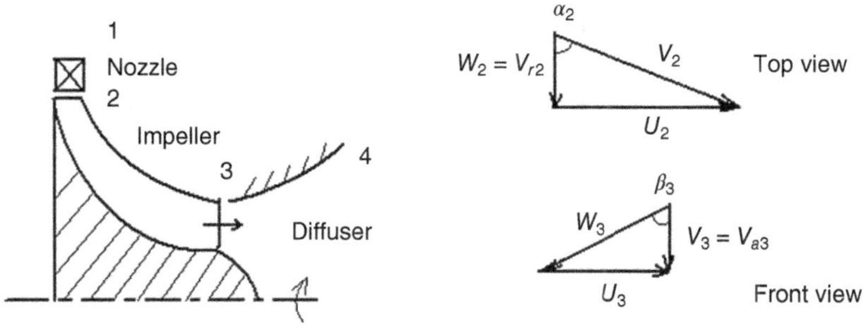

Figure 5.8 Radial-inflow turbine components and velocity diagram (with pure radial blades at inlet).

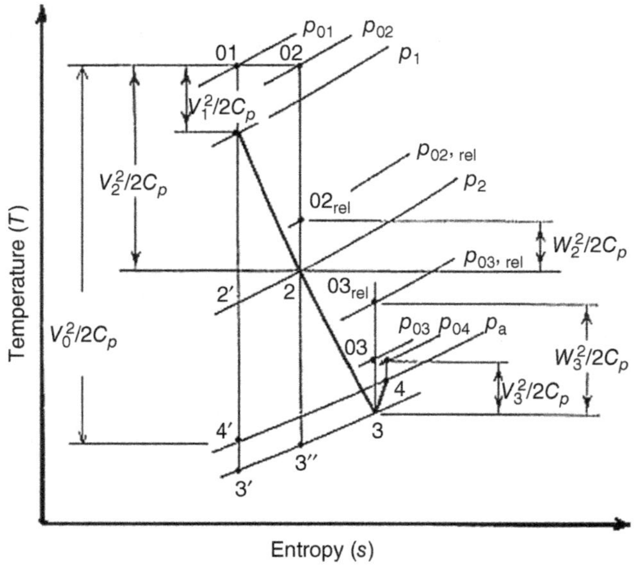

Figure 5.9 T–s diagram for a radial-inflow turbine.

ratio, V_0 is given by $V_0^2/2 = C_pT_{01}\left[1 - (p_a/p_{01})^{k-1/k}\right]$. The overall adiabatic efficiency of the turbine and diffuser is $\eta_{o/a} = (T_{01} - T_{03})/(T_{01} - T_{4'})$, where $T_{03} = T_{04}$. For the turbine alone, it is $\eta_t = (T_{01} - T_{03})/(T_{01} - T_{3'})$.

The turbine efficiency can also be expressed in terms of the nozzle and rotor loss coefficients, similar to the axial-flow turbine. They are defined as

$$\xi_N = \frac{T_2 - T_{2'}}{V_2^2/2C_p} \cong \frac{T_{3''} - T_{3'}}{V_2^2/2C_p}\frac{T_{2'}}{T_{3'}}, \qquad \xi_R = \frac{T_3 - T_{3''}}{W_3^2/2C_p}, \tag{5.1}$$

where

$$T_{2'}\,\Delta s \approx C_p\left(T_2 - T_{2'}\right), T_{3'}\,\Delta s \approx C_p\left(T_{3''} - T_{3'}\right), \text{or } T_2 - T_{2'} \approx \left(T_{3''} - T_{3'}\right)\left(T_{2'}/T_{3'}\right). \tag{5.2}$$

With

$$\begin{aligned} T_{01} - T_{3'} &= (T_{01} - T_{03}) + (T_{03} - T_3) + (T_3 - T_{3''}) + (T_{3''} - T_{3'}) \\ &= (T_{01} - T_{03}) + \frac{V_3^2}{2C_p} + \frac{\xi_R W_3^2}{2C_p} + \xi_N\left(\frac{V_2^2}{2C_p}\right)\left(\frac{T_{3'}}{T_{2'}}\right), \end{aligned} \tag{5.3}$$

and $V_2 = U_2 \operatorname{cosec} \alpha_2, W_3 = U_3 \operatorname{cosec} \beta_3, V_3 = U_3 \cot \beta_3, C_p(T_{02} - T_{03}) = U_2^2$, the turbine efficiency η_t can be written as

$$\begin{aligned} \eta_t &= \left[\frac{T_{01} - T_{3'}}{T_{01} - T_{03}}\right]^{-1} = \left[1 + \frac{1}{2C_p(T_{01} - T_{03})}\left(V_3^2 + \xi_R W_3^2 + \xi_N V_2^2\frac{T_{3'}}{T_{2'}}\right)\right]^{-1}. \\ &= \left\{1 + \frac{1}{2}\left[\left(\frac{r_3}{r_2}\right)^2(\cot^2\beta_3 + \xi_R \operatorname{cosec}^2\beta_3) + \xi_N\frac{T_{3'}}{T_{2'}}\operatorname{cosec}^2\alpha_2\right]\right\}^{-1}, \end{aligned} \tag{5.4}$$

where

$$\frac{T_{3'}}{T_{2'}} \cong \frac{T_{3''}}{T_2} = 1 - \frac{U_2^2}{2C_pT_2}\left[1 + \left(\frac{r_3}{r_2}\right)^2[(1 + \xi_R)\operatorname{cosec}^2\beta_3 - 1] - \cot^2\alpha_2\right] \tag{5.5}$$

and $T_2 = T_{01} - \left(U_2^2/2C_p\right)\operatorname{cosec}^2\alpha_2$. Eq. (5.5) is obtained from the following manipulation: From

$$
\begin{aligned}
T_{3''} &= T_2 - (T_3 - T_{3''}) - (T_2 - T_3),\\
U_2^2 &= C_p(T_{02} - T_{03}) = C_p(T_2 - T_3) + \frac{1}{2}\left(V_2^2 - V_3^2\right)\\
&= C_p(T_2 - T_3) + \frac{1}{2}\left(U_2^2 + W_2^2\right) - \frac{1}{2}\left(W_3^2 - U_3^2\right),
\end{aligned}
\tag{5.6}
$$

or

$$
T_2 - T_3 = \frac{\left(U_2^2 - U_3^2\right) + \left(W_3^2 - W_2^2\right)}{2C_p},
\tag{5.7}
$$

we have

$$
\begin{aligned}
\frac{T_{3''}}{T_2} &= 1 - \frac{T_3 - T_{3''}}{T_2} - \frac{\left(U_2^2 - U_3^2\right) + \left(W_3^2 - W_2^2\right)}{2C_pT_2}\\
&= 1 - \frac{\xi_R W_3^2}{2C_pT_2} - \frac{U_2^2}{2C_pT_2}\left[\left(1 - \frac{r_3^2}{r_2^2}\right) + \left(\frac{W_3}{U_2}\right)^2 - \left(\frac{W_2}{U_2}\right)^2\right]\\
&= 1 - \frac{U_2^2}{2C_pT_2}\left[\xi_R\left(\frac{r_3}{r_2}\right)^2\operatorname{cosec}^2\beta_3 + 1 - \left(\frac{r_3}{r_2}\right)^2 + \left(\frac{r_3}{r_2}\right)^2\operatorname{cosec}^2\beta_3 - \frac{V_2^2 - U_2^2}{U_2^2}\right]\\
&= 1 - \frac{U_2^2}{2C_pT_2}\left\{1 + \left(\frac{r_3}{r_2}\right)^2\left[(1 + \xi_R)\operatorname{cosec}^2\beta_3 - 1\right] - \cot^2\alpha_2\right\}.
\end{aligned}
\tag{5.8}
$$

Under the off-design condition, additional loss due to incidence or shock loss can be incurred at the rotor inlet. The process is depicted in Figure 5.10, where some additional increase in entropy is shown. The corresponding stagnation pressure drop can be accounted for by the expression $\Delta p_0 = (p_{02} - p_2)\cos^2\beta_2\,(\tan\beta_2 + 0.1)^2$, if β_2 is between $-65°$ and $+65°$, where β_2 is equal to the incidence for the pure radial inlet vanes.[6]

Example 5.1

A radial-inflow gas turbine with the dimensions shown is operated with the following parameters: $T_{01} = 2000\,\mathrm{R}$, $p_{01} = 50\,\mathrm{psia}$, and $N = 15{,}000\,\mathrm{rpm}$. If the estimated nozzle and rotor loss coefficients are $\xi_N = 0.08$ and $\xi_R = 0.45$, determine the expected (a) mass flow rate, (b) output shaft power, (c) turbine efficiency, and (d) exhaust pressure/temperature under the design condition. (Neglect the leakage and mechanical losses and the flow incidence and deviation at the blade leading and trailing edges.)

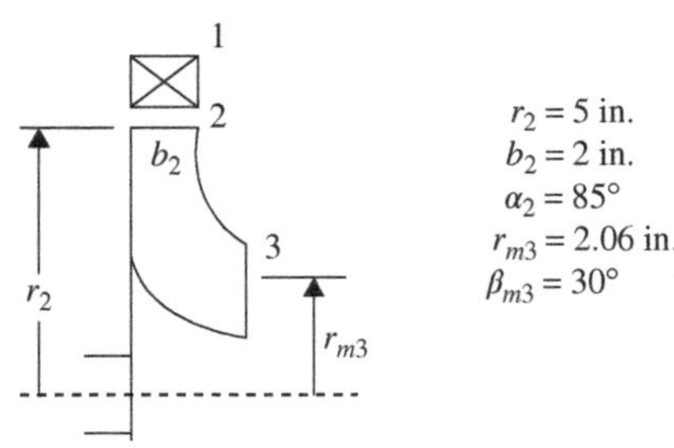

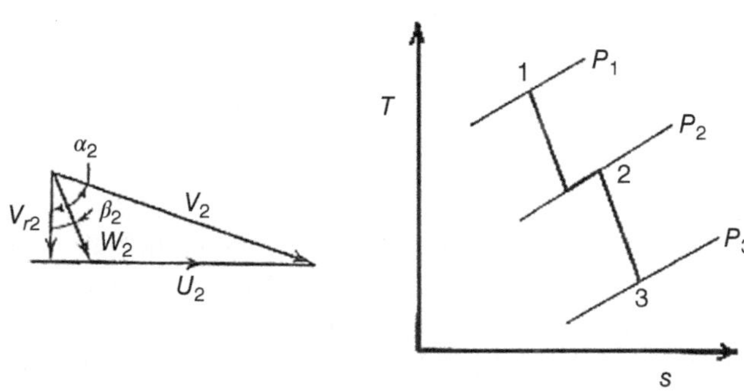

Figure 5.10 Effect of incidence loss.

Solution

Velocity diagrams at the rotor inlet and outlet are given as shown.

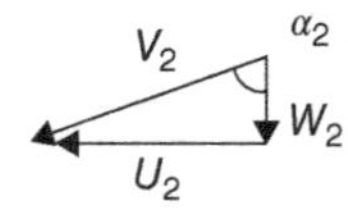

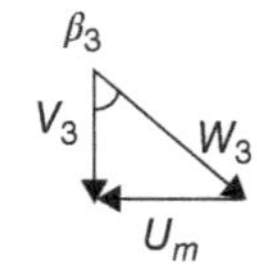

Velocities at the rotor inlet can be calculated as

$$U_2 = \frac{r_2 N\pi}{30} = \frac{(5/12)\times 15,000\pi}{30} = 654.5\text{ ft/s},$$

$$V_2 = \frac{U_2}{\sin\alpha_2} = \frac{654.5}{\sin 85°} = 657\text{ ft/s},$$

$$V_{r2} = W_2 = \frac{U_2}{\tan\alpha_2} = \frac{654.5}{\tan 85°} = 57.3\text{ ft/s}.$$

Hence from $C_p(T_{02} - T_2) = V_2^2/2$, where $T_{02} = T_{01}$, we have

$$T_2 = T_{02} - \frac{V_2^2}{2C_p} = 2000 - \frac{657^2}{2\times 0.24\times 778\times 32.2} = 1964.1\text{ R}.$$

From $\xi_N = (T_2 - T_{2'})/\left[V_2^2/(2C_p)\right]$, we have

$$T_{2'} = T_2 - \frac{\xi_N V_2^2}{2C_p} = 1964.1 - \frac{0.08\times 657^2}{2\times 0.24\times 778\times 32.2} = 1961.2\text{ R},$$

$$p_2 = p_{01}\left(\frac{T_{2'}}{T_{01}}\right)^{k/(k-1)} = 50\left(\frac{1961.2}{2000}\right)^{3.5} = 46.7\text{ psia},$$

$$\rho_2 = \frac{p_2}{RT_2} = \frac{46.7\times 144}{53.3\times 1964.1} = 0.064\text{ lb}_\text{m}/\text{ft}^3.$$

So the mass flow rate $m = \rho_2 V_{r2} A_2 = 0.064\times 57.3\times(2\pi\times 5\times 2/144) = 1.60\text{ lb}_\text{m}/\text{s}$. Assuming whirl-free flow at the rotor outlet under the design condition, we have $\Delta E = U_2 V_{u2} = U_2^2 = 654.5^2 = 428,370(\text{ft/s})^2 = 428,370/(32.2\times 550) = 24.2\text{ hp}/(\text{lb}_\text{m}/\text{s})$, or $P_s = m\,\Delta E = 1.60\times 24.2 = 38.7$ hp. Velocities at the rotor outlet are determined as

$$U_{m3} = \frac{(2.06/12)\times 15,000\times\pi}{30} = 269.6\text{ ft/s}, V_3 = \frac{U_{m3}}{\tan\beta_3} = \frac{269.6}{\tan 30°} = 467\text{ ft/s},$$

and $W_3 = U_{m3}/\sin\beta_3 = 269.6/\sin 30° = 539.2$ ft/s.

and $W_3 = U_{m3}/\sin\beta_3 = 269.6/\sin 30° = 539.2$ ft/s.

The turbine efficiency can be determined from Eqs. (5.4) and (5.5). Without detailed calculations, the result is given as $\eta_t = 0.691$.

The exhaust pressure/temperature can be determined from the following calculations with the help of Figure 5.9: From $C_p(T_{01} - T_{03}) = \Delta E = 428,370/(32.2\times 778) = 17.1\text{ Btu/lb}_\text{m}$ and $(T_{01} - T_{03})/(T_{01} - T_{3'}) = \eta_t = 0.691$, we have $T_{03} = 2000–17.1/0.24 = 1929$ R and $T_{3'} = T_{01} - (T_{01} - T_{03})/\eta_t = 2000 - (2000–1929)/0.691 = 1897$ R. Hence $p_3 = p_{01}(T_{3'}/T_{01})^{k/(k-1)} = 50(1897/2000)^{3.5} = 41.6$ psia, and $T_{3''} = T_2(p_3/p_2)^{(k-1)/k} = 1964.1(41.6/46.7)^{0.2857} = 1900.3$ R and $T_3 = T_{3''} + \xi_R W_3^2/(2C_p) = 1900.3 + 0.45\times 539.2^2/(2\times 0.24\times 32.2\times 778) = 1911.2$ R.

5.4 Thermodynamics of Axial-Flow Turbine

A Mollier diagram showing the thermodynamics processes of gas flow through a turbine stage, including the effects of irreversibility, is depicted in Figure 5.11. The subscripts 1, 2, and 3 designate the inlet/outlet of the stator and rotor, as shown in Figure 5.2. Across the stator, no work is done, and hence we have $h_{01} = h_{02}$. Across the rotor, the relative enthalpy $(h + W_2/2)$, sometimes called rothalpy* in the literature, remains constant, that is, $h_2 + W_2^2/2 = h_3 + W_3^2/2$, similar to the axial-flow compressor. The work done on a rotor by the flowing fluid equals

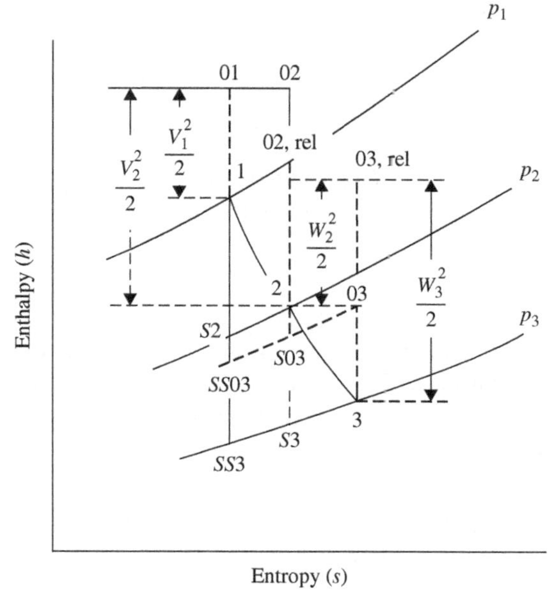

Figure 5.11 Mollier diagram for an axial-flow turbine stage.

the total enthalpy drop incurred by the fluid passing through the stage. It can be related to the flow velocity through Euler's Equation such that $\Delta E = h_{01} - h_{03} = U(V_{u2} + V_{u3})$, where V_{u2} and V_{u3} are opposite in direction.

The stage efficiency of a turbine stage can be defined as

$$\eta_s = \frac{h_{01} - h_{03}}{h_{01} - h_{ss03}} \cong \frac{C_p(T_{01} - T_{03})}{C_p(T_{01} - T_{ss03})} = \frac{\Delta T_{0s}}{T_{01}(1 - T_{ss03}/T_{01})}. \tag{5.9}$$

Hence we have

$$\Delta T_{0s} = \eta_s T_{01} \left[1 - \left(\frac{p_{03}}{p_{01}}\right)^{(k-1)/k}\right]. \tag{5.10}$$

For a normal stage, we have $V_3 = V_1$ and $\alpha_3 = \alpha_1$. If it is also assumed that $V_{ss3} = V_3$, then it can be written in the form

$$\begin{aligned}\eta_s &= \frac{h_1 - h_3}{h_1 - h_{ss3}} = \frac{h_1 - h_3}{(h_1 - h_3) + (h_3 - h_{s3}) + (h_{s3} - h_{ss3})} \\ &= \left(1 + \frac{h_3 - h_{s3}}{h_1 - h_3} + \frac{h_{s3} - h_{ss3}}{h_1 - h_3}\right)^{-1}.\end{aligned} \tag{5.10A}$$

To relate these static enthalpy changes with the component losses in the stator and rotor, we define $\xi_s = (h_2 - h_{s2})/(V_2^2/2)$ and $\xi_r = (h_3 - h_{s3})/(W_3^2/2)$ as the loss coefficients of the stator and rotor. They are usually available from empirical tests or have to be estimated in the design process. Also, from the slope of a constant-pressure line on a Mollier diagram, $\partial h/\partial s|p = T$, we obtain $h_{s3} - h_{ss3} \cong T_3(ss_3 - sss_3)$ and $h_2 - h_{s2} \cong T_2(s_2 - ss_2)$. Noting that $ss_3 - sss_3 = s_2 - ss_2$, these two equations can be combined into $h_{s3} - h_{ss3} = (T_3/T_2)(h_2 - h_{s2})$. Substituting these expressions into Eq. (5.10A), the stage efficiency becomes

$$\eta_s = \left(1 + \frac{\xi_r W_3^2 + \xi_s V_2^2(T_3/T_2)}{2(h_1 - h_3)}\right)^{-1}. \tag{5.11}$$

Again, substituting $W_3 = V_a \sec\beta_3$, $V_2 = V_a \sec\alpha_2$, and $h_1 - h_3 = UV_a(\tan\beta_3 + \tan\beta_2)$, the above equation becomes

$$\eta_s = \left(1 + \frac{\varphi}{2}\frac{\xi_r \sec^2\beta_3 + \xi_s(T_3/T_2)\sec^2\alpha_2}{\tan\beta_3 + \tan\beta_2}\right)^{-1}. \tag{5.12}$$

Here, the polytropic efficiency, defined as $\eta_p = dh/dh_s$, is close to the stage efficiency, similar to that discussed in Section 4.1.4 for the compressor. Analysis applied to an ideal gas expanding through a turbine results in

$$\frac{T_e}{T_i} = \left(\frac{p_e}{p_i}\right)^{\eta_p(k-1)/k}, \qquad \eta_{ad} = \frac{1-(p_e/p_i)^{\eta_p(k-1)/k}}{1-(p_e/p_i)^{(k-1)/k}}, \tag{5.13}$$

where p_i and p_e are the pressures at the inlet and exit of a multistage turbine. The relationship between η_{ad} and η_p for a turbine can also be plotted into curves similar to those in Figure 3.7. In contrast to a compressor, the adiabatic efficiency is always greater than the polytropic efficiency of a turbine. This is left as an exercise for the students in the problems.

In the case of steam turbines, the vapor cannot be treated as an ideal gas, except in highly superheated conditions, so the polytropic efficiency is related to the adiabatic efficiency with a parameter called *reheat factor* through numerical calculation such that $\eta_{ad} = \eta_p R_H$, where

$$R_H = \frac{\sum \Delta h_{s,i}}{h_i - h_{se}}. \tag{5.14}$$

This will be discussed in more detail in the next chapter.

5.5 Degree of Reaction

The velocity diagrams across an axial-flow turbine rotor are depicted in Figure 5.2 and again in Figure 5.12, where $\mathbf{V} = \mathbf{U} + \mathbf{W}$ and $U_1 = U_2 = U$. Similar to that discussed in Section 4.2, the degree of reaction for a turbine is defined as the ratio of static enthalpy change across the rotor to that of the whole stage, that is, $R = (h_2 - h_3)/(h_1 - h_3)$. For a normal stage, we have $V_3 = V_1$. Hence $h_1 - h_3 = h_{01} - h_{03} = h_{02} - h_{03} = U(V_{u2} + V_{u3})$ and $h_2 - h_3 = (1/2)(W_3^2 - W_2^2)$ across the rotor. So we have

$$R = \frac{W_3^2 - W_2^2}{2U(V_{u2} + V_{u3})}. \tag{5.15}$$

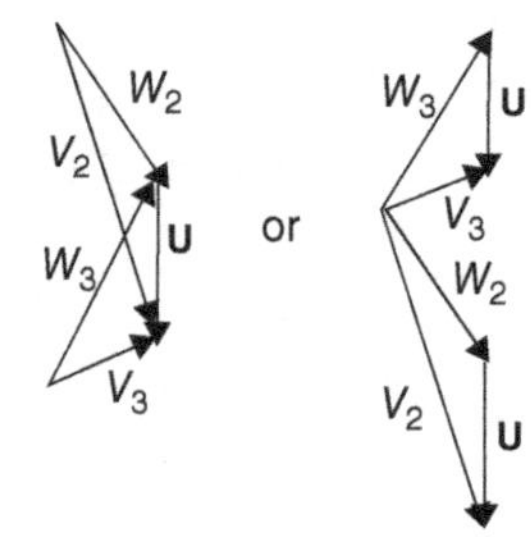

Figure 5.12 Velocity diagrams for an axial-flow turbine.

Assuming constant axial-flow velocity, it becomes

$$R = \frac{(W_{u3} - W_{u2})(W_{u3} + W_{u2})}{2U(V_{u2} + V_{u3})} = \frac{W_{u3} - W_{u2}}{2U},$$

since $V_{u2} = W_{u2} + U$ and $V_{u3} = W_{u3} - U$, from Figure 5.12. Hence, with the flow coefficient defined as $\varphi = V_a/U$, the degree of reaction can be expressed as

$$R = \begin{cases} \frac{1}{2}\phi(\tan\beta_3 - \tan\beta_2), \\ \frac{1}{2} + \frac{1}{2}\phi(\tan\beta_3 - \tan\alpha_2), \\ 1 + \frac{1}{2}\phi(\tan\alpha_3 - \tan\alpha_2), \end{cases} \tag{5.16}$$

which are similar to Eq. (7.2) for the compressor. Combining Eq. (5.16) with the expression of the loading coefficient, $\psi = 2C_p\,\Delta T_{os}/U^2 = 2\varphi(\tan\beta_3 + \tan\beta_2)$, we have

$$\tan\beta_3 = \frac{1}{2\varphi}\left(2R + \frac{\psi}{2}\right), \tag{5.17A}$$

$$\tan\beta_2 = \frac{1}{2\varphi}\left(\frac{\psi}{2} - 2R\right). \tag{5.17B}$$

For impulse turbines, $R = 0$; hence $\beta_2 = \beta_3$, as shown in Figure 5.2*a*. The Mollier and velocity diagrams corresponding to these conditions are sketched in Figure 5.13, which also shows that $h_2 = h_3$ since $W_2 = W_3$.

For 50% reaction, or simply reaction turbines, that is, $R = 0.5$, we have $\alpha_2 = \beta_3$ as obtained with Eq. (5.16). From the symmetry of the velocity diagram shown in Figure 5.14, it is obvious that $\alpha_3 = \beta_2$, too. The definition of the reaction implies that the static enthalpy drop in the stator is equal to that in the rotor, that is, $h_1 - h_2 = h_2 - h_3$ and $W_3 > W_2$, also shown in Figure 5.14. So, the rotor flow passage is convergent, as shown in Figure 5.2*b*.

For given rotating speed and rotor size, impulse turbines transfer more energy per unit stage, but the reaction turbines generally have higher adiabatic efficiency. These features will be discussed in more detail later.

Example 5.2 A multistage turbine consisting of all impulse stages with a stage efficiency of 85% is used to expand air from 1800 °F and 200 psia to 14.7 psia. If the mean blade speed and axial-flow velocity are constant at 750 and 400 ft/s, respectively, for each stage and the blade angle of the rotor is symmetric at 50° with respect to the axial direction, determine the overall adiabatic efficiency, exit air temperature, and number of stages.

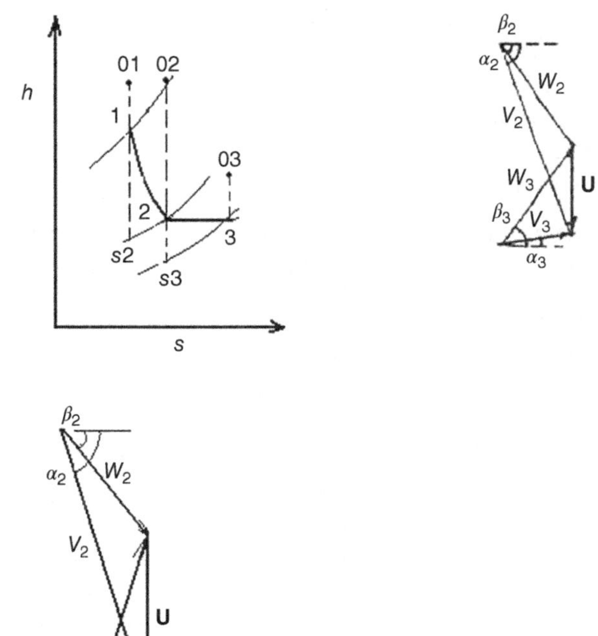

Figure 5.13 Mollier and velocity diagrams for an impulse turbine.

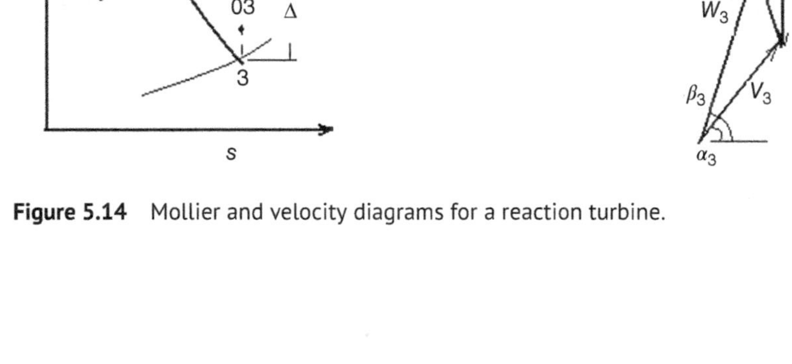

Figure 5.14 Mollier and velocity diagrams for a reaction turbine.

Solution

From

$$\eta_{ad} = \frac{1-(p_{0e}/p_{0i})^{\eta_s(k-1)/k}}{1-(p_{0e}/p_{0i})^{(k-1)/k}} = \frac{1-(14.7/200)^{0.85/3.5}}{1-(14.7/200)^{0.2857}}$$
$$= \frac{0.4695}{0.5256} = 0.893,$$
$$\frac{T_{soe}}{T_{0i}} = \left(\frac{p_{0e}}{p_{0i}}\right)^{(k-1)/k} = \left(\frac{14.7}{200}\right)^{0.2857} = 0.4743,$$
$$T_{soe} = 0.4743 \times (1800+460) = 1072\text{R},$$
$$\eta_{ad} = \frac{T_{0i}-T_{0e}}{T_{0i}-T_{soe}} = 0.893,$$

we have $T_{0e} = 1199$ R $= 739$ °F. Also, for impulse turbine, we have $\beta_2 = \beta_3$. Hence, from Figure 5.13 for each stage

$$\Delta h_{0s} = U_m(V_{u2}-V_{u3}) = U_m[U_m + V_a \tan\beta_2 - (U_m - V_a\tan\beta_3)]$$
$$= 2U_m V_a \tan\beta = 2\times 750 \times 400 \times \tan 50° = 7.15\times 10^5 (\text{ft/s})^2$$

or $0.24 \times 778 \times 32.2\ \Delta T_{0s} = 7.15 \times 10^5$, we have the temperature rise per stage, $\Delta T_{0s} = 119$ °F. Hence, the number of stages $n_s = (1800–739)/119 = 8.91 \approx 9$.

Example 5.3 An axial-flow turbine stage is tested with air. The result is reported as follows:

a) Total power developed $P_s = 375$ hp. with mass flow rate $m = 3\ \text{lb}_\text{m}/\text{s}$
b) Inlet condition: $p_{01} = 250$ psia, $T_{01} = 2860$ R
c) Axial flow velocity V_a is constant throughout the stage
d) No whirl at the inlet and outlet of the stage, that is, $V_1 = V_3 = V_a$
e) Condition between stator and rotor: $p_2 = 180$ psia, $\alpha_2 = 65°$
f) Estimated loss coefficients: $\xi_s = 7\%$, $\xi_r = 12\%$

Determine the axial-flow velocity, degree of reaction, static temperatures, and relative flow angles at the rotor inlet and outlet and stage efficiency.

Solution

Referring to Figure 5.11, we have $T_{s2} = T_{01}(p_2/p_{01})^{(k-1)/k} = 2604$ R. From $\xi_s = (h_2 - h_{s2})/(V_2^2/2)$, we have $C_p(T_2 - T_{s2}) = \xi_s V_2^2/2$. Combining with $C_p(T_{02} - T_2) = V_2^2/2$, where $T_{02} = T_{01}$, we have $C_p(T_{02} - T_{s2}) = (1 + \xi_s) V_2^2/2$, or $V_2 = [2C_p(T_{02} - T_{s2})/(1 + \xi_s)]^{0.5} = 1696$ ft/s. Hence we have $V_{u2} = V_2 \sin\alpha_2 = 1537$ ft/s, $V_a = V_2 \cos\alpha_2 = 716.8$ ft/s and $T_2 = T_{02} - V_2^2/(2C_p) = 2860–1696^2/(2 \times 0.24 \times 778 \times 32.2) = 2621$ R. Since $V_1 = V_a$, we have $T_1 = T_{01} - V_1^2/(2C_p) = 2817$ R. From $\Delta E = C_p\ \Delta T_{0s} = U(V_{u2} + V_{u3}) = UV_{u2}$, neglecting mechanical losses, we have $U = P_s/(mV_{u2}) = 375 \times 550 \times 32.2/(3 \times 1537) = 1440$ ft/s. Hence $\varphi = V_a/U = 0.498$, $\tan\beta_3 = U/V_a = 2.01$, $\beta_3 = 63.5°$, and $R = 1 + (\varphi/2)(\tan\alpha_3 - \tan\alpha_2) = 0.466$, where $\alpha_3 = 0$ and $\alpha_2 = 65°$. Also from the velocity diagram in Figure 5.12 with $V_3 = V_a$, we have $\tan\beta_2 = (V_a \tan\alpha_2 - U)/V_a = \tan\alpha_2 – 1/\varphi = 0.136$, and so $\beta_2 = 7.8°$. Similarly, we have $W_3 = V_a/\cos\beta_3 = 1606$ ft/s and $W_2 = V_a/\cos\beta_2 = 723.5$ ft/s. Across the rotor, we have $h_2 + W_2^2/2 = h_3 + W_3^2/2$. Hence $T_3 = T_2 + (W_2^2 - W_3^2)/(2C_p) = 2450$ R. From $\xi_r = C_p(T_3 - T_{s3})/(W_3^2/2)$, we have $T_{s3} = T_3 - \xi_r W_3^2/(2C_p) = 2424$ R, and $p_3 = p_2(T_{s3}/T_2)^{k/(k-1)} = 136.9$ psia. From Eq. (5.11), we have

$$\eta_s = \left(1 + \frac{\zeta_r W_3^2 + \zeta_s V_2^2 (T_3/T_2)}{2(h_1 - h_3)}\right)^{-1}$$
$$= \left(1 + \frac{0.12 \times 1606^2 + 0.07 \times 1696^2 (2450/2621)}{2 \times 0.24 \times 778 \times 32.2(2817 - 2450)}\right)^{-1} = 0.898,$$

or from Eq. (5.12), we have

$$\eta_s = \left(1 + \frac{\varphi}{2}\frac{\xi_r \sec^2\beta_3 + \xi_s(T_3/T_2)\sec^2\alpha_2}{\tan\beta_3 + \tan\beta_2}\right)^{-1}$$

$$= \left(1 + \frac{0.498}{2} \times \frac{0.12\sec^2 63.5° + 0.07(2450/2621)\sec^2 65°}{\tan 63.5° + \tan(7.8°)}\right)^{-1} = 0.898.$$

Also, η_s can be calculated from $\eta_s = (T_{01} - T_{03})/(T_{01} - T_{sso3})$, where

$$T_{03} = T_3 + \frac{V_3^2}{2C_p} = 2493\text{ R}, \qquad p_{03} = p_3\left(\frac{T_{03}}{T_3}\right)^{k/(k-1)} = 145.4\text{ psia},$$

$$T_{sso3} = T_{01}\left(\frac{p_{03}}{p_{01}}\right)^{(k-1)/k} = 2450\text{ R}.$$

Hence we have

$$\eta_s = \frac{2860 - 2493}{2860 - 2450} = 0.895.$$

5.6 Preliminary Design Procedure for Turbine Stage

A brief outline for designing an axial-flow turbine will be given in this section (a detailed discussion is beyond the scope of this book). The design process starts with a clear statement of the specifications.

1) the fluid medium to be used, steam, air, or other gases;
2) fluid stagnation pressure and temperature at the inlet, p_{01} and T_{01};
3) fluid stagnation or static pressure at stage outlet, p_{03} or p_3; and
4) either the fluid mass flow rate, m, or the mechanical power output P_s required and the design rotating speed N.

Depending on the applications, other considerations, such as maximum efficiency, first cost, minimum weight, and maximum life, should also be included.

Turbine performance at off-design operating conditions is then predicted. A cooling mechanism is designed, heat transfer calculations are performed, and stress and vibration analysis is completed. Several iterations are normally expected at different points before a final design is fixed.

For convenience, the relevant equations are listed here:

$$T_0 = T + \frac{V^2}{2C_p}, \qquad p = \rho RT, \qquad \frac{p}{p_0} = \left(\frac{T}{T_0}\right)^{k/(k-1)}, \tag{5.18A}$$

$$m = \rho V_a A, \qquad A = \pi r_t^2\left[1 - \left(\frac{r_h}{r_t}\right)^2\right],$$

$$r_m = \left[\frac{1}{2}\left(r_t^2 + r_h^2\right)\right]^{0.5}, \qquad b = r_t - r_h, \tag{5.18B}$$

$$\tan\alpha_3 = \tan\beta_3 - \frac{1}{\phi}, \tag{5.18C}$$

$$\tan\alpha_2 = \tan\beta_2 + \frac{1}{\phi}, \tag{5.18D}$$

$$\Delta T_{0s} = \eta_s T_{01}\left[1 - \left(\frac{p03}{p01}\right)^{k/(k-1)}\right] \qquad \text{[Equation (8.1)]}$$

$$\tan\beta_3 = \frac{1}{2\phi}\left(\frac{1}{2}\psi + 2R\right) \qquad \text{[Equation (8.6A)]}$$

$$\tan\beta_2 = \frac{1}{2\phi}\left(\frac{1}{2}\psi - 2R\right) \qquad \text{[Equation (8.6B)].}$$

The calculations at the mean radius can be summarized into the following steps:

1) Select a rotating speed N in (rpm) to match with the compressor and $U_m = r_m\pi N/30$ to satisfy the stress limitation.
2) Calculate ΔT_{0s} from Eq. (5.10) with assumed η_s, then ψ from $\psi = 2C_p\ \Delta T_{0s}/U_m^2$.
3) Select ϕ values of 0.8 ~ 1.0 and α_3 values of 0° ~ 20°.
4) Calculate β_3 from Eq. (5.18C), then R from Eq. (5.17A), and make sure R is not too small (say $R > 0.40$) to avoid a negative value at the root; otherwise, repeat step 3 with different ϕ and α_3.
5) Calculate β_2 with Eq. (5.17B) and α_2 with (5.18D).
6) Calculate V_{a2}, assuming $V_{a2} = V_{a3} = V_a$, with $\phi = V_a/U_m$, then V_2 from the velocity triangles.
7) Estimate the loss coefficient ξ_s, to be discussed in the next section, and then calculate ρ_2, p_2, T_2, and T_{s2} from Eq. (5.18A) and $\xi_s = (T_2 - T_{s2})/(V_2^2/2C_p)$, making sure that $p_2 > p_c$, p_c is the critical pressure corresponding to the choked condition and can be calculated from $p_c = p_{01}[(k+1)/2]^{-k/(k-1)}$.
8) Calculate A_2 and b_2 from Eq. (5.18B).
9) Assuming $V_1 = V_3$, which can be calculated from the velocity diagrams, then A_1 and b_1 from Eqs. (5.18A) and (5.18B).
10) With V_3 and p_{03}, a procedure similar to step 9 can be employed to determine A_3 and b_3. Calculate $\xi_{s,r}$ from $Y_{s,r}$ and compare them with the estimated values. Repeat item 7 if needed.

A constant V_a design is expected to result in an increasing flow passage area, that is, $A_3 > A_2 > A_1$, since we have $\rho_3 < \rho_2 < \rho_1$, in the expansion process. However, the flare-included angle should be limited to 25° to prevent boundary layer separation on the root surface, especially for impulse turbines.

With the mean radius configuration determined, the hub and tip radii and flow angles can be calculated with the radial equilibrium requirement, similar to the axial-flow compressor stage design discussed in Section 4.2.

Example 5.4 A gas turbine stage is designed with the following parameters:

a) Mass flow rate $\dot{m} = 50\ \text{lb}_\text{m}/\text{s}$, constant axial-flow velocity $V_a = 500$ ft/s
b) Peripheral velocity at mean radius, $U_m = 1200$ ft/s
c) Stator inlet condition: $T_{01} = 2000\,°\text{F}$, $p_{01} = 200$ psia
d) Rotating speed $N = 8000$ rpm
e) Stator exit angle $\alpha_2 = 75°$ and inlet angle $\alpha_1 = \alpha_3 = 10°$ with respect to the axial direction
f) Stator loss coefficient $\xi_s = 8\%$

Determine (a) rotor blade angle of inlet and outlet at mean radius, β_2, β_3; (b) stagnation temperature drop per stage, ΔT_{0s}; (c) shaft power output P_s; (d) degree of reaction; and (e) tip and hub radii at rotor inlet.

Solution

Use the velocity diagram shown in Figure 5.2 or 5.12. We have $V_a\ \tan\beta_2 = V_a\ \tan\alpha_2 - U_m$, or $\tan\beta_2 = \tan\alpha_2 - U_m/V_a = \tan 75° - 1200/500 = 1.33$, $\beta_2 = 53.1°$. Also $V_a\ \tan\beta_3 = V_a\ \tan\alpha_3 + U_m$, or $\tan\beta_3 = \tan\alpha_3 + U_m/V_a = \tan 10° + 1200/500 = 2.57$, $\beta_3 = 68.8°$. From

$$C_p\ \Delta T_{0s} = \Delta E = U_m(V_{u2} + V_{u3}) = U_m V_a(\tan\alpha_2 + \tan\alpha_3)$$
$$= 1200 \times 500\ (\tan 75° + \tan 10°) = 2.34 \times 10^6 (\text{ft/s})^2,$$

we have

$$\Delta T_{0s} = \frac{2.34 \times 10^6}{0.24 \times 778 \times 32.2} = 390\ \text{R}.$$

Hence neglecting leakage and mechanical losses, we have the shaft power output, $P_s = \dot{m}C_p\, \Delta T_{0s} = 50 \times 2.34 \times 10^6/(32.2 \times 550) = 6607\,\text{hp}$. The degree of reaction at mean radius can be determined from Eq. (5.16), $R = [V_a/(2U_m)](\tan\beta_3 - \tan\beta_2) = [500/(2 \times 1200)](\tan 68.8° - \tan 53.1°) = 0.259$. To determine the radii, the flow area A_2 can be determined from $m = \rho_2 A_2 V_a$. The density ρ_2 can be determined from p_2 and T_2, which can be calculated as follows. From $C_p T_{02} = C_p T_2 + V_2^2/2$ and $V_2 = V_a/\cos\alpha_2 = 500/\cos 75° = 1932\,\text{ft/s}$, we have

$$T_2 = T_{02} - \frac{V_2^2}{2C_p} = 2000 - \frac{1932^2}{2 \times 0.24 \times 778 \times 32.2}$$

$$= 2000 - 310 = 1690\,°\text{F} = 2150\ \text{R}.$$

From the definition of loss coefficient ξ_s, we have

$$T_{s2} = T_2 - \frac{\xi_s V_2^2}{2C_p} = 1690 - \frac{0.08 \times 1932^2}{2 \times 0.24 \times 778 \times 32.2}$$

$$= 1690 - 24.8 = 1665.2°\text{F} = 2125.2\ \text{R}$$

and $p_2/p_{01} = (T_{s2}/T_{01})^{k/(k-1)} = (2125.2/2460)^{1.4/0.4} = 0.60$, or $p_2 = 200 \times 0.60 = 120\,\text{psia}$. Hence the density and area can be calculated as

$$\rho_2 = \frac{p_2}{RT_2} = \frac{120 \times 144}{53.3 \times 2150} = 0.151\ \text{lb}_\text{m}/\text{ft}^3, \quad A_2 = \frac{\dot{m}}{\rho_2 V_a} = \frac{50}{0.151 \times 500} = 0.662\ \text{ft}^2.$$

The radii can be calculated from $r_t^2 - r_h^2 = A_2/\pi = 0.662/\pi = 0.211\ \text{ft}^2$ and $r_m = 30U_m/(N\pi) = 30 \times 1200/(8000\pi) = 1.43\,\text{ft}$, or $r_t^2 + r_h^2 = 2r_m^2 = 4.09\,\text{ft}^2$, and $r_t = 1.466\,\text{ft}$, $r_h = 1.393\,\text{ft}$, $b = r_t - r_h = 1.466–1.393 = 0.071\,\text{ft} = 0.88\,\text{in}$.

5.7 Determination of Turbine Stage Efficiency

Similar to the case of the compressor discussed in Section 4.2.4, loss coefficients for cascade/airfoil tests are defined in terms of pressure loss due to friction. The turbine stage efficiency and its component losses in the stator and rotor are defined in terms of the enthalpy differences as given in Eq. (5.11). The relationships between these parameters should be established.

In cascade tests, the stagnation pressure drop is measured instead of the temperature drop. Hence, the loss coefficient is defined in terms of the stagnation pressure drops across the stator and rotor, that is,

$$Y_s = \frac{p_{01} - p_{02}}{p_{02} - p_2}, \qquad Y_r = \frac{p_{02,\text{rel}} - p_{03,\text{rel}}}{p_{03,\text{rel}} - p_3}.$$

They can be related to ξ_s and ξ_r, respectively.

Referring to Figure 5.11, with

$$\frac{p_{01}}{p_{02}} = \frac{p_{01}}{p_2}\frac{p_2}{p_{02}} = \left(\frac{T_{01}}{T_{s2}}\right)^{k/(k-1)}\left(\frac{T_2}{T_{02}}\right)^{k/(k-1)} = \left(\frac{T_2}{T_{s2}}\right)^{k/(k-1)},$$

we can express

$$Y_s = \frac{(p_{01}/p_{02}) - 1}{1 - (p_2/p_{02})} = \frac{(T_2/T_{s2})^{k/(k-1)} - 1}{1 - (T_2/T_{02})^{k/(k-1)}} = \frac{[1 + (T_2 - T_{s2})/T_{s2}]^{k/(k-1)} - 1}{1 - [1 + (T_2 - T_{02})/T_{02}]^{k/(k-1)}}$$

$$\cong \frac{\frac{k}{k-1}\frac{T_2 - T_{s2}}{T_{s2}}}{\frac{k}{k-1}\frac{T_{02} - T_2}{T_{02}}} = \frac{T_2 - T_{s2}}{T_{02} - T_2}\frac{T_{02}}{T_{s2}} = \xi_s \frac{T_{02}}{T_{s2}} \cong \xi_s \frac{T_{02}}{T_2},$$

where $T_{02}/T_2 = 1 + (k - 1)M_2^2/2$, and the approximation is made with the Taylor expansion $(1 + \varepsilon)^x = (1 + x\varepsilon + \ldots) \cong 1 + x\varepsilon$ for a small number of ε.

Similarly, we have $Y_r \cong \xi_r T_{03,\text{rel}}/T_3$, where $T_{03,\text{rel}}/T_3 = 1 + (k - 1){M_{3,\text{rel}}}^2/2$. The stator can be a convergent or convergent-divergent nozzle. In general, the convergent nozzle is preferred. It is more adaptable to the off-design operation and more efficient when the pressure ratio is less than the critical pressure ratio.

If no specific cascade data are available, the loss coefficients ξ correlated by Soderberg[2] can be employed. They are based on the test data of some steam turbines of "good design." Zweifel[3] gave the good-design cascade criterion for minimum loss as $2(s/c_a)\cos^2\alpha_2(\tan\alpha_1 + \tan\alpha_2) = 0.8$, where s is the pitch and c_a is the axial length of the chord, $c_a = c\cos\gamma$.

Soderberg's correlation can be summarized as follows:

a) The parameter $\xi = 0.04 + 0.06(\varepsilon/100)^2$ is good for the deflection, $\varepsilon \leq 120°$ with $t_m/c = 0.2$, $b/c_a = 3$, and $\text{Re} = \rho_2 V_2 D_h/\mu = 10^5$, where the hydraulic diameter $D_h = 4(\text{throat area})/\text{wetted perimeter} = 2sb\cos\alpha_2/(s\cos\alpha_2 + b)$, t_m is the maximum thickness of the blade and $\varepsilon = \alpha_1 + \alpha_2$ in degrees.
b) Variation with t_m/c is roughly proportional to $[0.2/(t_m/c)]^{0.3}$.
c) For $b/c_a \neq 3$, the modified ξ_1 is given as

$$1 + \xi_1 = \begin{cases} (1+\xi)(0.993 + 0.021c_a/b) & \text{for stators,} \\ (1+\xi)(0.975 + 0.075c_a/b) & \text{for rotors.} \end{cases}$$

d) For $\text{Re} \neq 10^5$, we have $\xi_2 = (10^5/\text{Re})^{0.25}\,\xi_1$.

The flow deviation from the blade angle at the outlet is negligible for turbine cascades in general. These correlations can be used for the preliminary design unless some specific cascade data is available. Other types of loss correlation for turbines are also available in the literature.

5.8 Axial-Flow Turbine Performance

The basic turbine performance characteristics can be described in terms of the output shaft torque, power, and efficiency versus the rotating speed for a given mass flow rate of fluid supplied at a fixed inlet pressure and temperature to the turbine and discharged to a given ambient pressure. The ideal performance can be illustrated as follows. With the shaft rotating speed increasing from zero by reducing the load, the corresponding velocity diagram across the rotor can be sketched, as shown in Figure 5.15. It is seen that the axial-flow velocity V_a is kept constant as the flow rate is constant, and the peripheral velocity U increases as the rotating speed increases. The exit flow velocities from the stator and rotor, that is, V_2 and W_3, are relatively constant since the directions of these two flow velocities are guided by the blades. And the change of tangential velocities, $\Delta V_u = V_{u2} - V_{u3}$, decreases as the rotating speed increases. Hence, from Euler's Equation, the torque $\tau = \dot{m}r\,\Delta V_u$ decreases linearly. Or mathematically, it can be expressed as

$$\begin{aligned} \tau &= \dot{m}r(V_{u2} - V_{u3}) = \dot{m}r[V_a\tan\alpha_2 - (U - V_a\tan\beta_3)] \\ &= \dot{m}r[V_a(\tan\alpha_2 + \tan\beta_3) - r\omega] = a - b\omega, \end{aligned}$$

where $a = \dot{m}rV_a(\tan\alpha_2 + \tan\beta_3)$ and $b = \dot{m}r^2$ are constants. The shaft power $P_s = \omega\tau = a\omega - b\omega^2$ increases from zero to a maximum at the design condition, then drops to zero again at the so-called runaway speed. With the mass

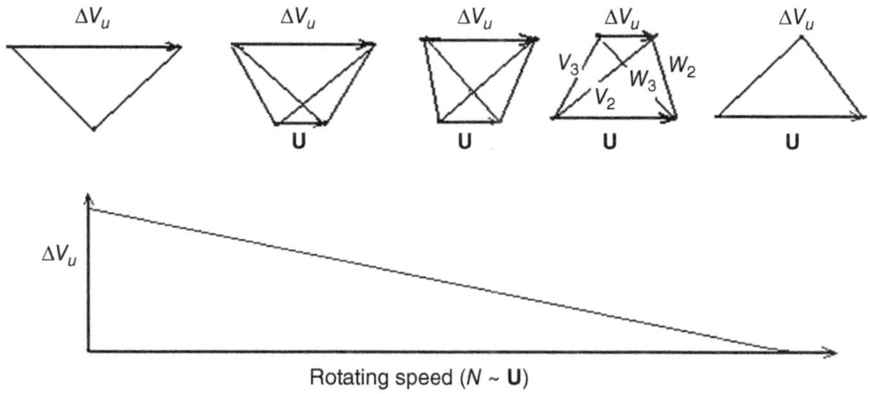

Figure 5.15 Velocity diagrams across turbine rotor with increasing rotating speed.

Figure 5.16 Theoretical shaft torque, power, and efficiency of a turbine versus rotating speed.

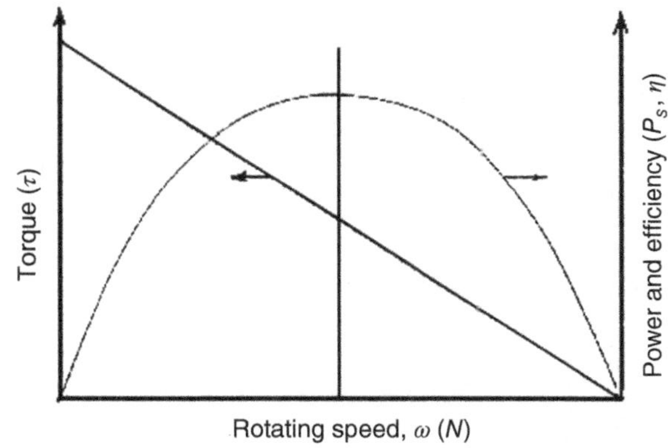

flow rate and inlet conditions kept constant, the efficiency will have a curve similar to that for the shaft power. They are shown in Figure 5.16. The runaway speed and maximum power/efficiency point can be calculated by setting $\tau = 0$ and $dP_s/d\omega = 0$, that is, $\omega_{r/a} = V_a(\tan\alpha_2 + \tan\beta_3)/r$ and $\omega_m = \omega_{r/a}/2$.

The actual performance curves in terms of the normalized parameters for the gas turbines are similar to the ideal curves, as shown in Figure 5.17*a*.

In order to match the turbine with a compressor, another form of the turbine performance curve is plotted in terms of the mass flow rate, shaft power, and efficiency versus the inlet/outlet total pressure or temperature ratio with the rotating speed as the parameter. Typical curves are shown in Figure 5.17*b*, where the horizontal axis is plotted in the reverse direction. The mass flow rate and rotating speed are normally expressed in terms of $\dot{m}\sqrt{T_{0i}}/p_{0i}$ and $N/\sqrt{T_{0i}}$, similar to those for the compressor discussed in Section 4.2. The mass flow rate may reach a constant choking condition at a low-pressure ratio.

For direct coupling of the compressor and turbine, the following conditions should be satisfied:

a) Compressor rpm = turbine rpm
b) Gas mass flow rate through turbine = air mass flow rate through compressor + fuel mass flow rate
c) Turbine output power = compressor input power + load

A more detailed analysis of performance matching will be discussed in the next section.

The performance characteristics of a turbine can also be illustrated with the utilization factor, defined as

$$\varepsilon = \frac{\Delta E}{\Delta E_{\text{avl}}} = \frac{\Delta E}{\Delta E + E_{\text{loss}}},$$

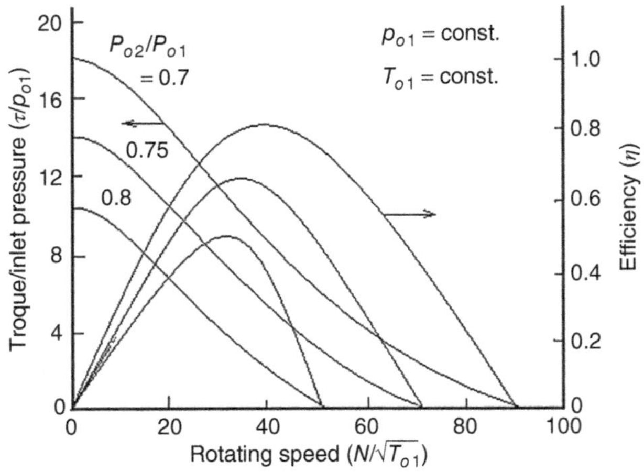

(a) Torque and efficiency versus rotating speed

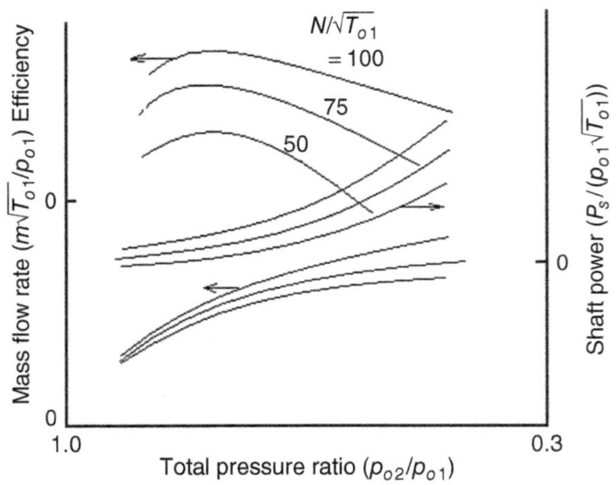

(b) Efficiency, shaft power, and mass flow rate versus pressure ratio

Figure 5.17 Typical performance curves of a reaction gas turbine with constant inlet condition of pressure and temperature (N: rpm, τ: ft-lbf, P_s: hp., $\dot{m}$: lbm/s, T_{01}: R, p_{01}, p_{02}: psia).

where ΔE is the energy transfer per unit mass of fluid defined in the Euler equation, and E_{loss} is part of the available energy in the fluid lost with the flow discharged from the turbine stage, that is, $E_{\text{loss}} = V_3^2/2$. Thus, from Eq. (2.42), we have

$$\varepsilon = \frac{(1/2)\left[\left(V_2^2 - V_3^2\right) + \left(W_3^2 - W_2^2\right)\right]}{(1/2)\left[V_2^2 + \left(W_3^2 - W_2^2\right)\right]}$$

since $U_2 = U_3$. With the degree of reaction expressed in the form

$$R = \frac{h_2 - h_3}{h_1 - h_3} = \frac{(1/2)\left(W_3^2 - W_2^2\right)}{(1/2)\left[\left(V_2^2 - V_3^2\right) + \left(W_3^2 - W_2^2\right)\right]},$$

the utilization factor can be written as

$$\varepsilon = \frac{\left(V_2^2 - V_3^2\right) + [R/(1-R)]\left(V_2^2 - V_3^2\right)}{V_2^2 + [R/(1-R)]\left(V_2^2 - V_3^2\right)} = \frac{V_2^2 - V_3^2}{V_2^2 - RV_3^2}. \tag{5.19}$$

This expression is valid for any degree of reaction except $R = 1$, which requires special analysis. Since $V_3^2/2$ represents the loss or part of available energy not utilized by the rotor, V_3 should be a minimum for the maximum utilization factor. The velocity diagram shows that V_3 is a minimum when it is in the pure axial direction. Thus, a general velocity diagram for maximum ε would appear as that shown in Figure 5.18, and Eq. (5.19) becomes

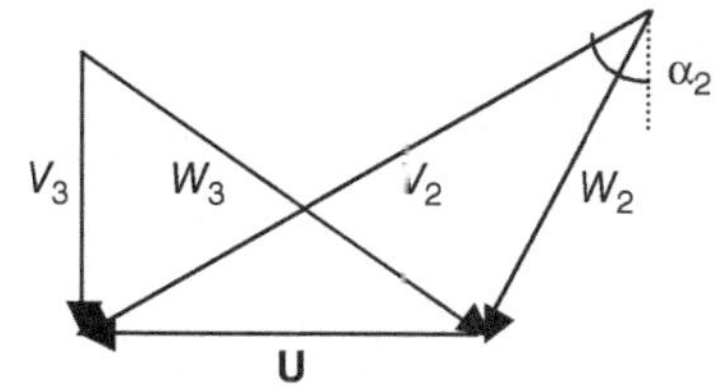

Figure 5.18 General velocity diagrams under maximum utilization factor condition.

$$\varepsilon_{max} = \frac{V_2^2 - V_2^2\cos^2\alpha_2}{V_2^2 - RV_2^2\cos^2\alpha_2} = \frac{\sin^2\alpha_2}{1 - R\cos^2\alpha_2}. \tag{5.20}$$

From this equation, it may be seen that for any degree of reaction, ε, is an absolute maximum and has a value of unity if $\alpha_2 = 90°$. Since $V_3 = V_2 \cos \alpha_2$, V_3 is zero, and the flow enters the rotor in a pure tangential direction. However, this is practically impossible. Normally, α_2 values of 65°–78° are used.

For an impulse turbine, $R = 0$, the velocity diagram under the maximum-utilization-factor condition is shown in Figure 5.19. It is seen that $W_2 = W_3$ and $\beta_2 = \beta_3$, and hence $2U = V_2 \sin \alpha_2$ or $U/V_2 = \sin \alpha_2/2$ and $\varepsilon_{max} = \sin^2 \alpha_2$. For a 50% reaction turbine, the velocity diagram under the maximum-utilization-factor condition is shown in Figure 5.20. It is seen that $\alpha_2 = \beta_3$ and $\beta_2 = \alpha_3 = 0$, and hence $U = V_2 \sin \alpha_2$ or $U/V_2 = \sin \alpha_2$ and $\varepsilon_{max} = \sin^2 \alpha_2/(1 - \cos^2 \alpha_2/2)$.

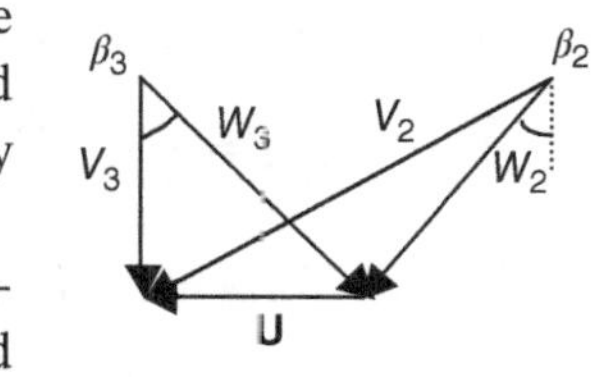

Figure 5.19 Velocity diagrams for impulse turbine.

The utilization factor versus U/V_2 for these two types of turbines is plotted in Figure 5.21 for a typical stator exit angle $\alpha_2 = 70°$.

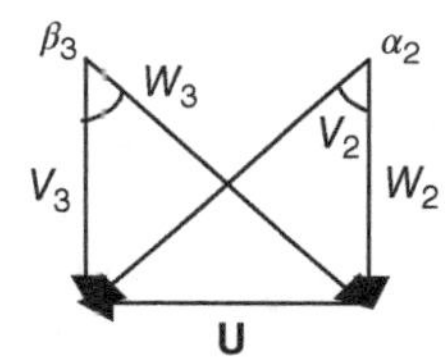

Figure 5.20 Velocity diagrams for 50 % reaction turbine.

The energy transfer per unit mass flow rate under the maximum-utilization-factor condition for these two types of turbines is compared as $\Delta E_i = UV_2 \sin \alpha_2 = 2U^2$ and $\Delta E_r = UV_2 \sin \alpha_2 = U^2$ from Figures 5.19 and 5.20. In other words, with all other conditions the same, the amount of energy transfer with an impulse turbine is double that with a 50% reaction turbine. But in general, the reaction turbine is more efficient than the impulse turbine because the flow in a reaction turbine rotor accelerates, and it is less likely to suffer from boundary layer separation, hence less loss.

In designing multistage axial-flow turbines, the impulse turbine stages are normally placed upstream to reduce fluid pressure and temperature quickly. The reaction turbine stages are placed downstream to take advantage of their higher adiabatic efficiency.

Example 5.5 Convert the data for $N/\sqrt{T_{01}} = 40$ in Figure 5.17*a* into the format of Figure 5.17*b*. Use the values $C_p = 0.24$ Btu/(lb$_m$R) and $k = 1.40$ for air.

Figure 5.21 Variation of utilization factor with U/V_2 for impulse turbine ($R = 0$) and reaction turbine ($R = 0.5$).

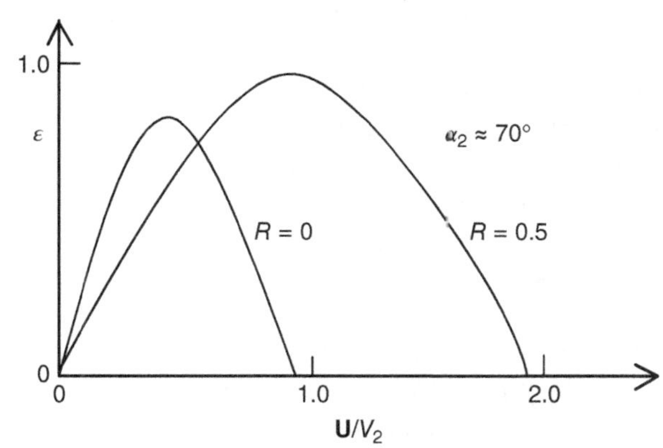

Solution

For $N/\sqrt{T_{01}} = 40$, the following data of p_{02}/p_{01}, τ/p_{01}, and η can be obtained from Figure 5.17*a*:

p_{02}/p_{01}	0.70	0.75	0.8
τ/p_{01}	8.7	5.3	2.2
η	0.81	0.64	0.41
$\frac{P_s}{p_{01}\sqrt{T_{01}}}$	0.066	0.040	0.017
$\frac{\dot{m}\sqrt{T_{01}}}{p_{01}}$	2.48	2.34	2.0

The power and mass flow rate have to be obtained with the following manipulations. From $P_s = \tau\omega$, we obtain

$$\frac{P_s}{p_{01}\sqrt{T_{01}}} = \frac{\tau N\pi}{30 \times 550 p_{01}\sqrt{T_{01}}} = \frac{\tau}{p_{01}} \times \frac{N}{\sqrt{T_{01}}} \times \frac{\pi}{30 \times 550}.$$

Also from $P_s/\dot{m} = \eta C_p T_{01}\,[1 - (p_{02}/p_{01})^{(k-1)/k}]$, we obtain

$$\frac{\dot{m}\sqrt{T_{01}}}{p_{01}} \equiv \frac{P_s/(p_{01}\sqrt{T_{01}})}{P_s/(\dot{m}T_{01})} = \frac{P_s/(p_{01}\sqrt{T_{01}})}{\eta C_p}\left[1 - \left(\frac{p_{02}}{p_{01}}\right)^{(k-1)/k}\right]^{-1},$$

where $(k - 1)/k = 0.40/1.4 = 0.2857$ and $C_p = 0.24\ \text{Btu}/(\text{lb}_\text{m}\text{R}) = 0.24 \times 778/550 = 0.339\ \text{hp-s}/(\text{lb}_\text{m}\text{R})$.

Example 5.6 A multistage axial-flow gas turbine is to be designed for the inlet air condition of 60 psia and 1400 R and to be exhausted to the ambient pressure of 14.7 psia. The rotating speed is selected to be 8000 rpm, blade velocity at mean radius is 850 ft/s, and blade length is 4 in. Determine (a) the blade radii at tip and hub and (b) the number of stages required if all stages selected are impulse or reaction type, assuming no friction loss.

Solution

From $U_m = r_m\omega = r_m N\pi/30$, we have $r_m = 30U_m/(N\pi) = 30 \times 850/(8000\pi) = 1.01$ ft $= 12.1$ in. $= [(r_t^2 + r_t^2)/2]^{0.5}$, or $r_t^2 + r_h^2 = 293.8$ in. Combined with $b = r_t - r_h = 4$ in., we have $r_t^2 - 4r_t - 138.9 = 0$, so $r_t = 13.95$ in. and $r_h = 9.95$ in. To find the number of stages required, the exhaust air temperature can be estimated as $T_{0e} = T_{0i}$ $(p_{0e}/p_{0i})^{(k-1)/k} = 1400(14.7/60)^{0.2857} = 936.7$ R. The overall energy available per unit mass of air is $\Delta H_s = C_p$ $(T_{0i} - T_{0e}) = 0.24(1400\text{–}936.7) = 111.2\ \text{Btu/lb}_\text{m}$. The maximum energy transfer per stage with an impulse turbine is $\Delta E_i = 2U_m^2 = 2 \times (850)^2/(32.2 \times 778) = 57.68\ \text{Btu/lb}_\text{m}$. Hence, the required number of stages is $n_{si} = 111.2/57.68 = 1.93 \approx 2$. With the reaction turbine stages, it will be $\Delta E_r = U_m^2 = 850^2/(32.2 \times 778) = 28.84$ Btu/lb$_\text{m}$ and $n_{sr} = 3.85 \approx 4$.

5.9 Compressor and Turbine Matching

The design point performance of a complete gas turbine engine is obtained from cycle analysis. Each component, such as the compressor, combustor, and turbine, is then designed based on the conditions specified for the design point. Their off-design performances are obtained from calculations with empirical formulas or laboratory tests. When the components are combined into a complete engine, the off-design performance range for the system will be reduced.

The system's off-design point is to find the corresponding operating points on the characteristics of each component when the engine is running at a steady speed or in equilibrium. The equilibrium running points for several speeds are plotted on the compressor characteristic diagram and joined to form an equilibrium line. Then, the power output or thrust curves can be obtained from the turbine's or nozzle's characteristics. All of these establish an equilibrium diagram, which shows the proximity of the operating line or zone to the compressor surge line and the compressor's corresponding efficiency. The complete gas turbine engine becomes unstable if the operating line intersects with the surge line.

5.9.1 Component Characteristics

The performance characteristics of the nondimensional parameters for individual compressors and turbines are discussed in the previous sections. For the compressor, expressing the efficiency and pressure ratio versus mass flow rate at a constant rotating speed is normal. A sample map is shown in Figure 5.22*a*. The flow could be choked at high rotating speeds, and both the efficiency and pressure ratio result in the vertical lines on the mass flow base. Hence, plotting the efficiency curves against the pressure ratio is convenient. For the impulse turbine, the nondimensional mass flow is choked at the high-pressure ratio, and its variation concerning the nondimensional rotational speed is small, even for the non choking condition. Hence, a single mass flow curve is approximated in the preliminary calculations.

Fine-tuning with the rotating speed effect under the nonchoking condition can be considered later. A simplified diagram is shown in Figure 5.22*b*. It should be noted that subscripts 3 and 4 are the turbine inlet and outlet in this figure. The load characteristics for a shaft power engine should also be available. A typical curve for general machinery is shown in Figure 5.22*c*.

5.9.2 Off-Design Operation of Single-Shaft Gas Turbine

The following steps describe the procedure to establish the equilibrium running lines of the complete single-shaft gas turbine driving a typical mechanical load, which varies with the rotating speed, with T_{01}, p_{01} given.

1) Select a constant-speed line on the compressor characteristic diagram and choose any point on this line. The values of $\dot{m}\sqrt{T_{01}}/p_{01}, p_{02}/p_{01}, \eta_c$, and $N/\sqrt{T_{01}}$ are then set for this particular point.
2) Since the compressor and turbine are directly coupled, their rotational speeds should be the same, that is, $N_t = N_c = N$. Assume the same mass flow rate, that is, $\dot{m}_t = \dot{m}_c = \dot{m}$, since the fuel flow rate is much less than the airflow rate, and also neglect inlet and exhaust pressure losses, that is, $p_{04} = p_{01} = p_a$. With the pressure loss in the combustor Δp_b given, we have $p_{03}/p_{02} = 1 - \Delta p_b/p_{02}$, and hence $p_{03}/p_{04} = (p_{03}/p_{02})(p_{02}/p_{01})$.
3) From the turbine characteristics, we can obtain the approximate value $m\sqrt{T_{03}}/p_{03}$, which mainly depends on the pressure ratio only, as shown in Figure 5.22*b*. Since it also can be related to $\dot{m}\sqrt{T_{01}}/p_{01}$ in the form

$$\frac{\dot{m}\sqrt{T_{03}}}{p_{03}} = \left(\frac{\dot{m}\sqrt{T_{01}}}{p_{01}}\right)\left(\frac{p_{01}}{p_{02}}\right)\left(\frac{p_{02}}{p_{03}}\right)\sqrt{\frac{T_{03}}{T_{01}}}, \tag{5.21}$$

   the turbine inlet temperature T_{03} can be determined. Then, the parameter $N/\sqrt{T_{03}}$ and η_t can be determined. Or the more accurate values $\dot{m}\sqrt{T_{03}}/p_{03}$ and T_{03} can be obtained by iteration for the non-choking condition.
4) The corresponding temperature changes and the net power output of the turbine can then be obtained from

$$\Delta T_{034} = \eta_t T_{03}\left[1 - \left(\frac{p_{04}}{p_{03}}\right)^{(k-1)/k}\right] \quad \text{[Equation (8.1)]},$$

$$\Delta T_{012} = \frac{T_{01}}{\eta_c}\left[\left(\frac{p_{02}}{p_{01}}\right)^{(k-1)/k} - 1\right] \quad \text{[Equation (7.4)]},$$

   and

$$P_0 = mC_{pg}\,\Delta T_{034} - \frac{1}{\eta_m} mC_{pa}\,\Delta T_{012}, \tag{5.22}$$

   where η_m is the mechanical efficiency of the compressor–turbine combination, C_{pa} and C_{pg} are the specific heat for air and combustion gas, respectively, and m can be obtained from the value of $\dot{m}\sqrt{T_{01}}/p_{01}$.
5) The net output power P_0 is compared with the required load P_L at the chosen speed. Iteration is needed until they match closely.

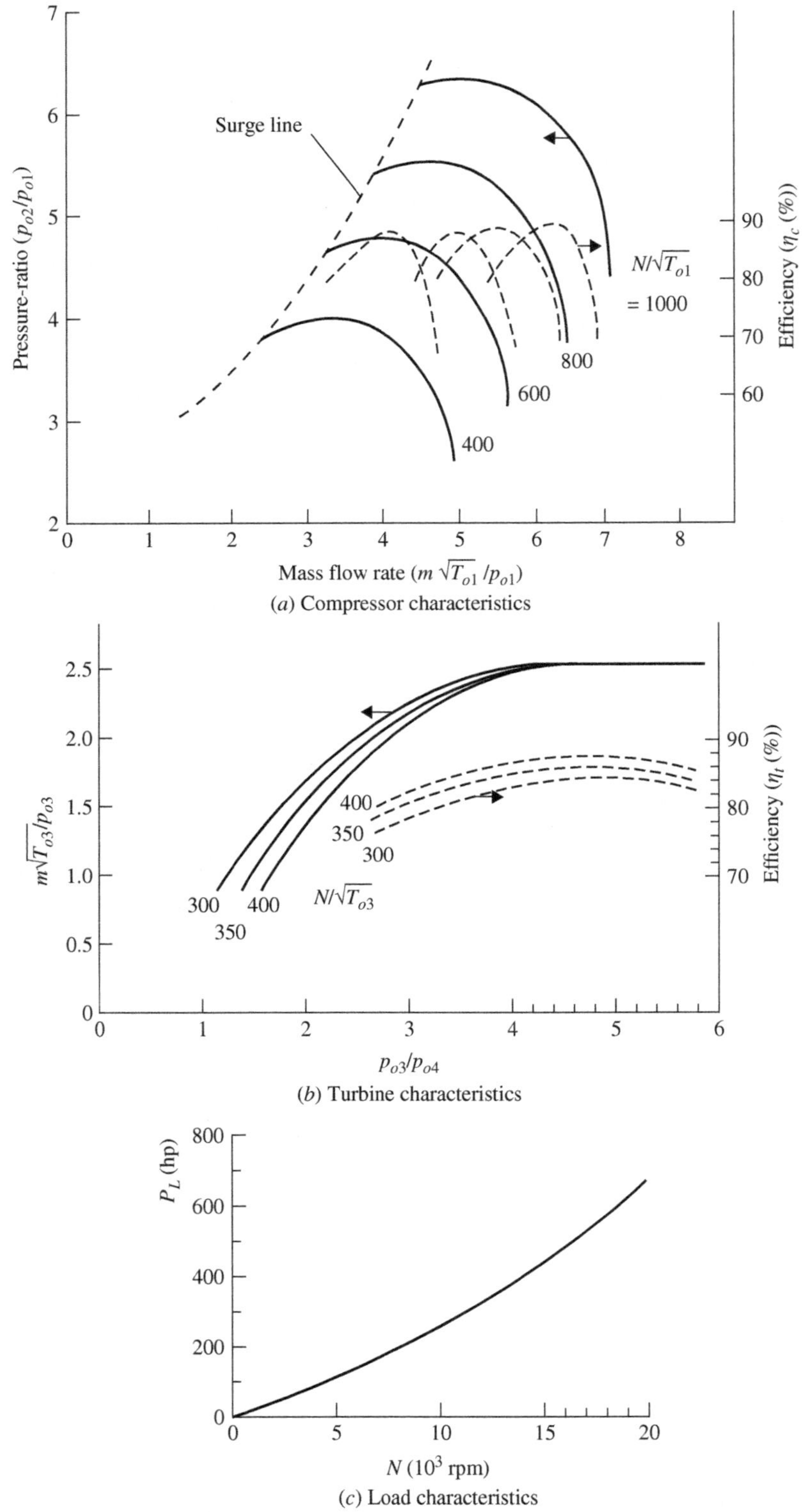

Figure 5.22 Typical component characteristics for compressor/turbine matching [$\dot{m}$(lbm/s), T_0(R), p_0(psia), N(rpm), and P_L(hp)].

Calculations are repeated for several rotating speeds. The equilibrium points for these speeds are located on the compressor characteristic diagram and connected to an equilibrium running line. Its proximity to the surge line can then be investigated. At each point, other parameters, such as T_{02}, T_{03}, and m, can also be evaluated. If the load is a constant-speed device, such as a generator or variable-pitch propeller, the calculations can proceed along a constant-speed curve on the compressor characteristics. Each point corresponds to an output power for the load. The no-load conditions, that is, $P_0 = 0$, can also be evaluated. A more detailed discussion can be found in Ref. 5.

Example 5.7 A shaft power gas turbine engine, of which the compressor and turbine characteristics are shown in Figures 5.22*a* and *b* is used to drive a machine of which the load characteristics are shown in Figure 5.22*c*. If the engine is operated at the ambient condition $p_a = 14.7$ psia, $T_a = 70\,°\text{F}$, locate its running point on the compressor when the machine is operated at a rotating speed of 18,400 rpm. Assume the following data:

a) Mechanical efficiency for the compressor–turbine combination is $\eta_m = 95\%$.
b) Air and gas properties are $k_a = 1.4$, $k_g = 1.33$, $C_{pa} = 0.24\ \text{Btu/lb}_\text{m}\text{R}$, $C_{pg} = 0.28\ \text{Btu/lb}_\text{m}\text{R}$.
c) Pressure drop in the combustor is $\Delta p_b = 2$ psi.

Solution
From Figure 5.22*c* we have $P_L = 620$ hp. at $N = 18{,}400$ rpm. Pick a point on the curve of $N/\sqrt{T_{01}} = 18,400/\sqrt{530} = 800$ in Figure 5.22*a*, say $p_{02}/p_{01} = 5$. So we have

$\dot{m}\sqrt{T_{01}}/p_{01}$	η_c	p_{02} (psia)	$\dot{m}$ (lb$_\text{m}$/s)	p_{03} (psia)	p_{03}/p_{04}
5.7	0.85	73.5	3.64	71.5	4.86

where $p_{02} = 5 \times 14.7, m = 5.7 \times 14.7/\sqrt{530}$, $p_{03} = (73.5 - 2) = 71.5$ psia, $p_{03}/p_{04} = 71.5/14.7 = 4.86$. Then from Figure 5.22*b*, with $p_{03}/p_{04} = 4.86$ and $m_3 = m_2$, we have

$\dot{m}\sqrt{T_{03}}/p_{03}$	T_{03} (R)	$N/\sqrt{T_{03}}$	η_t
2.56	2528	366	0.87

where $T_{03} = (2.56 \times 71.5/3.64)^2, N/\sqrt{T_{03}} = 18,400/\sqrt{2528}$. So, from Eqs. (5.10), (4.20), and (5.22), we have

$$\Delta T_{034} = \eta_t T_{03}\left[1 - \left(\frac{p_{04}}{p_{03}}\right)^{(k-1)/k}\right] = 0.87 \times 2528[1 - (4.86)^{-0.248}] = 713\ \text{R},$$

$$\Delta T_{012} = \left(\frac{T_{01}}{\eta_c}\right)\left[\left(\frac{p_{02}}{p_{01}}\right)^{(k-1)/k} - 1\right] = \left(\frac{530}{0.85}\right)[(5)^{0.2857} - 1] = 364\ \text{R},$$

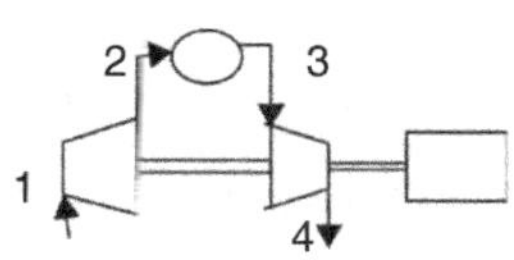

and $P_0 = 3.64(0.28 \times 713\text{–}0.24 \times 364/0.95) = 392$ Btu/s $= 554$ hp, which is less than P_L. So we pick another point on the same curve, say $p_{02}/p_{01} = 5.2$, and repeat the calculations:

$m\sqrt{T_{01}}/p_{01}$	η_c	p_{02} (psia)	$\dot{m}$ (lb$_\text{m}$/s)	p_{03} (psia)	p_{03}/p_{04}	$\dot{m}\sqrt{T_{03}}/p_{03}$	T_{03} (R)	$N/\sqrt{T_{03}}$	η_t
5.6	0.88	76.4	3.57	74.4	5.06	2.55	2824	346	0.85

$$\Delta T_{034} = 0.85 \times 2824[1 - (5.06)^{-0.248}] = 794 \text{ R}, \qquad \Delta T_{012} = \left(\frac{530}{0.88}\right)[(5.2)^{0.2857} - 1] = 362 \text{ R},$$

and net power output $P_0 = 3.57(0.28 \times 794–0.24 \times 362/0.95) = 467$ Btu/s = 660 hp, which is too much greater than P_L. Pick another point, say $p_{02}/p_{01} = 5.15$. Then we have

$\dot{m}\sqrt{T_{01}}/p_{01}$	η_c	p_{02} (psia)	$\dot{m}$ (lb$_m$/s)	p_{03} (psia)	p_{03}/p_{04}	$\dot{m}\sqrt{T_{03}}/p_{03}$	T_{03} (R)	$N/\sqrt{T_{03}}$	η_t
5.65	0.87	75 0.7	3.61	73.7	5.01	2.55	2710	353	0.86

$$\Delta T_{034} = 0.86 \times 2710[1 - (5.01)^{-0.248}] = 768R, \Delta T_{012} = \left(\frac{530}{0.87}\right)[(5.15)^{0.2857} - 1] = 364 \text{ R},$$

and $P_0 = 3.61(0.28 \times 768–0.24 \times 364/0.95) = 444.3$ Btu/s = 628 hp, which is close to P_L. So the running point is around $p_{02}/p_{01} = 5.15$, $\dot{m}\sqrt{T_{01}}/p_{01} = 5.65$, and $N/\sqrt{T_{01}} = 800$ on the compressor characteristics. It is not too close to the surge line and hence it is safe.

5.10 Other Topics Related to Gas Turbine Engine

5.10.1 Combustion Chamber (Combustor)

The combustor is an important component in the gas turbine engine. Its design is a complicated task involving knowledge of fluid dynamics, combustion, and mechanical design. A detailed discussion is beyond the scope of this book. Only a brief discussion will be provided here.

The combustors can be classified according to their arrangements around the engine:

1) In a can-type (or tubular) combustor, as shown in Figure 5.23*a*, air from the compressor is split into several streams entering the individual can of combustor. Fuel is also injected into the individual chamber from a common line.
   The advantage of this type of combustor is that development can be carried out for a single can with a small-scale test rig. The overall system can be designed by arranging the multiple cans around the axis. Its disadvantages are the heavy weight and larger frontal area for the overall system.
2) The annular type shown in Figure 5.23*b* is more compact and has lower pressure loss. Its disadvantages are (1) fuel-air ratio and temperature distributions are less uniform, (2) lower structural integrity, and (3) higher development cost.
3) The cannular type modifies the annular combustor where the individual flame tubes are uniformly spaced around an annular casing. Both annular and cannular designs can have a reverse-flow arrangement and result in a reduced overall shaft length.
4) The silo-type combustor is a vertical cylindrical chamber, as shown in Figure 5.23*c*. It is relatively large and hence has lower flow velocity and pressure loss. It is primarily used in ground-based industrial gas turbines.

The following parameters characterize combustor performance:

1) Pressure Loss: The stagnation pressure loss across a combustor is due to the flow friction and stagnation temperature rise associated with combustion.
2) Combustion Efficiency: The combustion efficiency is defined as either $\eta_b = (f/a)_{\text{th}}/(f/a)_{\text{ac}}$ or $\eta_b = (\Delta T)_{\text{ac}}/(\Delta T)_{\text{th}}$, where $(f/a)_{\text{th}}$ and $(f/a)_{\text{ac}}$ are the theoretical and actual fuel-air ratios required to raise a certain temperature across the combustor; $(\Delta T)_{\text{th}}$ and $(\Delta T)_{\text{ac}}$ are the theoretical and actual temperature rises for a certain fuel-air ratio. Both definitions result in approximately the same magnitude under the same conditions.

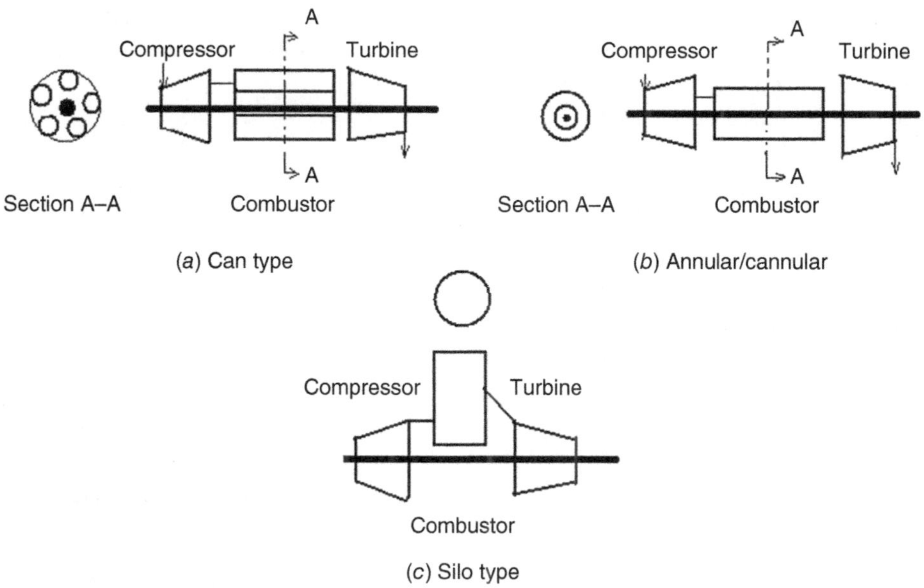

Figure 5.23 Types of combustors.

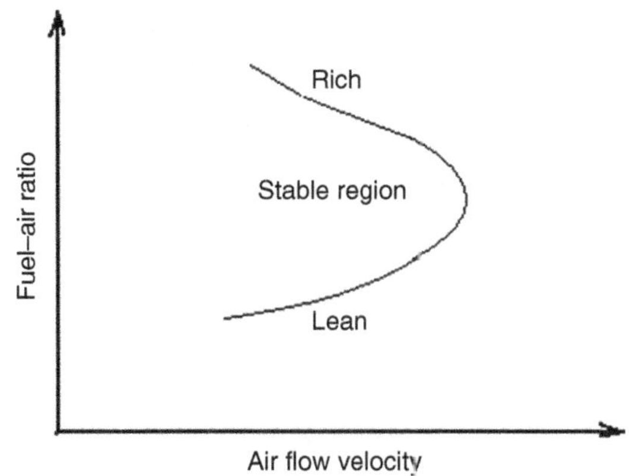

Figure 5.24 Combustion stability limits.

3) Combustion Stability Limits: The combustion process is possible for a certain range of fuel-air ratios. It becomes unstable and flames when the fuel-air ratio is too rich or too lean. This range is a function of the chamber gas pressure and flow velocity for a given combustor. As the pressure decreases, the stability range will decrease. It is shown qualitatively in Figure 5.24 that the stability range of the fuel-air ratio decreases and eventually flames out as the airflow velocity increases.

5.10.2 Control of Emission Pollutants

The air pollution from gas turbines is relatively minor compared with that from conventional power plants burning coal or residual fuel. The typical fuel used in a gas turbine, either natural gas or liquid distillate, is relatively clean because the combustion product gas has to flow through the turbine. However, with the increasing widespread usage of gas turbines, their pollution problems are receiving more attention.

The main types of pollutants emitted from a gas turbine are CO (carbon monoxide), UHC (unburned hydrocarbon), and NO_x (oxides of nitrogen). Carbon monoxide and UHC harm human health, while NO_x will deplete the ozone in the upper atmosphere. Carbon monoxide and UHC are formed due to incomplete combustion when the machine is operated in off-design conditions. Nitrogen oxides are formed due to the high flame temperature required to achieve high efficiency at the design condition.

The effects of the fuel-air ratio on the emission of these pollutants can be plotted qualitatively, as shown in Figure 5.25. The residence time of combustion (related to the gas flow velocity or cross-sectional area of the combustor) also has some minor effects on forming these pollutants. Longer residence time will promote the formation of NO_x but will improve combustion efficiency and reduce the amount of CO and UHC. The effect of flame temperature on NO_x emission is shown in Figure 5.26. The NO_x concentration increases exponentially concerning the flame temperature range of 1500–2000 K, as reported in Ref. 10.

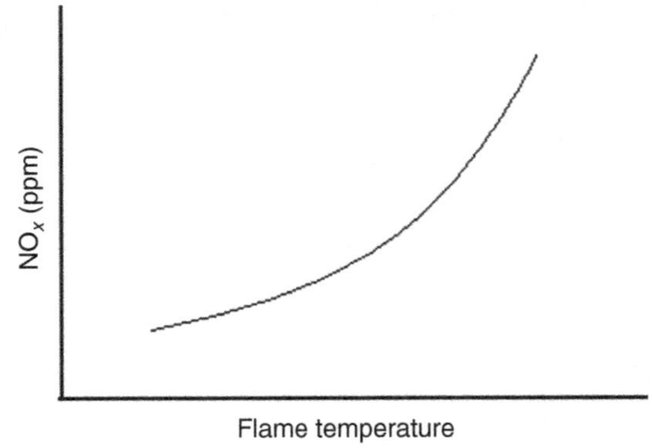

Figure 5.26 Qualitative effects of flame temperature on NO_x emission.

These pollutant emissions can be controlled with the following approaches:

1) Water or Steam Injection: The NO_x concentration can be reduced by injecting water or steam into the combustor to decrease the flame temperature. This will decrease the plant's thermal efficiency, although some additional output power can be gained from the increased mass flow rate. Also, this will increase CO and UHC concentration, and the water/steam required is significant (about 50–100% of fuel flow rate)
2) Selective Catalytic Reduction: Further reduction of NO_x concentration can be achieved by installing some selected catalyst and ammonia injection at the combustor exhaust to convert NO_x to N_2 and H_2O. However, these catalysts only work at a gas temperature between 285 °C and 400 °C. Hence, this approach can only be employed conveniently for the system with an HRSG (heat recovery steam generator), and additional equipment costs will be incurred.

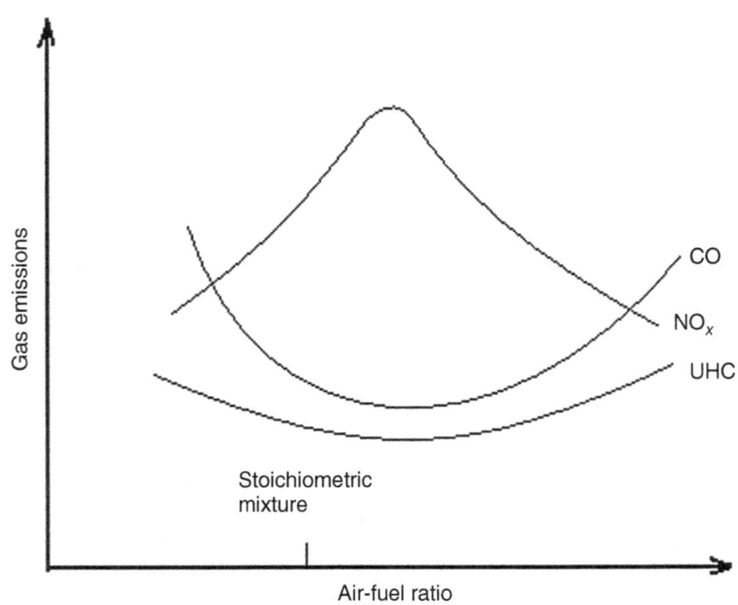

Figure 5.25 Qualitative effects of fuel-air ratio on emissions.

3) Dry Low-NO_x Combustor: Recently, some innovative combustor designs have been pursued. The approach is to control the combustion process in the combustor without water/steam injection. Typically, the fuel is supplied into two zones with two stages. The first (or primary) zone is the pilot stage for starting and idling. The second zone is for the main stage of combustion. Different fuel-air ratios are supplied for different operating conditions.
4) Other new clean fuels, such as syngas and hydrogen-rich fuel, are being developed. A more detailed discussion of these topics is beyond the scope of this book, and the interested reader should consult Refs. 10–14

5.11 Applications of Gas Turbine Engines

Gas turbine engines can be classified according to their applications and power levels. They can be divided into three major groups: (1) aircraft engines, (2) ground-based industrial power units, and (3) surface vehicle power plants, both ground and marine vehicles.

Aircraft engines have received most of the attention and have been the major thrust for gas turbine development in the past few decades. The main concerns in their designs are high reliability, high thrust/weight ratio, and high efficiency. Four types of aircraft gas turbines have been designed for different applications according to flying speed and other considerations. The schematic diagrams are shown in Figure 5.1. The turboprop is for subsonic flight, and the ramjet is for supersonic flight. In a turbofan, the bypass air supplied by the upstream fan is used to cool the combustor and turbine, maintaining the same thrust power with a higher exhaust gas mass flow rate and lower jet velocity. Hence, it is quieter than the other jet engines. Detailed thermodynamic cycle analyses are given in Appendix A.

The main differences between the ground-based industrial gas turbines and aircraft gas turbines are as follows:

a) The lifetime for industrial gas turbines is around 100,000 h, while that for aircraft is much shorter.
b) The size and weight are not so important for the industrial turbines; hence, various accessories can be added to improve overall efficiency and performance.
c) The kinetic energy of exhaust gas is useful for producing thrust in aircraft gas turbines, but it is wasted in industrial ones.

Hence, some modifications and innovations have been developed for industrial applications other than stand-alone gas turbine systems to improve overall plant performance, including reheating, regenerating, combined cycle, and cogeneration. More discussions of combined cycle and cogeneration will be given in the next chapter on steam turbines.

These modifications can be classified as follows.

a) Industrial heavy-duty gas turbines have up to 350 MW output power. They have reached a pressure ratio of 35, turbine inlet temperature of 2500 °F, and thermal efficiency of 50%. Typical units have heavy wall casing, thick airfoil section blades, and large frontal areas. The axial-flow compressor, turbine, and can-annular or silo-type combustor are mostly employed in these units.
b) Aero-derivative gas turbines have output power between 2.5 and 50 MW. The gas generator is the aircraft engine modified to burn industrial fuel, and a power turbine is added downstream. An additional low-pressure compressor for the turbofan engine replaces the fan. The power and petrochemical industries use them for remote-area power generation, gas transmission, and reinjection in off-shore platforms. The advantages of these units are as follows: (1) the right size and weight for packaging and testing within the manufacturer's plant before shipping to job sites for installation; (2) convenience for maintenance and repair. Gasifier that can be removed and sent back to the factory, a power turbine section that is less likely to fail since its inlet temperature is lower; and (3) easy to adapt for remote control since their auxiliary systems are simpler and no water cooling is needed (only air or oil/air cooling).

c) Medium-range gas turbines have output power from 5 to 15 MW and a pressure ratio between 5 and 15. They are split-shaft designs, with the gasifier operating at peak efficiency and the power turbine running at variable speeds to match the load. They are also used in petrochemical plants and off-shore platforms.
d) Small gas turbines have an output power of less than 5 MW. They can be aero-derivative engines with centrifugal (or combined axial/centrifugal) compressors and radial-inflow turbines. With their smaller sizes, the turbine blades are not cooled, so the turbine inlet temperature has to be lower (around 1800 °F), resulting in a lower efficiency.
e) Microturbines have output power between 20 and 350 kW. They employ a centrifugal compressor and radial-inflow turbine and normally burn diesel fuel or natural gas. With their compact size and relatively quiet operation, they are finding applications for APU (auxiliary power unit), cogeneration, and absorption chillers in commercial and medium-sized industrial sites. One example is shown in Figure 5.27.[14]

Other applications of gas turbines for stationary power plants are being developed:

1) Integrated gasification combined cycle (IGCC)—To expand the industrial applications of gas turbines, low-quality coal and heavy fuel oil with high sulfur content can be used with coal gasification or fluidized-bed systems. The coal and heavy fuel oil can be converted into clean gaseous fuel through a gasification unit before being burnt in the combustor, as shown in Figure 5.28. The exhaust hot gas from the gas turbine is used to generate steam for the steam turbine and process.
2) Another gas turbine application being developed is the compressed-air energy storage (CAES) scheme shown in Figure 5.29.[15] It can be a stand-alone plant integrated with a grid system, or a hybrid system integrated with wind or solar power. During the off-peak hours when electricity is abundant or wind/solar power is available, the electric generator is run reversibly as an electric motor (or a separate motor is installed) to drive the compressor, pumping high-pressure air into the empty cavern. A regenerator with pebbles of alumina or silica can be added to store heat. During peak load hours or when wind/solar power is unavailable, high-pressure hot air in the cavern is released to drive the turbine and electric generator (or replace the compressor for regular gas turbine engine operation). The compressor and turbine can be separated or connected to a common shaft and disconnected when necessary.

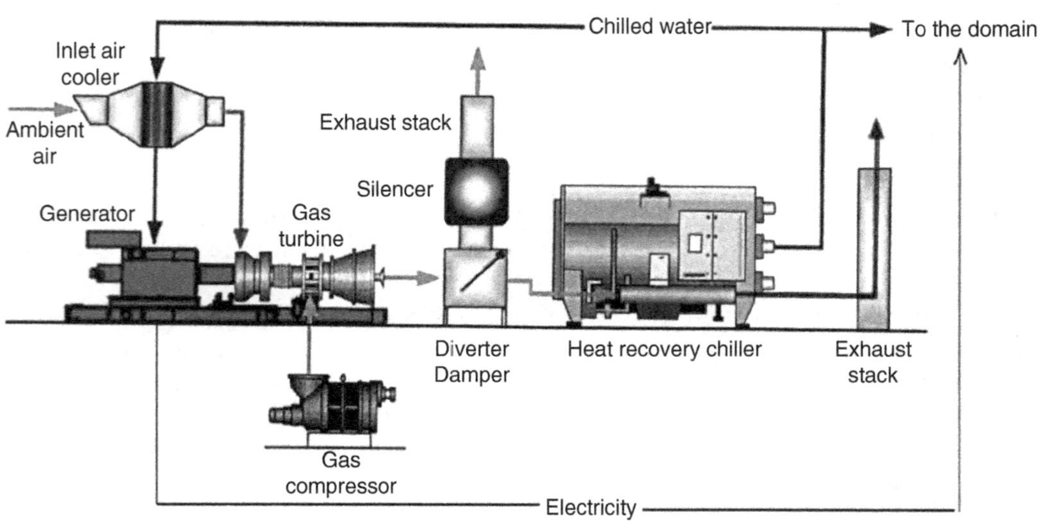

Figure 5.27 Microturbine in an integrated system to produce electricity, hot air, and chilled water. (*Source:* Oakridge National Laboratory.[14])

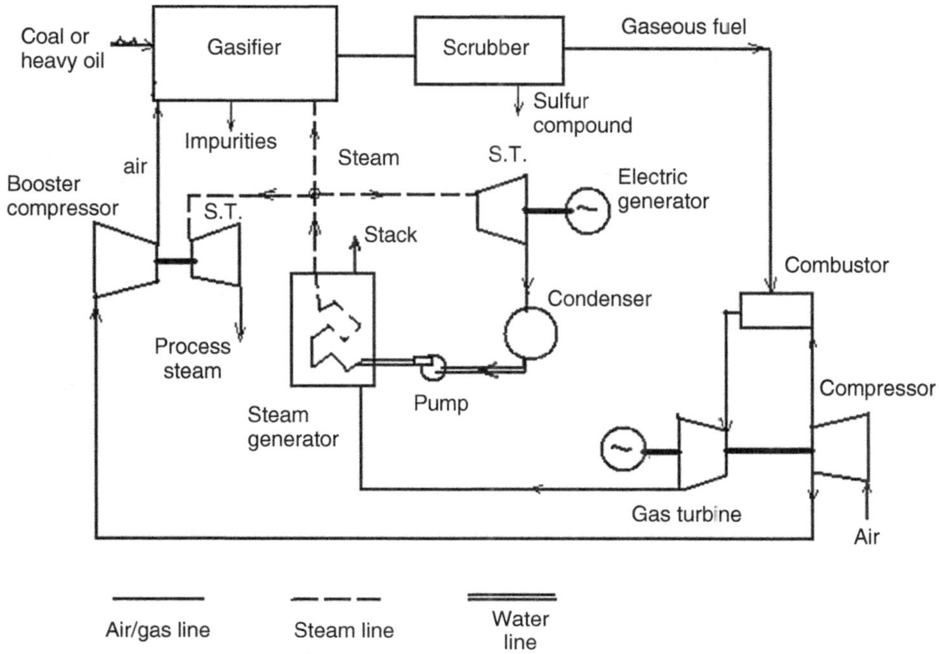

Figure 5.28 Gas turbine in an integrated gasification combined cycle (S.T. = steam turbine).

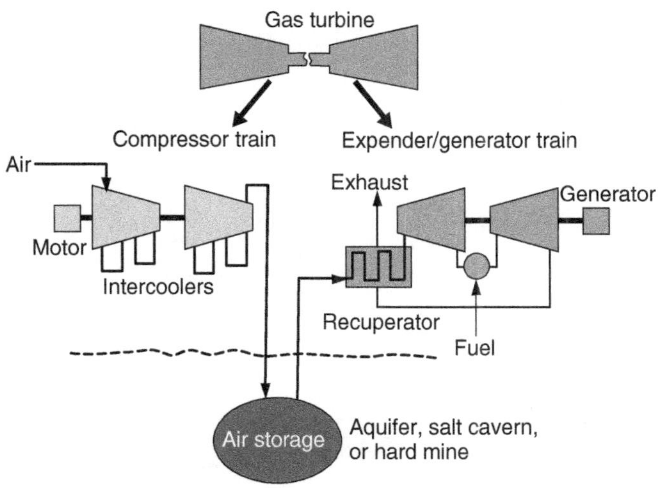

Figure 5.29 Gas turbine converted into a compressed air energy storage system (CAES). (*Source:* Reproduced from Greenblatt.[15])

Three different modes of operation are possible:

a) Electricity is generated by fuel supplied to the combustor by a regular gas turbine.
b) The compressor is driven with electricity from a grid or wind/solar power to store energy in the air storage (the common shaft is disconnected).
c) Electricity is generated by releasing the high-pressure air from CAES to drive the turbine (the common shaft is disconnected). Some extra power can be generated by burning fuel in the combustor and replacing the compressor with CAES.

Applying the gas turbine for automotive power plants has potential advantages of improving fuel efficiency and using alternative fuels. But some additional requirements have to be satisfied:

a) Fuel efficiency is required over the duty cycle, which includes a significant portion of operation under off-design conditions for both off-speed and off-power. This may require turbomachines of variable geometric design.
b) Low initial manufacturing and maintenance costs are required to compete with other types of engines. This means that low-cost materials have to be used for a relatively simple design.
c) Acceptable acceleration, emission, and noise level are required as well as engine size and weight.

Three engine configurations are currently considered, shown in Figure 5.30. All employ a regenerator to improve efficiency:

1) Single shaft with continuously variable transmission (CVT)—The compressor and both gasifier and power turbines are mounted on a single shaft. So, a complicated CVT mounted on the same shaft has to be employed to change the driving speed.

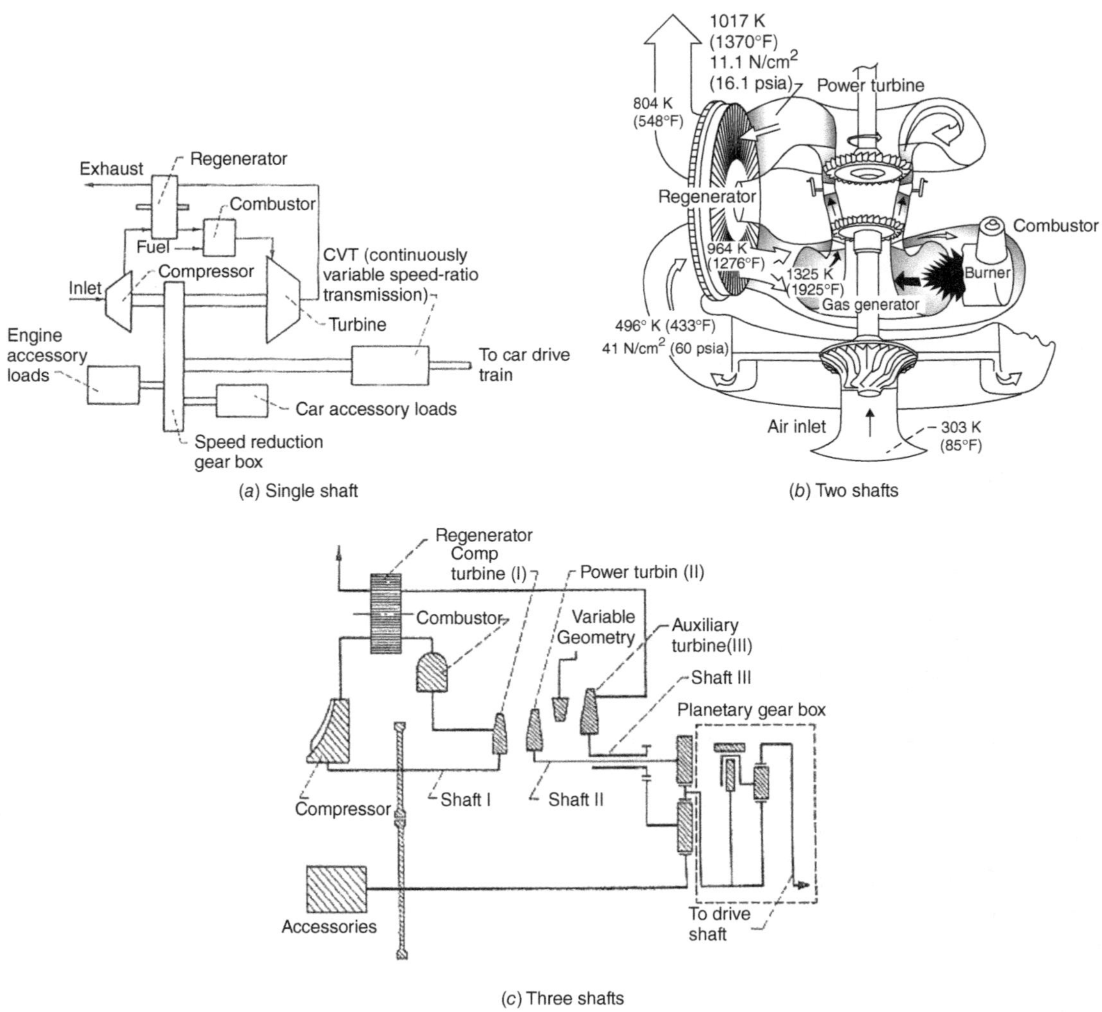

Figure 5.30 Automotive gas turbine configurations. (*Source:* O'Brien[16]/with permission of Elsevier.)

2) Two shafts with the power turbine uncoupled from the gasifier mechanically—A conventional three-speed transmission can be coupled with the power turbine.
3) Three shafts with two power turbines.

The second power turbine (auxiliary turbine) is mounted on a separate shaft to assist in driving the compressor, vehicle, and accessories. A relatively simple planetary gear transmission is coupled with the main power turbine.

Automotive gas turbine development has been a long-term project pursued by automotive manufacturers and government agencies. More detailed information can be found in Ref. 17.

5.12 Manufacturing Challenges on High-Temperature Gas Turbine Components

5.12.1 Gas Turbine Cooling Technologies

Since the thermal efficiency of a gas turbine is strongly related to the gas temperature at the turbine inlet, it is common practice to cool the turbine blades for the high-power units. So the gas temperature at the turbine inlet can be raised. It can be either air cooling or liquid cooling. Liquid or steam cooling is used for the ground-based industrial gas turbine. Air cooling with a 1.5–2% flow rate from the compressor can reduce the turbine blade temperature by 300–500 °F. Hence, the gas temperature at the turbine inlet can be increased significantly, or a less expensive material can be employed. Today, the highest turbine entry temperature is about 1750 K for the G and H class industrial gas turbine with 60% thermal efficiency in combined cycle mode (Ref. 18). However, 1950 K is under development with potentially 65% thermal efficiency (Ref. 19).

Two types of air cooling are possible, as shown in Figure 5.31. It can be either (a) internal cooling with coolant flowing through the internal channel from the blade root to tip or (b) external cooling by either injecting coolant

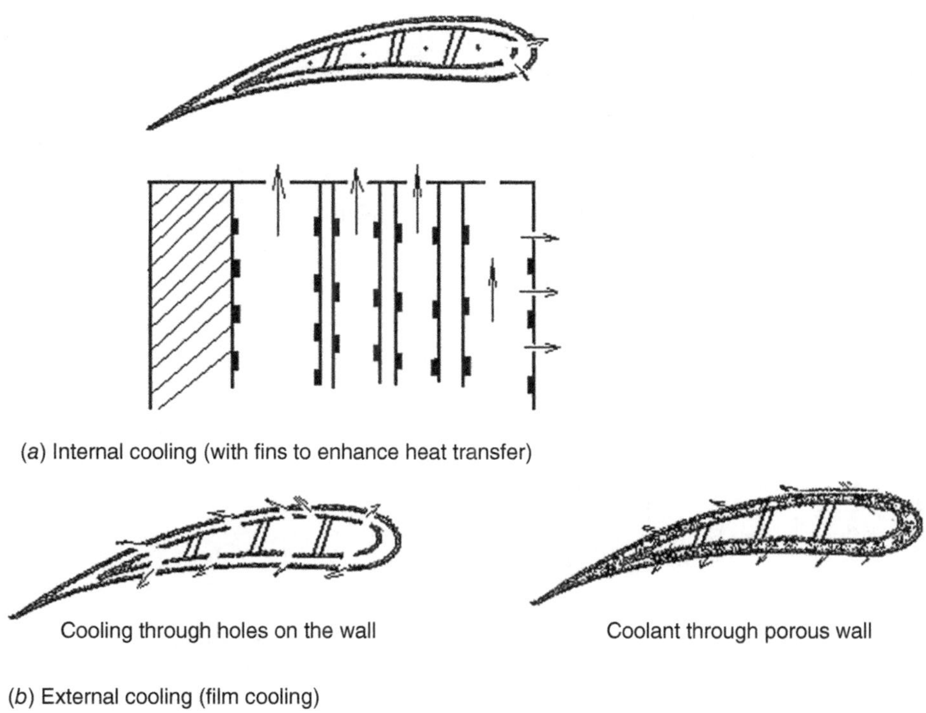

Figure 5.31 Turbine blade cooling.

onto the blade surface to form a film or forcing the coolant from the inside through the blade porous wall to the outside of the blade. Or a combination of both types can be employed.

A simplified one-dimensional heat transfer model for the internal cooling of air will be illustrated. Temperature distributions along the blade height from root to tip for the coolant, blade, and gas around the blade (T_c, T_b, and T_g) are shown in Figure 5.32. Here, T_g is assumed to be constant. A parameter $(T_b - T_{ci})/(T_g - T_{ci})$, called the blade relative temperature coefficient, is used to indicate the cooling effectiveness, where T_{ci} is the inlet coolant temperature at the blade root. The heat balance for an elemental length dl can be written as

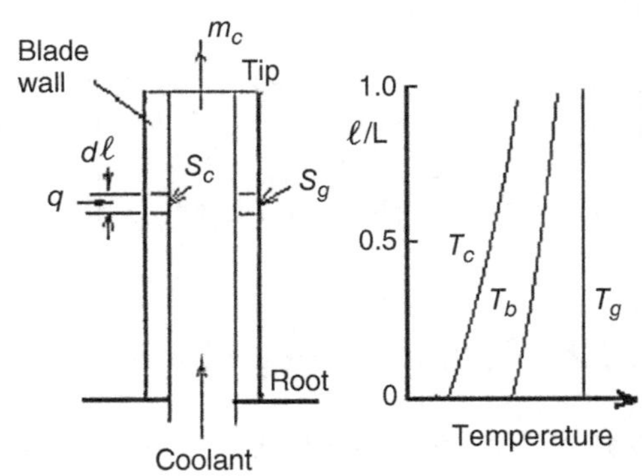

Figure 5.32 Simplified internal cooling model.

$$h_g S_g (T_g - T_b) = h_c S_c (T_b - T_c), \tag{5.23}$$

where h_g and h_c are the heat transfer coefficients and S_g and S_c are the wetted perimeters of the gas side and coolant side, respectively. For the coolant flow m_c, we also have

$$\frac{\dot{m}_c C_{pc} dT_c}{dl} = h_c S_c (T_b - T_c). \tag{5.24}$$

Equation (5.23) can be rewritten as

$$T_c = T_b - \frac{h_g S_g}{h_c S_c}(T_g - T_b) \tag{5.25}$$

or

$$\frac{dT_c}{dl} = \left(1 + \frac{h_g S_g}{h_c S_c}\right)\frac{dT_b}{dl}, \tag{5.26}$$

since T_g is constant. Combining Eqs. (5.23), (5.24), and (5.26) to eliminate T_c, we have

$$\left(1 + \frac{h_g S_g}{h_c S_c}\right)\frac{d(T_g - T_b)}{dl} + \frac{h_g S_g}{\dot{m}_c C_{pc}}(T_g - T_b) = 0$$

Integrating with $T_b = T_{br}$ at $l = 0$, assuming constant h_g, h_c, S_g, and S_c, results in

$$T_g - T_b = (T_g - T_{br})e^{-kl/L}, \tag{5.27}$$

where

$$\frac{k}{L} = \frac{h_g S_g}{\dot{m}_c C_{pc}[1 + (h_g S_g / h_c S_c)]}.$$

Also, Eq. (5.25) can be written as

$$T_g - T_c = (T_g - T_b)\left(1 + \frac{h_g S_g}{h_c S_c}\right) = (T_g - T_{br})\left(1 + \frac{h_g S_g}{h_c S_c}\right)e^{-kl/L} \tag{5.28}$$

or

$$T_g - T_{ci} = (T_g - T_{br})\left(1 + \frac{h_g S_g}{h_c S_c}\right). \tag{5.29}$$

Combining Eqs. (5.28) and (5.29), we have

$$T_g - T_c = (T_g - T_{ci})e^{-kl/L}. \tag{5.30}$$

Subtracting Eqs. (5.27) from (5.29), we have

$$T_b - T_{ci} = (T_g - T_{br})\left(1 + \frac{h_g S_g}{h_c S_c} - e^{-kl/L}\right).$$

By dividing this by Eq. (5.29), we have the blade relative temperature coefficient,

$$\frac{T_b - T_{ci}}{T_g - T_{ci}} = 1 - \frac{e^{-kl/L}}{1 + (h_g S_g / h_c S_c)}. \tag{5.31}$$

With the heat transfer coefficients h_c and h_g available from the empirical correlation, the one-dimensional blade temperature distribution T_b along the spanwise direction can be evaluated, as shown in Figure 5.32 qualitatively. Then, using the finite-element method, two-dimensional heat transfer and thermal stress analysis, including the rotational effect, can be performed at the critical blade section to locate the maximum temperature and stress.

In the more sophisticated design, the cooling air can be extracted from a single stage or two stages of the compressor; the flow can be single pass or multi-pass to provide film cooling and internal cooling. A more detailed discussion on this topic can be found elsewhere.[7, 8]

Other techniques to increase the allowable turbine inlet temperature involve using special materials such as ceramic or TBC (thermal barrier coating) for the turbine blades.

5.12.1.1 Necessity of Innovative Gas Turbine Cooling Techniques

Advanced heat transfer and cooling techniques form a major pillar supporting the development of high-efficiency, high-power output gas turbine engines.[19, 20] Conventional gas turbine thermal management technology is composed of six main elements: (1) internal convective cooling, (2) external surface film cooling, (3) materials selection, (4) thermal-mechanical design at the component and system levels, (5) selection and/or pre-treatment of the coolant fluid, and (6) material coating. This subsection summarizes specific cooling technologies representing innovative methods expected to enhance the aero-thermal-mechanical performance of turbine engines further. The techniques discussed include forced convective cooling with unconventional turbulators and concavity surface arrays, swirl-cooling chambers, latticework cooling networks, augmentations of impingement heat transfer, synergistic approaches using mesh networks, and film cooling.

The technology of cooling gas turbine components, primarily via internal convective flows of single-phase gases and external surface film cooling with air, has developed over the last several decades into complex geometries involving many differing surfaces, architectures, and fluid-surface interactions. The fundamental aim of this technology area is to obtain the highest overall cooling effectiveness with the lowest possible penalty on the thermodynamic cycle performance. As a thermodynamic Brayton cycle, the efficiency of the gas turbine engine can be raised by increasing the firing temperature of the turbine. Modern gas turbine systems are fired at temperatures far over the material melting temperature limits. This is made possible by the aggressive cooling of the hot gas path components using a portion of the compressor discharge air, as depicted in Figure 5.33.

Using 15% to 25% of this compressed air to cool the high-pressure portions of the turbine presents a severe penalty on thermodynamic efficiency unless the firing temperature is sufficiently high for the gains to outweigh the losses. The efficiency gain is significant enough to justify the added complexity and cost of the cooling technologies employed in all properly operating cooled turbine systems. Actively or passively cooled regions in power-generating gas turbines include the stationary vanes or nozzles and the rotating blades or buckets of the high-pressure stages, the shrouds bounding the rotating blades, and the combustor liners and flame-holding segments (fuel nozzles, splash plates). These engines also cool the interfaces and secondary flow regions around the immediate hot gas path. A more detailed schematic of the cooling for an aircraft engine combustor and turbine first stage, which may also be considered an aero-derivative power turbine, is shown in Figure 5.34.

Cooling technology, as applied to gas turbine components, is composed of five main elements: (1) internal convective cooling, (2) external surface film cooling, (3) materials selection, (4) thermal-mechanical design, and

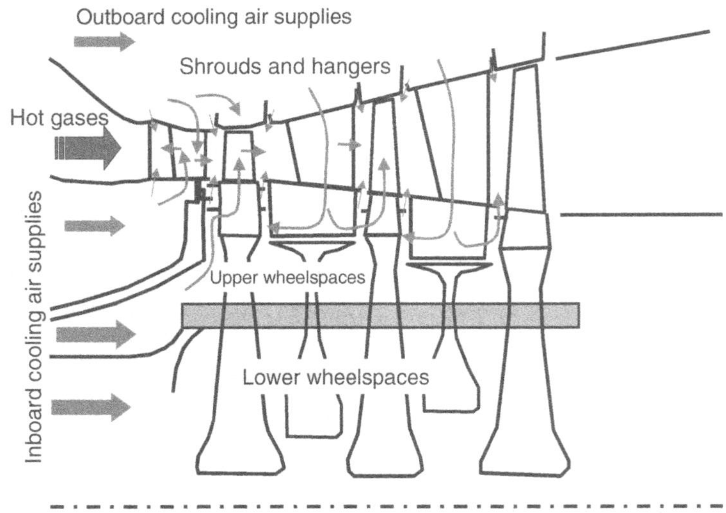

Figure 5.33 Schematic of turbine with cooling flows.

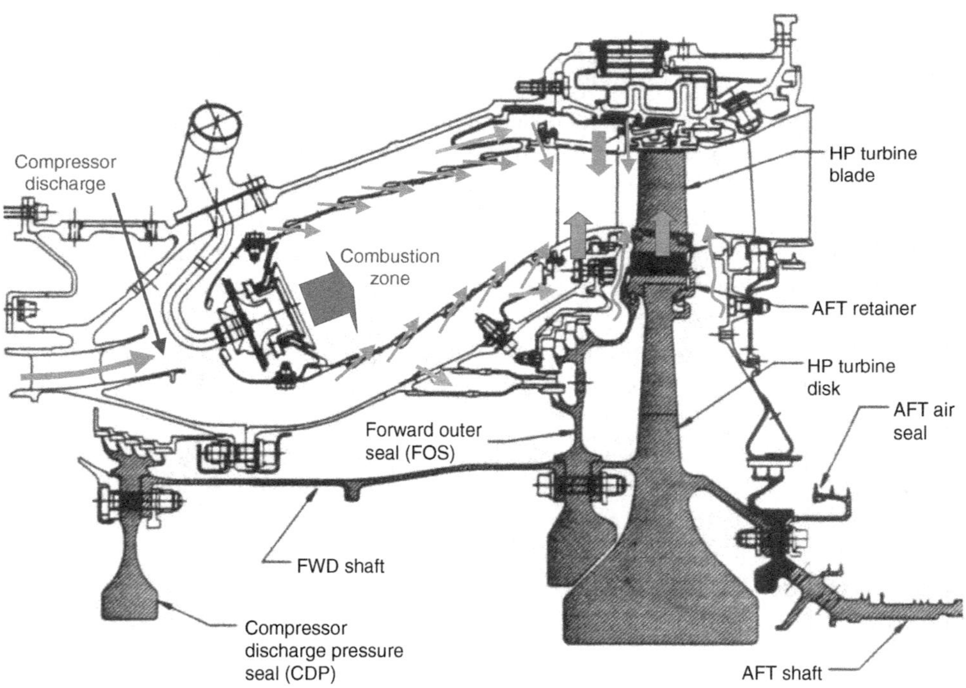

Figure 5.34 Cooling flows for a combustor and high-pressure turbine.

(5) selection and conditioning of the coolant fluid. Cooled turbine components are merely highly specialized and complex heat exchangers that release the cold side fluid in a controlled fashion to maximize work extraction. The enhancement of internal convective flow surfaces for the augmentation of heat transfer was initially improved some 25–30 years ago by introducing rib-roughen or turbulators and pin-banks or pin-fins. Figure 5.35 shows an example schematic of a blade cooling circuit that utilizes many tubulated passages, a pin bank in the trailing edge, and impingement in the leading edge (coolant is released via film holes, tip holes, and trailing edge).

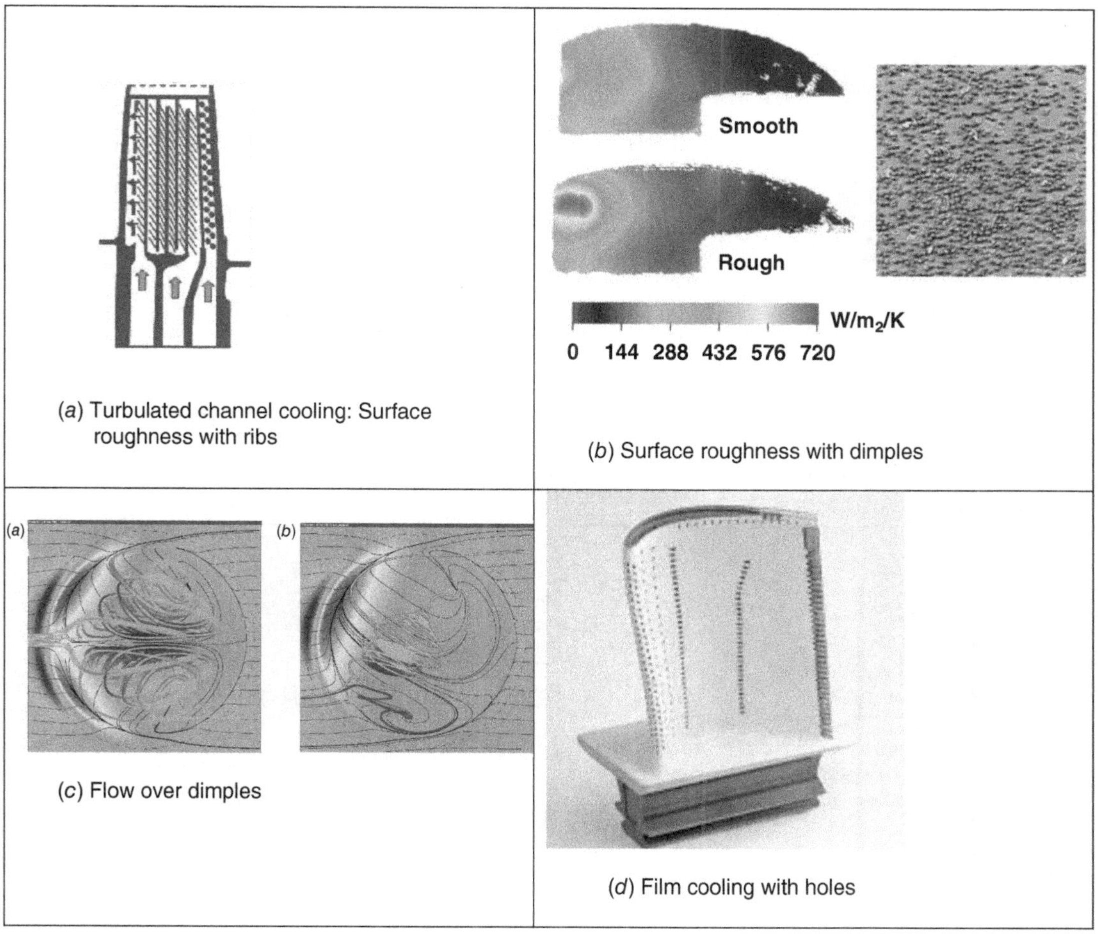

Figure 5.35 Different turbine blade cooling cases.

5.12.2 Different Turbine Blade Cooling Cases

Before describing any specific technologies, several characteristics of gas turbine cooling are worth noting. Most highly cooled regions of the high-pressure turbine components involve turbulent convective flows and heat transfer (Re ∼ 10,000–60,000). This is true of both stationary and rotating components. The enhancement of heat transfer coefficients for turbine cooling makes full use of the turbulent flow nature by seeking to generate mixing mechanisms in the coolant flows that actively exchange cooler fluid for the heated fluid near the walls. These mechanisms include shear layers, boundary layer disruption, and vortex generation. In a marked difference from conventional heat exchangers, most turbine cooling means do not rely on an increase in cooling surface area since the available surface area to volume ratios are very small. Surface area increases are beneficial but are not the primary objective of enhancements. Various enhancement techniques typically result in at least a 50% and as much as 300% increase in local heat transfer coefficients over that associated with fully developed turbulent flow in a smooth duct.

One of the most common means for enhancing heat transfer coefficients within internal cooling passages, especially the serpentine passages of many turbine blade designs, is turbulators (see Figure 5.35*a*), also known as rib roughness (Figures 5.35*b* and *c*). Turbulators in the form of trip strips placed transverse to the bulk flow direction were one of the first improvements made to the cooling of blades. Hence, many researchers have been

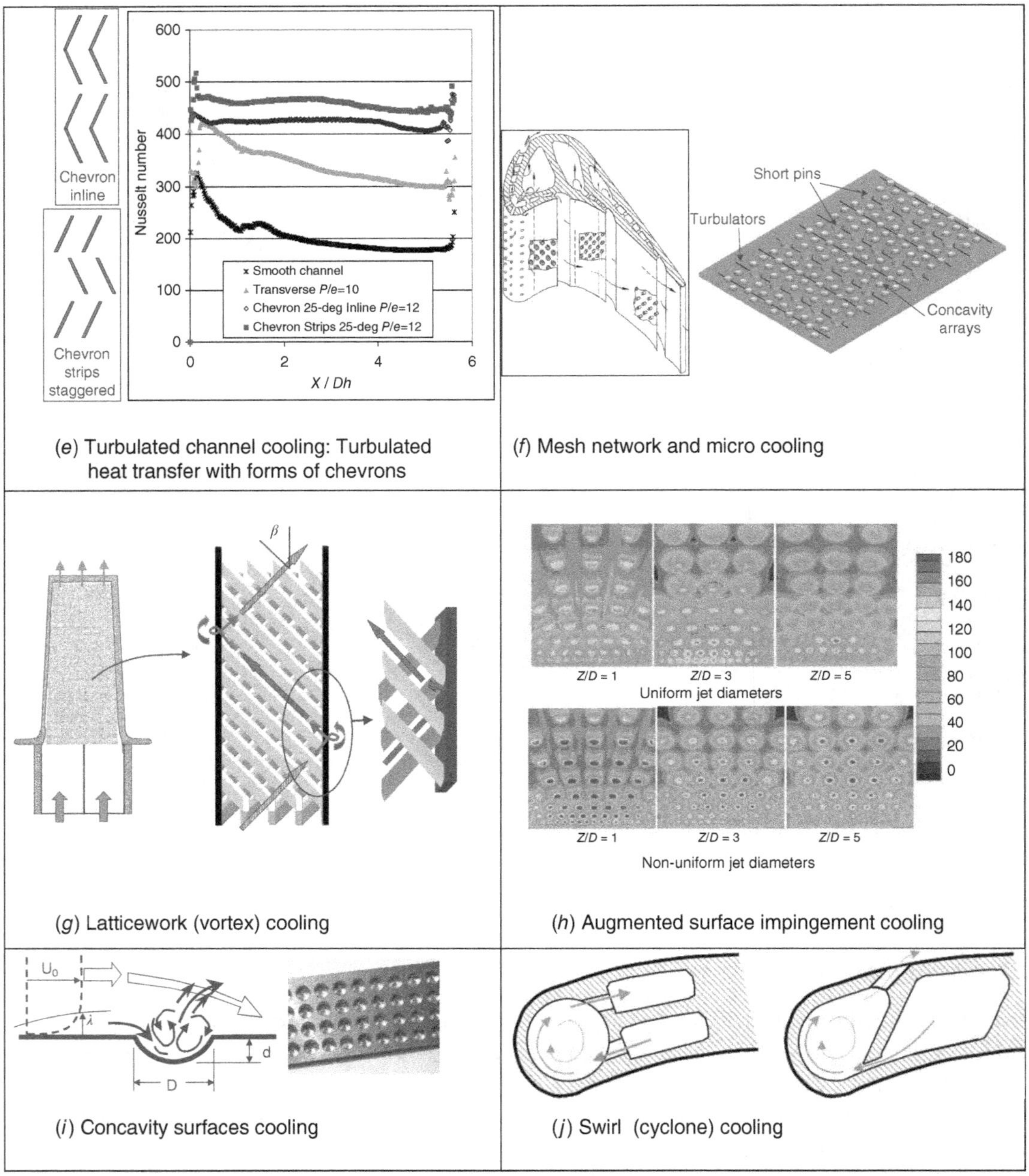

(*e*) Turbulated channel cooling: Turbulated heat transfer with forms of chevrons

(*f*) Mesh network and micro cooling

(*g*) Latticework (vortex) cooling

(*h*) Augmented surface impingement cooling

(*i*) Concavity surfaces cooling

(*j*) Swirl (cyclone) cooling

Figure 5.35 (*Continued*)

concerned with heat transfer and friction characteristics. Basic transverse turbulator research, which resulted in widely used data and correlations, was performed by Webb et al.,[21] Burggraf,[22] and Han et al.,[23, 24] With the last decade's advancements in materials and manufacturing technologies, a drastically larger realm of surface enhancement techniques has become cost-effective for use in cooling turbine airfoils. Turbulators may now be of varying shape, orientation, segmentation, and size, providing a continuous spectrum of possible geometries for achieving flow-surface interactions that enhance local and global heat transfer coefficients. Most researchers concentrated on turbulators of relative height e/D less than 0.2 but greater than that relative roughness typically

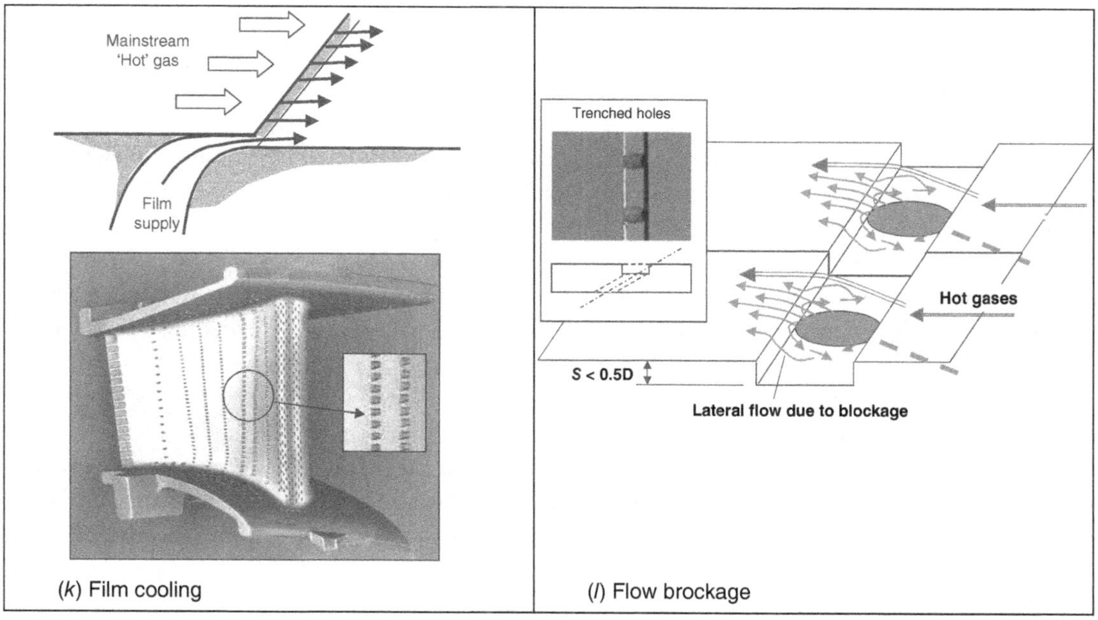

Figure 5.35 *(Continued)*

associated with uniform surface roughness, where e and D stand for roughness and duct diameter, respectively. The general findings of all such research have been that surface averaged heat transfer coefficients within stationary, tubulated passages may be enhanced by factors from 1.8 to 2.8, while factors of 3–10 increase the friction factors or require pumping power. Many parameters have been investigated in tubulated passages (angle to the bulk flow, roughness, e/D, shaping, taper, etc.) and many more in serpentine circuits. However, the range of effects has remained the same over the years.

For example, Figure 5.35*a* shows that surface enhancement methods continue to play a key role in today's turbine cooling designs. Film cooling is the practice of bleeding internal cooling flows onto the exterior skin of the components to provide a heat flux-reducing cooling layer, as shown by the many holes placed over the airfoil in Figure 5.35*a*.

The roughness on the blade can frequently be tailored to address cooling needs in specific localities. Effects that tend to thin the boundary layer relative to the roughness element heights, such as increased Reynolds number or decreased jet diameter, lead to increased heat transfer. Chakroun et al.[25] examined impingement heat transfer for a single normal air jet on a deterministic patterned rough surface (Figure 5.35*b*). Heat transfer augmentations of up to 28% were noted, with a measured increase in the turbulence intensity of the flow around the roughness elements. El-Gabry and Kaminski[26] measured peak and average heat transfer coefficients for an array of normal or angled jets impinging on a randomly rough surface. Figure 5.35*b* provides the smooth and rough surface heat transfer coefficient distributions for Bunker and Bailey's turbine blade platform cooling model study.[27] Also shown is a photo of the random roughness surface, which, in this case, is not close-packed. This study uses a single nearly normal impingement jet with a Reynolds number of 130,000, spaced 2.5 jet diameters from the surface. The immediate stagnation region of the jet exhibits a 50% increase in heat transfer coefficient, while this benefit is very quickly lost as flow transitions to a convective condition over the surface.

Several forms of patterned surface augmentation, more amenable to investment casting, have also received increasing attention for use under impingement jets. These include pin arrays in the form of discrete-shaped bumps, regularly spaced turbulators, and even arrays of concavities. These surface augmentation methods aim to derive a greater benefit than simple wet surface area increase by creating additional vortices and turbulence.

The study of Son et al.[28] examined the use of cylindrical surface pins under impingement jets for a flat surface, while that of Taslim et al.[29] investigated conical bumps to augment airfoil leading edge impingement cooling. Both studies found heat transfer to be augmented by roughly an amount equal to the increased surface area ratio factor (Figure 5.35*e*).

Cooling meshes were tested using round pins and rounded-diamond-shaped pins with height-to-diameter (H/D) ratios of 0.2 and center spacing (S/D) ratios of 1.5, as well as less dense rounded diamond pins of smaller H/D of 0.3 and S/D of 2.14. These geometries differ substantially from conventional pin fin arrays in which $H/D > 1$ and S/D ratios are typically about 2.5. Special attention was paid to the combination of techniques, including pin meshes, turbulators, and dimple/concavity arrays, which have been shown to provide multiple design solutions for heat transfer and pressure loss objectives. Figure 5.35*c* shows one such combined geometry using pins, concavities, and turbulators. Figure 5.35*c* also shows the potential use of these new in-wall mesh networks as improvements in cooling designs for turbine blades, as per Lee and Bunker.[30]

In the bulk of the literature, improvements in the use of transverse turbulators within channels have focused on angled turbulators, segmented turbulators, and many other combinations of these in such forms as chevrons and segmented discrete strips. Many researchers have noted significant heat transfer coefficient enhancements for surfaces with upstream pointing chevrons and staggered segments of chevrons. For example, Han and Zhang[31] investigated four cases of various segmented chevron geometry, chevrons, segmented angled strips, and conventional turbulators in a simple duct flow. They found heat transfer coefficient enhancements as high as 3 to 4 compared to smooth surfaces for Re from 15,000 to 80,000. The friction coefficient enhancements were also generally higher than those for more conventional turbulators. Taslim et al.[32] and Kiml et al.[33] obtained similar results at lower Re conditions. While the beneficial effects of turbulator angle were maintained, the number of segmentations did not show a consistent improvement trend. Chyu and Natarajan[34] presented detailed mass transfer data for transverse segmented turbulators where the segments were aligned or staggered. The alignment of the segments lowered the average mass transfer. While the segmentation gaps increased vertical mixing locally, these also disrupted the benefits of reattaching flows. On the other hand, staggered segments produced higher average mass transfer coefficients by providing a more periodic nature to the disturbances. All preceding studies examined enhancement features inside square or nearly square channels, so the features were not repeated laterally, as would be the case between parallel plates or in ducts of higher aspect ratio.

Applications of concavity surfaces involving heat transfer were largely unknown until the recent dissolution of the Soviet Union, at which time Russian research from the 1980s began to surface. The basic fluid dynamic condition for flow over concavities of spherical or cylindrical shape is well described in the study of Afanas'yev et al.,[35] A flow with boundary layer thickness less than the concavity surface diameter reacts with the cavity by flowing into the "bowl," experiencing a separated region of some extent on the entry side as depicted in Figure 5.35*i*. The spherical shape, or one nearly approximating it, creates a pressure field within the bowl acting to collapse or concentrate the flow in the downstream portion of the recess, creating a vortex structure. In a steady flow over a symmetric spherical dimple, a pair of symmetric, counter-rotating vortices is ideally formed, as shown by the computational work of Isaev et al.[36] in Figure 5.35*c*. In most real flow cases, only one vortex is created, and this vortex may move side-to-side with some frequency. This condition is said to "expel" flow from the cavity as an organized vortical structure. As the vortex penetrates into and interacts with the mainstream flow, it provides a scrubbing action, which brings fresh core fluid to the surface for enhanced heat transfer. Because the motion is organized, rather than the more dissipative effect of shearing layers, the pressure loss is less than that observed with projecting obstructions such as turbulators. In fact, some cases of concavity surfaces have friction nearly the same as smooth surfaces. Studies focusing on the heat transfer coefficients on a surface with a single hemispherical concavity show the fundamental potential of this method for thermal enhancement. Kesarev and Kozlov[37] show the detailed shear stress and heat transfer coefficient distributions relative to a flat surface for a single dimple, observing overall heat transfer augmentations of up to 1.5.

The full potential for concavity surface heat transfer comes from applying full-surface arrays of ordered concavities. Wind tunnel surface studies of Afanas'yev and Chudnovskiy[38] and also Afanas'yev et al.[39] for arrays of spacings from about 1 to 1.7 dimple diameters showed heat transfer enhancements of 30% to 40% with no increase in friction.

As turbine blade cooling advances beyond the conventional methods, cooling will migrate from a largely internal bulk flow medium within hollow airfoils (e.g., Figure 5.35*d*) to one that may be built into or "within" the exterior walls of an airfoil. This change moves the heat sink (the coolant and its wetted surface area) closer to the heat source (the hot gas path surfaces), producing a more effective heat exchanger. This approach has the benefit of retaining a cooler internal bulk structure for the blade, which results in a longer rupture life. In extreme cases, this type of design might be conceived as a microcooling method. Many small cooling passages near the surface have potential benefits, but the risks associated with dirt plugging, manufacturing tolerances, and cost must also be considered. The nearer-term version of in-wall cooling is sometimes called double-walled or mesh network cooling, in which the passages are of sizes amenable to investment casting. A greater degree of uniformity in internal cooling is key to many current thermal stress–strain limitations. Turbine airfoils can generally withstand higher bulk temperatures as the local thermal gradients are reduced. Zhang et al.[40] addressed this type of cooling network for different open flow areas, or alternately different solidity of channel meshes, using straight channels that intersect with an included angle of 60°, showing 2 to 2.5 times local heat transfer enhancement compared to a smooth channel. Heat transfer in small pre-film impingement chambers within airfoil wall sections was studied by Gillespie et al.,[41] showing the contributions to cooling of both wall chamber interior surfaces.

Microcooling for airfoils is the natural extension of the more macroscopic mesh network cooling to reach the limits of thermodynamics and heat transfer. Microcooling spreads out the cooling network in a series of smaller and highly distributed channels, or sub-channels, providing better uniformity of cooling and lesser in-plane thermal gradients. See Figure 5.35*f*. This is analogous to the distribution of blood vessels in the human body. Microcooling may be considered a complete airfoil cooling solution or a regional cooling device (e.g., leading edges or trailing edges). The most notable developments in micro-cooling are those of the Allison. The airfoil walls in Lamilloy are formed as two or more sheets of metal bonded together with distributed pins, while those of CastCool are investment cast and limited to a double-wall type. The cooling flow is introduced via the staggered impingement jets concerning the exiting film holes. Full cooling distribution is obtained, and transpiration-like film cooling results from the normal holes in the outer layer. Both overall cooling effectiveness and film cooling are shown in Nakamata et al.[42] and Hale et al.,[43] The challenges of micro-cooling include hole plugging, wall strength, film cooling, manufacturing, and cost. However, no commercial use of these micro-cooled solutions has yet appeared.

Recent research into further tubulated channel or tubulated surface cooling is taking the direction of even more complex geometries such as perforated turbulators, turbulators detached from the wall, and turbulators with other vortex generators added on top. Care must be taken, however, to maintain robustness and high-yield manufacturing also. Among the most promising geometries are variations on chevron turbulators, including broken or segmented chevrons, and the staggering of these features to increase surface flow interactions. Figure 5.35*e* shows an example of heat transfer coefficient data from several surfaces placed in a parallel plate channel flow at a Reynolds number of 100,000. Among the geometries tested are a smooth surface, transverse turbulators, unbroken chevron arrays, and broken/staggered counter-angled strips formed by removing the apex of each chevron, as shown in Bunker et al.,[44] The parallel plate channel is large enough in width to have about a dozen such pairs of features in a full adjacent surface array. The laterally averaged smooth surface Nusselt number distribution agrees with literature values for a developing flow between parallel plates. The Nusselt number distribution for the transverse turbulators is elevated by a factor of about 1.75 and exhibits the characteristic trait of decay with axial distance from the channel entry. The somewhat surprising behavior of the chevrons is that no decay characteristic is present; the heat transfer coefficient is enhanced to a level of more than two times the developed smooth surface value and remains there throughout. The broken and staggered chevron strip heat transfer coefficients improve more, which might be attributed to the increased free edges generating additional mixing. Friction coefficients also increase with these changes.

Most research has focused on the modification of the surfaces, but the jet arrays may also be modified from the conventional uniform in-line or staggered spacing. Bailey and Bunker[45] examined heat transfer with jet array spacings outside that of prior literature, both smaller and larger. Significant deviations from past data were noted, most especially in cases where jet arrays become very dense, leading to large variations in jet Reynolds numbers. As an extension to this work, Gao et al.[46] studied the effects of linearly stretched jet arrays in which the row-to-row spacing is increased with distance inside the channel. Figure 5.35*f* shows an example of the Nusselt number distribution for an average jet Reynolds number of 10,000. The streamwise averaged heat transfer coefficients decline as the jets become less numerous.

Latticework cooling (Figure 5.35g), also known as vortex cooling or bounded vortical duct cooling, in its application to high-temperature gas turbine components originated within the former Soviet design bureau engineering system 25 years ago (Nagoga[47]). This method of vane, blade structure, and cooling developed into the most recognizable standard within the many Soviet designs (Russian and Ukrainian designs today). Its application history parallels the Western use of serpentine cooling for turbine blades.

Figure 5.35*h* shows an example of the Nusselt number distribution for an average jet Reynolds number of 10,000. The streamwise averaged heat transfer coefficients decline as the jets become less numerous. Figure 5.35*h* also shows a second modification to compensate for this decline by increasing the jet diameter on each row.

The use of impingement jets for cooling various regions of modern gas turbine engines is widespread, especially within the high-pressure turbine. Since the cooling effectiveness of impingement jets is very high, this method of cooling provides an efficient means of component heat load management, giving sufficient available pressure head and geometrical space for implementation. Regular arrays of impingement jets are used within turbine airfoils and end walls to provide relatively uniform and controlled cooling of open internal surface regions. Such regular impingement arrays are generally directed against the target surfaces using sheet metal baffle plates, inserts, or covers, which are fixed relative to the target surface. These arrangements allow for the design of a wide range of impingement geometries, including in-line, staggered, or arbitrary patterns of jets. In more confined regions of airfoils, such as the leading edge or trailing edge, spanwise lines of impingement jets are sometimes used to focus cooling on one primary location of high external heat load, like the airfoil aerodynamic stagnation region. Many other applications for individual impingement jets exist on selected stationary and rotating surfaces. Vane end walls, blade platforms, unattached shrouds, and combustor liners may all have specific local cooling requirements that are well suited to individual jet cooling. Impingement jets are also used on rotor disk cavity faces and, in some applications, may provide additional sealing functions.

A wealth of information exists on the basic cases of individual and array jet impingement heat transfer. The heat transfer coefficient distributions due to a single axisymmetric jet impinging normally on a smooth, flat plate with free spent air discharge were investigated by Gardon and Cobonpue.[48] This study determined the basic effects of jet-to-target spacing and jet Reynolds number on stagnation region and radial heat transfer. Perry[49] measured single-air jet impingement heat transfer characteristics on a free surface for normal and oblique impingement angles, determining the local and averaged effects of jet angling.

Another class of surface enhancements results from the depression of features into the cooling channel or surface walls, forming recesses rather than projections. Generically, such features are known as concavities (or dimples) and may be formed in an infinite variation of geometries with various resulting heat transfer and friction characteristics. Concavity surfaces are commonly known for their drag-reduction characteristics in external flows over bodies. The most famous example is golf balls, which delay the boundary layer separation point, thereby reducing overall drag for the sphere, as shown by Bearman and Harvey.[50] Application of such concavity surfaces to external flows for marine vessels and airframes has been suggested by Kiknadze et al.,[51] in which defined arrays of shallow concavities may cover the main hull or fuselage surfaces to prevent a thick boundary layer from forming, thereby reducing the drag.

As with many other cooling technologies, care must be taken to examine the use and application of the augmentation means. For example, Bunker and Donnellan[52] investigated using dimpled surface arrays in circular cooling passages with Reynolds numbers up to 90,000.

Film cooling is a major technology that allows today's gas turbines to obtain extremely high turbine firing temperatures, subsequent high efficiencies, and longer life parts. The art and science of film cooling concerns the bleeding of internal components, cooling air through the external walls to form a protective cooling layer between the hot gases and the component's external surfaces. Applying effective film cooling techniques provides the first and best line of defense for hot gas path surfaces against the onslaught of extreme heat fluxes, directly reducing the incident convective heat flux on the surface. Fundamental early research concentrated on nearly ideal film cooling formed by a two-dimensional layer, as Figure 5.35*j* depicts. Because of its high importance and widespread application, research into the many aspects of film cooling has seen a tremendous increase in the last 10–15 years. The publications relating directly or indirectly to film cooling deal with the major effects of film hole internal fluid dynamics, interactions with the mainstream gas flow, turbulence, and vorticity production, effects of approach flow before the hole entry, hole shape, orientation, and spacing, hole length-to-diameter ratio, density ratio, blowing strength, momentum flux ratio, effects of mainstream turbulence intensity, mainstream acceleration, external surface curvature, and external surface roughness.

Over the past 30+ years, investigations have been performed by a broad spectrum of researchers to understand the fundamental physics of film cooling and to improve the state-of-the-art. The primary focus of most research has been on the use of discrete film holes, or rows of film holes, on the hot gas path surfaces of the turbine since mechanical constraints dictate this format. Figure 5.35*k* shows an example of a typical high-pressure turbine inlet guide vane with many film cooling rows of varying spacing and orientation.

5.12.2.1 Summary

This summary has presented several cutting-edge, innovative cooling methods that are expected to enhance the aero-thermal-mechanical performance of turbine engines used for power systems further. These methods are by no means an exhaustive or comprehensive summary, however. Many other variations and combinations of these techniques are anticipated as manufacturing advances become reality. Further improvements and new techniques may become feasible as materials, systems integration, and controls also advance.

5.12.3 Active Cooling of Turbine Components

5.12.3.1 Active Cooling

Gas turbines play a significant role in the lives of all of us. They provide electricity or fuel to our homes and transportation via aircraft and ships. Despite over fifty years of operation and their current ubiquity, gas turbines continue to evolve, becoming more fuel-efficient, reliable, and durable. Significant engineering challenges remain with the present emphasis on improving efficiency and operation with alternative fuels. The gas turbine did not come about under a single, momentous breakthrough but rather through numerous small advances on many fronts, including heat transfer, aerodynamics, materials, dynamics, lubrication, and fuels. The efforts of tens of thousands of engineers have produced the present, extensive body of knowledge that provides our basis for further development. Given the complexity of the unit, it is quite easy to become overwhelmed when trying to become acquainted with the technology on which it is based.

The main objective sought by the gas turbine heat transfer engineer is to provide an airfoil and gas path design that is both efficient and can meet life goals. This requires a broad understanding of the system. First, an understanding of the constraints associated with the material to be used must be developed. This information (often proprietary and carefully protected) includes parameters such as the material's maximum use temperatures associated with oxidation and corrosion life, creep, and thermal-mechanical fatigue (all functions of temperature and stress). Given the projected usage and life goals, the designer applies life equations to determine metal temperature limits throughout the airfoil and the passage walls. The internal and film cooling schemes are designed to maintain the metal temperatures below these limits.

The heat transfer engineer has several tools that can be applied to provide a cooling scheme that meets the temperature constraints with minimal negative impact on engine aerodynamic and structural performance. These tools generally fall into two broad categories: those that cool the airfoil and end walls inside and those that reduce the external heat load. Note that, concerning heat transfer from the passage flow to the airfoil, the two broad categories have opposite effects on heat transfer. Internal cooling generally increases airfoil surface heat transfer rates, while external cooling aims to reduce them. They both tend to reduce airfoil and end wall temperatures. The complexity of choices for the various techniques to be discussed requires the gas turbine heat transfer engineer to understand the entire heat transfer problem thoroughly.

Internal cooling schemes generally employ one or more of the following techniques: impingement of cooling flow upon the wall to be cooled, convective cooling by channeling coolant flow through passages within the airfoil or end wall, and conduction through pin fin arrays to augment heat transfer within the solid components of the airfoil and enhance convection by increasing both flow turbulence and fluid-to-metal contact area. The choices for flow conditions and cooling path configurations associated with the cooling scheme are influenced by the pressure of the cooling flow supply, the temperature of the coolant fluid, the heat load distribution, and the stress distribution within the airfoil.

One or more of the following methods may control the external heat load. First, there is the geometry. Although aerodynamic considerations generally define the airfoil's shape, there are regions, such as the leading and trailing edges, where some improvements may be affected while maintaining acceptable aerodynamic performance. As necessary to accommodate cooling, the airfoil is thicker than the aerodynamic design engineer would have it. Also, there are thermal barrier coatings. These are generally ceramic-based coatings that have low thermal conductivity. They are used to reduce the heat flux through the airfoil wall. Finally, film cooling can be applied effectively on the external surface of the airfoil and on the end wall to reduce the temperature of the gas flowing immediately adjacent to the airfoil and end wall surfaces. It should be noted that film cooling injections into the gas path is usually done with an aerodynamic penalty. Effective cooling schemes must consider all impacts on the engine design.

Designers quickly realize that airfoil design is truly complex. The internal configuration directly influences the external heat load distribution, while the external heat load defines what the internal thermal design must accomplish. Finding the final configuration of the airfoil requires a systematic approach to the analysis, either beginning with the external heat load and then moving into the coolant fluid (the direction of energy flow) or beginning with the internal coolant and then documenting the external heat load (essentially moving in the general direction of the coolant). Although the choice is arbitrary, the discussion herein will move from the inside of the airfoil to the outside. As with any design process, multiple iterations through the analysis are required.

The thermal capacity of the coolant used for internal cooling is finite, defined by the mass flow rate, the coolant's specific heat, and the coolant temperature. The heat transfer engineer must carefully balance coolant flow rates and pressure drops while augmenting heat transfer coefficients with various devices to affect proper cooling.

The external heat transfer situation is different. Energy flows past and into the airfoil from an infinite source, the passage stream. The influence on the passage energy flow of the airfoil's cool surface is minor. The main concerns are (1) reducing the external effective heat loads between the airfoil wall or end wall and the bulk passage flow and (2) reducing the external area. The heat transfer engineer cannot effect significant changes in the second concern. Film cooling offers the possibility of introducing a film of cooler fluid (spent from the internal cooling functions) to cover the external surface of the airfoil to help effect the first concern listed above. This film offers a cooler source temperature for convective heat transfer through the thermal boundary layer from the passage flow to the surface and, as such, reduces heat flux across the convective layer. Over the years, film cooling has become a very effective method for reducing airfoil and end wall surface temperatures. A second opportunity for reducing the surface heat flux is introducing a solid-material coating of high thermal resistivity on the surface. In conjunction with the reduced heat flux due to film cooling and effective internal cooling, this gives lower temperatures on the airfoil external surfaces and load-bearing regions of the airfoil and end wall.

The internal geometries of high-performance engine airfoils tend to be very complex, incorporating many features designed to strategically cool the internals so that the entire external surface of the airfoil remains generally at a uniform temperature at a level suitable for sustained operation. Complicating the problem is getting sufficient coolant supply pressure and dealing with the thin trailing edge, the exposed tip, and the high heat transfer stagnation region at the airfoil's leading edge. A rise in convective heat transfer coefficient in a channel flow by some devised method or another is generally accompanied by a rise in pressure drop, giving a trade-off between enhancement and flow rate. Cooling flow circuit designs optimize the cooling effectiveness distribution for a given pressure drop. Manufacturing considerations, such as accommodating component thermal growth, maintaining clearances, and controlling cost (usually reflected in avoiding very fine feature sizes), are also considered. Designing a cooling scheme for a high-performance engine gas path becomes a delicate balancing act between designs for internal and external heat transfer performance, among many other design considerations. Tools include a fundamental understanding of flow and heat transfer and a working knowledge of the many other design considerations. Codes and design methods are available to the heat transfer engineer, but a fundamental understanding is paramount.

5.12.3.2 Hot Gas Path Heat Transfer Characteristics

The flow entering the turbine gas path comes from the combustor, as shown in Figure 5.36. Within the combustor, fuel (either a liquid similar to kerosene or diesel fuel or a gas, such as methane, syngas, or High Hydrogen and Syngas [HSG] processed from coal) and air combine in a highly exothermic reaction that leads to high-temperature products of combustion at the high combustor pressure. A modern dry, low-NO_x combustor operates lean and without water injections. A second type, catalytic combustor, has a selective catalytic reduction in the exhaust stream. In the low-NO_x combustor, excess air is used to maintain the combustor exit temperature at a suitable limit consistent with the engine design and turbine durability limits. Combustor designs often have strong swirls in the combustion zone to enhance mixing and shorten the reaction zone. About 30% of the airflow enters the primary combustion zone either through a swirl section or through holes in the primary zone of the combustion chamber. The remainder of the air, about 70% of the total airflow, is supplied through holes in the combustor liner as dilution air to mix and cool the flow and to protect the liner from reaching excessive temperature levels. In the case of a catalytic combustor, a catalytic reactor section is put downstream of the pre-burner's fuel/air mixing section. This reactor consists of fine channels coated with a catalyst, such as palladium oxide. The catalytic reaction

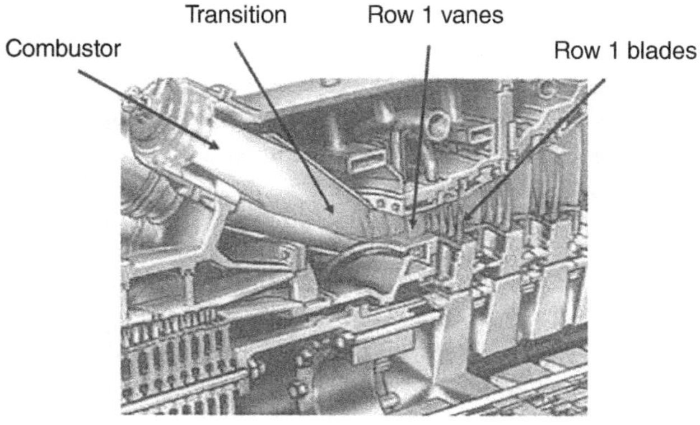

Figure 5.36 Hot gas path of a Siemens SGT6-5000F gas turbine. (*Source:* Brummel et al.[53]/Siemens Power Generation.)

occurs at temperature levels below the turbine inlet temperature, so a downstream combustor is needed to raise the gas temperature to the turbine inlet temperature. This combustor is similar to the dry low-NO_x combustor.

The flow passes through a transition section from the combustor exit to the turbine. Depending on the combustor and turbine designs, this section generally has a decreasing area in the streamwise direction and some wall curvature. The flow departs the transition section to enter the vane section of the first-stage turbine. To accommodate the thermal growth of components, the junction between the transition section and the turbine is fabricated with gaps in the wall, see Figure 5.37. These gaps are sealed against the ingression of hot gases with compressor discharge air that has been routed through a cavity in the center of the engine.

The flow then is turned through the vane, or nozzle, row of the first stage of the turbine. In this row, the flow is given a swirl (a circumferential velocity component) and is accelerated to a higher velocity. With the rise in velocity, there is a drop in the static temperature of the passage flow. This temperature change in the passage core flow can be estimated by analyzing it as a one-dimensional isentropic flow. Local velocities within the vane row can be at sonic velocity or mildly supersonic, and shock waves can stand within the passage. Generally, a shock from the trailing edge of one vane crosses the passage to impinge on the suction surface of the neighboring vane. This shock will affect wall heat transfer rates on the suction surface at the location of impingement and downstream. Upon entry to the vane row, the fluid temperature exceeds the maximum allowable materials temperature; thus, this row must be aggressively cooled. The gas path heat transfer engineer spends most of their time designing the first stage vane and rotor sections. The techniques for thermal control mentioned above are applied to this section: film cooling, impingement cooling, internal convection, pin fins augmentation, and trailing edge and tip cooling. The airfoil surfaces and the end walls are actively cooled using the appropriate techniques. Because of the thermal expansion of components, the airfoils/end wall assemblies are segmented, having one or two (doublet) airfoils per

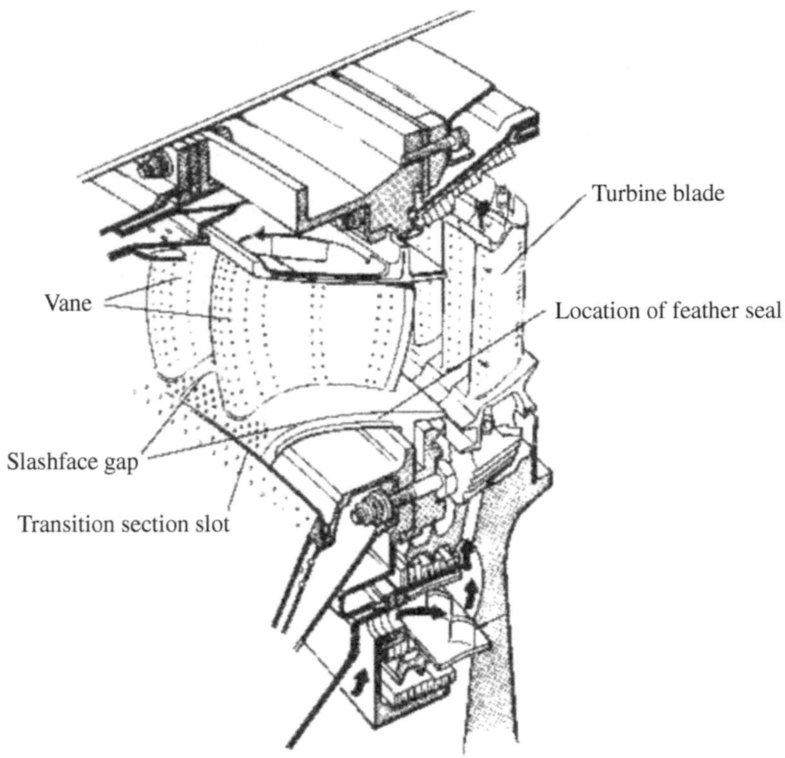

Figure 5.37 The first stage vane and rotor section showing the transition section to the vane slot and the slash-face or gutter gap. (*Source:* Cumpsty[54]/Cambridge University Press.)

assembly. A gap separates the segments to allow for thermal expansion, and the interface gap (also known as the slash-face or gutter, see Figure 5.37) is sealed by air taken from the engine cavity. This sealant, or leakage, the air is cool and mixes with the passage flow near the end wall surface, so it has important effects on the gas path cooling. Because of mixing with passage flow, which is of high velocity in the downstream half of the passage, there may be strong aerodynamic losses associated with this leakage.

The flow next enters the first stage rotor. In this stage, the swirl is removed from the flow while the velocity decreases, and work is extracted. This work extraction results in a drop in stagnation temperature. Though the fluid velocity drops, the net effect of velocity change and work interaction is a drop in static temperature across the rotor passage (within the passage core flow). This stage is rotating; thus, the end wall, or platform, moves relative to the end wall of the upstream vane row. The flow in the near-end wall region thus is skewed as it passes onto a moving platform from a stationary end wall. To accommodate the relative motion of the two rows and the thermal growth of the components, there is a gap between the vane and rotor sections (see Figure 5.37). This gap is sealed with airflow from the cavity, again with effects on gas path heat transfer and aerodynamic losses. The degree of skewing mentioned above can be modified by introducing the sealant flow with pre-swirl. Because of rotation, the flow and heat transfer are affected by buoyancy forces within the centripetal force field and Coriolis forces wherever there is a radial flow component, such as with the internal cooling channels.

As the flow proceeds through the remainder of the turbine, it passes through a series of vane and rotor components of subsequent stages. In each stage, work is extracted, and the total temperature is reduced. In contrast, the loss penalty for the system due to cooling air usage continues to accumulate (the coolant air used in the downstream stage is not available to do work within an upstream stage). The reduction in stage temperature significantly impacts the choice of materials and cooling schemes. Parts with lower external heat loads (such as the second blade) tend to require less aggressive internal cooling, resulting in reduced thermal gradients within the part material (compared to the gradients in the first stage blade). Also, fewer internal ribs are used to divide internal cooling passages, and since they tend to remain much cooler than the external walls, they carry much of the centrifugal stress load. This reduces the significance of local surface failure modes, such as oxidation, and increases the relative importance of creep, which may be understood as a section average temperature and stress-related failure mode. The change in the life-limiting failure mode and the reduced external heat loads may allow the designer to choose fewer exotic materials or apply less aggressive thermal control, thereby reducing the cost of the part. The greater cost benefit may be realized by choosing fewer exotic materials in the large castings found in industrial gas turbine engines. Regarding the smaller castings in aircraft engines, machining costs may dominate the part price, driving the designer to choose a less complex cooling scheme. In both cases, the thermal design techniques are employed in the first stage but are applied more sparingly in the downstream stages.

Other components demanding thermal design attention in the gas path are support struts and the engine propulsion nozzle (in the case of the aircraft gas turbine). Thermal loads on these components are more severe if the engine has an afterburner and a combustor in the gas path upstream of the nozzle. Our more detailed discussions will be restricted to the vane and rotor stages of the turbine.

5.12.3.3 Active Cooling of the Gas Turbine Components in the Gas Path

As discussed in the introduction, stage temperature profiles of gas turbine engines often exceed the temperature capability of the turbine materials. To correct the problem, designers incorporate passages and heat transfer augmentation features on the internals of airfoil castings. When relatively cool fluid drawn from the compressor is passed through internal passages, the metal temperature of the airfoil can be maintained below the limits defined by material capability. The design of these internal passages has evolved over many years, leading to complex and highly effective configurations. Some of the most popular schemes are:

5.12.3.4 Impingement Cooling

Impingement cooling can be a very effective way to cool the internal surfaces of a turbine airfoil. Properly designed impingement cooling schemes tend to have high heat transfer coefficients and large pressure drops. As impingement cooling systems require some rather unique geometries; a feed plenum, hole plate, and an impingement target and plenum, the use of impingement cooling systems is typically confined to the first stage vane (where the high heat loads and relatively low cooling supply pressure to passage dump pressure ratio precludes the use of long-channel cooling circuits) and the leading edge and trailing edge regions of other airfoils. There are several mechanisms associated with the impinging jet that improve heat transfer. These will be discussed in the following sections.

5.12.3.5 Pin Fin Cooling

Pin fins are useful tools for enhancing heat transfer coefficients. They are often found in the trailing edge region of turbine airfoils where the high heat transfer coefficients and relatively large pressure drops, they produce are advantageous. Pins can also be designed to span from one side of the channel to the other, providing a conduction path that helps maintain the uniformity of airfoil metal temperatures. Pins also provide additional surface area, which further enhances cooling performance.

5.12.3.6 Channel Cooling Techniques

In the mid-region of turbine airfoils, some form of channel flow cooling is often employed. This cooling may be directed through a simple channel that spans the airfoil from the root to the tip (a radial flow configuration) or it may be configured in a more complex arrangement that directs the flow outward toward the tip, then inward toward the root, then repeating, through a series of channels and turns (a serpentine configuration). The number of passes that a serpentine scheme may incorporate will vary with the specific application and will depend on several factors including the external heat load, available supply pressure, internal heat transfer coefficient that can be attained, the cross-sectional area of the airfoil, and manufacturing considerations. Actual turbine applications typically employ three or five passes, although patents for schemes with additional passes have been filed. The channels can be either smooth or include some form of roughening elements which typically increase both the heat transfer coefficient and the pressure loss. The turns between the passages tend to exhibit complex flow fields making heat transfer coefficient and pressure loss predictions difficult. Examples were already explained in Section 5.12.1, but one is listed as follows:

- Rough channel cooling
- Channels roughened with perpendicular ribs
- Effect of rib spacing and height
- Channels Roughened with Ribs, Heat Transfer Correlations
- Channels roughened with V and W-shaped ribs
- Dimpling

5.12.3.7 External Heat Transfer

External heat transfer applies to surfaces adjacent to the passage flow, whether the airfoil external surfaces or the passage walls (end walls or platforms). As noted, the airfoil resides in a flow in which the temperature exceeds the maximum materials temperature for a suitable design; thus, the surfaces adjacent to the passage flow must be aggressively cooled. This can be by conduction to the internal surfaces, which are cooled by the means discussed above, or by film cooling, which will be discussed in later sections. The first step in the design process is to describe the boundary conditions. The entry flow conditions to the first stage of the turbine have been discussed. The maximum surface temperature limits that the design must meet are determined by materials limitations, such as yield, oxidation, creep, and Thermal Mechanical Fatigue. Establishing these limits requires empirical data from the materials and is, therefore, outside of the scope of this chapter. As mentioned, the design for external thermal control must be done in conjunction with the internal cooling design, as air from the internal cooling operations is used for film cooling, and the external heat load affects the coolant heat up. Thus, the coolant temperature and pressure available for film cooling are determined by the internal cooling design. The examples are (1) airfoil external heat transfer and (2) film cooling.

5.12.3.8 Concluding Remarks

Important aspects of gas turbine hot gas path heat transfer technology have been discussed. In some respects, preliminary designs can proceed with rather simple correlations, such as that presented for heat transfer on the leading edge. In most design aspects, the flow is too complex, and heat transfer is influenced by too many factors for the application of any general correlations. Thus, in support of engine design, (1) information in the literature is studied and interpreted; (2) measurements in cascades or other flow rigs that simulate the engine geometry and flow conditions of interest are taken and interpreted; (3) detailed calculations are made and the computed results are interpreted and; ultimately, (4) prototype engines are constructed, instrumented, operated, and examined for indications of excessive thermal loading. All these steps require a strong understanding of the fundamental principles that govern flow and heat transfer in the gas path. Experiments and computations never include all effects; thus, an understanding of physics is needed in making experimental or computational setup choices.

Though reliance on computations is rising, the art of computation is still encumbered with unanswered questions regarding computation of transition to turbulence and estimation of turbulent transport.

The objective of this chapter is to acquaint the designer with many important features of flow and heat transfer in the gas path that are to be considered when entering the literature, preparing experiments and computer models, and interpreting the information gathered.

5.12.4 Heat Transfer for Sustainable Energy Systems

5.12.4.1 Introduction

The concept of sustainable development dates back several decades ago, which refers back to the World Commission on Environment and Development in 1987.[55] At the same time, it also provided the henceforth most used definition of sustainable development that satisfies the present needs without compromising the ability of future generations to satisfy their needs. This concept has indeed expressed people's concern about the growing crisis of the Earth's life support systems, that is, the use of the available resources on our planet. Energy is a very important part of the world, and sustainable energy systems have become a world concern among scientific and political communities and ordinary people.

Today, the production of electricity and heat is mainly based on finite primary energy sources. Fossil fuels are combusted in such large amounts that flue gas emissions have affected the environment, e.g., the greenhouse effect and toxic pollutants. A general approach to improving the degree of sustainability of the energy supply lies in the following three aspects: reducing final energy consumption, improving overall conversion efficiency, and using renewable sources.[56] To reduce final energy consumption is an obvious approach, which requires more energy-efficient process components and systems. The energy source requirement for the same energy output can be brought down by improving overall conversion efficiency. Renewable energy sources, such as hydropower, biomass, wind, and solar energy, are attractive because they are sustainable.

In all three aspects, it was found that heat transfer and heat recovery play an important role. For instance, increasing the efficiency in thermal processes for heat and power generation requires an increase of the highest temperature in the process, which must be increased further in the future. Cooling is needed to enable the equipment materials in gas turbine units to withstand such high temperatures. This paper will illustrate several examples to stress the importance of the relevant heat transfer and heat exchangers in developing sustainable energy systems. Examples will also be given to illustrate that heat transfer and heat exchanger technologies can bring down the emissions of greenhouse gases and other pollutants. It can finally be concluded that an attempt must be made to provide efficient, compact, and cheap heat transfer methods and heat exchangers to develop sustainable energy systems.

It is known that sustainable energy development must be achieved through three parallel approaches: reducing final energy consumption, improving overall conversion efficiency, and making use of renewable energy sources. Applying advanced heat transfer and heat recovery technologies in these three areas is important. In addition, heat transfer and heat exchangers are also important in protecting the environment by reducing emissions and pollutants. The attempt to provide efficient heat transfer methods and heat exchangers is a real challenge for turbomachinery technology.

5.12.4.2 Reduction of Energy Consumption

The process industry remains one of the largest sectors in consuming energy. A typical process, shown in Figure 5.38, consists of three parts: a chemical plant, a utility plant, and a heat recovery network.[57] The purpose of the chemical plant is to produce products from raw materials with the supply of energy from both the utility plant and heat recovery network. The utility plant produces power, hot utility, and cold utility. The heat recovery network, which consists of many heat exchangers, aims to recover heat from hot streams to heat cold streams. Maximizing the heat recovery network can reduce energy consumption and flue gas emissions from the utility plant. Therefore, reducing energy consumption requires optimizing the heat recovery network, that is, heat exchanger networks. Advanced heat exchanger technologies can improve the efficiency of heat exchanger networks. Such technologies include compact heat exchangers, multi-stream heat exchangers, heat transfer enhancement, and mini- and micro-heat exchangers.[58] These technologies can improve current processes, and the final energy demands can be reduced.

Conventional heat exchangers in process industries are shell-and-tube heat exchangers. There are several disadvantages in using such units, e.g., low ratio of surface to volume, the tendency to severe fouling, use of multi-pass design, low efficiency due to a relatively high-pressure drop per unit of heat transfer in the shell side, etc. Most of these advantages are due to the relatively large hydraulic diameter. To overcome these, compact heat exchangers have been developed. A compact heat exchanger incorporates a heat transfer surface with area density (or compactness) above 700 m^2/m^3 on one side of the fluids.[59] The common types include plate heat exchangers, plate-fin

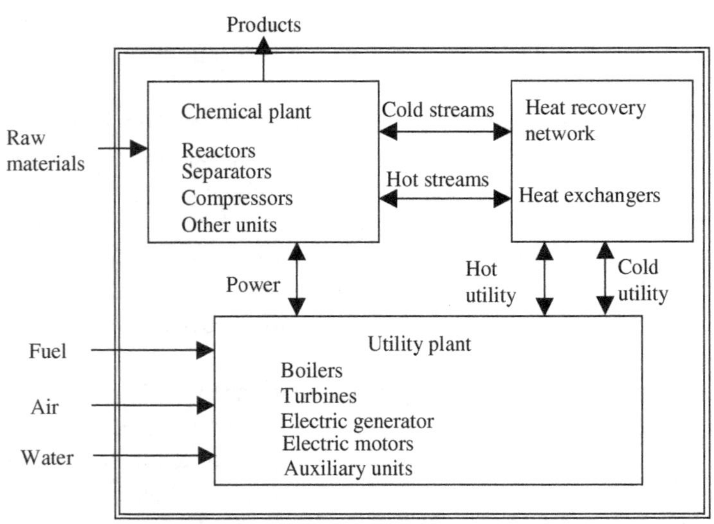

Figure 5.38 Total processing system.

heat exchangers, tube-fin heat exchangers, etc. In the process industries using compact heat exchangers, energy consumption can be reduced in addition to the plant's reduced capital cost and complexity.

Compact heat exchangers usually have a small hydraulic diameter, which results in high heat transfer coefficients. This will reduce the unit size and weight, hence the unit capital cost. In addition, the high heat transfer coefficients permit compact heat exchangers to operate under conditions with close temperature approaches. This is significant in the optimization of heat exchanger networks. In the pinch analysis method for the design of heat exchanger networks,[60] the minimum temperature difference is the decisive parameter to construct the so-called composite curves, which are shown schematically in Figure 5.39. By using compact heat exchangers, the minimum temperature difference can be reduced significantly compared to shell-and-tube heat exchangers. This makes the two lines in the composite curves approach very close to each other, which means that the heat recovery is enlarged, and at the same time, the external utility requirements are reduced. Therefore, the utility consumption in the entire plant is reduced. Due to the high heat transfer coefficients and low unit capital costs, the heat recovery system's total cost can still be lower than that of shell-and-tube heat exchangers.

A multi-stream heat exchanger is a good option when too many heat exchanger units are required. In optimizing heat exchanger networks using pinch technology, many exchangers are often required when the network is designed for two-stream exchangers. This not only increases the capital cost but also increases the complexity of the network. Therefore, it may challenge the optimal solution, and relaxation must be made. Using multi-stream heat exchangers might be a good way to circumvent this problem, and it offers several potential benefits, including large savings in capital and installation costs, reduction in physical weight and space, better integration of the process, etc. However, the streams connected to them should not be too far away in physical space to save piping costs. Common multi-stream heat exchangers include multi-stream plate-fin heat exchangers, multi-stream plate heat exchangers, etc.[61]

Heat transfer enhancement for shell-and-tube heat exchangers should also be considered when optimizing heat exchanger networks. It reduces the capital cost because of the small size needed for a given duty. It also reduces the temperature driving force, reducing entropy generation and increasing the second law efficiency. In addition, heat transfer enhancement enables heat exchangers to operate at smaller velocities but achieve the same or even higher heat transfer coefficient. This means that a reduction of pressure drops, corresponding to less power utilization, may be achieved. All these advantages have made heat transfer enhancement technology attractive in heat exchanger applications. For the tube side, different geometries (e.g., low-finned tubes, twisted tubes, grooved tubes) and tube inserts (e.g., twist-taped inserts, wire coil inserts, extended surface inserts) have been developed.[62] Improvements have been made for the shell side, e.g., helical baffles and twisted tube heat exchangers.[58]

More heat transfer and heat exchanger technologies are available to improve the process and reduce the final energy consumption. These may include micro- and mini-heat exchangers, integrated chemical reactor heat exchangers, etc. Due to the limited space in this article, these are not further explored in detail. However, the possibilities of their application in process industries should not be underestimated.

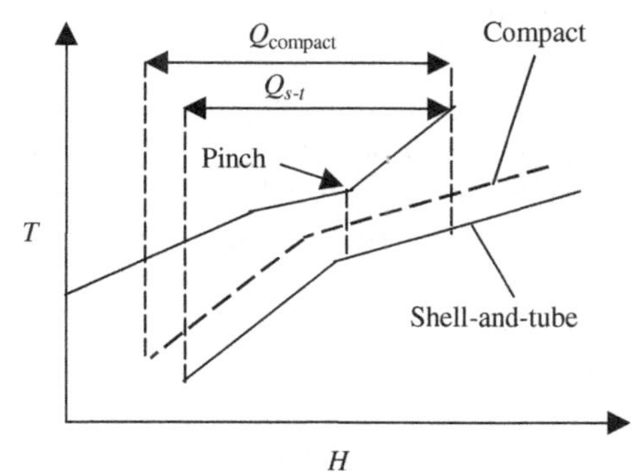

Figure 5.39 Composite curves.

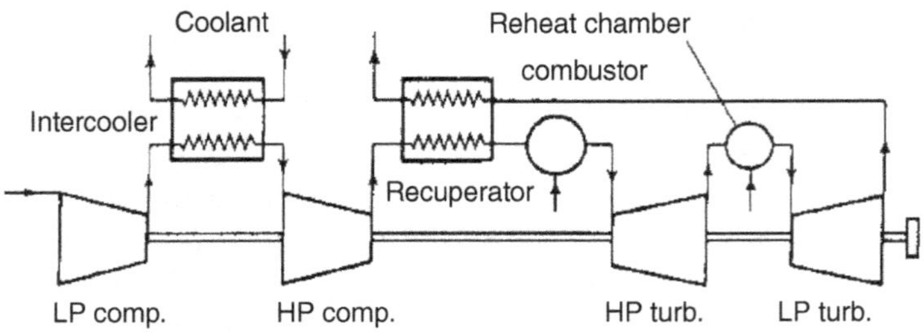

Figure 5.40 Power plant with recuperator and reheater.

5.12.4.3 Improved Efficiency of Energy Conversion

There are many ways to improve the efficiency of thermal power plants, but heat transfer and heat exchangers play a significant role in all means. This can be highlighted by considering a power plant using gas turbines as an example. The original Brayton cycle for the power plant only needs a compressor, a combustion chamber, and a power turbine, and this concept can be found in any textbook on thermodynamics, e.g., Cengel and Boles.[63] However, the thermal efficiency is usually very low in such systems, and improvement can be made by employing the concept of intercooling, recuperation (regeneration), and reheating. Such a flow sheet is illustrated in Figure 5.40, and the corresponding thermodynamic cycle is shown in Figure 5.41. Two stages of gas compression are provided to reduce the power consumption for compression due to the lower inlet temperature of the gas in the second compression stage by using an intercooler. The net power output is increased because the compression power required is reduced. Recuperation uses energy in the turbine exhaust gases to heat the air entering the combustion chamber, thus saving a certain amount of fuel in the combustion process. This will certainly increase overall thermal efficiency as well.

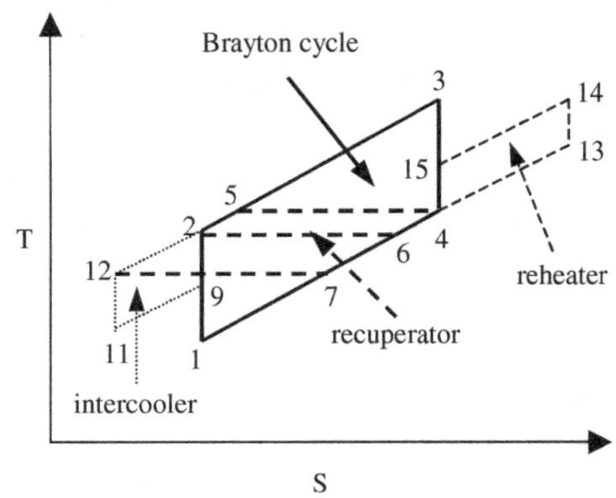

Figure 5.41 Thermodynamic cycle of a gas turbine power plant with intercooler, recuperator, and reheater.

In addition, the turbine output can be increased by dividing the expansion into two or more stages and reheating the gas to the maximum permissible temperature between the stages. Although the power output is improved, the cost of additional fuel will be heavy unless a heat exchanger is also employed. These concepts can also be seen in the thermodynamic cycle in Figure 5.41. The cycle 1-2-3-4-1 corresponds to the simple Brayton cycle. 9-11-12-2 represents the intercooling, and 15-14-13-4 represents the reheating. 4-7-12-5 and 4-6-2-5 represent recuperation in the case of intercooling and no intercooling, respectively. This concept has already been incorporated into some real gas turbines, e.g., LMS100 from G.E., which has the application of an intercooler. Mercury 50 from Solar Turbine has the application of recuperator, and GT24/26 from ASLTOM has the application of sequential Combustion. These features significantly increase the efficiency of gas turbines, and a lot of work has been done to design reliable heat exchangers operated at higher temperatures.

In addition, the thermal efficiency and power output of gas turbines will increase with an increasing turbine rotor inlet temperature, which corresponds to the temperature at point 3 in Figure 5.41. This is the reason why modern advanced gas turbine engines operate at high temperatures (ISO Turbine Inlet Temperature [TIT] in the range of 1200 °C–1400 °C), and the trend is to operate at even higher temperatures. To enable this, cooling technologies and material innovation must be developed for combustion chambers, turbine blades, guide vanes, etc. Over the years, film cooling, convection cooling, and impingement cooling have been developed for both combustion chamber (see Figure 5.42) and turbine blade (see Figure 5.43), and transpiration cooling is still under

Figure 5.42 Cooling concepts of the combustion chamber.

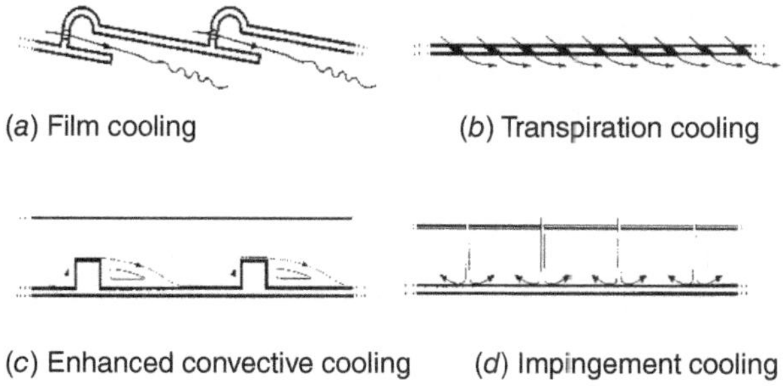

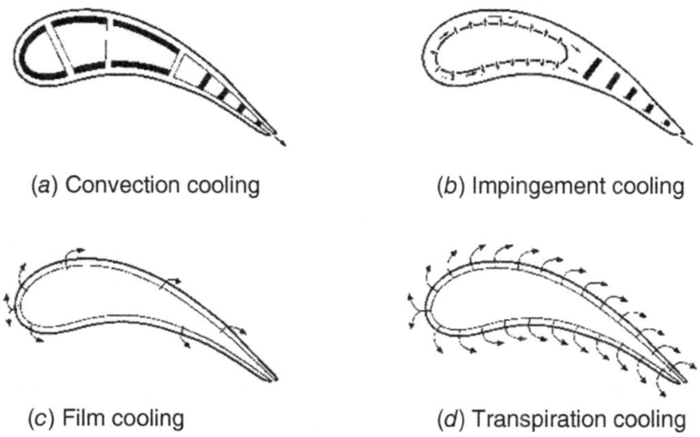

Figure 5.43 Cooling concepts of gas turbine blade.

development due to engineering difficulty. In addition, more advanced high-temperature materials, such as superalloys of single crystal and ceramic coating also significantly contribute to the high TIT operation. These advanced cooling technologies can sustain reliable and highly efficient power plants.

The above blade cooling technologies are for air-cooled gas turbines, but a new turbine cooling concept is available, that is, steam-cooled gas turbines. Steam provides several benefits over air as a cooling medium. Firstly, steam provides higher heat transfer characteristics because its heat capacity is higher than air. Secondly, using steam as a cooling medium reduces the use of cooling air, which means that more cooling air is available for the combustion process, which contributes to improving emissions. Thirdly, reducing cooling air results in less temperature dilution of the hot gas caused while mixing with the cooling air. This increases the turbine inlet temperature, which results in more power being available. Finally, no ejection of cooling air to the main gas flow minimizes aerodynamic loss. With this technology, the efficiency of the gas turbine is greatly enhanced, and G.E.'s H-class gas turbine is the best example. It is the first gas turbine to break 60% efficiency in the combined cycle power plants. However, to design such turbines, the heat transfer characteristics of steam as a cooling medium must be thoroughly understood, which requires extensive research. For the gas side, the different fuels can lead to significant changes in the properties of the gas in the turbine part. The IGCC (Integrated Gasification Combined Cycle) application operates on hydrogen. Consequently, the syngas will increase the heat transferred to rotating and stationary airfoils due to increased moisture content and mass flow. Thus, research will provide a better understanding of heat transfer mechanisms in a syngas environment.

Another way to improve energy conversion efficiency is to use combined cycles incorporating steam turbines, fuel cells, etc. A combined cycle with steam turbines is a relatively old but still very effective approach, and heat transfer and heat exchangers play a significant role there, without any doubt. Here, a brief discussion is given about the heat transfer issues associated with fuel cells. Figure 5.44 shows a typical configuration for a combined cycle using a gas turbine and a fuel cell. As is well known, fuel cells can convert the chemical energy stored in

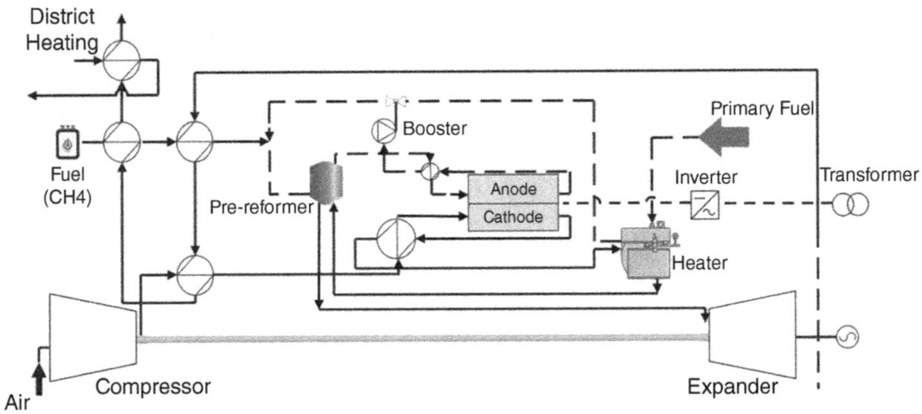

Figure 5.44 Reference fuel cell and gas turbine system layout. (*Source:* Adapted from Selimovic.[65])

the fuel into electrical and thermal energy through electrochemical processes. High electrical efficiencies can be obtained because these processes are not subject to the Carnot cycle limitation. Typical fuel cell types include Phosphoric Acid Fuel Cells (PAFC), Proton Exchange Membranes (PEM), Solid Oxide Fuel Cells (SOFC), and Molten Carbonate Fuel Cells (MCFC) etc.[64]

The operation principle indicates that heat and mass transfer are important in fuel cells.[64] One typical fuel cell construction is the flat plate design for SOFC, shown in Figure 5.45. As can be seen, the fuel in fuel ducts has both heat and mass transfer on the top wall with the anode, and the air in air ducts has both heat and mass transfer on the bottom wall with the cathode. In addition, two-phase flows exist in fuel ducts after a part of the fuel is consumed. Therefore, fluid flow and heat transfer conditions in air and fuel ducts greatly affect fuel cells' performance and, consequently, the entire power cycle. Most current designs are based on constant Nusselt number and friction factor values. Such rough estimations cannot meet future developments, and considerable research efforts must be made to address this complex heat and mass transfer problem.

The above analysis demonstrates that high power conversion efficiency can be reached with the help of relevant heat transfer and heat exchanger technologies. Therefore, attempts to provide compact, efficient heat transfer methods and heat exchangers while allowing a cheap and relatively simple manufacturing technique are real challenges for research.

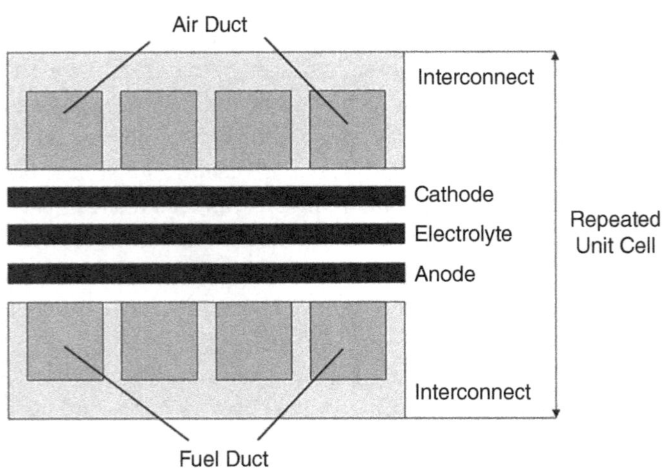

Figure 5.45 Structure of a unit of a fuel cell.

5.12.4.4 Use of Renewable Energy

Hydropower, biomass, wind, and solar energy are the most important renewable and sustainable energy sources. Hydropower depends on the earth's contour, which is not substantial for those countries with flat earth surfaces. Biomass appears to be an attractive option for many countries, and technologies for converting biomass into electricity and heat are highly similar to those for other solid fossil fuels. Wind and solar energy sources are strongly fluctuating sources, but they are very clean with no pollutant emissions and have received great attention. Heat transfer and heat exchangers play an important role in these renewable energy systems, just as in those systems described earlier.

Consider now a simple solar energy system as an example. Figure 5.46 is a schematic view of a typical domestic hot water heating system designed for residential applications. When there is sun, the photovoltaic (P.V.) module produces power, which runs a small circulating pump. Antifreeze is pumped and heated through the solar collectors. The fluid then returns to a reservoir in the heat exchanger module. Water coils in the reservoir absorb the heat from the solar fluid. Domestic water flows through these heat exchanger coils through natural thermosiphon action. As the water is heated, it rises and returns to the top of the tank, drawing cold water from the bottom of the tank and transferring it into the heat exchanger.

No external heat exchanger, as shown in Figure 5.46, was used historically. Instead, the heat exchangers were coils of copper pipes located in the bottom of the solar storage tank. The current design shown in Figure 5.46 has some advantages. Firstly, the system performance is enhanced. External heat exchangers can be configured so that the potable water side circulates by natural convection (i.e., thermosiphons), allowing excellent temperature stratification in the storage tank. With the hot water remaining in the top of the tank, usable hot water is available more rapidly with an external heat exchanger. Secondly, the thermodynamic efficiency is improved with the external heat exchanger configuration. The heat transfer rate is directly proportional to the difference in temperature between the water being heated and the antifreeze from the solar collectors. With the external heat exchanger configuration, the heat exchanger coil is always surrounded by very cold water, so thermal efficiency is greatly improved. Thirdly, low cost can be achieved due to the long lifetime of the external heat exchanger compared to the solar tank. The external heat exchanger can be saved when the solar tank develops a leak; thus, cost saving is achieved. However, the heat transfer mechanism involved in the external heat exchanger is highly complex. Both forced convection and natural convection have important impacts. Shell-and-tube heat exchangers may serve well, but compact heat exchangers (such as plate heat exchangers) claim superior operating conditions. Because this practical application is still in its infancy, considerably more research is expected in the future.

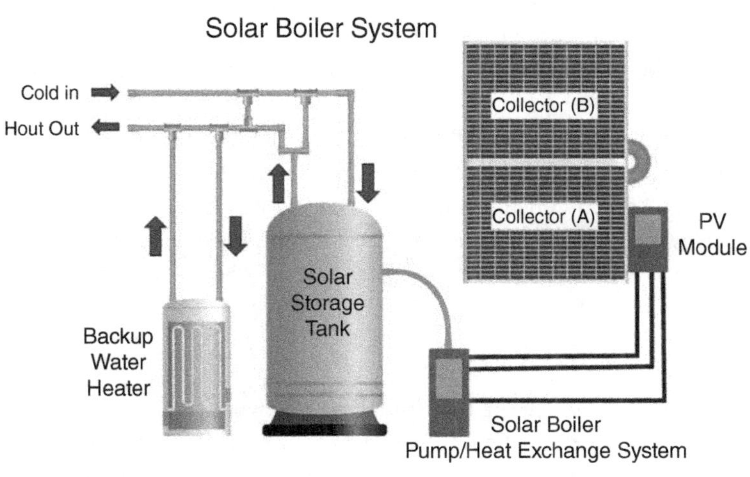

Figure 5.46 A solar energy system (from Solar-Works Inc.).

In addition, the solar collector using the photovoltaic (P.V.) module is a special heat exchanger. Solar energy (radiant energy) is absorbed on one side of the surface. This energy is transferred to the second side of the coolant. This is quite a complex heat transfer problem, not only because it involves both radiant and convection heat transfer but also because it is time dependent. Solar energy varies with time and location, which must be considered when using this renewable energy.

The importance of heat transfer and heat exchangers has been illustrated for the solar energy system. A similar conclusion can be reached when dealing with other renewable energy systems. However, they are not fully exploited here due to limited space.

5.12.4.5 Reduction of Emission and Pollutants

Heat transfer and heat exchangers are also important in reducing emissions and pollutants. As has been illustrated, they play an important role in the development of sustainable energy systems. Reducing final energy consumption means less prime energy (e.g., fossil fuels) consumption, which results in an overall reduction in emissions and pollutants. The improved efficiency of power plants also reduces primary energy consumption and consequent emissions. Alternative fuels like biofuels (including biomass and waste utilization) are said to be neutral regarding CO_2. The other renewable energy sources, solar, hydropower, and wind, are clean enough, and no emissions exist. In addition, by considering the pressure drop and associated pressure losses (work loss) in the heat transfer processes and attempting to reduce them, electricity consumption will decrease, which is also beneficial. Therefore, heat transfer and heat exchangers are important for protecting the environment and their role in developing a sustainable energy system.

The above influences on emissions and pollutants are the indirect effect. However, heat transfer and heat exchangers can also directly affect reducing emissions and pollutants in many situations. One example is their presence in internal combustion engines. In diesel combustion engines, exhaust gas recycling (EGR) has been used for a while because this is an efficient method to reduce NO_x. However, particle emissions have increased, and the engine performance is reduced. It has been recognized that if the exhaust gas is cooled down in a heat exchanger, the above problems can be avoided or at least partially avoided. In addition, the NO_x emissions will be further reduced, as seen in Figure 5.47. In this situation, several factors must be considered. Firstly, an EGR cooler must be compact and lightweight due to the limited automobile space. Secondly, because the cooling water is taken from the total engine cooling water, the amount of cooling water for the EGR cooler is limited and must be kept as small as possible. This means that the EGR cooler must be of high thermal efficiency. Thirdly, the EGR cooler is always subject to unsteady or oscillatory operation and is also severely affected by fouling. This means operating reliability and lifetime are extremely important when selecting the heat exchanger type. Therefore, a compact heat exchanger (e.g., a brazed plate heat exchanger) may be a better option, although shell-and-tube heat exchangers are often used in automobiles. Further research must be carried out to design an EGR cooler that gives very reliable performance and durability.

Another example is the combustion chamber in gas turbine systems. It is well known that the production of NO_x is related to high flame temperatures. One way to reduce the flame temperature is to use high air-to-fuel ratios.[67] This means that much more compressor air is needed for combustion, and consequently, less is available for cooling combustion chambers and turbine blades. However, low-temperature zones lead to unburned hydrocarbons. Thus, the emission control and the cooling system are coupled and need careful attention. This is another evidence that heat transfer design directly reduces emissions and pollutants.

5.12.4.6 Conclusions

To develop sustainable energy systems, one must minimize energy use, improve energy conversion efficiency, and use renewable energy sources. In all these aspects, heat transfer and heat exchangers play a significant role, which has been reviewed and illustrated throughout the paper. In addition, heat transfer and heat exchangers also have a great influence on the reduction of emissions and pollutants directly and indirectly. Therefore, the

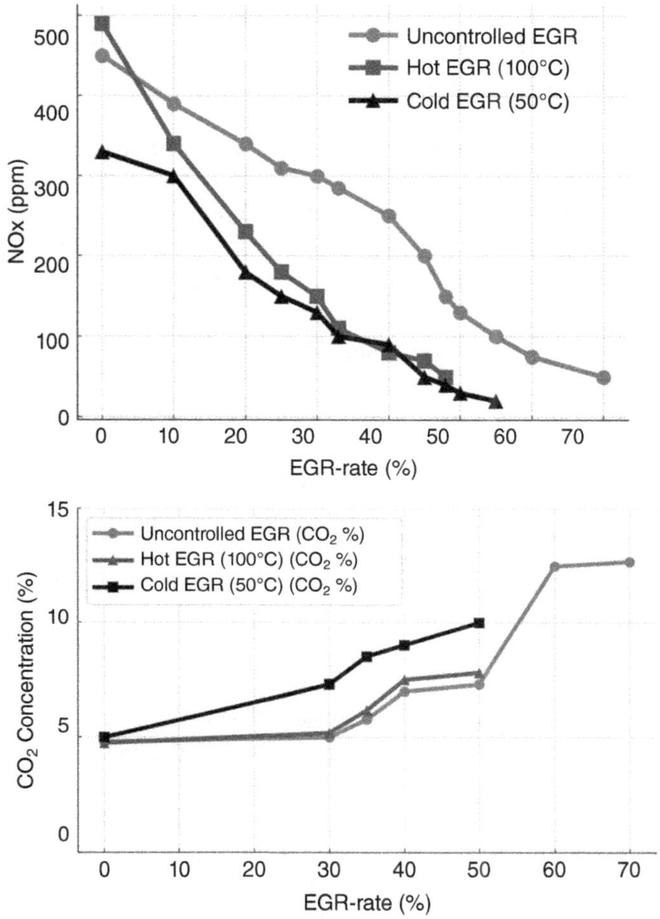

Figure 5.47 Effect of cooled EGR on NO_x and CO_2 emissions. (*Source:* Adapted from Arcoumanis.[66])

attempt to provide efficient, compact, and cheap heat transfer methods and heat exchangers is a real challenge for research. Both theoretical and experimental investigations must be conducted, and modern scientific techniques must be adopted, such as computational fluid dynamics, laser techniques, liquid crystal thermography, and many other technological resources. By doing so, sustainable energy systems can be established, contributing to global sustainable development.

References

1 Moran, M. J., and Shapiro, H. N., *Fundamentals of Engineering Thermodynamics*, 3rd ed., John Wiley & Sons, New York, 1995.

2 Horlock, J. H., "Losses and Efficiencies in Axial-flow Turbines," *Int. J. Mech. Sci.*, Vol. 2, pp. 48–75, 1960.

3 Zeifel, O., "The Spacing of Turbomachine Blading, Especially with Large Angular Deflection," *Brown Boveri Rev.*, Vol. 32, No. 12, pp. 436–444, Dec. 1945.

4 Saravanamuttoo, H., Roger, C., and Cohen, H., *Gas Turbine Theory*, 5th ed., Prentice-Hall, New York, 2001.

5 *Garrett/Ford AGT101 Advanced Gas Turbine Program Summary*, Garrett Turbine Engine Co., Phoenix, AZ, 1985.

6 Bridle, E. A., and Boulter, R. A., "A Simple Theory for the Prediction of Losses in the Rotors of Inwards Radial Flow Turbines," *Proc. Instn. Mech. Engrs.*, Vol. 182, No. Part 3H, 1968.

7 Ainley, D. G., *Internal Air-Cooling for Turbine Blades—A General Design Survey, R&M Aeronautical Research Council, 3013*, HMSO, London, 1957.

8 Han, J. C., Dutta, S., and Ekkad, S., *Gas Turbine Heat Transfer and Cooling Technology*, Taylor & Francis, New York, 2000.

9 Leonard, G., and Stegmaier, J., "Development of an Aero-Derivative Gas Turbine Dry Low Emissions Combustion System," Paper No. 93-GT-288, American Society of Mechanical Engineers, 1993.

10 Lipfert, F. W., "Correlation of Gas Turbine Emission Data," Paper No. 72-GT-60, American Society of Mechanical Engineers, 1972.

11 Davis, L. B., and Washam, R. M., "Development of a Dry Low NO_x Combustor," Paper No. 89-GT-255, American Society of Mechanical Engineers, 1989.

12 Maghon, H., Berenbrink, P., Termuehlen, H., and Gartner, G., "Progress in NO_x and CO Emission Reduction of Gas Turbines," Paper No. 90-IPGC/GT-4, American Society of Mechanical Engineers, 1990.

13 Sattelmeyer, T., Felchlin, M. P., Hauman, J., and Tyner, D., "Second Generation Low-Emission Combustors for ABB Gas Turbines: Burner Development and Tests at Atmospheric Pressure," *ASME Trans.*, Vol. 114, pp. 118–124, 1992.

14 Oakridge National Laboratory *Distributed Energy Program FY04 Accomplishments Report*, Oakridge National Laboratory, Oakridge, TN, 2004.

15 Greenblatt, J. B., *Using Gas Turbines to Enhance the Value of Wind Power, Combined Cycle J., 1Q/2005*, PSI Media, Las Vegas, NV, 2005.

16 O'Brien, J. P. (Ed.), *Gas Turbines for Automotive Use*, Noyes Corporation, Park Ridge, NJ, 1980.

17 Ai, T, Koeneke, C, Arimura, H, & Hyakutake, Y. "Development of an Air Cooled G Class Gas Turbine (the M501GAC)." Proceedings of the ASME Turbo Expo 2009: Power for Land, Sea, and Air. Volume 1: Aircraft Engine; Ceramics; Coal, Biomass and Alternative Fuels; Controls, Diagnostics and Instrumentation; Education; Electric Power; Awards and Honors. Orlando, Florida, USA. June 8–12, 2009. pp. 961–968. ASME. https://doi.org/10.1115/GT2009-60321.

18 Ito, E, Okada, I, Tsukagoshi, K, Muyama, A, & Masada, J. "Development of Key Technologies for the Next Generation 1700C-Class Gas Turbine." Proceedings of the ASME Turbo Expo 2009: Power for Land, Sea, and Air. Volume 1: Aircraft Engine; Ceramics; Coal, Biomass and Alternative Fuels; Controls, Diagnostics and Instrumentation; Education; Electric Power; Awards and Honors. Orlando, Florida, USA. June 8–12, 2009. pp. 919–929. ASME. https://doi.org/10.1115/GT2009-59783.

19 Amano, R. S., and Sunden, B., *Thermal Engineering in Power Systems (Developments in Heat Transfer)*, WIT Press, 2008, ISBN978-1-84564-062-0.

20 Amano, R. S., and Sunden, B., *Impingement Jet Cooling in Gas Turbines*, WIT Press, 2014, ISBN978-1-84564-906-7.

21 Webb, R. L., Eckert, E. R. G., and Goldstein, R. J., "Heat Transfer and Friction in Tubes with Repeated-Rib Roughness," *Int. J. Heat Mass Transf.*, Vol. 14, pp. 601–617, 1971.

22 Burggraf, F., "Experimental Heat Transfer and Pressure Drop with Two-Dimensional Turbulence Promoter Applied to Two Opposite Walls of a Square Tube," in *Augmentation of Convective Heat and Mass Transfer*, Bergles, A. E. and Webb, R. L. (Eds), ASME, pp. 70–79, 1970.

23 Han, J. C., Glicksman, L. R., and Rohsenow, W. M., "An Investigation of Heat Transfer and Friction for Rib-Roughened Surfaces," *Int. J. Heat Mass Transf.*, Vol. 21, pp. 1143–1156, 1978.

24 Han, J. C., Park, J. S., and Lei, C. K., "Heat Transfer Enhancement in Channels with Turbulence Promoters," *J. Eng. Gas Turbines Power*, Vol. 107, pp. 628–635, 1985.

25 Chakroun, W. M., Al-Fahed, S. F., and Abdel-Rehman, A. A., "Heat Transfer Augmentation for Air Impinged on Rough Surface," Paper 97-GT-436, IGTI Turbo Expo, Orlando, FL, 1997.

26 El-Gabry, L. A., and Kaminski, D. A., "Experimental Investigation of Local Heat Transfer Distribution on Smooth and Roughened Surfaces Under an Array of Angled Impinging Jets," *J. Turbomach.*, 2004.

27 Bunker, R. S., and Bailey, J. C., "Turbine Blade Platform Impingement Cooling with Effects of Film Extraction and Roughness," Paper No. GT-2005-68415, IGTI Turbo Expo, Reno-Tahoe, NV, 2005.

28 Son, C., Gillespie, D., Ireland, P., and Dailey, G. M., "Heat Transfer Enhancement Strategy for an Impingement Cooling System," Proceedings of the 8th International Symposium on Transport Phenomena and Dynamics of Rotating Machinery, Hemisphere Pub. Co., Vol. 2, pp. 721–729, 2000.

29 Taslim, M. E., Setayeshgar, L., and Spring, S. D., "An Experimental Evaluation of Advanced Leading Edge Impingement Cooling Concepts," *ASME J. Turbomach.*, Vol. 123, pp. 147–153, 2002.

30 Lee, C. P., and Bunker, R. S., "Thermal Shield Turbine Airfoil," US Patent 7011502, 2006.

31 Han, J. C., and Zhang, Y. M., "High Performance Heat Transfer Ducts with Parallel Broken and V-Shaped Broken Ribs," *Int. J. Heat Mass Transf.*, Vol. 35, No. 2, pp. 513–523, 1992.

32 Taslim, M. E., Li, T., and Kercher, D. M., "Experimental Heat Transfer and Friction in Channels Roughened with Angled, V-Shaped and Discrete Ribs on Two Opposite Walls," *J. Turbomach.*, Vol. 118, pp. 20–28, 1996.

33 Kiml, R., Mochizuki, S., and Murata, A., "Heat Transfer Enhancement Mechanism in a Rectangular Passage with V- and L-Shaped Ribs," *J. Flow Vis. Image Process.*, Vol. 8, pp. 51–68, 2001.

34 Chyu, M. K., and Natarajan, V., "Effects of a Slit in a Rib on the Heat Transfer from a Rib-Mounted Wall," *J. Heat Transf.*, Vol. 115, pp. 792–796, 1993.

35 Afanas'yev, V. N., Veselkin, V.Yu., Leont'ev, A. I., Skibin, A. P., and Chudnovskiy, Ya.P., "Thermohydraulics of Flow Over Isolated Depressions (Pits, Grooves) in a Smooth Wall," *Heat Transf. Res.*, Vol. 25, No. 1, pp. 22–56, 1993.

36 Isaev, S. A., Leont'ev, A. I., and Baranov, P. A., "Identification of Self-Organized Vortexlike Structures in Numerically Simulated Turbulent Flow of a Viscous Incompressible Liquid Streaming Around a Well on a Plane," *Tech. Phys. Lett.*, Vol. 26, No. 1, pp. 15–18, 2000.

37 Kesarev, V. S., and Kozlov, A. P., "Convective Heat Transfer in Turbulized Flow Past a Hemispherical Cavity," *Heat Transf. Res.*, Vol. 25, pp. 156–160, 1993.

38 Afanas'yev, V. N., and Chudnovskiy, Ya.P., "Heat Transfer and Friction on Surfaces Contoured by Spherical Depressions," *Heat Transf. Res.*, Vol. 24, pp. 24–104, 1992.

39 Afanas'yev, V. N., Chudnovskiy, Ya.P., Leont'ev, A. I., and Roganov, P. S., "Turbulent Flow Friction and Heat Transfer Characteristics for Spherical Cavities on a Flat Plate," *Exp. Thermal Fluid Sci.*, Vol. 7, pp. 1–8, 1993.

40 Zhang, N., Yang, W. J., and Lee, C. P., "Heat Transfer and Friction Loss Performance in Flow Networks with Multiple Intersections," *Exp. Heat Transf.*, Vol. 6, pp. 243–257, 1993.

41 Gillespie, D., Wang, Z., Ireland, P., and Kohler, S. T., "Full Surface Local Heat Transfer Coefficient Measurements in a Model of an Integrally Cast Impingement Cooling Geometry," *J. Turbomach.*, Vol. 120, pp. 92–99, 1996.

42 Nakamata, C., Okita, Y., Matsuno, S., Mimura, F., Matsushita, M., Yamana, T., and Yoshida, T., "Spatial Arrangement Dependence of Cooling Performance of an Integrated Impingement and Pin Fin Cooling Configuration," Paper No. GT-2005-68348, IGTI Turbo Expo, Reno-Tahoe, NV, USA, 2005.

43 Hale, C. A., Plesniak, M. W., and Ramadhyani, S., "Film Cooling Effectiveness for Short Film Cooling Holes Fed by a Narrow Plenum," *J. Turbomach.*, Vol. 122, pp. 553–557, 1999.

44 Bunker, R. S., Bailey, J. C., Widener, S. K., and Johnson, T. E., "Combustor Cooling with Angled Segmented Surfaces," US Patent Application US2006/0042255A1, 2006.

45 Bailey, J. C., and Bunker, R. S., "Local Heat Transfer and Flow Distributions for Impinging Jet Arrays of Both Sparse and Dense Extent," GT-2002-30473, IGTI Turbo Expo, Amsterdam, 2002.

46 Gao, L., Ekkad, S. V., and Bunker, R. S., "Impingement Heat Transfer Part I: Linearly Stretched Arrays of Holes," *AIAA J. Thermophys. Heat Transf.*, Vol. 19, No. 1, pp. 57–65, 2005.

47 Nagoga, G. P., "Personal Communication," Moscow, 2000.

48 Gardon, R., and Cobonpue, J., "Heat Transfer Between a Flat Plate and Jets of Air Impinging on It," International Developments in Heat Transfer, Proceedings 2nd International Heat Transfer Conference, ASME, New York, pp. 454–460, 1962.

49 Perry, K. P., "Heat Transfer by Convection from a Hot Gas Jet to a Plane Surface," *Proc. Inst. Mech. Eng. (London)*, Vol. 168, pp. 775–784, 1954.

50 Bearman, P. W., and Harvey, J. K., "Golf Ball Aerodynamics," *Aeronaut. Q.*, pp. 112–122, May 1976.

51 Kiknadze, G. I., Gachechiladze, I. A., and Oleinikov, V. G., "Method and Apparatus for Controlling the Boundary or Wall Layer of a Continuous Medium," US Patent No. 6119987, 2000.

52 Bunker, R. S., and Donnellan, K., "Heat Transfer and Friction Factors for Flows Inside Circular Tubes with Concavity Surfaces," *J. Turbomach.*, Vol. 125, pp. 665–672, 2003.

53 Drummel, H.-G., LeMeiux, D. H., Voigt, M., and Zombo, P. J., "Online Monitoring of Gas Turbine Power Plants," Siemens Power Generation Technical Paper, 2005.

54 Cumpsty, N., *Jet Propulsion*, Cambridge University Press, Cambridge, 1997.

55 http://www.un.org/esa/sustdev/agreed.htm

56 Woudstra, N., "Towards Sustainable Energy Systems," International Conference on Efficiency, Cost, Optimisation, Simulation and Environmental, Enschede, the Netherlands, pp. 1381–1392, 2000.

57 Grossmann, I. E., "Mixed-Integer Programming Approach for the Synthesis of Integrated Process Flowsheets," *Comput. Chem. Eng.*, Vol. 9, No. 5, pp. 463–482, 1985.

58 Wang, L., and Sundén, B., "On Heat Exchanger Technologies in Process Integration, accepted by Process Innovation and Process Intensification (P.I.)2 Conference," Edinburgh, Scotland, UK, Sep. 9–13, 2002.

59 Shah, R. K., "Compact Heat Exchanger Surface Selection Methods," *Proc. Sixth Int. Heat Transfer Conf.*, Vol. 4, pp. 193–199, 1978.

60 Linnhoff, B., Townsend, D. W., Boland, D., Hewitt, G. F., Thomas, B. E. A., Guy, A. R., and Marsland, R., *User Guide on Process Integration for the Efficient Use of Energy, Institution of Chemical Engineers*, Pergamon Press, Oxford, 1982.

61 Wang, L., "Performance Analysis and Optimal Design of Heat Exchangers and Heat Exchanger Networks," Ph.D. thesis, Lund Institute of Technology, Sweden, 2001.

62 Wang, L., and Sundén, B., "Performance Analysis of Some Tube Inserts," *Int. Comm. Heat Mass Transf.*, Vol. 29, No. 1, pp. 45–56, 2002.

63 Cengel, Y. A., and Boles, M. A., *Thermodynamics: an Engineering Approach*, 3rd ed., McGraw-Hill, Boston, 1998.

64 Larminie, J., and Dicks, A., *Fuel Cell Systems Explained*, Wiley, New York, 2000.

65 Selimovic, A., "Modelling of Solid Oxide Fuel Cells Applied to the Analysis of integrated Systems with Gas Turbines," Ph.D. thesis, Lund Institute of Technology, Sweden, 2002.

66 Arcoumanis, C., Nagwaney, A., Hentschel, W., and Ropke, S., "Effect of EGR on Spray Development, Combustion and Emissions in a 1.9L Direct-Injection Diesel Engine," SAE paper 952356, 1995.

67 Cohen, H., Rogers, G. F. C., and Saravanamuttoo, H. I. H., *Gas Turbine Theory*, 4th ed., Longman Group Limited, Essex, 1996.

Problems

5.1 A 50% reaction turbine with blades as shown is used for a 10-stage gas turbine. The exit velocity of air from the stator blades is 480 ft/s, the rotor blade speed is 455 ft/s, and the axial flow velocity is constant. Determine the work done by air per stage with a mass flow rate of 8 lb_m/s. If the total-to-total efficiency is 80%, $p_{01} = 1600$ psia, and $T_{01} = 1000\,°\text{F}$ at the first-stage stator inlet, determine the air total temperature and pressure at the outlet of the last stage.

5.2 Pressures in psia are measured at various stations of an impulse gas turbine stage, all at the mean blade radius $p_{01} = 60$ psia, $p_{02} = 55$ psia, $p_2 = 27$ psia, and $p_3 = 24$ psia. The mean blade speed is 850 ft/s, $T_{01} = 1900$ R, and the flow angle at the stator exit is 70° with respect to the axial direction. Determine the total-to-total efficiency of the stage.

5.3 An axial-flow turbine stage with $r_t = 8.0$ in., $r_h = 6.5$ in., and $\alpha_1 = 72°$ is designed for a 50% degree of reaction at the mean radius while rotating at $N = 8500$ rpm. If the free-vortex condition is to be satisfied across the blade radius with a constant axial-flow velocity of 550 ft/s, determine the rotor blade angles at mean, hub, and tip for both leading and trailing edges. Also determine the degree of reaction at hub and tip.

5.4 A multistage axial-flow gas turbine is used to produce power with air supplied from a combustor at $p_0 = 140$ psia, $T_0 = 2160$ R, and exhausted to the atmosphere of 14.7 psia. If each stage is designed to produce the same amount of power with a constant mean radius of 5 in., rotating at 12,000 rpm, axial-flow velocity of 600 ft/s, flow angle at the stator outlet of 70°, and the stage efficiency of 86%, determine the total number of stages required, the blade height of the first stage, and the total power produced if the mass flow rate is 30 lb_m/s. (use room temperature air properties).

5.5 Air is expanded in a multistage axial-flow turbine, and the pressure drop across each stage is very small. If the expansion can be assumed to be reversible, but (a) the heat loss in each stage is a constant fraction J of the enthalpy drop in that stage or (b) the heat loss is proportional to the absolute temperature, derive the corresponding pressure–temperature relationship of the inlet and outlet.

5.6 Plot Eq. (5.11) to show the relationship between η_{ad} and η_p for a gas turbine similar to Figure 4.9. Set $\eta_p = 0.6, 0.7, 0.8, 0.9$ and $p_i/p_e = 1, \ldots, 10$.

5.7 A single-stage 50% reaction gas turbine with a mean blade radius of 3 in. and annular flow area of 20 in^2. is tested with the following data: discharge temperature $T_3 = 35\,°\text{F}$, discharge pressure $p_3 = 14.7$ psia, rotating speed $N = 45{,}000$ rpm, constant axial flow velocity through the rotor $V_a = 185$ ft/s, and $V_{u2} = 1100$ ft/s, at the rotor inlet. Determine (a) stator outlet angle, (b) air mass flow rate, and (c) output horsepower.

5.8 A turbine stage is tested with air to obtain the following data: (a) mass flow rate $m = 3.5$ lb_m/s, (b) constant axial-flow velocity through the stage, $V_a = 480$ ft/s, (c) blade speed at mean radius, $U_m = 980$ ft/s, and (d) absolute flow angles at stator and rotor exits, $\alpha_2 = 65°$, $\alpha_3 = 10°$. Determine the relative flow angles at the rotor inlet and outlet, degree of reaction, stage temperature drop, total shaft power output, and utilization factor.

5.9 Based on the turbine performance curves of Figure 5.17, plot the shaft power P_s and efficiency versus the rotational speed N for the inlet pressure/temperature of 100 psia/2000 R and the exhaust pressure of 70 psia.

5.10 Repeat the calculations given in Example 5.5 for $N/\sqrt{T_{01}} = 30$.

5.11 Repeat the calculations given in Example 5.7 to locate the other running point for $N = 13{,}800$ rpm.

6

Steam Turbines

6.1 Introduction

Most of the basic theories for gas turbines, discussed in the last chapter, are also applicable to steam turbines. But the steam can only be treated as an ideal gas when it is highly superheated. The criteria can be expressed in terms of the compressibility factor $Z \equiv p/\rho RT$. The compressibility factor is a function of p_R and T_R, that is, $Z = f(p_R, T_R)$, where the reduced pressure and temperature are defined as $p_R \equiv p/p_c$ and $T_R \equiv T/T_c$. The critical pressure p_c and critical temperature T_c of steam are 3207 psia and 1165 R, respectively. When Z is close to unity, under low p_R and high T_R, steam can be treated as an ideal gas with a specific heat capacity ratio $k = 1.30$. Otherwise, the chart (Mollier diagram), steam table, or other empirical formula has to be used.

Another special characteristic of the steam turbine is that the fluid flowing through it can be a two-phase mixture of vapor and liquid droplets, especially in low-pressure turbines, if it expands into the saturated region. The state of the mixture is identified with a parameter called the moisture content $y = m_l/m_t$ (ratio of liquid mass over total mixture mass per unit volume). The effects of liquid droplets on steam turbine performance are twofold. First, the liquid droplets move slower than the vapor and may impinge on the blade surface, causing additional friction loss and damaging the blades in the long run. Second, due to the fast expansion (pressure/temperature reduction) through the turbine, some vapor condensation can be delayed, even when the mixture drops into the saturated condition. This phenomenon is called a *nonequilibrium process* or *supersaturation*. Less energy in the vapor is released to do work than under the equilibrium process.

In central electricity-generating power plants, steam turbines are predominantly employed because steam can be conveniently generated with different types of fuel. Since the water is separated from the combustion gas in the steam generator/boiler, the fuel can be fossils, nuclear, solid waste, geothermal, or others. The pollutants from combustion can be treated separately.

In process plants where steam is needed, steam can be produced economically at pressures and temperatures higher than needed. Sometimes, steam at different pressure/temperature levels is needed. In these cases, steam turbines can be used to reduce the pressure and produce mechanical power simultaneously instead of using other pressure-reducing devices. In addition to these advantages, steam turbines are more flexible for power and speed variations compared to electrical motors, especially for medium- and high-power applications. Also, steam turbines are spark-proof in hazardous environments, have higher starting torque, and are inherently self-limiting against overloading.

The disadvantage of the steam turbine compared to the gas turbine at high power and the electric motor at low power is its bulky size. Since it is generally a closed-loop machine, a condenser and its accessories are required. Also, the steam generator/boiler is heavier than the combustion chamber in a gas turbine. The pressure of superheated steam is higher than that of air at comparable temperatures and densities, as indicated from the ideal gas equation $p = \rho(R_o/\mathrm{Mw})T$, where Mw is the molecular weight.

Fundamentals of Turbomachinery: Theory and Applications, Second Edition. Ryoichi S. Amano and William W. Peng.

Companion website: www.wiley.com/go/amano/turbomachinery2e

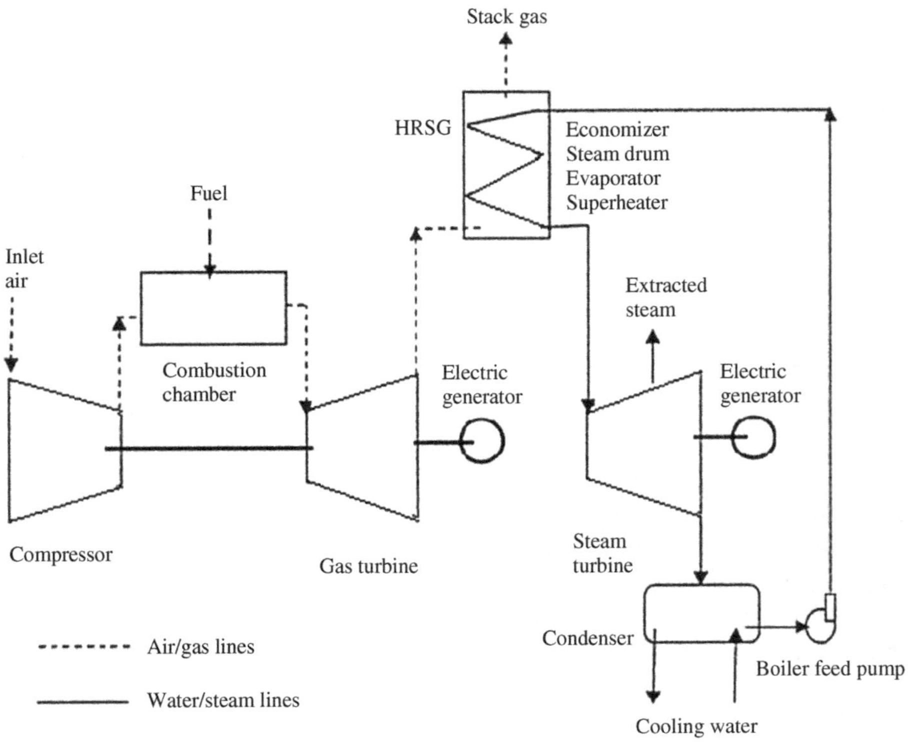

Figure 6.1 Combined cycle and cogeneration plant.

Cogeneration and combined-cycle plants have become more popular in recent years due to their higher overall plant efficiency. Some of them have been discussed in the sections related to gas turbines. A detailed schematic is shown in Figure 6.1. In this plant, natural gas or other fuel is burnt in the gas turbine. The residual heat in its exhaust gas is used to produce steam in a heat recovery steam generator (HRSG), also called a waste heat boiler (WHB). The steam is used to power a steam turbine. Both turbines drive generator sets or other machinery. Some steam is extracted for space heating, industrial processes, or gas turbine blade cooling. Some sample calculations are given in later sections and Appendix A.

6.2 Velocity-Compounded Stage

As mentioned in Section 5.8, for the same size and rotating speed, an impulse turbine stage converts more energy per stage than a reaction turbine stage. Hence, in multistage turbines, the impulse stages are placed upstream to reduce the pressure faster, thereby reducing the number of stages required. This is especially advantageous for the steam turbine because the steam boiler pressure is much higher. Facilitate the pressure drop even further, a velocity-compounded arrangement is sometimes used for the first stage of a multistage steam turbine. It can be either a Curtis stage or a re-entry stage, as shown in Figure 6.2.[1] More pressure drop occurs in the nozzle (normally a convergent–divergent nozzle), and hence, higher velocity is obtained at the rotor inlet.

In a Curtis stage, two rows of impulse rotors are used with a stator in between to change the flow direction. The velocity and h–s diagrams across both rotors are shown in Figure 6.3.

The required conversion of enthalpy to kinetic energy takes place in the nozzle, and the static enthalpies downstream of station 1 are constant, as shown. For the ideal case of flow without friction losses, the static pressure will

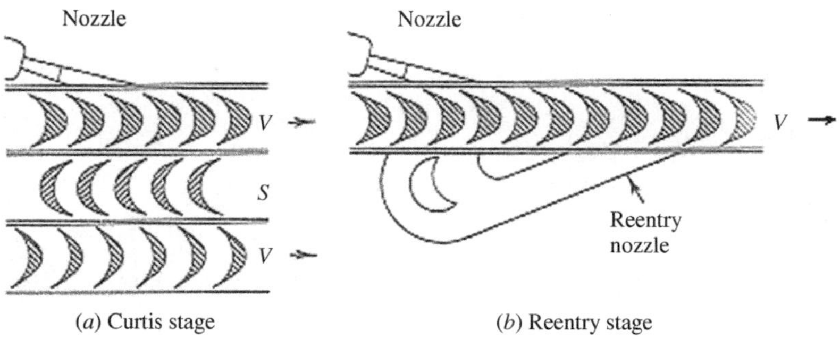

Figure 6.2 Curtis (velocity-compounded) stages for steam turbine. (Reproduced from Gartman, H. (Ed.), *De Laval Engineering Handbook*, 3rd ed., McGraw-Hill, Inc., New York, 1970.)

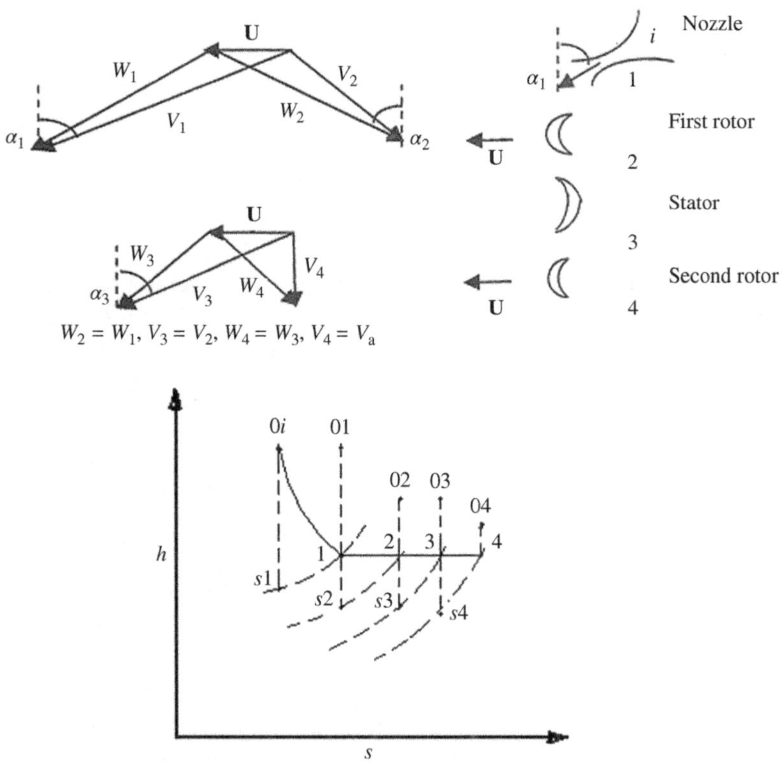

Figure 6.3 Curtis stage velocity and h–s diagrams.

also be constant. This can be shown in the h–s diagram when all points 1, 2, 3, 4 and $s2$, $s3$, $s4$ collapse into one single point $s1$.

The energy transfer of both rotors can be written as

$$\begin{aligned}\Delta E_1 &= U(V_1 \sin\alpha_1 + V_2 \sin\alpha_2) = U(V_1 \sin\alpha_1 + W_2 \sin\beta_2 - U)\\ &= U[V_1 \sin\alpha_1 + (V_1 \sin\alpha_1 - U) - U] = 2U(V_1 \sin\alpha_1 - U),\\ \Delta E_2 &= U(V_3 \sin\alpha_3) = 2U(W_3 \sin\beta_3) = 2U(V_3 \sin\alpha_3 - U) = 2U(V_2 \sin\alpha_2 - U)\\ &= 2U(W_2 \sin\beta_2 - 2U) = 2U(V_1 \sin\alpha_1 - 3U).\end{aligned}$$

Hence

$$\Delta E_c = \Delta E_1 + \Delta E_2 = 4U(V_1 \sin \alpha_1 - 2U).$$

Expressed in terms of the utilization factor, we have.

$$\varepsilon = \frac{\Delta E_c}{\Delta E_c + V_4^2/2} = \frac{\Delta E_c}{V_1^2/2} = 8\left(\frac{U}{V_1}\right)\left[\sin \alpha_1 - 2\left(\frac{U}{V_1}\right)\right],$$

where $V_1^2/2 = \Delta E_c + V_4^2/2$ is the total available energy. To find the condition of maximum ε, we set $d\varepsilon/d(U/V_1) = 0$, that is,

$$8 \sin \alpha_1 - 32\left(\frac{U}{V_1}\right) = 0 \quad \text{or} \quad \left(\frac{U}{V_1}\right)_{\text{max}} = \frac{\sin \alpha_1}{4}. \tag{6.1}$$

The corresponding maximum ε and ΔE_c are obtained as

$$\varepsilon_{\text{max}} = 2 \sin \alpha_1 \left(\sin \alpha_1 - \frac{1}{2} \sin \alpha_1\right) = \sin^2 \alpha_1, \tag{6.2}$$

$$\Delta E_{c,\text{max}} = 4U(2U) = 8U^2. \tag{6.3}$$

For the regular impulse and reaction turbines, the maximum energy transfers per stage are $\Delta E_{i,\text{max}} = 2U^2$ and $\Delta E_{r,\text{max}} = U^2$. Hence, the advantage of a Curtis stage is obvious.

Example 6.1 Referring to Figure 6.3, determine the flow velocities and energy transfer for each rotor of a Curtis stage under the condition of maximum utilization factor. The following data are given: $\alpha_1 = 70°$ with respect to the axial direction, $r_m = 8$ in., and $N = 7500$ rpm. Neglect all friction losses. What is the exit pressure if the steam is supplied to the nozzle at 3000 psia and 950°F?

Solution

The peripheral velocity is calculated as

$$U = \frac{r_m N\pi}{30} = \frac{8 \times 7500\pi}{12 \times 30} = 523.6 \text{ ft/s}.$$

From Equation (6.1), we have $U/V_1 = \sin \alpha_1/4 = 0.2349$, and hence $V_1 = 523.6/0.2349 = 2228.8$ ft/s. From the velocity diagram, at station 1, we have $V_1 \sin \alpha_1 - W_1 \sin \beta_1 = U$, and $V_1 \cos \alpha_1 = W_1 \cos \beta_1$ or $W_1 \sin \beta_1 = 2228.8 \sin 70° - 523.6 = 1570.8$ ft/s and $W_1 \cos \beta_1 = 2228.8 \cos 70° = 762.3$ ft/s. Hence $\tan \beta_1 = 2.06$, $\beta_1 = 64.1°$, and $W_1 = 1746$ ft/s. At station 2, we have $W_2 \sin \beta_2 - V_2 \sin \alpha_2 = U$ and $V_2 \cos \alpha_2 = W_2 \cos \beta_2$, with $W_2 = W_1 = 1746$ ft/s and $\beta_2 = \beta_1 = 64.1°$, or $V_2 \sin \alpha_2 = 1746 \sin 64.1° - 523.6 = 1047$ ft/s, $V_2 \cos \alpha_2 = 1746 \cos 64.1° = 762.6$ ft/s. Hence $\tan \alpha_2 = 1.373$, $\alpha_2 = 53.9°$, and $V_2 = 1295.2$ ft/s. At station 3, we have $V_3 \sin \alpha_3 - W_3 \sin \beta_3 = U$ with $V_3 = V_2 = 1295.2$ ft/s and $\alpha_3 = \alpha_2 = 53.9°$ or $W_1 \sin \beta_3 = V_1 \sin \alpha_3 - U = U = 523.6$ ft/s and $W_3 \cos \beta_3 = V_3 \cos \alpha_3 = 1295.2 \cos 53.9° = 763.1$ ft/s. Hence $\tan \beta_3 = 0.685$, $\beta_3 = 34.4°$, $W_3 = 925.1$ ft/s; also $W_4 = W_3 = 925.1$ ft/s, $\beta_4 = \beta_3 = 34.4°$, and $V_4 = V_a V_1 \cos \beta_4 = 763.3$ ft/s, $\alpha_4 = 0°$.

From these velocities, the energy transfer of the rotors can be calculated as

$$\begin{aligned}
\Delta E_1 &= U(V_1 \sin \alpha_1 + V_2 \sin \alpha_2) = 523.6(2228.8 \sin 70° + 1295.2 \sin 53.9°) \\
&= 523.6(2094.4 + 1046.5) = 1.643 \times 10^6 \text{ (ft/s)}^2 = \frac{1.643 \times 10^6}{32.2 \times 778} = 65.6 \text{ Btu/lb}, \\
\Delta E_2 &= U(V_3 \sin \alpha_3) = 523.6(1295.2 \sin 53.9°) = 0.547 \times 10^6 \text{ (ft/s)}^2 \\
&= \frac{0.546 \times 10^6}{32.2 \times 778} = 21.8 \text{ Btu/lb}.
\end{aligned}$$

Hence, the total energy transfer is $\Delta E_c = 65.6 + 21.8 = 87.4\,\text{Btu/lb}$. To compare with that calculated with Equation (6.3), we have

$$\Delta E_c = 8U^2 = 8 \times (523.6)^2 = 2.19 \times 10^6\ (\text{ft/s})^2 = \frac{2.19 \times 10^6}{32.2 \times 778} = 87.5\ \text{Btu/lb}.$$

The difference is due to round-off error. The static enthalpies and pressures at stations 1, 2, 3, and 4 are the same for the ideal case and can be calculated from $h_1 = h_{01} - V_1^2/2$, where $h_{01} = h_{oi} = 1405\,\text{Btu/lb}$ from the Mollier diagram for $p_{oi} = 3000\,\text{psia}$, $T_{oi} = 950°\text{F}$, and $V_1^2/2 = 2228.8^2/(2 \times 32.2 \times 778) = 99\ \text{Btu/lb}$. Hence, we have $h_1 = 1306\,\text{Btu/lb}$ and $p_1 \cong 1400\,\text{psia}$.

6.3 Classification of Steam Turbines

Steam turbines can be classified according to several different schemes:

a) Based on steam flow direction, a steam turbine can be an axial-flow, radial-inflow, radial-outflow (also called Ljungstrom turbine), or tangential-flow machine.
   1) The construction of an axial-flow steam turbine is similar to that of a gas turbine. The flow through the stator and rotor is substantially in the axial direction. This type of design is predominantly adapted in medium- and high-power units. Any reasonable number of stages can be used for the expansion of high-pressure/temperature steam in a compact arrangement.
   2) The radial-inflow turbine, as discussed in Chapter 5 for the gas turbine, is primarily designed for lower power units. As indicated in the Euler equation for turbines, radial inflow is a logical arrangement to maximize the energy transfer for radial-flow turbines.
   3) The radial-outflow turbine, also called a Ljungstrom turbine, is shown in Figure 6.4. The steam enters the blades near the axis radially. Two rotors, with blades arranged alternatively, rotate in opposite directions. There is no stator between the rotor blades. The velocity diagrams shown are for the first two inner rings of blades. It is noted that the peripheral velocity U changes direction and the absolute velocity $V_3 = V_2$. This type of turbine has high efficiency but is complicated and expensive to construct. Hence, its development has not been widely received in the United States.
   4) The tangential-flow turbine, as shown in Figure 6.5 is like the Pelton wheel of the hydraulic turbine, to be discussed in Chapter 7. The steam exiting nozzles that are arranged almost tangent to the periphery of

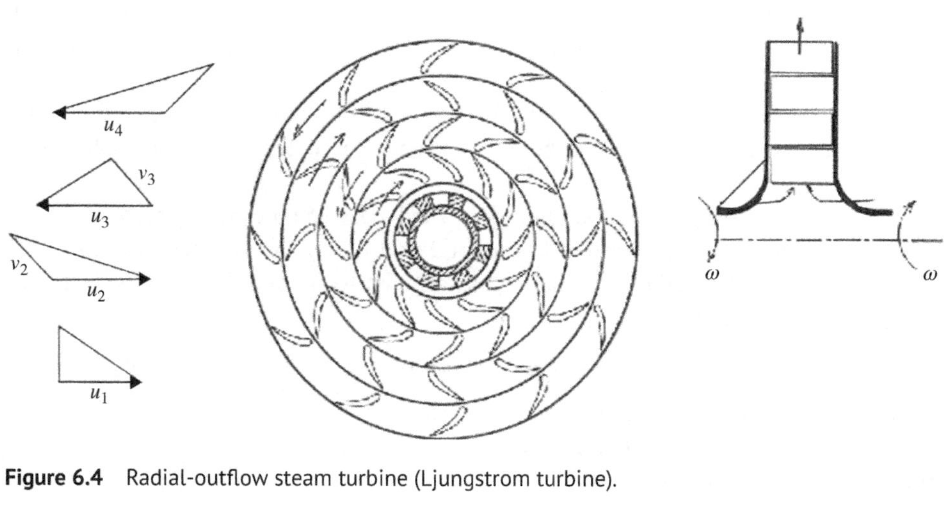

Figure 6.4 Radial-outflow steam turbine (Ljungstrom turbine).

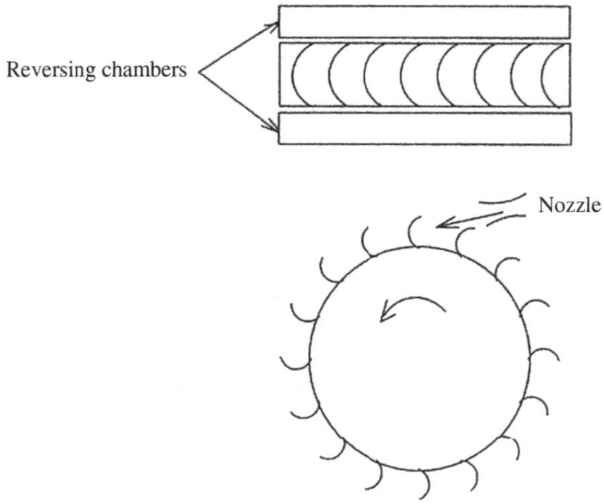

Figure 6.5 Tangential-flow steam turbine.

the rotor impinges on a semicircular rotor blade (or bucket). It then discharges to a reversing chamber and reenters the rotor blades.

b) Based on the rotor blade type, axial-flow steam turbines can be classified as (1) impulse, (2) reaction (also called Parsons turbine), or (3) combined impulse and reaction in the multistage turbine. Most of the principles for the gas turbine discussed in Chapter 5 are also applicable to the steam turbine.
c) Based on the number of stages that the steam pressure drop is divided into from inlet to exhaust, steam turbines can be classified as single-stage or multistage machines. In a single-stage turbine, one row of nozzles is used to convert all pressure drops into kinetic energy. Then, one or two rows of bladed rotors are used to convert the flow kinetic energy into mechanical energy. In a multistage turbine, the pressure drop is divided into two or more stages. In each stage, one row of nozzles is followed by sets of rotors. Schematic diagrams of these two types of steam turbines are shown in Figure 6.6.
   In a single-stage turbine or the first few stages of a multistage machine, typically, impulse-type blades are employed since they can convert more energy than the reaction type for the same diameter and rotating speed, as discussed earlier. Or a Curtis stage with a few rows of rotors can be used.
d) Based on the exhaust condition, steam turbines can be classified as (1) condensing and (2) noncondensing or topping. In the noncondensing turbine, the exhaust steam is used for processing or heating, and hence it is an open-cycle machine. The topping turbine exhausts steam to another steam turbine.
e) Based on the arrangement of the first-stage nozzles, steam turbines can be classified as (1) full admission and (2) partial admission. Turbines with full admission have nozzles arranged around the complete circle upstream of the first-stage rotor.
f) Steam turbines can also be classified according to (1) orientation of their shaft, either horizontal or vertical, and (2) connection with the load, either directly connected or geared. The choice will depend on the application, installation, and space available.
g) For high-power systems, more than one casing or one shaft may be needed. Based on the arrangement of casing and shaft, steam turbines can be classified as (1) tandem-compound unit or (2) cross-compound unit. The tandem-compound unit has all turbines driving the same shaft, while the cross-compound unit has the turbines driving the separate shafts, as shown in Figure 6.7.

In the large plants, steam turbines may be divided into three sections according to their pressure ranges:

1) high-pressure (HP) turbine with pressure ranging from 400 to 4500 psia,
2) intermediate-pressure (IP) turbine with pressure ranging from 300 to 1300 psia, and
3) low-pressure (LP) turbine with pressure below 300 psia.

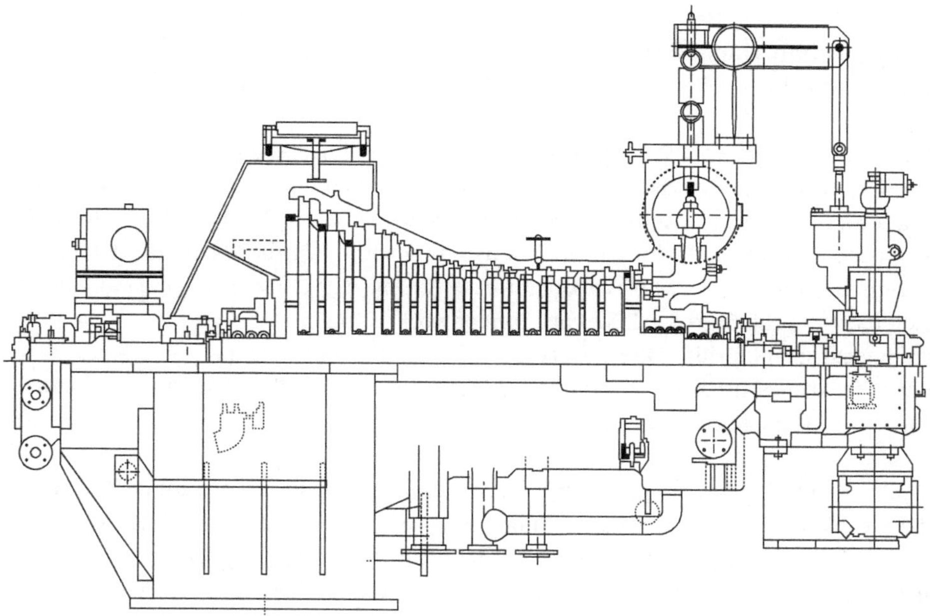

Figure 6.6 Typical axial-flow multistage steam turbine installation. (Reproduced from Gartman, H. (Ed.), *De Laval Engineering Handbook*, 3rd ed., McGraw-Hill, Inc., New York, 1970.)

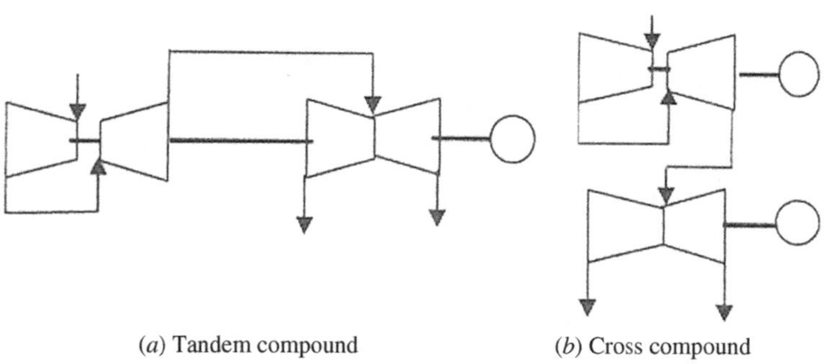

Figure 6.7 Arrangement of steam turbine casing and shaft.

6.4 Steam Turbine Parameters and Performance

Like other types of turbomachines, the overall efficiency of a steam turbine can be attributed to several mechanisms of energy losses in converting input power to output power. The overall efficiency (also called engine efficiency or brake engine efficiency) is defined as $\eta_{\mathrm{o/a}} \equiv P_s/(m_t \Delta H_s)$, where P_s is the output shaft power, m_t is the steam mass flow rate, and ΔH_s is the ideal total enthalpy drop per unit mass of steam from inlet to exhaust condition.

The energy losses can be divided into

1) losses due to the friction and secondary flow through the flow passages.
2) losses due to the flow leakage over the rotors; and
3) losses due to the friction of the rotor disk, bearings, seals, and transmission gear and the power to drive accessory devices, if there is any.

The efficiencies associated with these losses are defined as adiabatic efficiency η_a, volumetric efficiency η_v, and mechanical efficiency η_m. The overall efficiency can, therefore, be related as $\eta_{o/a} \equiv \eta_a \eta_v \eta_m$. Sometimes, the internal efficiency, defined as $\eta_i \equiv \eta_a \eta_v$, is used. A detailed discussion on this topic was given in Section 3.1.1 for centrifugal pumps.

Sometimes, a parameter called *steam rate* SR is used instead of efficiency. It is defined as the steam mass flow rate required to produce one unit of shaft power, expressed in units of lb_m/hp-h or kg/kW-h. It is inversely proportional to the overall efficiency for a given set of steam inlet/outlet conditions, since SR $= m_t/P_s \equiv 1/(\Delta H_s \eta_{o/a})$ from the definition. This is true only for the basic turbine without extraction and reheating. Or the "*heat rate*" HR can be defined as the heat supply required to produce a unit of work in terms of Btu/hp-h or Btu/kW-h. It is a more useful measure of performance, as it indicates the cost of fuel for each unit of work output.

The average efficiencies of typical steam turbines are correlated with a parameter called quality factor QF. The situation is similar to the specific speeds used in other turbomachines discussed earlier. The quality factor is defined as QF $\equiv \sum U^2/\Delta H_s$, where U is the rotor tip velocity of the stages. It can be related to the other parameters as follows.

Across each stage, we have $\Delta h_s \sim V_2^2/(2g_c) = U^2/(2g_c\lambda^2)$, or $U^2/\Delta h_s \sim 2g_c\lambda^2$, where V_2 is the absolute flow velocity at the nozzle outlet and $\lambda \equiv U/V_2 = \sin\alpha_2/2$ for impulse blading or $\lambda = \sin\alpha_2$ for reaction blading. Assuming identical stages, the above expression can be summed over all stages in the form $\sum U^2 = 2g_c\lambda^2 \sum \Delta h_s = 2g_c\lambda^2 R_H \Delta H_s$, where R_H is the reheat factor for steam mentioned in Section 5.4. Hence, the quality factor can be related as QF $\equiv \sum U^2/\Delta H_s = 2g_c\lambda^2 R_H$.

A typical chart relating the average engine efficiency and quality factor for multistage steam turbines is given in Figure 6.8. Here the conversion factor is $g_c = 778 \times 32.2 = 25{,}052$ ft-lb_f-lb_m/(Btu-slug).

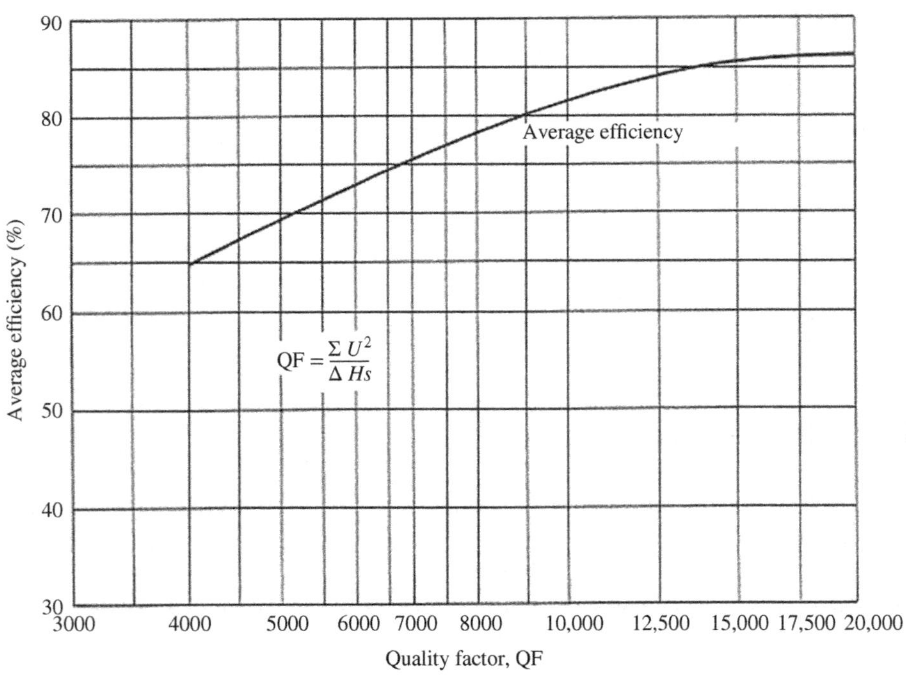

Figure 6.8 Typical correlation of average efficiency with quality factor for multistage steam turbines. (Reproduced from Gartman, H. (Ed.), *De Laval Engineering Handbook*, 3rd ed., McGraw-Hill, Inc., New York, 1970.)

Example 6.2 A 10-stage steam turbine running at 3600 rpm is proposed for the steam available at 300 psia and 1000°F and to be discharged to a condenser at 10 psia. Estimate the efficiency and size of the proposed turbine.

Solution

To use Figure 6.8, with $\text{QF} \equiv \sum U^2/\Delta H_s = 2g_c\lambda^2 R_H$, the value of R_H can be estimated with Equation (5.13). Using $k = 1.3$ for steam and assuming $\eta_p \approx 0.90$, we have $\eta_{\text{ad}} = [1 - (p_e/p_i)^{\eta_p(k-1)/k}]/[1 - (p_e/p_i)^{(k-1)/k}] = 0.931$, and hence $R_H = \eta_{\text{ad}}/\eta_p = 0.931/0.90 = 1.035$.

For impulse stages, the optimal efficiencies occur at $\lambda = U/V_2 = \sin\alpha_2/2 \approx 0.47$ with $\alpha_2 \approx 70°$. So QF can be calculated as $\text{QF} \approx 2 \times 25{,}052 \times 0.47^2 \times 1.035 = 11{,}455$. From Figure 6.8, efficiency can be estimated to be $\eta \approx 83\%$.

From the Mollier diagram in Figure A.1, we have $h_i = 1525$ Btu/lb$_\text{m}$, $h_{se} = 1150$ Btu/lb$_\text{m}$, with $s_i = s_{es} = 1.8$ Btu/lb$_\text{m}$-R, and hence $\Delta H_s = 1525 - 1150 = 375$ Btu/lb$_\text{m}$. So we have $\sum U^2 = 11{,}455 \times 375 = 4{,}295{,}625$ (ft/s)^2. With 10 identical stages, we have $U^2 = 429{,}562$ or $U = 655$ ft/s, and the turbine diameter $D = 2U/\omega = 2 \times 655/(3600\pi/30) = 3.47$ ft.

6.4.1 Part-Load Operation

Most turbines are designed at the high-power condition (but not the maximum power) and are required to operate at a constant speed for some variation of output power. For gas turbines, the output power can be changed by injecting different amounts of fuel, hence changing the air–fuel ratio and turbine inlet temperature. For steam turbines, the boiler condition and, hence, the steam pressure/temperature at the turbine inlet are usually kept constant. The variation of output power must be affected by either (1) reducing the steam pressure with throttling or (2) partial admission, that is, admitting steam into only some of the nozzles upstream of the first stage of the turbine. If the turbines must be operated with overload for a short period of time, the steam can bypass the first or second stage with full pressure. This will cause the mass flow rate, and hence output power increase since the flow passage area normally increases from upstream to downstream in order to accommodate the increasing specific volume of steam. Of course, all these operations under off-design conditions will result in a slight decrease in plant efficiency.

Throttling is a process that causes pressure reduction with flow friction, while enthalpy is constant and entropy increases, so the available energy in steam drops. It is simple but inefficient, so it is primarily used in small turbines, where cost is more important than efficiency.

Partial admission is more complicated but causes less efficiency. The steam supply following the main valve is split into two or more streams, each with its own control valve. Under part-load operation, the steam is admitted to only some nozzles, with the design pressure maintained. No extra energy loss is incurred, except that due to the windage and turbulence in the rotors as they pass "idle nozzles." Partial admission is commonly used in large turbines.

In both methods, however, the parameter that changes significantly with output power is the steam mass flow rate. The efficiency and enthalpy changes are relatively minor. Hence, under the off-design operation, shaft power and steam flow rate are related almost linearly. This relationship, shown in Figure 6.9, is called the Willans line. With this simple relationship, the complete off-design performance of a steam turbine can be obtained with tests at two conditions. Then, a straight line is plotted through them. The intercept on the abscissa (point 1) indicates the no-load loss, P_l, while that on the ordinate (point 2) represents the flow required for no load, m_o. The steam rate $\text{SR} = \dot{m}_t/P_s$ is also plotted. They can be expressed mathematically as

$$\dot{m} = \dot{m}_o\left(1 + \frac{P_s}{P_l}\right), \quad \text{SR} \equiv \frac{\dot{m}_t}{P_s} = \dot{m}_o\left(\frac{1}{P_s} + \frac{1}{P_l}\right)$$

The heat rate ($\text{HR} \equiv m_t Q_h/P_s$) curve is similar to the steam rate curve since the inlet steam condition is kept constant, and the heat required to produce one unit mass of steam Q_h is constant. The thermal efficiency is reciprocal

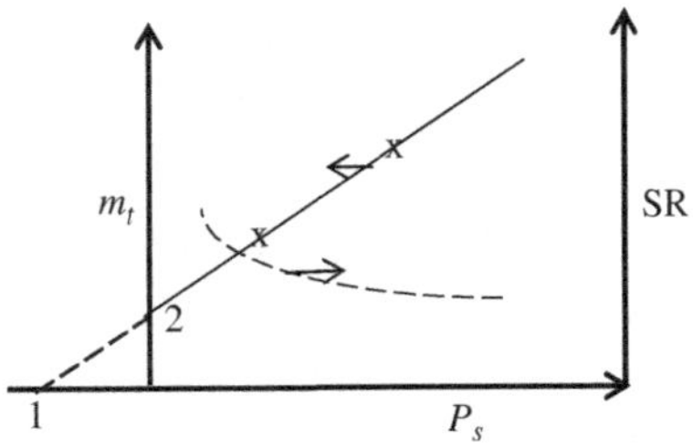

Figure 6.9 Willans line.

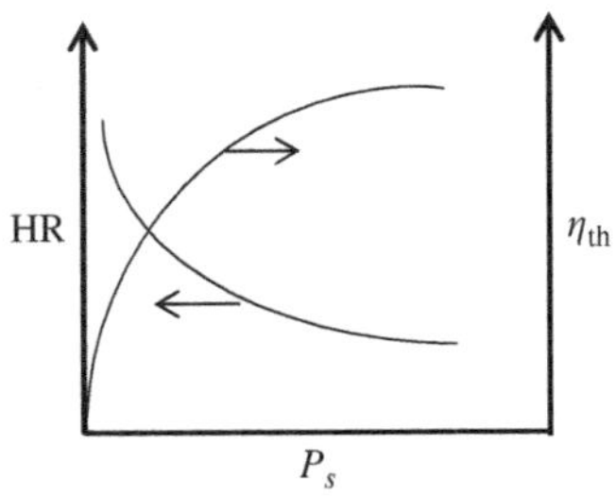

Figure 6.10 Qualitative heat rate and thermal efficiency versus shaft power for part load operation (with constant rotating speed and steam inlet condition).

to the heat rate with a conversion factor since we have $\eta_{th} = P_s/(\dot{m}_t Q_h)$. These two curves are shown in Figure 6.10 qualitatively.

Typical performance curves of axial-flow multistage steam turbines of both condensing and noncondensing types are shown in Figure 6.11. The effects of steam inlet pressure and degrees of superheat and vacuum at the condenser on turbine performance are also shown.

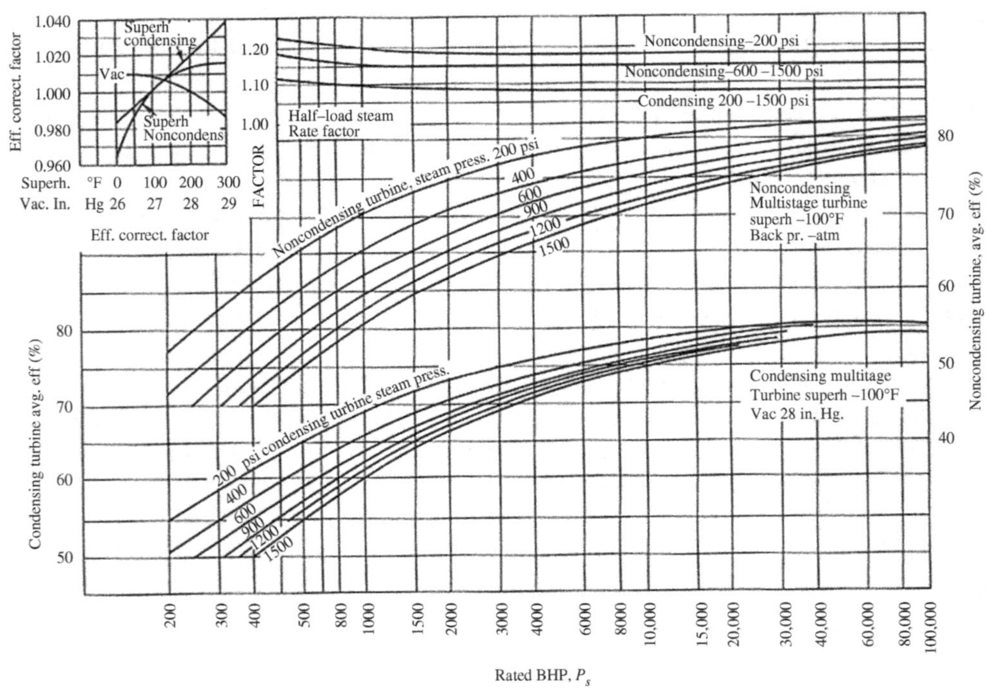

Figure 6.11 Typical axial-flow multistage steam turbine performance curves. (Reproduced from Gartman, H. (Ed.), *De Laval Engineering Handbook*, 3rd ed., McGraw-Hill, Inc., New York, 1970.)

With a fixed exhaust pressure, the steam mass flow rate is approximately proportional to the turbine inlet pressure. For choked flow, as is likely in an impulse turbine stage, the mass flow rate can be calculated as

$$\dot{m}_t = \rho_1 V_1 A_1 = \rho_1 a_1 A_1 \sim \left(\frac{p_1}{RT_1}\right)\sqrt{KRT_1} \sim \frac{p_1}{\sqrt{T_1}}$$

if the superheated steam can be treated as an ideal gas.

For non-choking flow, the manipulation given below also shows that linear relationship. Across each stage, we have $\Delta h \sim \left(V_1^2 - V_2^2\right) \sim V_1^2(1 - C_1)$, where C_1 is a constant, or $V_1^2 \sim \Delta h \sim \Delta p/\rho$. And from $\dot{m}_t = \rho VA$ across each stage,

$$\dot{m}_t^2 = (\rho VA)^2 = \frac{(\rho_1 V_1)^2}{(1/A_1)^2} = \frac{(\rho_2 V_2)^2}{(1/A_2)^2} = \cdots = \frac{(\rho_n V_n)^2}{(1/A_n)^2},$$

we have

$$\dot{m}_t^2 = \frac{\sum(\rho_i V_i)^2}{\sum(1/A_i)^2} = \frac{\sum(\rho_i \Delta p_i)}{\sum(1/A_i)^2}.$$

Also from $\Delta(\rho p) = \rho\Delta p + p\Delta\rho$ and $p = C\rho^k$, or $\rho\Delta p = kp\Delta\rho$, we have $\Delta(\rho p) = \rho\Delta p + \rho\Delta p/k = [(k+1)/k]\rho\Delta p$. Or substituting it back, we have

$$\begin{aligned}\dot{m}_t^2 &= \frac{[k/(k+1)]\sum[\Delta(\rho_i p_i)]}{\sum(1/A_i)^2}\\ &= \frac{[k/(k+1)][(\rho_1 p_1 - \rho_2 p_2) + (\rho_2 p_2 - \rho_3 p_3) + \cdots - \rho_n p_n]}{\sum(1/A_i)^2}\\ &\cong \frac{[k/(k+1)](\rho_1 p_1)}{\sum(1/A_i)^2},\end{aligned}$$

or $\dot{m}_t^2 \sim \rho_1 p_1 = p_1^2/(RT_1)$. So, we have $\dot{m}_t \sim p_1$.

Example 6.3 A 12-stage steam turbine with impulse blades and dimensions as given is rotating at 3600 rpm. If the inlet steam condition is 1000 psia and 1100°F, determine the exhaust steam condition and total shaft power under design condition with a steam flow rate of 7 lb_m/s (assuming the efficiencies as $\eta_{ad} = 0.85$, $\eta_v = 0.92$, and $\eta_m = 0.90$; neglect all flow deviations).

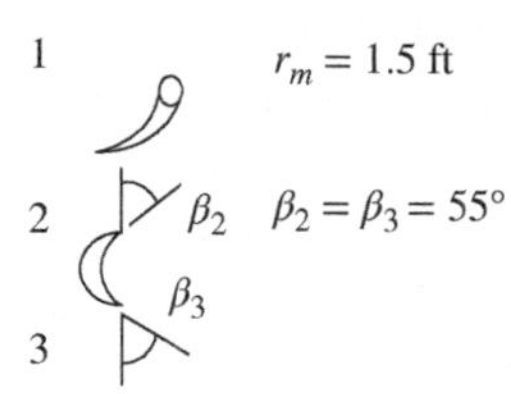

Solution

The tangential velocity at the rotor mean radius is $U_m = r_m N\pi/30 = 1.5 \times 3600\pi/30 = 565.5$ ft/s. From the velocity diagram in Figure 5.19 for the impulse stage, we have

$$\begin{aligned}\Delta h_o = \Delta E_t &= \frac{U_m V_{u2} - U_m V_{u3}}{g_c} = \frac{U_m V_{u2}}{g_c} = \frac{2U_m^2}{g_c} = \frac{2 \times 565.5^2}{32.2}\\ &= 19{,}863\ \text{lb}_\text{f}\text{-ft/lb}_\text{m} = 25.5\ \text{Btu/lb}_\text{m}.\end{aligned}$$

From the Mollier diagram in Appendix A, we have $h_{oi} = 1565\ \text{Btu/lb}_\text{m}$. For the stages with constant mean radii, we have $h_{oi} - h_{oe} = n_s\Delta h_o$ or $h_{oe} = h_{oi} - n_s\Delta h_o = 1565 - 12 \times 25.5 = 1259\ \text{Btu/lb}_\text{m}$. Also, from

$\eta_{ad} = (h_{oi} - h_{oe})/(h_{oi} - h_{soe})$, we have $h_{soe} = h_{oi} - (h_{oi} - h_{oe})/\eta_{ad} = 1565 - 306/0.85 = 1205\,\text{Btu/lb}_\text{m}$. From $s_{soe} = s_{oi} = 1.69\,\text{Btu/(lb}_\text{m}\text{-R)}$, we have $p_{oe} = 50\,\text{psia}$ and $T_{oe} = 450°\text{F}$. Also from the given efficiencies, we have $P_s = \eta_v \eta_m m n_s \Delta h_o = 0.92 \times 0.90 \times 7 \times 12 \times 19{,}863/550 = 2512\,\text{hp}$.

Example 6.4 In the last example, if the steam mass flow rate is estimated to be 0.5 lb_m/s for the no-load condition, determine the steam rate and heat rate for the full-load and 50% load conditions (if it requires $Q_h = 1750$ Btu of heating to generate 1 lb_m of steam).

Solution

Based on Willan's law, we have the steam mass flow rate $\dot{m}_t = 0.5 + CP_s$, where $C = (7 - 0.5)/2512 = 2.6 \times 10^{-3}\,\text{lb}_\text{m}/\text{(hp-s)}$. So we have $\text{SR} = \dot{m}_t/P_s = 0.5/P_s + 2.6 \times 10^{-3}$ and $\text{HR} = Q_h \times \text{SR}$ or at full load,

$$\begin{aligned} \text{SR}_1 &= \frac{0.5}{2512} + 2.6 \times 10^{-3} = 2.8 \times 10^{-3}\ \text{lb}_\text{m}/\text{(hp-s)} \\ &= 10.1\ \text{lb}_\text{m}/\text{(hp-h)}, \\ \text{HR}_1 &= 1750 \times 10.1 = 17{,}675\ \text{Btu/(hp-h)}; \end{aligned}$$

at 50% load

$$\begin{aligned} \text{SR}_2 &= \frac{0.5}{1256} + 2.6 \times 10^{-3} = 3.0 \times 10^{-3}\ \text{lb}_\text{m}/\text{(hp-s)} \\ &= 10.8\ \text{lb}_\text{m}/\text{(hp-h)}, \\ \text{HR}_2 &= 1750 \times 10.8 = 18{,}900\ \text{Btu/(hp-h)}. \end{aligned}$$

6.5 Cogeneration and Combined-Cycle Plants

Cogeneration is a system by which fuel energy is used to generate shaft power and steam thermal energy simultaneously. So, the overall efficiency of the plant can be improved.

It can be either a topping-cycle or a bottom-cycle system. In the topping cycle, fuel is burnt to produce high-pressure/temperature gas or steam to drive a gas/steam or combined-cycle turbine. Then, the exhaust gas from the gas turbine is used to produce steam through a HRSG (also called a waste heat recovery boiler in the literature). Or the steam is extracted from the intermediate stages of the steam turbine or from its final exhaust.

In the bottom-cycle system, the fuel is burnt to produce steam for industrial plant processes. Then, the waste heat from the exhaust steam is used to drive a steam turbine for shaft power production. They are shown schematically in Figure 6.12.

The high-temperature exhaust gas from a shaft power gas turbine can be used to generate steam to drive a steam turbine. This is a combined-cycle power plant. It can generally be designed so that the appropriate equipment can be selected from existing models (including gas turbine, steam turbine, and HRSG). Additional fuel can be burnt in HRSG to make up the discrepancy between the gas turbine exhaust condition and the design inlet condition for the steam turbine.

The definition of thermal efficiency (also called utilization factor in the literature) for a cogeneration system is not straightforward since the mechanical energy of the shaft is worth more than the steam heating energy from a thermodynamics point of view. Hence, two definitions are devised, namely,

$$\eta_{th} = \frac{P_s + Q_{th}}{Q_{in}} \quad \text{or} \quad \eta_{th} = \frac{A_s + A_{th}}{A_{fuel}},$$

where P_s is the mechanical shaft power, Q_{in} is the heating rate of fuel, A_{th} is the availability of generated steam, Q_{th} is the heating rate of steam produced, A_s is the equivalent availability of shaft power ($\approx P_s$), and A_{fuel} is the availability of the fuel required.

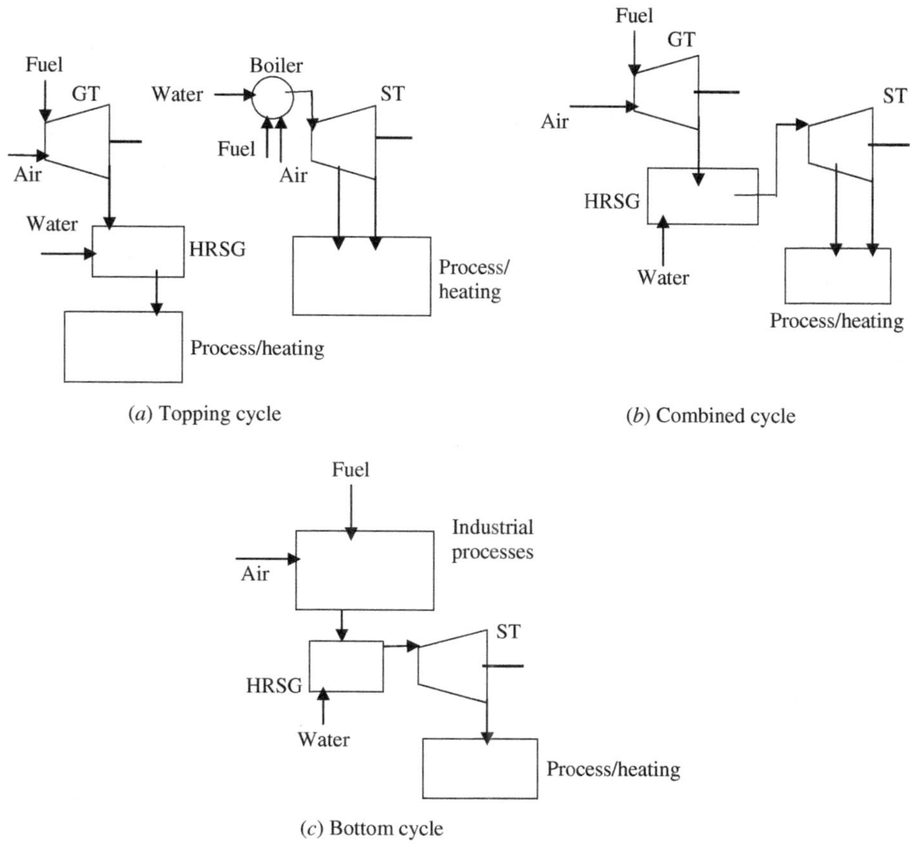

Figure 6.12 Cogeneration systems.

Detailed discussions of the concept of availability and thermodynamic cycles of cogeneration or combined-cycle plants can be found in typical thermodynamics texts, such as Ref. 2.

Example 6.5 If the exhaust steam from the steam turbine in Example 6.4 is used for processing/heating, what is the thermal efficiency of the cogeneration system under full load Compare it with the simple shaft power system.

Solution

To calculate thermal efficiency, the units have to be consistent. With 1 hp = 0.707 Btu/s = 2545 Btu/h, we have

$$\eta_{\text{th}} = \frac{P_s + Q_e}{Q_{\text{in}}} = \frac{2512 \times 0.707 + 1259 \times 7}{7 \times 1750} = 0.864.$$

For the simple shaft power system, we have $\eta_{\text{th}} = 2512 \times 0.707/(7 \times 1750) = 0.145$, or from the heat rate, $\eta_{\text{th}} = 2545/\text{HR} = 2545/17{,}675 = 0.144$.

Example 6.6 The combined-cycle plant shown has the following parameters given:

a) Ambient atmospheric condition, $p_5 = 14.7\,\text{psia}$, $T_5 = 80°\text{F}$
b) Compression ratio, $p_6/p_5 = 15$
c) Gas turbine inlet temperatures $T_7 = 2100°\text{F}$, exhaust temperature $T_9 = 450°\text{F}$
d) Adiabatic efficiencies for compressor and gas turbine: $\eta_c = 82\%$, $\eta_{\text{gt}} = 90\%$

e) Steam maximum pressure and temperature: $p_3 = 1000$ psia, $T_3 = 900°$F
f) Condenser pressure $p_1 = p_4 = 1.5$ psia
g) Adiabatic efficiencies for pump and steam turbine: $\eta_p = 80\%$, $\eta_{st} = 85\%$

Determine (1) the ratio of the mass flow rates for steam and combustion gas and (2) the thermal efficiencies of the gas turbine cycle, steam turbine cycle, and combined-cycle plant (neglect all friction and heat losses in the connecting pipes).

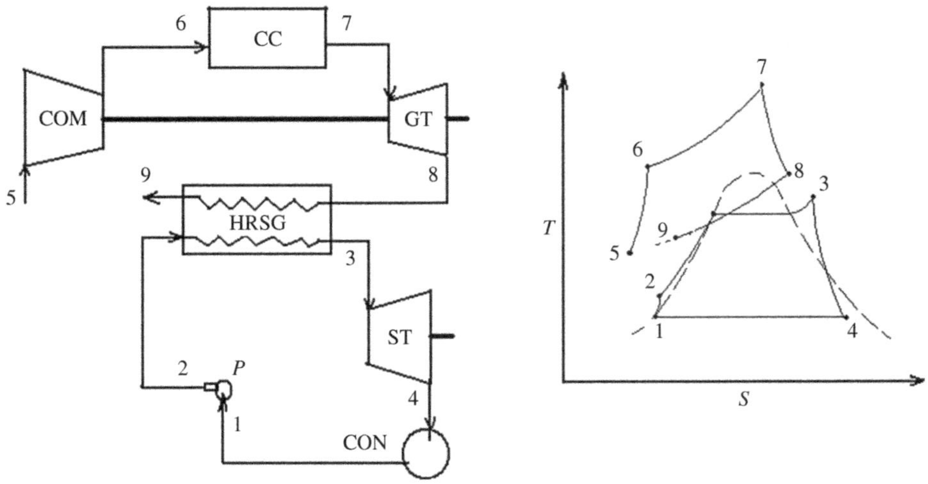

Solution
The enthalpies at various points must be determined first. For the steam turbine cycle, from the Mollier diagram or steam tables, we have

$$h_1 = h_{f1} = 83.6\ \text{Btu/lb}_\text{m},$$

$$h_2 \approx h_1 + \frac{v\Delta p}{\eta_p} = 83.6 + \frac{0.0185(1000 - 1.5) \times 144}{778 \times 0.80} = 83.6 + 4.27 = 87.9\ \text{Btu/lb}_\text{m},$$

$$h_3 = 1447\ \text{Btu/lb}_\text{m},$$

$$s_{s4} = s_3 = 1.61\ \text{Btu/(lb}_\text{m}\text{-R)}, \quad h_{s4} = 925\ \text{Btu/lb}_\text{m}.$$

Hence from $\eta_{st} = (h_3 - h_4)/(h_3 - h_{s4})$, we have

$$h_4 = h_3 - (h_3 - h_{s4})\eta_{st} = 1447 - (1447 - 925) \times 0.85 = 1003\ \text{Btu/lb}_\text{m}.$$

The thermal efficiency of the steam turbine cycle is then obtained as

$$\eta_{th,s} = \frac{W_{st} - W_p}{Q_{in,s}} = \frac{(h_3 - h_4) - (h_2 - h_1)}{h_3 - h_2}$$
$$= \frac{(1447 - 1003) - (87.9 - 83.6)}{1447 - 87.9}$$
$$= \frac{439.7}{1359.1} = 0.323 = 32.3\%.$$

For the gas turbine cycle, an ideal gas with constant C_p is assumed for the working gas. With $C_p = 0.24$ Btu/(lb$_\text{m}$-R), and $k = 1.4$, we have

$$T_6 = T_5 + \frac{T_5\left[(p_6/p_5)^{(k-1)/k} - 1\right]}{\eta_c} = 540 + \frac{540(15^{0.2857} - 1)}{0.82}$$
$$= 540 + 769 = 1309\ \text{R} = 849°\text{F},$$

$$T_8 = T_7 - T_7\left[1 - \left(\frac{1}{15}\right)^{0.2857}\right]\eta_{gt}$$
$$= 2560 - 2560(1 - 0.461) \times 0.85 = 1388\ \text{R} = 928°\text{F}$$

(which should be greater than T_3). The thermal efficiency of the gas turbine cycle is obtained as

$$\eta_{th,g} = \frac{W_{gt} - W_c}{Q_{in,g}} = \frac{C_p(T_7 - T_8) - C_p(T_6 - T_5)}{C_p(T_7 - T_6)}$$
$$= \frac{(2560 - 1388) - (1309 - 540)}{2560 - 1309} = 0.322 = 32.2\%.$$

From the energy balance equation across the HRSG, we have $\dot{m}_g C_p(T_8 - T_9) = \dot{m}_s(h_3 - h_2)$, or

$$\frac{\dot{m}_s}{\dot{m}_g} = \frac{C_p(T_8 - T_9)}{h_3 - h_2} = \frac{0.24(928 - 450)}{1447 - 87.9} = \frac{114.7}{1359} = 0.084.$$

Hence, the thermal efficiency of the combined cycle is obtained as

$$\eta_{th,c} = \frac{(W_{gt} - W_c) + (\dot{m}_s/\dot{m}_g)(W_{st} - W_p)}{C_p(T_7 - T_6)}$$
$$= \frac{0.24(1172 - 769) + 0.084 \times 439.7}{0.24 \times 1251}$$
$$= \frac{96.7 + 36.9}{300.2} = 0.445 = 44.5\%.$$

References

1 Gartmann, H. (Ed.), *De Laval Engineering Handbook*, 3rd ed., McGraw-Hill, New York, 1970.
2 Moran, M. J., and Shapiro, H. N., *Fundamentals of Engineering Thermodynamics* 3rd ed., John Wiley & Sons, New York, 1995.

Problems

6.1 A 50% reaction turbine with blades as shown is used for a 10-stage steam turbine. The exit velocity of steam from the stator blades is 480 ft/s, the blade speed is 455 ft/s, and the axial flow velocity is constant. Determine the work done by steam per stage with mass flow rate of 8 lb_m/s. If the total-to-total efficiency is 80%, $p_{01} = 1600$ psia, and $T_{01} = 1000°$F at the first-stage stator inlet, determine the steam total temperature and pressure at the outlet of the last stage.

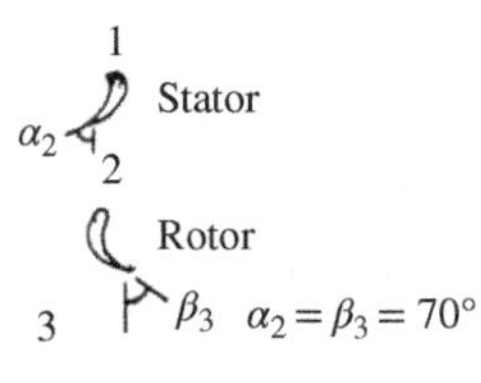

6.2 A multistage high-pressure steam turbine is supplied with steam at a stagnation pressure of 7 MPa absolute and a stagnation temperature of 500°C. The corresponding specific enthalpy is 3410 kJ/kg. The steam exhausts from the turbine at a stagnation pressure of 0.7 MPa absolute. The superheated steam is assumed to behave like an ideal gas with $k = 1.3$. If the small stage efficiency is 82%, determine (a) the temperature and density of the steam at the exit of the turbine and (b) the reheat factor. The ideal gas equation for this steam is represented by $p/\rho = 0.23(h - 1943)$, where p is in kPa, ρ is in kg/m^3, and h is in kJ/kg.

6.3 A cogenerating steam power plant operates with a boiler steam at 100 psia, 900°F, with a steam rate of 50 lb_m/s. The condenser operates at 4 psia. The steam is extracted from the turbine at 60 psia with 20 lb_m/s for process heating and returned to the feedwater as saturated liquid at 12 psia. If the turbine efficiency is 85%, determine the total shaft power and the process heat rate available. If the efficiencies of both condensate and boiler feed pumps are 78%, determine the total plant efficiency.

6.4 Convert the performance curve in Figure 6.11 to a Willian line (steam rate versus shaft power) for the condensing multistage turbine. Select the line for inlet pressure of 400 psi, 100°F superheat, and discharge condition of 28 in. Hg vacuum.

6.5 A steam turbine is needed to produce 3000 hp. Use the data in Figure 6.11 to calculate the steam rates for the boiler condition of 600 psi (consider both condensing and noncondensing types with 100°F superheat and 200°F superheat).

7

Hydraulic Turbines

7.1 Introduction

7.1.1 Introduction to Hydropower

Hydropower is a significant energy source as it accounts for 16.4% of the total electricity generation all over the world.[1] It is considered as a leading renewable energy resource, supplying 71% of the global renewable-energy electricity generation.[2, 3] In 2017, hydropower contributed to 7.5% of the total U.S. electricity generation accounting for 44% of electricity generated by renewable energy.[4, 5] The United States is ranked fourth in the annual hydroelectric power production after China, Canada, and Brazil, as indicated in Figure 7.1.[6]

There are 2198 active hydropower plants in the United States with a total capacity of 79.64 GW in which half of this capacity is located in Washington, California, and Oregon.[7] The majority of the installed hydropower plants in terms of capacities are located in the Northwest, whereas the Northeast has the highest number of facilities. The largest hydroelectric plant in the United States is located at the Grand Coulee Dam, WA, with a capacity of 6809 MW.

Hydro turbines are classified into three main categories depending on their application: Kaplan, Francis, and Pelton. Kaplan turbines are applicable for low heads and high flow rates scenarios, whereas Francis turbines for medium heads and flow rates applications, and Pelton turbines are used in low flow rates and high heads cases. Francis turbines are the most common type and used in the hydropower plants since they possess a wide range of flow rate and head combinations. From 1996 to 2011, 9455 MW identified turbine capacity were installed. 74% of the total installed capacity was Francis turbines, Kaplan with 23%, and 83 MW for Pelton turbines.[8] Hydraulic turbines are used to convert hydropower stored in water at high elevation or in flowing rivers into mechanical power to drive an electrical generator or other machinery. Hydropower is a type of renewable energy derived from solar energy. Water is evaporated by solar radiation, then rises to the sky as cloud, is transported by wind, condenses as rain, and falls onto elevated ground, as shown in Figure 7.2.

According to the U.S. National Hydropower Association (www.hydro.org), hydropower accounts for approximately 7% of U.S. electricity in terms of actual generation (2.75×10^8 MW-h). It represents about 9% in terms of generating capacity (7.9×10^4 MW). Hydropower accounts for 83% of U.S. renewable energy capacity and approximately 77% of actual renewable electricity generation. According to the Department of Energy, approximately 2.1×10^4 MW of new hydropower capacity could be developed without building new dams or impoundments. This is enough power for 6.9 million homes.

In recent years, pumped storage plants have become a popular way to store energy. The turbine is run reversibly (or a separate pump is used) to pump water from a low reservoir to a high reservoir during off-peak hours. During peak hours, it is run by allowing the water to flow back to the low reservoir to supplement the extra demand. A typical plant is shown in Figure 7.3*a*. The hydroelectric plant can be started and brought up to full capacity quickly.

Fundamentals of Turbomachinery: Theory and Applications, Second Edition. Ryoichi S. Amano and William W. Peng.

Companion website: www.wiley.com/go/amano/turbomachinery2e

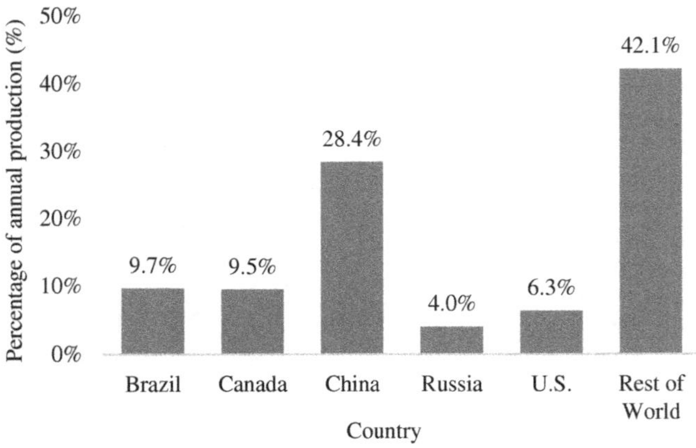

Figure 7.1 Percentage of annual hydropower production per country.

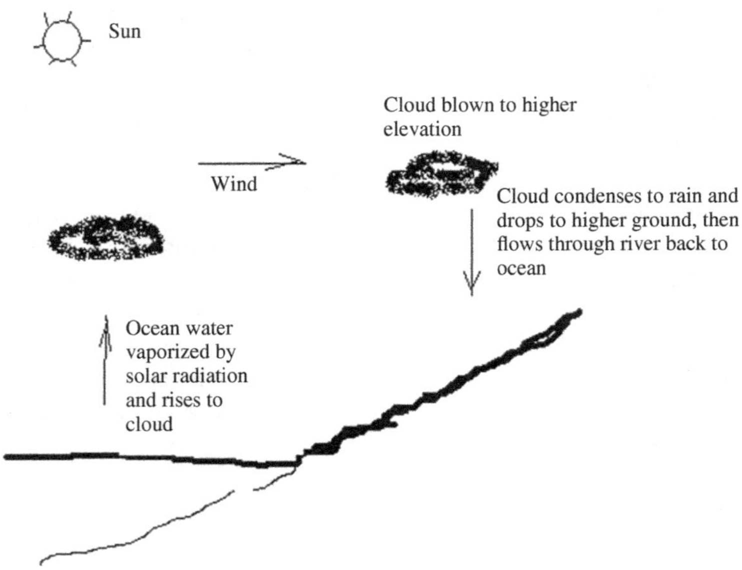

Figure 7.2 Conversion of solar energy to hydropower.

It can also be designed to be started without the use of an external power source. Advantages of pumped storage hydropower with a thermal base-load plant include the following: (a) improved total system reliability, (b) more efficient use of available base-load units, and (c) ability for rapid change in load.

In a large pumping system, energy can be saved with a small hydraulic turbine served as a power recovery device. As shown in Figure 7.3*b*, in a lower discharge pressure line, a power recovery turbine can be used instead of a pressure-reducing valve, or a turbine can be installed in the recirculated line to keep the main pump operating around the best efficiency point for the lower flow requirement condition.

Increasing energy cost in recent years has sparked interest in using pumps as hydraulic turbines in small hydropower projects. A typical multistage vertical pump operated as a hydraulic turbine is reported in Ref. 9. It was shown that pumps of the similar specific speed and similar performance characteristics at pump mode may have completely different turbine characteristics.

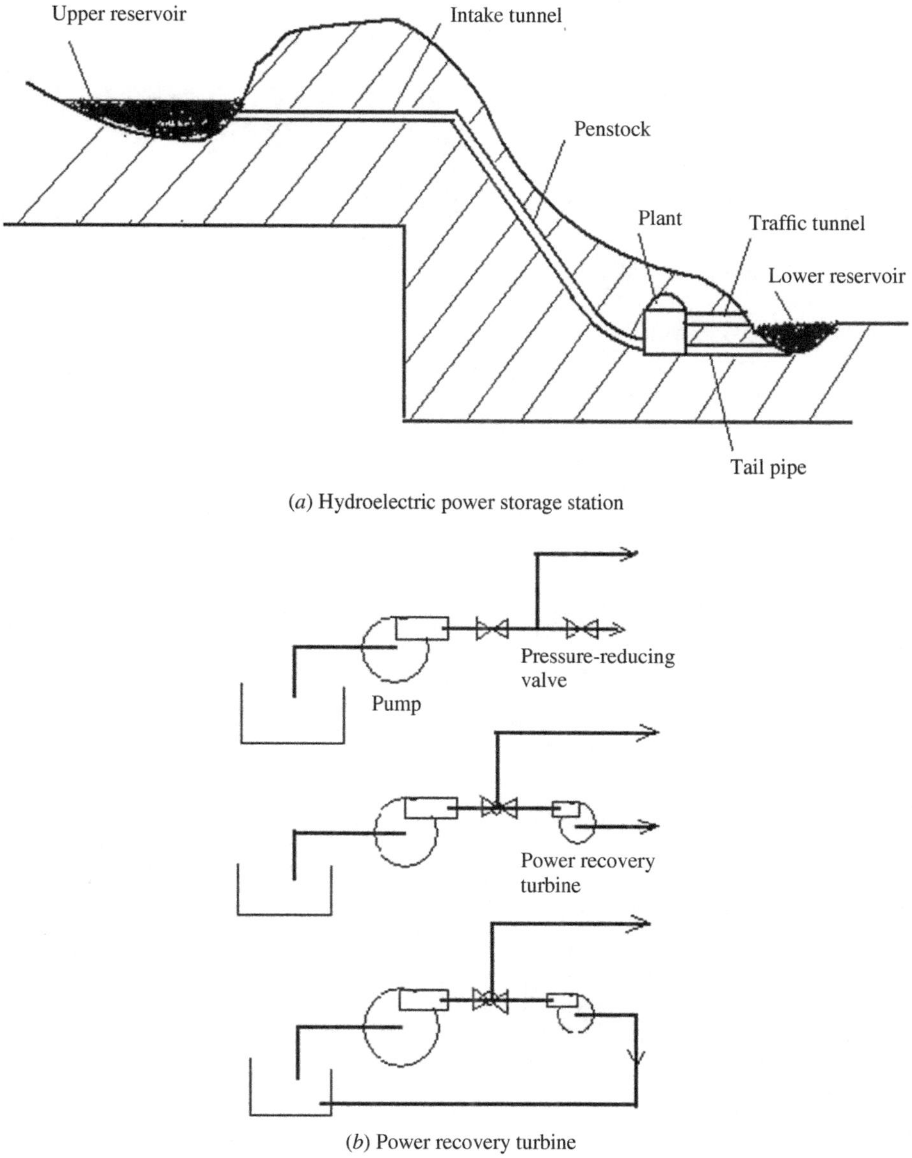

Figure 7.3 Special application of hydraulic turbines.

7.1.2 Types of Hydraulic Turbines

The types of hydraulic turbines can be classified into three groups: (a) Pelton turbines (impulse type) for low flow, high head, (b) Francis and pump turbines (mixed-flow type) for medium flow and medium head, and (c) Kaplan and bulb turbines (axial-flow type) for high flow and low head. They can be either horizontal or vertical axis. Examples are shown in Figure 7.4.

Two Pelton turbine installations are shown in Figure 7.5. There can be one or more nozzles operating simultaneously on the turbine runner. It is also possible to have some backward-pointing nozzles for braking. The load is changed by controlling the water flow with needle valves, deflector, or shutting off some valves. The Pelton turbine

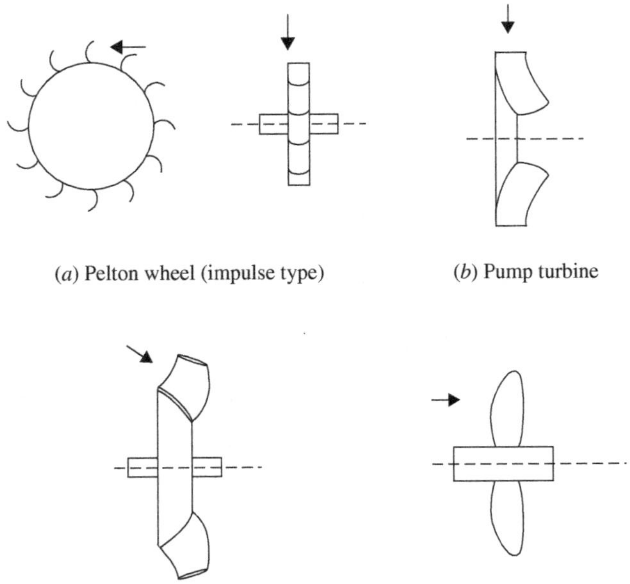

(*a*) Pelton wheel (impulse type)   (*b*) Pump turbine

(*c*) Francis turbine (mixed flow)   (*d*) Kaplan turbine (axial flow)

Figure 7.4 Typical types of hydraulic turbines.

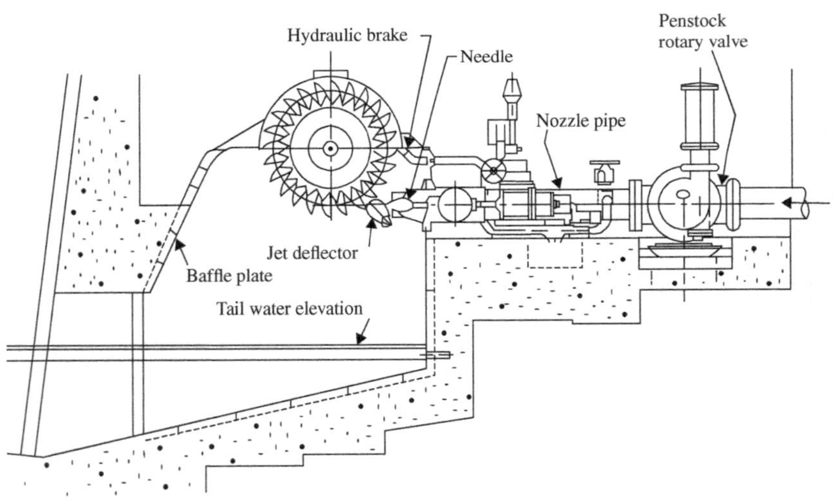

(*a*) Horizontal-axis turbine with single nozzle (side view)

Figure 7.5 Pelton turbine installation. (Reproduced from Wilbur, L. C. (Ed.), 1985/with permission of John Wiley & Sons.)[10]

is an impulse-type turbine since all hydraulic head is converted into kinetic energy through the nozzles. The water is under constant atmospheric pressure through the runner buckets.

Reaction turbine impellers can be either mixed-flow or axial-flow type. They are similar to centrifugal pumps in construction, except for the wicket gates at the inlet around the outer periphery to control the flow rate. The mixed-flow type can be a Francis turbine or a pump turbine, which is a pump operated as a turbine. A typical installation is shown in Figure 7.6. A variation of the mixed-flow turbine is the Deriaz turbine (also called diagonal-flow turbine), shown in Figure 7.7. The blades are shorter like a propeller and adjustable.

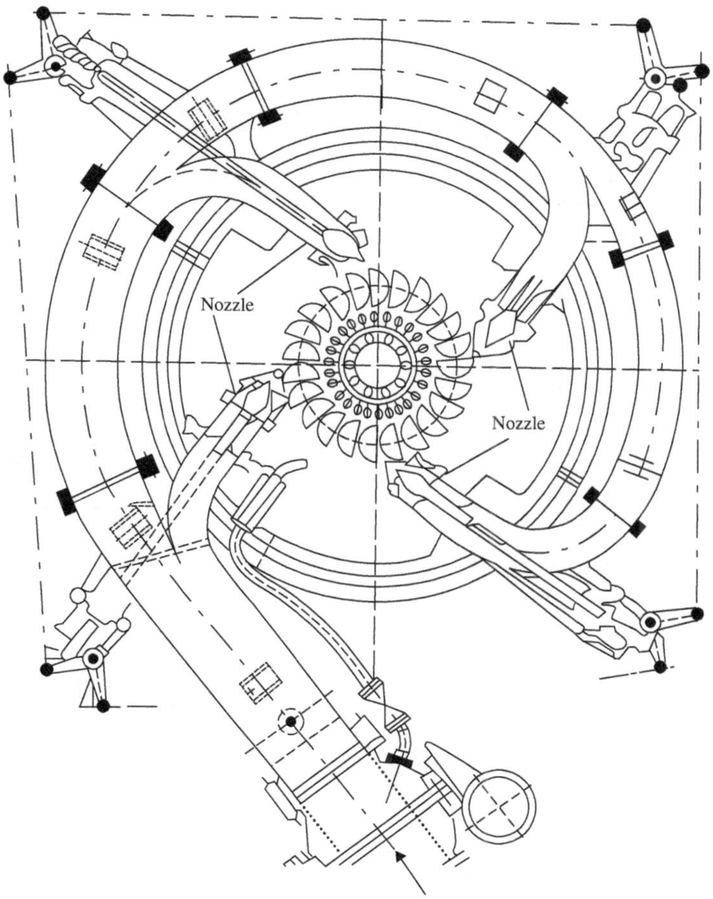

(*b*) Vertical-axis turbine with multiple nozzles (top view)

Figure 7.5 *(Continued)*

The axial-flow turbine runner can be either a fixed-blade propeller or adjustable, the latter called a Kaplan turbine, shown in Figure 7.8. Turbines with adjustable blades, gates, or needles can be operated for constant high efficiency as expected. Some variations of the axial-flow turbines are as follows:

a) Bulb turbine—The generator is enclosed and submerged in the water passage.
b) Tubular turbine—The generator is located outside of the water passage and is connected to the turbine shaft either directly or through a set of angle gears.
c) Rim generator turbine—The generator rotor is mounted on the outer tips of the turbine runner blades (seals are needed to keep the water away from the generator).

A schematic of these arrangements is shown in Figure 7.9.

7.1.3 Selection of Hydraulic Turbines

The nondimensional parameters used to describe the performance of a hydraulic turbine are similar to those for a pump: (a) flow coefficient $\phi = Q/(\omega D^3)$, (b) energy coefficient (equivalent to head coefficient for pump) $\psi = gH/(\omega^2 D^2)$, (c) power coefficient $\Pi = P_s/(\rho\omega^3 D^5)$, and (d) efficiency $\eta = P_s/(\rho QgH) = \Pi/(\phi\psi)$. The specific speed and specific diameter are also defined and can be nondimensional or dimensional, such as specific

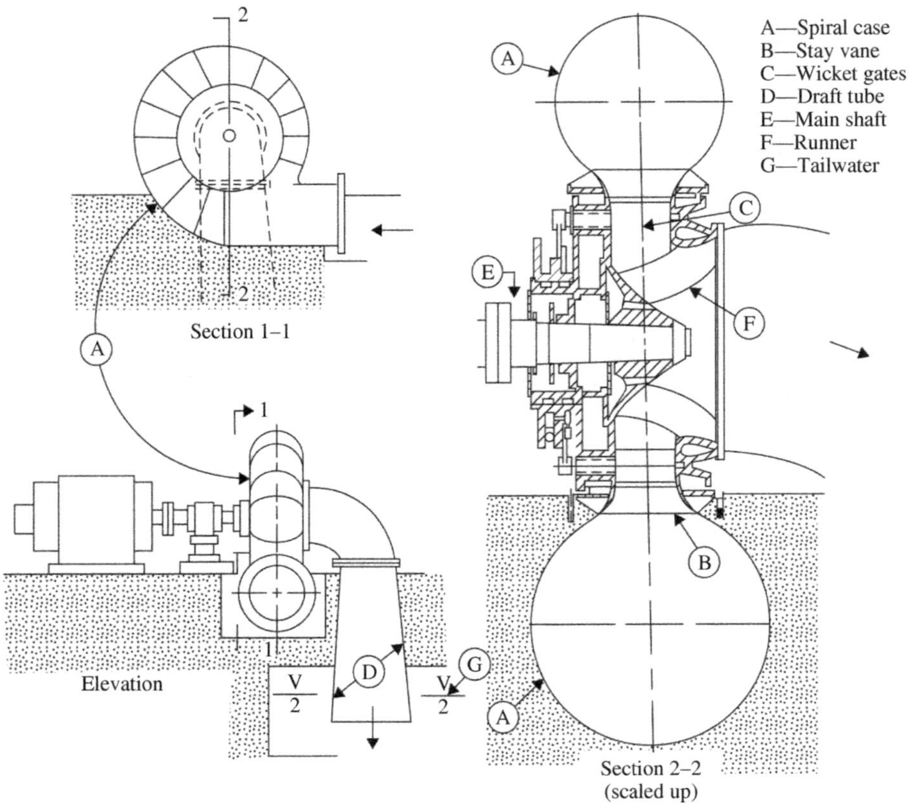

Figure 7.6 Mixed-flow turbine installation. (Reproduced from Wilbur, L. C. (Ed.), 1985/with permission of John Wiley & Sons.)[10]

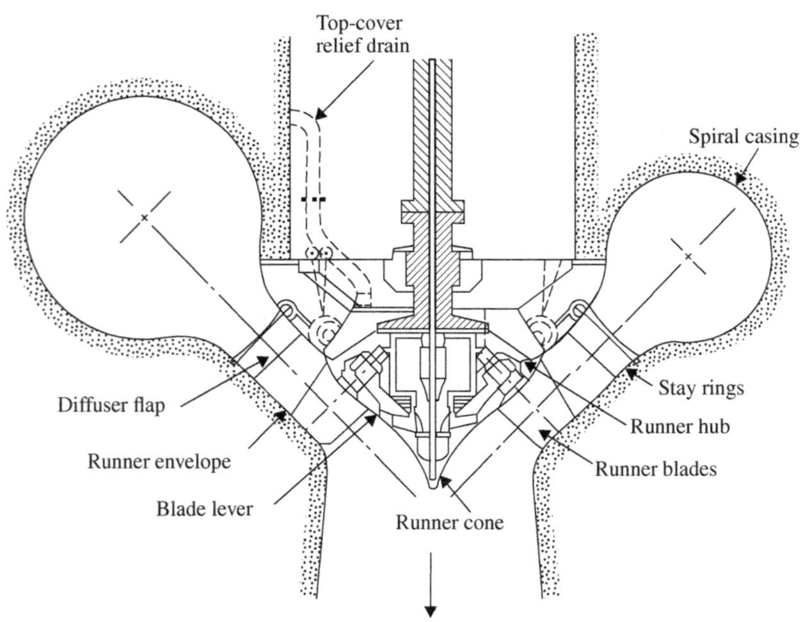

Figure 7.7 Deriaz turbine (diagonal-flow turbine). (Reproduced from Wilbur, L. C. (Ed.), 1985/with permission of John Wiley & Sons.)

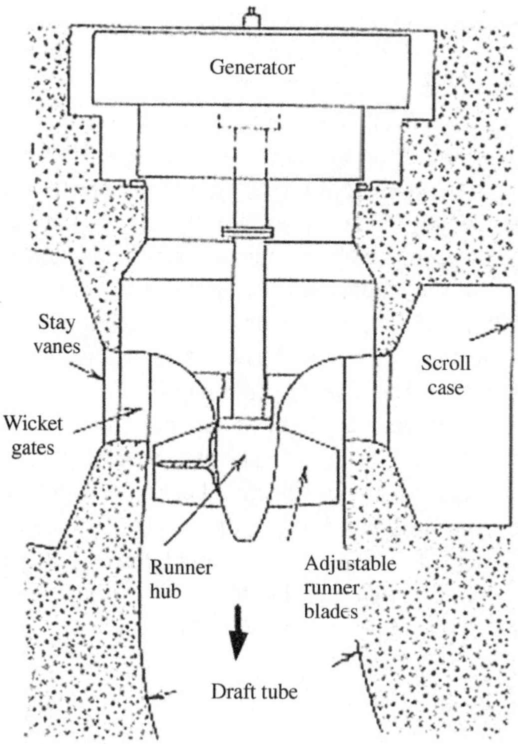

Figure 7.8 Axial-flow turbine installation. (Reproduced from Wilbur, L. C. (Ed.), 1985/with permission of John Wiley & Sons.)

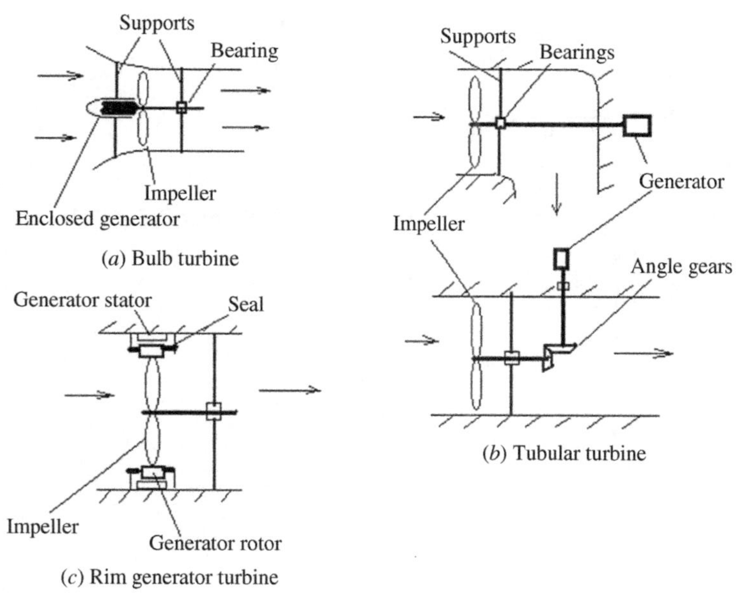

Figure 7.9 Variation of axial-flow turbine installations.

speed $\omega_s = \omega Q^{0.5}/(gH)^{0.75}$ and specific diameter $\Delta_s = D(gH)^{0.25}/Q^{0.5}$ in nondimensional forms or $N_s = NQ^{0.5}/H^{0.75}$ (sometimes, the power specific speed $N_{sp} = NP_s^{0.5}/H^{1.25}$ is used) and $D_s = DH^{0.25}/Q^{0.5}$ in dimensional forms, where D is the runner diameter.

The peak efficiency of the turbines can be correlated with the specific speed, similar to those for the pumps, fans, and compressors discussed earlier. They are shown in Figure 7.10.

To select an appropriate type of turbine for a given hydropower site, we start with the head and flow capacity available. Then, the possible output power can be estimated with $P_o = \eta\rho QgH$ with approximate efficiency $\eta = 0.90$. Charts prepared by turbine manufacturers (such as the one shown in Figure 7.12[13]) can be used to select the type of turbine. Then, the rotating speed N (rpm) can be selected such that the specific speed N_s or ω_s calculated will fall in the range of the turbine type given in Table 7.1, based on Ref. 12, and be close to the maximum efficiency shown in Figure 7.10.[11] The obtained efficiency can be used to calculate the revised output power and to confirm or change the selected turbine type.

If the generator is to be directly coupled with the turbine, the speed N should be the synchronous speed $N = 120f/n_p$, where f is the electrical frequency in hertz, which is 60 in United States, and n_p is the number of generator poles, which should be multiples of two or four.

Empirical data relating the specific diameter with specific speed for different types of hydraulic turbines are presented in Figure 7.11, also adapted from Ref. 12. They can be used to estimate the impeller diameter. The specific diameter used in this chart is nondimensional, defined as $\theta = K_\theta D(gH)^{0.25}/Q^{0.5}$, and the specific speed is defined as $\sigma = K_\sigma nQ^{0.5}/(gH)^{0.75}$ for nondimensional forms, where $K_\theta = \pi^{0.5}/2^{0.75} = 1.054$ and $K_\sigma = \pi^{0.5}\,2^{0.25} = 2.11$,

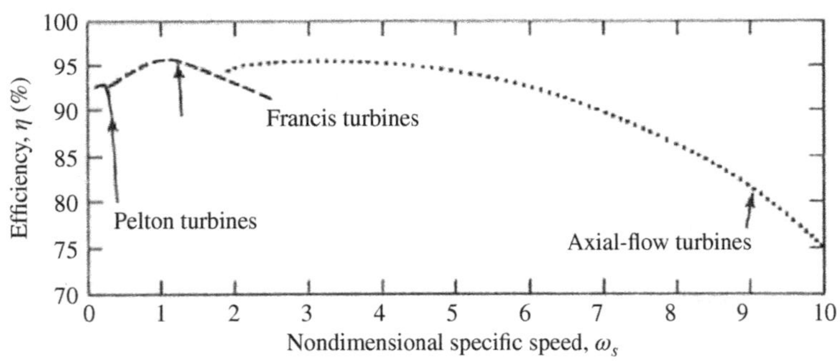

Figure 7.10 Efficiency versus specific speed of hydraulic turbines. (Reproduced from Gulliver, J. S. & Arndt, R. E., 1991/with permission of McGraw-Hill Education.)

Table 7.1 Range of specific speed.

	Specific Speed
Turbine type	N_s [rpm(m^3/s)$^{0.5}$/m$^{0.75}$]
Pelton	1–20 (with one jet)
Pump	25–120
Francis	20–140
Kaplan	100–300
Bulb	150–400

Source: From Ref. [12]/Reed Business Publishing.

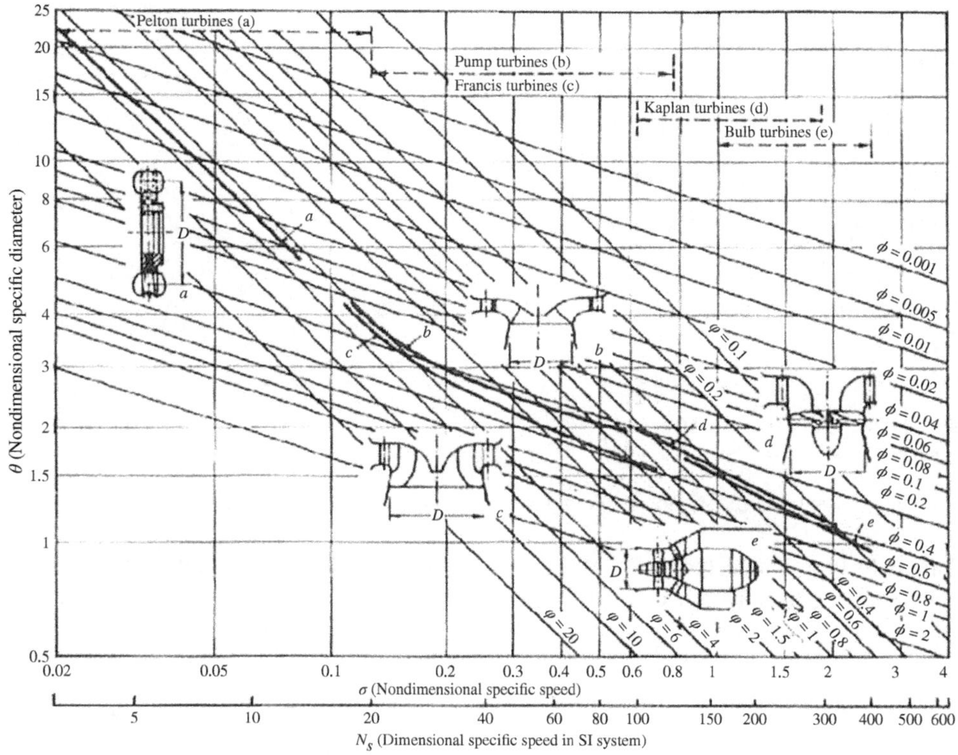

Figure 7.11 Relationship between specific diameter and specific speed of hydraulic turbines. (Reproduced from Schweiger, F. & Gregor, J., 1989/with permission of Reed Business Publishing.)

n is in revolutions per second. The other parameters can be in either SI or English unit. On the horizontal axis, the dimensional-specific speed N_s in rpm$(m^3/s)^{0.5}/(m^{0.75})$ is also given.

The constant energy coefficient $\psi = gH/(k_\psi n^2 D^2)$ and flow coefficient $\phi = Q/(k_\phi n D^3)$ curves are also shown in Figure 7.11, where $K_\psi = \pi^2/2$, $K_\phi = \pi^2/4$.

If no rotational speed is specified, the selection procedure can be summarized as follows:

1) Given available head H, capacity Q, and estimated efficiency $\eta = 0.90$, the output power is calculated as $P_o = \eta\rho QgH$.
2) With H and P_o, the type of turbine can be selected from Figure 7.12.
3) Select the appropriate rotating speed N so that the specific speed $\omega_s = \omega Q^{0.5}/(gH)^{0.75}$ and $N_s = NQ^{0.5}/H^{0.75}$ will result in an efficiency close to the maximum value in Figure 7.10 and within the range specified in Table 7.1.
4) Estimate the impeller diameter using Figure 7.11 if needed.

Example 7.1 If a hydropower site has a head of 85 m and flow rate of 16 m^3/s, suggest the type of hydraulic turbine to be used, its operating speed, and the approximate size.

Solution

From $H = 85$ m and $Q = 16$ m^3/s, the output power can be estimated as

$$P_o = \eta\rho QgH = \frac{0.90 \times 998 \times 9.8 \times 16 \times 85}{1000} = 11,971 \text{ kW}.$$

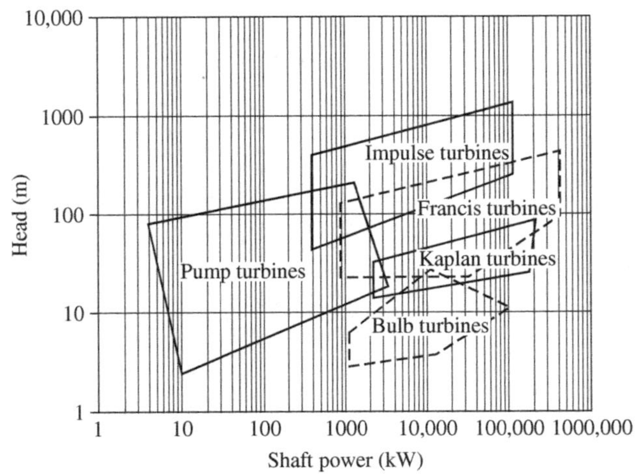

Figure 7.12 Application range for hydraulic turbines. (*Source:* Warnick, C. C., et al. 1984/Pearson PLC.)

From Figure 7.12, a Francis turbine is selected. Then with the selected synchronous speed of 16-pole generator, $N = 120 \times 60/16 = 450$ rpm, we have $N_s = NQ^{0.5}/H^{0.75} = 64.3\,\text{rpm}(\text{m}^3/\text{s})^{0.5}/\text{m}^{0.75}$, or $\sigma = K_\sigma nQ^{0.5}/(gH)^{0.75} = 2.11 \times (450/60) \times 4/(9.8 \times 85)^{0.75} = 0.41$, or

$$\omega_s = \frac{450\pi\sqrt{16}}{30(9.8 \times 75)^{0.75}} = 1.33,$$

which is appropriate for the Francis turbine based on Figures 7.10 and 7.11. From Figure 7.11, we also have $\theta = 1.9$ for $N_s = 64.3$. Hence, $K_\theta D(gH)^{0.25}/Q^{0.5} = 1.9$ or $1.054D(9.8 \times 85)^{0.25}/16^{0.5} = 1.9$, and $D = 1.34$ m. From Figure 7.10, we have the efficiency $\eta \cong 0.95$, which is close to the original estimation.

The flow coefficient and energy coefficient at this point can also be calculated as $\phi = 16/(2.47 \times 7.5 \times 1.34^3) = 0.359$ and $\psi = 9.8 \times 75/(4.93 \times 7.5^2 \times 1.34^2) = 1.47$, which check with the point on Figure 7.11.

7.1.4 Hydraulic Turbine Performance Characteristics

The basic performance characteristics of a hydraulic turbine can be obtained from Euler's equation with a procedure similar to the one described in Section 5.8 for the gas turbine. The typical relationships of shaft torque, power, and efficiency versus impeller rotating speed expressed in terms of speed ratio $U/\sqrt{2gH}$ are shown in Figures 7.13 and 7.14 with a constant head. The effects of flow rate or gate opening are also shown in Figure 7.14. These nondimensional parameters were discussed in Chapter 2.

The similarity laws (or affinity laws) applied to pumps are also applicable to hydraulic turbines, that is, $Q \sim ND^3$, $H \sim N^2D^2$, and $P_s \sim N^3D^5$.

A more complete map of performance characteristics in terms of nondimensional parameters ψ, φ, Π, and η called hill curves is shown in Figure 7.15. It is developed from extensive laboratory tests of model turbines and can be used to predict the performance of a prototype machine based on similarity laws, or for the convenience of applications, the efficiency can be plotted versus the percentage of full load, as shown in Figure 7.16.

Example 7.2 Convert the data for 80% gate opening in Figure 7.15 into a variable-speed performance characteristic for a head of 75 m and a runner diameter of 1.5 m. Show the variations of shaft power, flow rate, and efficiency versus rotating speed.

phi	psi	eff (%)	pai	N(rpm)	P_s(mw)	Q(m^3/s)	eff (%)
0.158	0.093	55	0.0078	566.3056	12.32442	31.58513	55
0.151	0.083	56.5	0.0067	599.4505	12.55608	31.95251	56.5
0.14	0.071	58	0.0058	648.1316	13.73842	32.03066	58
0.127	0.06	62.5	0.0045	705.0448	13.72088	31.60786	62.5
0.108	0.048	69	0.0034	788.264	14.48816	30.05178	69
0.092	0.04	71.5	0.0025	863.5	14.00381	28.04303	71.5
0.076	0.03	67.5	0.0015	997.0839	12.93617	26.74977	67.5
0.066	0.025	60	0.001	1092.251	11.33669	25.44726	60

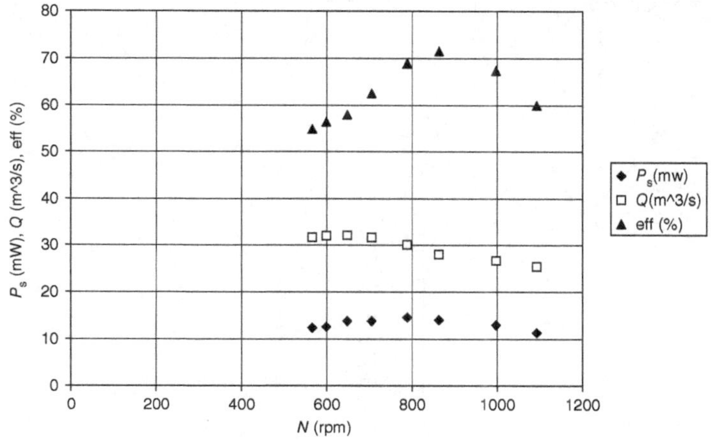

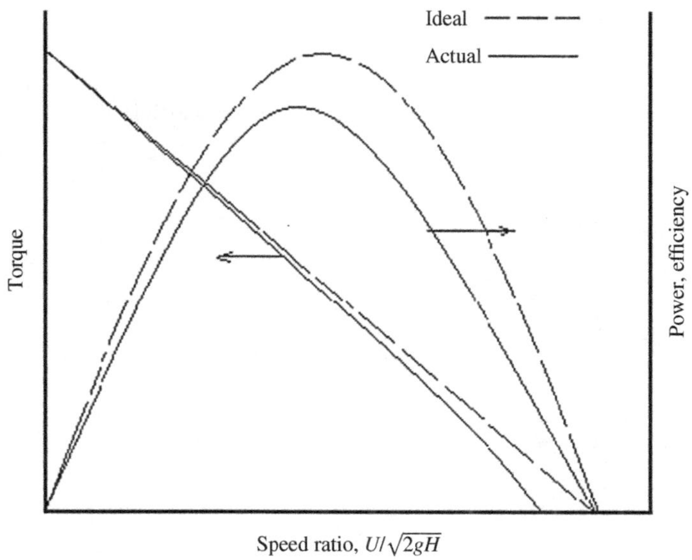

Figure 7.13 Variable-speed performance for a hydraulic turbine.

Solution

From $\psi = gH/(\omega^2 D^2)$, we have $N = 30\omega/\pi = 9.55[9.81 \times 75/(1.5^2\psi)]^{0.5}$ or $N = 172.7/\sqrt{\psi}$ (rpm). Also, from $\phi = Q/(\omega D^3)$ and $\Pi = P_s/(\rho\omega^3 D^5)$, we have $Q = (\pi N/30)D^3\phi = 0.353N\phi$ and $P_s = 0.998 \times (\pi N/30)^3 \times (1.5)^5\Pi = 0.0087N^3\Pi$. Pick the points along the 80% gate opening curve, read the values for φ, ψ, Π, and efficiency, where Π = efficiency $\varphi\psi$ and calculate N, P_s, and Q with the above equations. The calculations and plot are carried out with Excel.

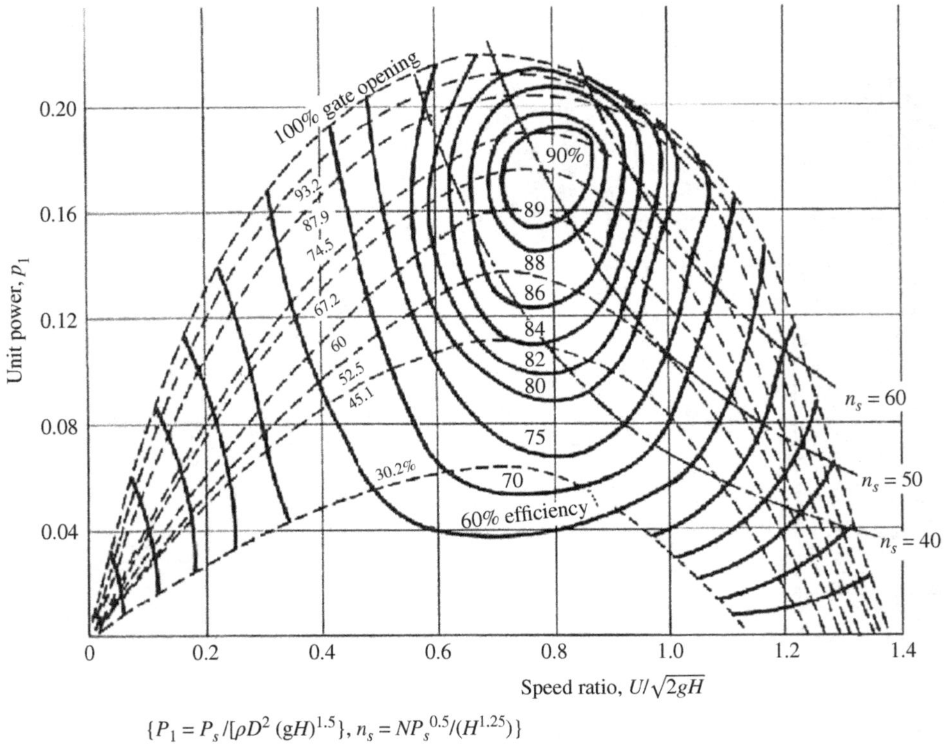

Figure 7.14 Typical reaction turbine performance characteristics. (*Source:* Warnick, C. C., et al. 1984/Pearson PLC.)

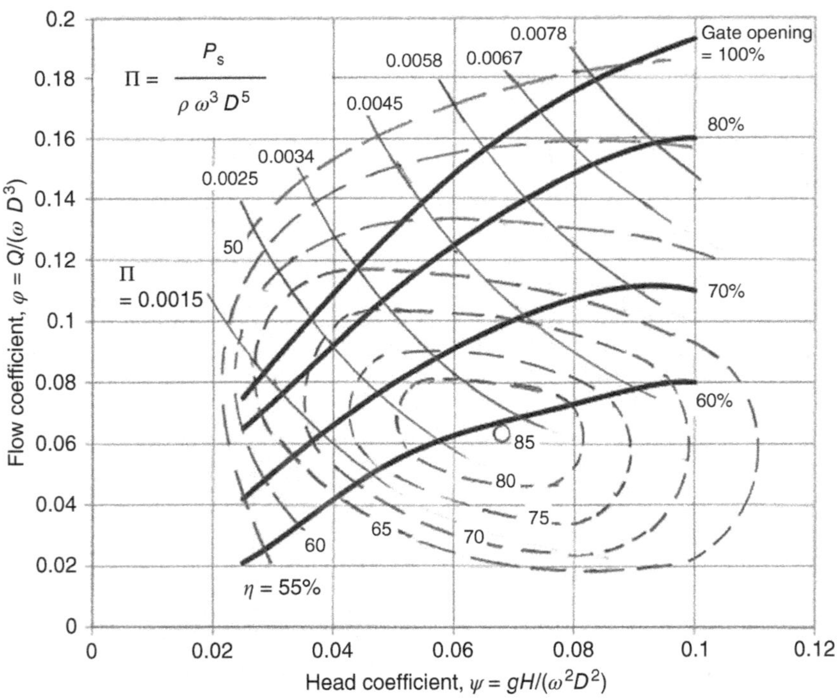

Figure 7.15 Typical turbine hill curves (in nondimensional parameters).

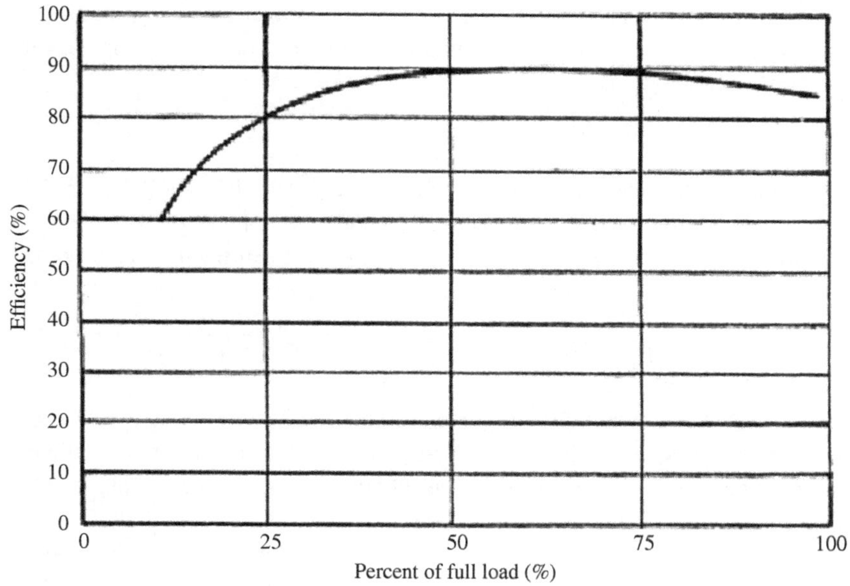

Figure 7.16 Typical variation of hydraulic turbine efficiency with load.

7.2 Pelton Turbines

A Pelton turbine is a pure impulse machine used in a high-head, low-flow installation. The hydraulic energy is converted through a nozzle (stator) into kinetic energy of a high-speed jet. It then strikes the buckets on the wheel (rotor) to produce mechanical energy, as shown in Figure 7.17. It is an impulse machine since all pressure drop takes place in the stator, none in the rotor.

Figure 7.17 Velocity diagrams for Pelton wheel turbine.

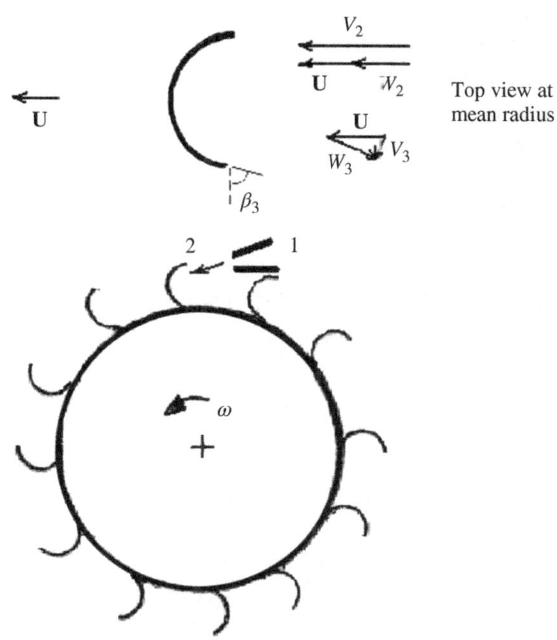

It was shown in the last chapter that for an impulse turbine, the maximum utilization factor occurs at $U/V_2 = 1/2$. This can be demonstrated in a different way. From the Euler equation and the velocity diagram in Figure 7.17, the energy transfer per unit mass flow rate, ΔE, is

$$\Delta E = U(V_{u2} - V_{u3}) = U[(U + W_2) - (U - W_3 \sin \beta_3)] = U(W_2 + W_3 \sin \beta_3).$$

Since $W_3 = W_2$ for an ideal flow without friction, we have $\Delta E = UW_2\,(1 + \sin \beta_3) = U(V_2 - U)(1 + \sin \beta_3)$. We have shown that ΔE can also be expressed in the form $\Delta E = \left[\left(U_2^2 - U_3^2\right) + \left(V_2^2 - V_3^2\right) + \left(W_3^2 - W_2^2\right)\right]/2$ for a turbine. So, for the Pelton turbine without friction, we have $\Delta E = \left(V_2^2 - V_3^2\right)/2$. Hence, the utilization factor ε can be written as

$$\varepsilon = \frac{\Delta E}{\Delta E + V_3^2/2} = \frac{\Delta E}{V_2^2/2} = 2(1 + \sin \beta_3)\left(\frac{U}{V_2}\right)(1 - U/V_2).$$

For the maximum ε, we set $d\varepsilon/d(U/V_2) = 0$, and hence, $(U/V_2)_{\max} = 1/2$. Accordingly, we have $\varepsilon_{\max} = (1 + \sin \beta_3)/2$ and $\Delta E_{\max} = U^2(1 + \sin \beta_3)$. To compare these expressions with those given in the last chapter, it should be noted that we have adopted $\alpha_2 = 90°$ here and assumed $\beta_3 = 90°$ in the last chapter. The final expressions will be the same if both α_2 and β_3 are set to 90° in both cases.

When frictional losses in the nozzle and buckets are included, with the nozzle efficiency and velocity coefficient of the bucket defined as $\eta_n = V_2^2/(2gH)$ and $W_3 = \mathrm{CV}_b W_2$, the hydraulic efficiency of the turbine, defined as $\eta_h = \Delta E/gH$, becomes

$$\eta_h = \frac{2\Delta E \eta_n}{V_2^2} = 2\eta_n\left(\frac{U}{V_2}\right)\left(1 - \frac{U}{V_2}\right)(1 + \mathrm{CV}_b \sin \beta_3), \quad \text{where}$$

$$\Delta E = U(W_2 + W_3 \sin \beta_3) = UW_2(1 + \mathrm{CV}_b \sin \beta_3) = U(V_2 - U)(1 + \mathrm{CV}_b \sin \beta_3)$$

Or under the optimum condition, we have $\eta_{h,\max} = 0.5\eta_n(1 + \mathrm{CV}_b \sin \beta_3)$. The shaft power, including mechanical and leakage losses, becomes $P_s = \eta_m \eta_v m\ \Delta E = \eta_m \eta_v m \eta_h gH = \eta_{o/a} mgH$, and the flow rate can be calculated as $Q = (\pi/4)d_n^2\sqrt{\eta_n(2gH)}$, where d_n is the nozzle diameter.

Example 7.3 A Pelton turbine wheel of 3.0 m pitch diameter with a single jet of 8 cm nozzle diameter is installed to produce power from a water head of 350 m. If the nozzle efficiency, including the loss in penstock, is $\eta_n = 0.82$ and the velocity coefficient in the wheel bucket is $\mathrm{CV}_b = 0.95$ and mechanical and volumetric efficiencies are $\eta_m = 0.90$, $\eta_v = 0.96$, respectively, determine the optimum rotating speed, the shaft power output, and the corresponding specific speed.

Solution

The jet flow velocity can be calculated as $V_2 = (2\eta_n gH)^{0.5} = (2 \times 0.82 \times 9.81 \times 350)^{0.5} = 75$ m/s. Hence, the optimum wheel tangential velocity is $U_m = 0.5V_2 = 37.5$ m/s, and the rotating speed $N = 60 \times U_m/(\pi D) = 238.7$ rpm. Under the maximum utilization-factor condition, we have $\beta_3 = 90°$, and

$$\Delta E_{\max} = (1 + \mathrm{CV}_b)U(V_2 - U) = 1.95U^2 = 1.95 \times 37.5^2 = 2742.2\ \text{N - m/kg}.$$

The flow rate is calculated as $Q = \mathrm{V}_2 A_n = 75.0 \times \pi \times 0.08^2/4 = 0.377\ \mathrm{m^3/s}$, and hence, the total shaft power output is

$$P_s = \eta_m \eta_v m \Delta E = \frac{0.90 \times 0.96 \times 998 \times 0.377 \times 2742.2}{1000} = 891.4\ \text{kW}.$$

The specific speed can be calculated as

$$N_s = \frac{NQ^{0.5}}{H^{0.75}} = \frac{238.7 \times (0.377)^{0.5}}{(350)^{0.75}} = 1.81\ \mathrm{rpm(m^3/s)^{0.5}/m^{0.75}},$$

or in nondimensional form:

$$\omega_s = \frac{\omega Q^{0.5}}{(gH)^{0.75}} = \frac{(238.7\pi/30)(0.377)^{0.5}}{(9.81 \times 350)^{0.75}} = 0.034.$$

7.3 Francis Turbines

A Francis turbine is a radial-inflow turbine with either a pure radial or a mixed-flow impeller as shown in Figure 7.18. The flow enters the impeller radially inward and discharges in the axial direction, reversed to that of a typical radial or mixed-flow impeller pump. The inlet guide vanes enclosing the impeller circumferentially are used to direct the flow in the proper direction for different flow rates.

A Kaplan (or propeller) turbine is an axial-flow type, as shown in Figure 7.19, along with the velocity diagrams across the impeller. It can be either a fixed or adjustable blade type. With the adjustable blades, a flat efficiency

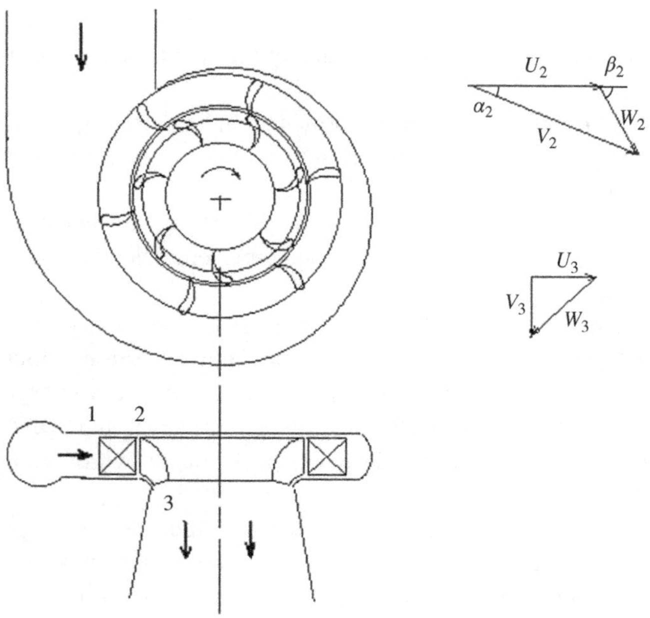

Figure 7.18 Velocity diagrams for Francis-type turbine.

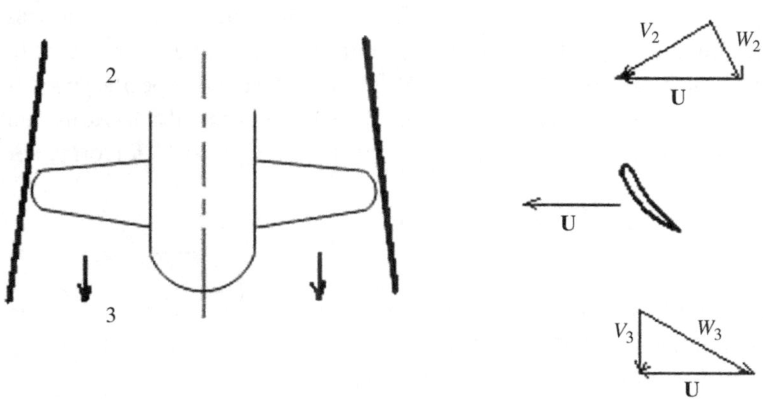

Figure 7.19 Velocity diagrams for Kaplan turbine (propeller type).

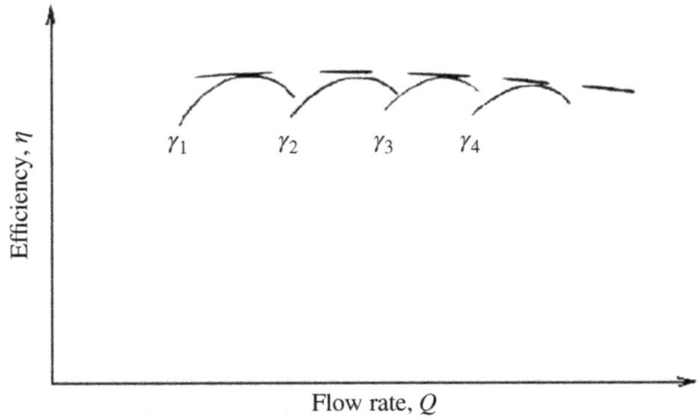

Figure 7.20 Efficiency versus flow rate of a Kaplan turbine with adjustable pitch angle.

curve can be obtained, as shown in Figure 7.20. Each solid curve is for a certain blade setting, and the envelope of these curves is the overall-efficiency curve.

The basic design analysis for both Francis and Kaplan turbines also can be performed with the Euler's equation such that $\Delta E = U_2V_{u2} - U_3V_{u3} = U_2V_{u2} = U_2V_2 \cos \alpha_2$, assuming no whirl at the impeller exit under design condition. Hence, we have the output shaft power $P_s = \eta_m\eta_v \dot{m}$

ΔE, and the flow rate $Q = V_2 \sin \alpha_2 A_2$, where $V_2 = \sqrt{2\eta_n gH}$, $A_2 = b_2\pi D_2$ for the Francis turbine and $A_2 = \pi\left(D_t^2 - D_h^2\right)/4$ for the Kaplan turbine. Here, η_n is to account for the losses through the penstock and nozzle as that for the Pelton turbine.

Since the pressure decreases in the direction of flow in the turbine, there is no likelihood of boundary layer separation in the flow passage. Hence, the turbine has higher efficiency compared to the pump of corresponding type. It is not unusual to have an efficiency of 90% or higher. So, in a pumped storage system, it takes more energy to pump water back to a higher elevation reservoir than that to be produced by the turbine in the reversed process. Even so, it is still a worthwhile scheme because the energy stored and available during peak hours is worth more than the excess energy produced in off-peak hours.

The cavitation phenomenon is also observed in reaction turbine operation because the water flow passage is enclosed in this type of turbine. It also causes noise, vibration, and erosion on the structure, similar to the situation in the pump. However, it takes place at the turbine discharge, where the pressure is minimum. Both NPSH and Thoma's parameter σ are also used for turbines, that is, $\text{NPSH} = H_a - Z_1 + V_e^2/2g + H_L - H_v$ and $\sigma = \text{NPSH}/H$. The heads are defined in Figure 7.21, where H_a is the atmospheric pressure head, Z_1 is the height of the turbine center above the tail water level, V_e is the flow velocity at the draft tube exit, H_L is the head loss in the draft tube, and H_v is the water vapor pressure head. The typical minimum required values of σ related to the specific speed for Francis and propeller turbines are given in Figure 7.22,[14] where the power-specific speed is defined as $N_{\text{sp}} = NP_s^{0.5}/H^{1.25}$ and N is in rpm, P_s in horsepower, and H in feet. These curves can also be expressed analytically as $\sigma = 0.006 + 0.55(N_{\text{sp}}/100)^{1.8}$ for Francis turbines and $\sigma = 0.10 + 0.30(N_{\text{sp}}/100)^{2.5}$ for propeller turbines.

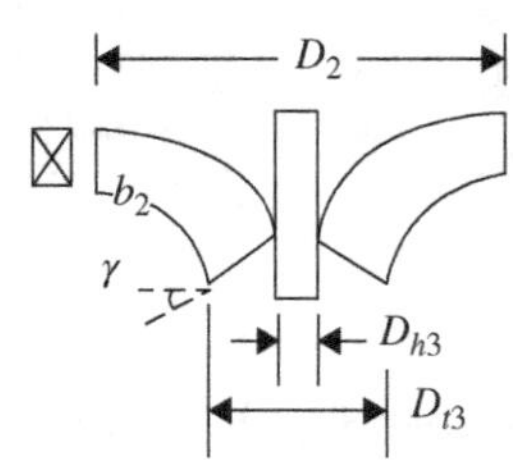

Example 7.4

A Francis-type impeller turbine is designed for a hydropower site with a head of 80 ft, flow capacity of 63 ft^3/s, and estimated stator efficiency of 97%. If the impeller rotating speed is set as 400 rpm, discharge flow velocity is limited to 25 ft/s for cavitation reasons, and outlet hub diameter is picked as 2 in. to accommodate the shaft size, determine the other impeller dimensions and the blade angles at the mean radius.

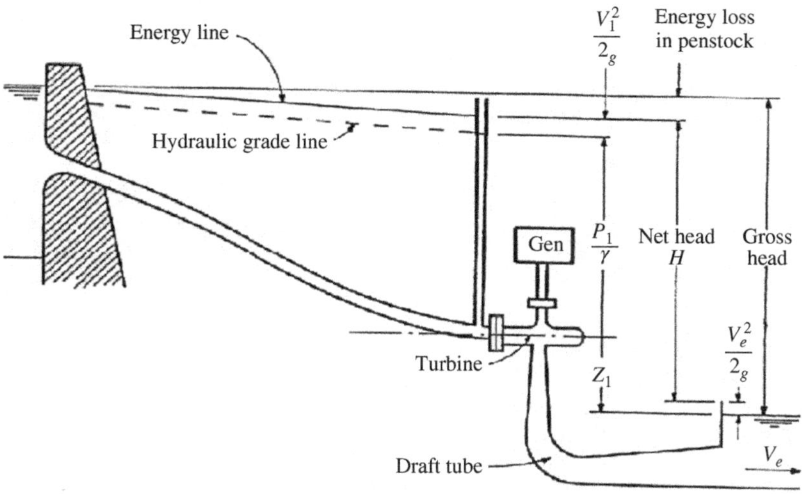

Figure 7.21 Definitions of heads in hydraulic turbine installation. (Reproduced from Gulliver, J. S. & Arndt, R. E., 1991/with permission of McGraw-Hill Education.)

Solution

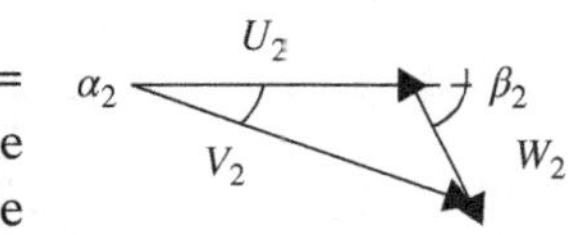

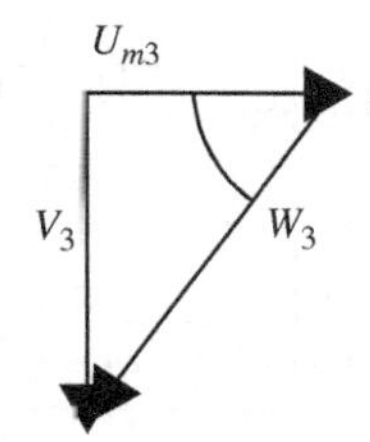

From Figure 7.11, with $\sigma = K_\sigma n Q^{0.5}/(gH)^{0.75} = 2.11 \times (400/60)(63^{0.5})/(32.2 \times 80)^{0.75} = 0.31$, we have $\theta = 2.4 = K_\theta D_2(gH)^{0.25}/Q^{0.5}$, or $1.054D_2(32.2 \times 80)^{0.25}/63^{0.5} = 2.4$ or the inlet diameter $D_2 = 2.5$ ft. Hence, $U_2 = D_2 N\pi/60 = 2.5 \times 400\pi/60 = 52.3$ ft/s. Also, the inlet flow velocity V_2 can be calculated as $V_2 = (2gH\eta_s)^{0.5} = (2 \times 32.2 \times 80 \times 0.97)^{0.5} = 70.6$ ft/s. With the inlet velocity diagram and $\alpha_2 = 20°$ selected, we have $V_{r2} = V_2 \sin \alpha_2 = 70.6 \sin 20° = 24.1$ ft/s. Therefore, $\tan \beta_2 = V_{r2}/(V_2 \cos \alpha_2 - U_2) = 24.1/(70.6 \cos 20° - 52.3) = 1.72$ and $\beta_2 = 59.8°$. Selecting the incidence $i = 2.2°$, we have $\beta_{b2} = 62°$. From $A_2 = Q/V_{r2} = 63/24.1 = 2.61$ ft^2 $= \pi D_2 b_2$, we have $b_2 = 2.61/(2.5\pi) = 0.33$ ft $= 4.0$ in. At the outlet, with $r_{h3} = 1$ in., setting $\gamma = 15°$, and $V_3 = 25$ ft/s, we have $A_3 = \left(r_{t3}^2 - r_{h3}^2\right)\pi/\cos\gamma = 63/25 = 2.52$ ft^2. Hence, $r_{t3} = [2.52 \cos 15°/\pi + (1/12)^2]^{0.5} = 0.88$ ft $= 10.6$ in. At the mean radius, we have $r_{m3} = [(10.6^2 + 1)/2]^{0.5}/12) = 0.627$ ft, $U_{m3} = 26.3$ ft/s, and hence, $\tan \beta_{m3} = V_3/U_{m3} = 25/26.3 = 0.95$, or $\beta_{bm3} = \beta_{m3} = 43.6°$, assuming no deviation.

Example 7.5 Estimate the maximum height Z_1 shown in Figure 7.21 for the turbine in Example 7.4 to be cavitation free. Use the curve in Figure 7.22, assuming $V_e = 5$ ft/s, $H_L = 0.7$ ft, and water temperature is 70 °F.

Solution

To use Figure 7.22, we need the dimensional power-specific speed $N_{sp} = NP_s^{0.5}/H^{1.25}$. So, the shaft power/efficiency has to be estimated from Figure 7.10, where the nondimensional specific speed ω_s is needed.

From $\omega_s = \omega Q^{0.5}/(gH)^{0.75} = (400\pi/30)(63)^{0.5}/(32.2 \times 80)^{0.75} = 0.92$, we have $\eta = 0.95$, or $P_s = \eta\rho gQH = 0.95 \times 62.4 \times 63 \times 80/550 = 543.2$ hp, or $N_{sp} = 400(543.2)^{0.5}/80^{1.25} = 38.9$ rpm(hp)$^{0.5}$/ft$^{1.25}$. From Figure 7.22, we obtain $\sigma \cong 0.1$ or $\text{NPSH}_{avail}/H \geq 0.1$, where $\text{NPSH}_{avail} = H_a - Z + H_L + V_e^2/(2g) - H_v$ and $\text{NPSH}_{avail} \geq 8$ ft. At sea level and $T = 70°$F, we have $H_a = 14.7 \times 144/62.4 = 33.9$ ft, $H_v = 0.363 \times 144/62.4 = 0.84$ ft, and $V_e^2/(2g) = 5^2/(2 \times 32.2) = 0.39$ ft. So, the maximum height Z_{max} should be 26.1 ft.

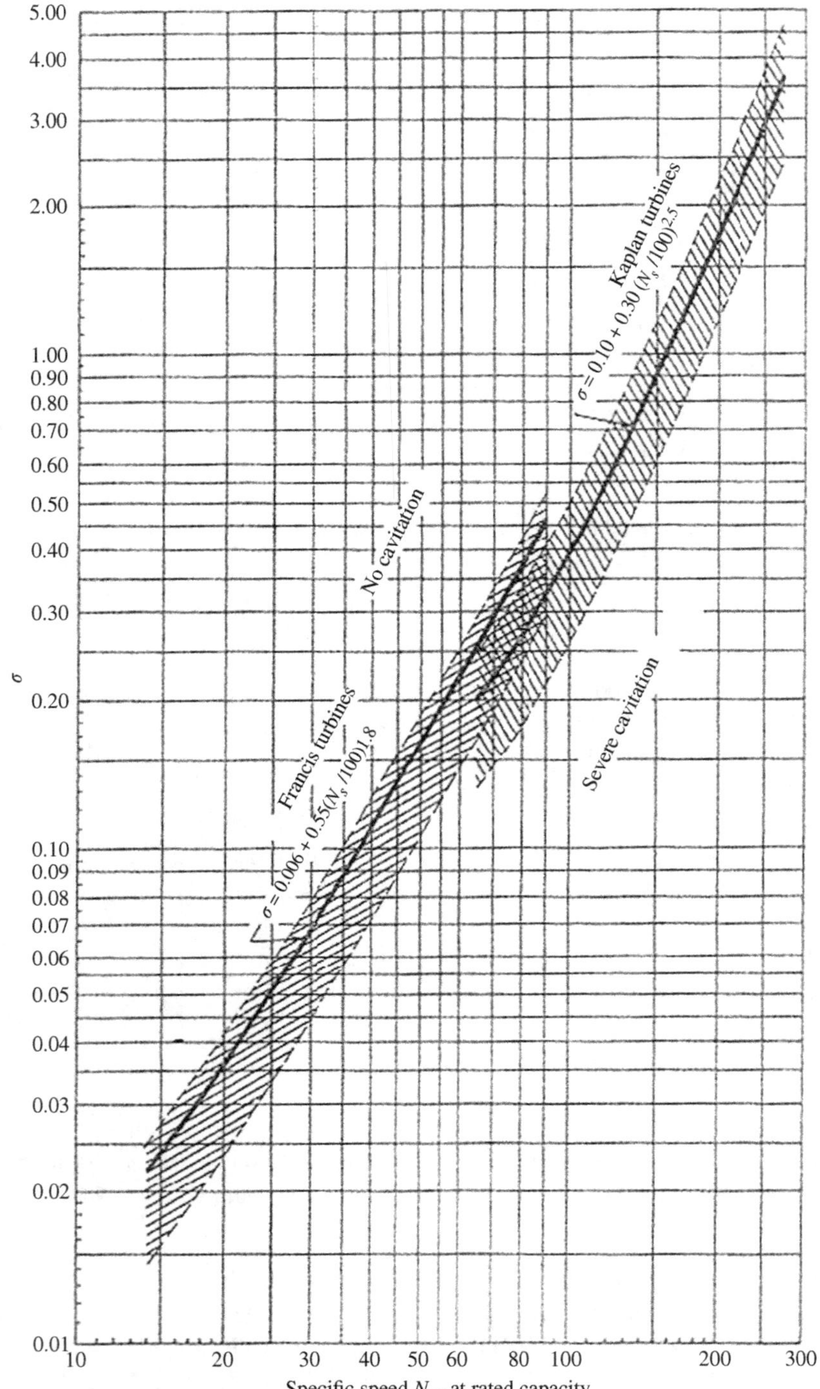

Figure 7.22 Required minimum Thoma's cavitation parameter versus power-specific speed N_{sp} [rpm $(hp)^{0.5}/(ft)^{1.25}$] for Francis and Kaplan turbines. (Reproduced from Moody, L. F., 1969/with permission of McGraw-Hill Education.)

7.4 Kaplan Turbines

7.4.1 Design and Performance

7.4.1.1 Introduction

The Kaplan turbine is a propeller-type water turbine that has adjustable blades. It was developed in 1913 by Austrian professor Viktor Kaplan,[15] who combined propeller blades with automatically adjusted wicket gates to achieve efficiency over a wide range of flow and water levels.

The Kaplan turbine was an evolution of the Francis turbine. Its invention allowed efficient power production in low-head applications, impossible with Francis turbines.

A study by Abeykoon and Hantsch[16] discussed the design of a Kaplan turbine rotor, aiming to improve the turbine's efficiency by optimizing the rotor design. An investigation of the design and the velocity distribution of a Kaplan runner blade was conducted by Permana and Rudianto,[17] and the design was verified. An airfoil profile from the Gottingen (GOE) series was selected for the runner's blade. This study focused on defining the major characteristics of the turbine with a given head and flow rate.

7.4.1.2 Multidisciplinary Design Optimization

The process of design optimization for hydro turbines is complicated due to the complexity of the rotor and the stator geometries. Several variables in the design optimization process entail an engineering optimization technique called multidisciplinary design optimization (MDO). The MDO method is followed in this research to achieve the optimized design of the turbine.[18]

Several fields of engineering, such as aerospace and automotive, have extensively considered MDO in their applications, and this technique has paved the way for designers to integrate all connected disciplines simultaneously with time and cost. Aerospace engineering applications (i.e., aircraft and spacecraft) were the leading implementers of MDO.[19–23] MDO helps avoid complications accompanying sequential design or partial optimization, providing more effective and vigorous convergence than simple iteration and supporting the management of the design procedure.[24]

Herein, multiple variables involved in the design optimization process of the turbine's rotor, main hub and blade dimensions, meridional contour, number of blades, number of spans, blade's leading edge, and trailing edge angles, and blade wrap angle are some examples of the design variables. This study limits the optimization process to the number of blades and blade wrap angle. Figure 7.23 shows the optimization algorithm for the design of Kaplan turbine rotor.

Figure 7.24 shows the velocity triangles of the inlet and the outlet at the mid-span of the blade. Fixed blades Kaplan turbine is considered for the study to avoid the complexity in manufacturability, mainly where 3D-printed parts are used in this project. The CAD of the baseline design is shown in Figure 7.25.

The design is generated to fit the available boundary conditions of the turbine; however, this design might not be the optimum one as some of the design variables depend on the designer's knowledge. Also, some variables have a range of applicable values for the given conditions, contributing to the turbine design's development, improvement, and optimization.

The first variable selected for the optimization process was the number of blades. The baseline design is a 6-blade rotor, as shown in Figure 7.25. Three, four, five, and seven blades are also considered. Evaluation for a matrix of thirty simulations is conducted to locate the maximum power output (in watts) for the five different numbers of blade rotors at the rotational speed range of 500–3000 rpm for each. Yen et al. studied the number of blades for the stator, and based on their study, the optimum number of blades was nine.[25]

The blade wrap angle ($\Delta\varphi$) is the second variable to examine for further optimization, where the wrap angle is defined as the angle between the leading and trailing edges of tangent lines. Also, this angle is recognized by two angles at the trailing edge: the starting angle at the hub side and the ending angle at the shroud side.

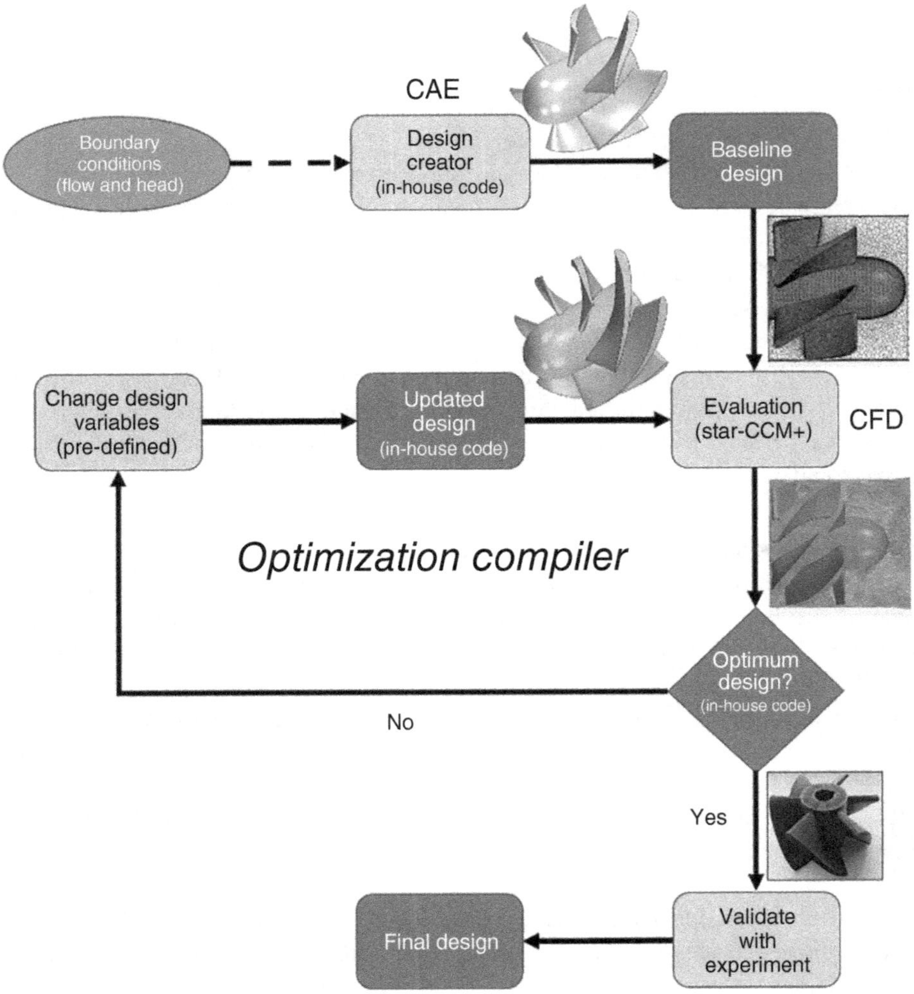

Figure 7.23 Optimization algorithm for the design of Kaplan rotor.

In Figure 7.25, for example, the blade wrap angle is 80° at the hub side of the blade and 60° at the shroud side. Thus, the blade wrap angle is 80°–60° with a linear relation to describe another four angles between the starting and ending angles. The total number of sub-angles for a wrap angle is equal to the number of spans of the blade. Figure 7.26 shows the design details of the baseline rotor; six spans are presented at the suction side of the blade.

Another matrix of 121 simulations is established to evaluate the power output of the rotor at 121 different wrap angles, from 60° to 110° with a 5° increment for both sides (the hub and the shroud).

Experimental Setup An experimental setup was developed in a laboratory managed by the University of Wisconsin-Milwaukee in Milwaukee's Global Water Center. It is a system with 2.6 m (8.5 ft) of the head and a maximum flow rate of 28 L/s (1.00 cfs). The turbine rotor outside diameter was set at 7.6 cm (3.0 in.) and 3D printed in the laboratory, as shown in Figure 7.27.

The setup included a tank with a head of up to 2.8 m (9.0 ft), a pump to circulate the water, two flow meters (one before the tank and one in the overflow pipe), and a turbine. The turbine had a transparent housing for observation (see Figure 7.28). The rotor is connected to a 5/8-in. diameter stainless steel shaft for the speed and torque

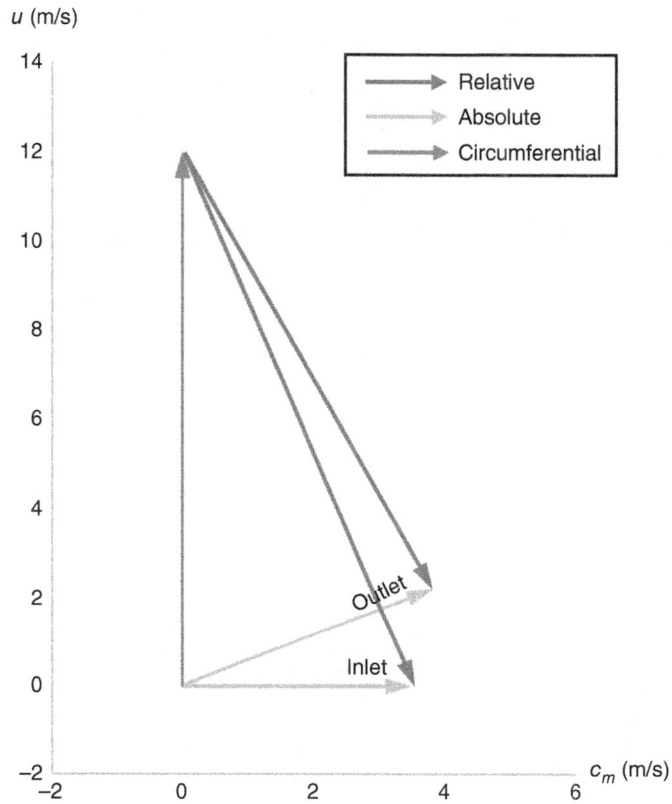

Figure 7.24 Velocity triangles at the mid-span for the inlet and the outlet.

measurement. The shaft was connected to a torque meter with an accuracy of less than 0.1% and a DC generator. The RPM and torque values were displayed using a device with an accuracy of 0.01% and 0.02%, respectively. Electric power is dissipated in a resistive load bank. An electromagnetic flow meter measured the flow rate with 0.25% accuracy.

The turbine's optimum rotational speed is not limited to 1500 rpm, as the trend lines in Figure 7.29 show that the turbine can keep the highest power output within the range of 1450–1700 rpm. Error bars represent a fixed error percentage of 5% for all data points.

Based on the previous first iteration of optimization, the 6-blade rotor is considered for the wrap angle simulation matrix. Figure 7.30 presents the geometry of the baseline design with an 80°–60° wrap angle versus the optimized design with a 100°–100° angle; the latter showed nearly 23% improvement in the power output (137 W) of the turbine over the baseline design (111 W). One reason for this increment is that the optimized design's long blade increases the blade passage area.

The complete solution of the matrix is shown in Figure 7.31. The figure illustrates the effect of the blade wrap angle on the turbine power output; 0% represents the baseline design where no improvements exist. Regardless of whether the five wrap angles would decrease the power output below the baseline for the cases 60°–60°/65°, 65°–60°, 70°–60°, and 75°–60° (hub-shroud), most of the angles caused improvement in the turbine performance. The four wrap angles 100°–100°, 100°–105°, 105°–100°, and 105°–105° marked the highest positive effect of 23% on the turbine power output with up to 25% additional power, compared to the lowest power output at the wrap angle of 60°–60° (−2%).

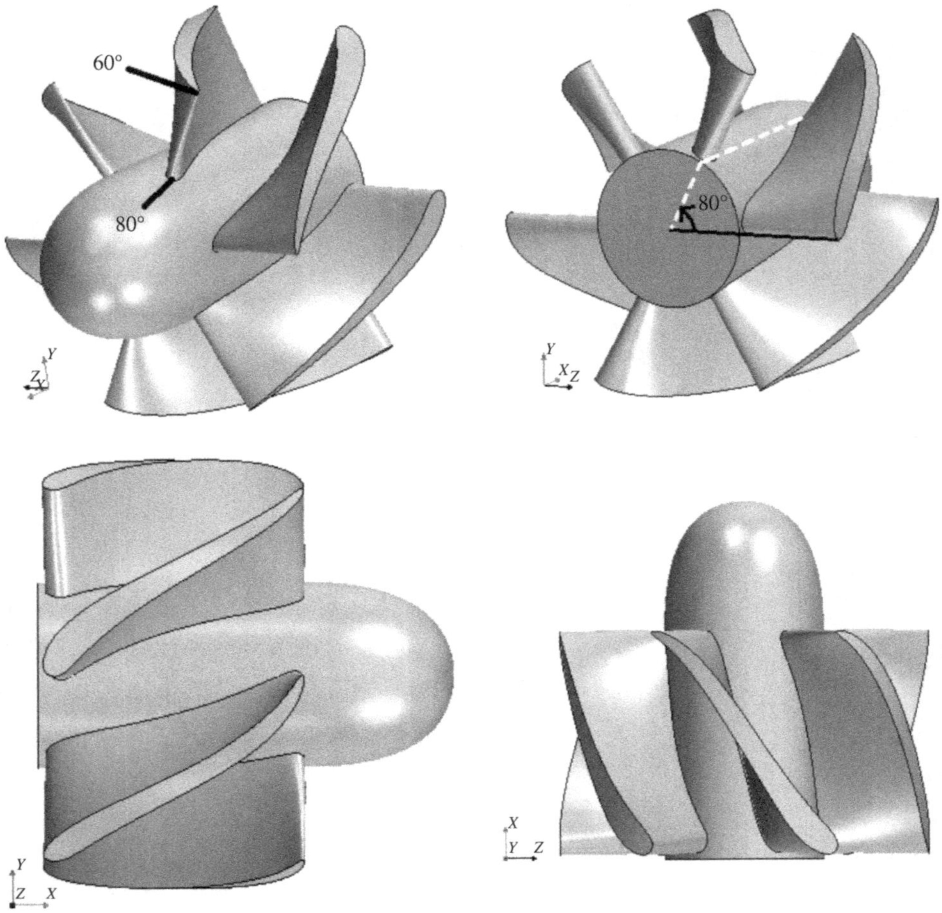

Figure 7.25 CAD for the proposed preliminary rotor design.

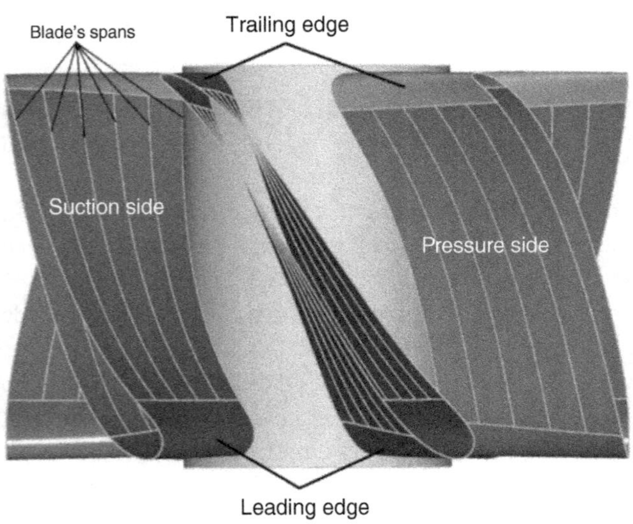

Figure 7.26 Significant design parts of the baseline rotor. (Trailing edge, leading edge, suction side, pressure side, and blade's spans).

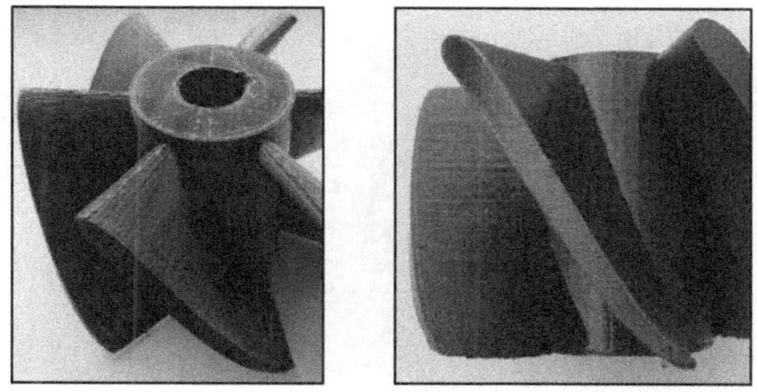

Figure 7.27 The 3D-printed 7.6 cm (3-in.) 6-blade rotor for the experimental testing.

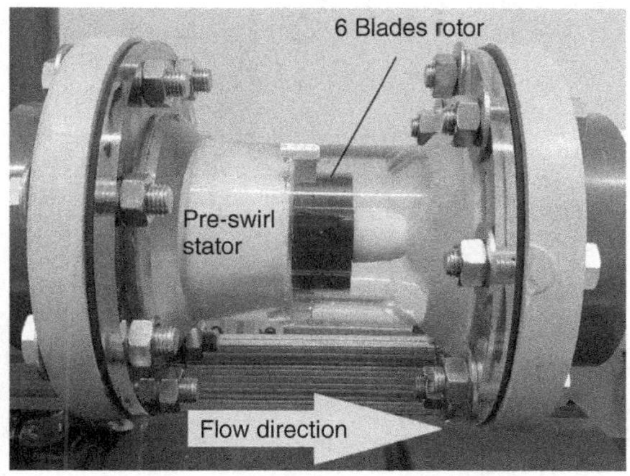

Figure 7.28 Turbine housing includes a 3D-printed pre-swirl stator and the 6-blade rotor under operation.

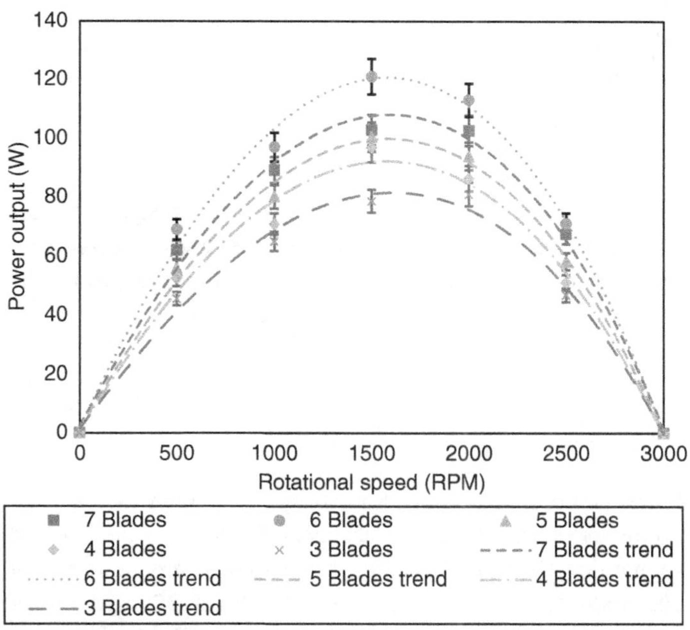

Figure 7.29 Performance curve of the baseline rotor design at a different number of blades and different rotational speeds.

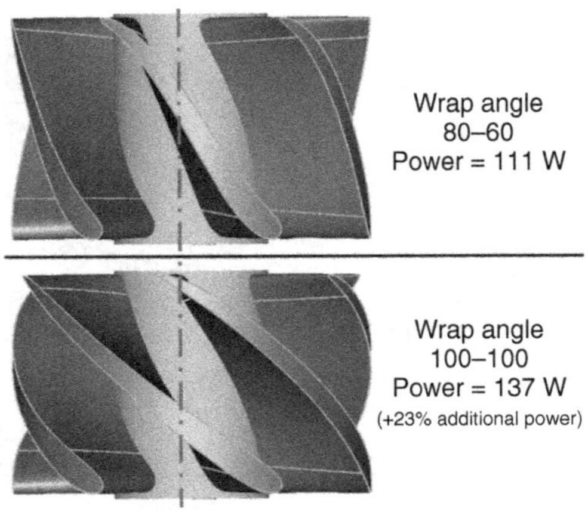

Figure 7.30 Effect of the blade wrap angle (hub to shroud) on the turbine power output.

Angle at the hub (°)	Angle at the shroud (°)										
	60	65	70	75	80	85	90	95	100	105	110
60	-2%	-1%	2%	3%	4%	6%	8%	10%	12%	13%	15%
65	-1%	2%	3%	4%	6%	8%	9%	11%	13%	14%	15%
70	-1%	3%	5%	6%	8%	9%	9%	12%	13%	15%	16%
75	-1%	4%	6%	7%	9%	10%	11%	13%	14%	16%	17%
80	0%	5%	8%	9%	11%	12%	13%	15%	15%	17%	18%
85	1%	5%	9%	11%	12%	13%	14%	16%	17%	18%	19%
90	2%	6%	10%	12%	13%	14%	15%	17%	19%	20%	20%
95	3%	7%	10%	13%	14%	16%	17%	20%	22%	21%	21%
100	4%	8%	11%	13%	15%	17%	19%	21%	23%	23%	21%
105	4%	8%	12%	13%	16%	18%	20%	22%	23%	23%	22%
110	5%	9%	13%	15%	17%	19%	21%	21%	21%	22%	22%

Figure 7.31 Effect of the blade wrap angle on the power output of the turbine.

For all angles larger than 105°–105°, it was anticipated that the improvement would be the largest as the blade was supposed to have a larger passage area. However, the larger the angles, the higher the losses occurred near the trailing edge, negatively affecting the performance. Eventually, for optimum power output for the turbine, the blade's wrap angle range should be within 100°–105° at both the hub and the shroud sides.

7.4.2 Rim-Drive Design

7.4.2.1 Introduction

The objective of this subsection is to study the design and optimization of Kaplan hydro turbines for very low head (less than 3 m), with a particular emphasis on the use of rim-drive electrical generators. The work is based on an experimental setup and computation fluid dynamics (CFD) analysis of a variety of design parameters for maximum output power and efficiency. Two designs are presented in the paper. One is a 90-cm (35-in.) diameter vertical-oriented Kaplan hydro turbine system as an intended product capable of generating over 50 kW. The other is a smaller, 7.6-cm (3-in.) diameter horizontal-oriented system for prototyping and laboratory verification. Specific designs for the runner and the stator, intake tube shape, as well as guide vanes upstream of the turbine, were studied to get the most from the available head.

Currently, renewable energy plays a significant role in new power generation worldwide. Hydropower, wind, biomass, and photovoltaic are the leading renewable energy streams with 99% of total renewable sources. These streams add up to hundreds of gigawatts in global energy generation. Furthermore, these mitigate tons of greenhouse gases emissions.[26, 27] Based on REN21's 2017 report, renewables contributed 19.3% to energy consumption and 24.5% to electricity generation in 2015 and 2016, respectively.[28]

Currently, hydropower supplies 16.6% of the global electricity production[29] and 86% of the renewable energy sector.[30] The global installed hydropower capacity is 1171 GW, and the generated power from all different sources is around 3983 TWh/year. In the United States, with about 2400 dams are utilized for hydropower represents 9.25% of overall generating capacity, accounts for 83% of U.S. renewable energy capacity and 77% of actual renewable electricity generation.[31] However, most of the installed hydropower consists of large plants. Much potential hydro generation remains untapped, particularly at lower power and head levels.[32] This subsection focuses on a study in finding a power-efficient solution to this situation, precisely, at heads, less than 3 m (10 ft), where more energy is available on the aggregate but with a lack of viable solutions both technically (efficiency wise) and economically. Specifically, the authors envision leveraging 3D-printing techniques to produce units that are both low cost and optimized for specific sites. Such a design methodology, however, requires a thorough design optimization, which this paper addresses.

Bijukchhe[33] conducted a comparison between experimental and CFD results of 8-in. diameter, horizontal Kaplan turbines. The study described the numerical simulation of the prototype turbine at full and partial load and compares the results with experimental values for a range of 6–15 m (20–50 ft) of the head, flow rates of 200–340 L/s (7–12 cfs), and six different runner rotational speeds in the range of 1333–3350 rpm. The simulations match experimental results best for the optimum point but overpredict for the overload conditions.

Yen et al.[25] presented a study of a 30-cm (12-in.) diameter Kaplan hydro turbine geometry with performance optimization through CFD method based on large eddy simulation (LES) of transient turbulence modeling. The study included a selection of horizontal versus vertical setup, tailwater head test, draft tube length test, draft tube angle test, intake tube geometry test, and blade geometry combination test. All of these geometric parameters were used in the computation and the results were based on power output, mass flow rate, and efficiency.

ElGammal et al.[34] focused on the turbine's intake tube design based on numerical investigations to maximize energy extraction. Optimizing the efficiency of a system with less material cost was the primary goal of the study. The results showed the merit of higher slopes per side length in enhancing output power, with an average of 2.7% of the output power by full expansion from minimum to the maximum angle. A bell-mouth profile was also investigated and provided an additional 2.2% increase in turbine efficiency.

The current study expands on these earlier results in various ways, particularly with the use of a rim-driven electrical generator similar in concept to recently introduced ship propellers.[35, 36] It consists of a rotating ring around the outside of the runner blades, (see Figure 7.32). The generator stator is mounted outside the water conduit. This approach has some advantages: No shaft or other part is going from inside the water to the outside. The system has a benefit as a compact unit for easy installation including retrofits in existing plants, maintenance, and eventual replacement. With fewer requirements for the center of the runner, one can minimize the overall system diameter, making for a more compact design. The smaller size can make better use of the low head and will facilitate manufacturing, especially with 3D printing where the cost goes up non-linearly with size. Furthermore, it is possible to use a hub less runner, which even if less efficient may be advantageous regarding allowing debris, or fish, to go through.

Figure 7.32 Turbine model—runner attached to rim-generator ring.

For the current study, numerical optimization was based on flow simulations with LES for transient turbulence modeling. Both the larger and smaller systems were investigated with a maximum head of 2.7 m (9 ft) and a constant number of blades: five and nine blades for the runner and the stator, respectively.

7.4.2.2 Methodology

For this project, the propeller type (Kaplan) axial flow hydro turbine configuration was selected based on the maximum low-head available.[37] Such axial flow turbines are applicable with either horizontal or vertical orientation. The 7.6-cm system was designed horizontal, while the 90-cm system is vertical. Kaplan runners can feature fixed or adjustable blades. With low cost being a requirement for this application, fixed-blade Kaplan turbines were exclusively investigated.

Experimental Setup An experimental setup was developed in a laboratory managed by the University of Wisconsin-Milwaukee in Milwaukee's Global Water Center. It is a system with 2.7 m of the head and a maximum flow rate of 28 L/s (1.00 cfs). The turbine runner outside diameter was set at 7.6 cm, capable of a mechanical output on the order of 400 W. The setup is shown in Figure 7.33. It was used as a testbed for computational tools, for various ideas being considered, and for validation.

The setup included a tank to provide a head of up to 3 m (10 ft), a pump to circulate the water, two flow meters (one before the tank and one in the overflow pipe), and a turbine. The turbine had an embedded steel ring around the runner to emulate the electrical generator rotor. At this stage, the runner also had a shaft for measurement of speed and torque. As shown in Figure 7.33, this shaft was connected to a torque meter and a DC generator. Electric power is dissipated in a resistive load bank. An electromagnetic flow meter measured the flow rate.

The literature conjugated with same project conditions led to the conclusion that a combination of a nine-vane stator paired with a five-blade runner provided an efficient turbine while also providing sufficient flow to power the system.[25] Also, the blade thickness fraction should be 0.07, and the tip thickness fraction should be 0.03 to provide a safety factor of 2 with the materials available for 3D printing.

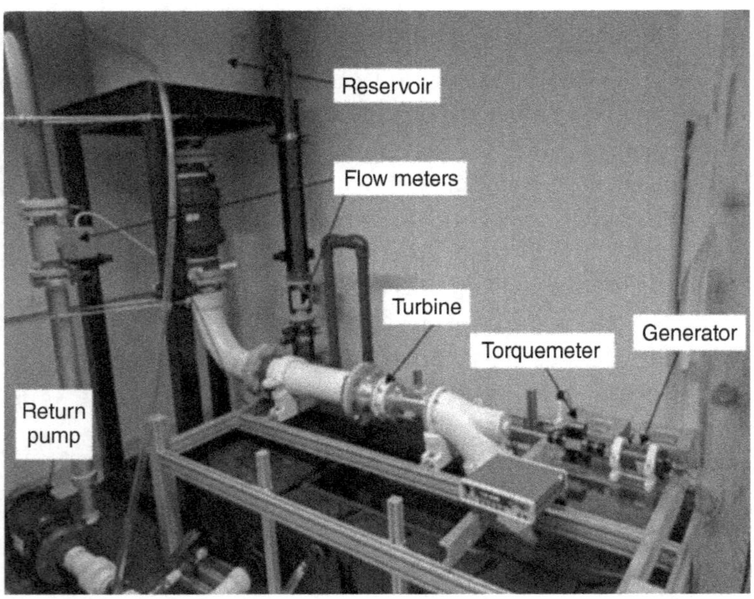

Figure 7.33 Experimental setup with the 7.6-cm turbine at University of Wisconsin-Milwaukee.

The turbine includes a 9-mm thick steel ring to emulate the electrical generator rotor, which will eventually be made of permanent magnets mounted on a steel ring. The computations led to an estimation that an eventual generator rotor would be made of a 5.5-mm steel ring supporting 3.5-mm thick magnets, with a length of 10 mm. Such a generator should be able to absorb 2 N-m and generate 1 kW (electrical, assuming a generator efficiency of 88%) at 5400 rpm. Since magnets have a specific mass close to that of steel, doing the experiments with a dummy rotor of 9-mm thickness was realistic regarding inertia and mechanical integrity of the rotor-runner assembly.

Figure 7.34a shows the tested turbine runner with a slot to slide the electrical generator rotor in place. This slot is integral with the runner, manufactured by 3D printing in a single step. Figure 7.34b shows both turbine stator and runner before assembly. Figure 7.34c shows a turbine runner/rim assembly in the process of being 3D printed (the different color is immaterial—several units were built during the project from the various source material).

The theoretical power equation for a water stream can be calculated as follows. The formulation sets the systems boundary conditions and makes it possible to verify the results regarding power and efficiency[38]:

$$P = \eta \times \rho \times g \times H \times Q \tag{7.1}$$

The use of two different modeling scales for the hydraulic system with variables like head, flow rate, rotational speed, diameter, and power requires a computational bridge between the two. The modeling is based on the affinity laws (similarity laws). Such laws should be considered for the estimation, and the comparison between the two systems and the most relevant law based on the given variables in this study is[39]:

$$\Pi_{\text{Prototype}} = \Pi_{\text{Model}} \tag{7.2}$$

$$\frac{P_p}{\rho_p D_p^5 N_p^3} = \frac{P_m}{\rho_m D_m^5 N_m^3} \tag{7.3}$$

A test facility consists of a water loop in a closed system with two reservoirs. The water is steadily circulated using a variable frequency drive (VFD) motor to pump. The flow rate, power, and torque of the turbine rotor are monitored through devices. The preliminary design for the rim-drive turbine and the results are shown in Figure 7.35. The design of the rim-drive is being improved using additive manufacturing technology. The data shows that the rim-driven turbine enhances efficiency by 10 % over a typical Kaplan turbine with a shaft.

The United States presently has 43 pumped storage hydropower (PSH) plants with more than 21 GW installed storage capacity. PSH dominates—by far—all other utility-scale storage nationwide, and the installed capacity can be potentially doubled. Generally, most of the installed capacity is in the form of reversible single-stage

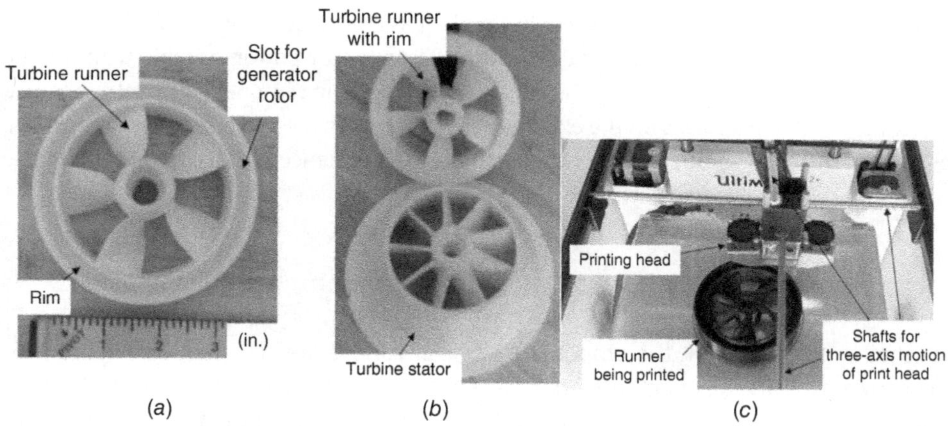

Figure 7.34 Turbine prototype made from 3D-printing process.

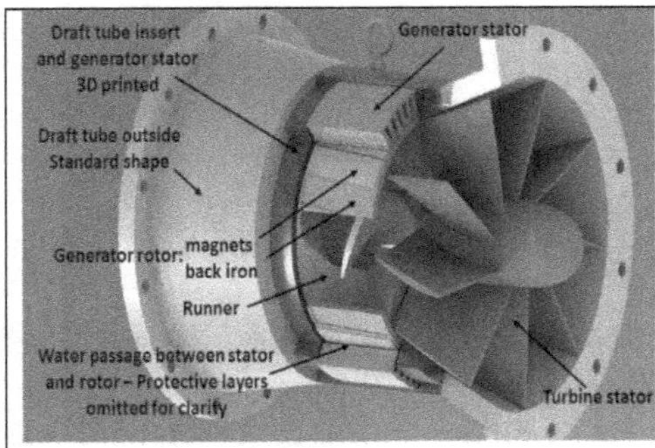	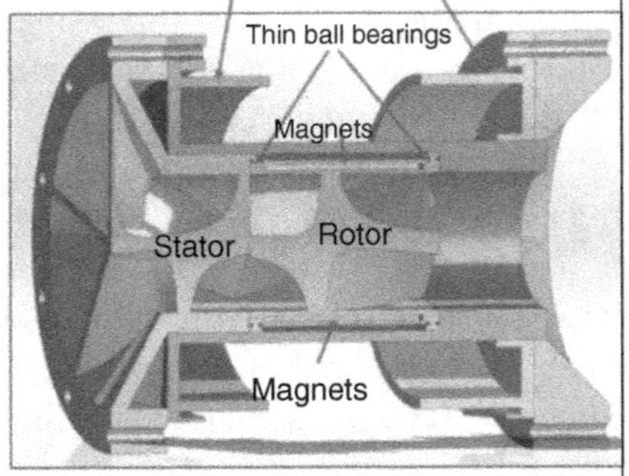
Currently using rim-driven turbine	Design of a new rib-driven turbine
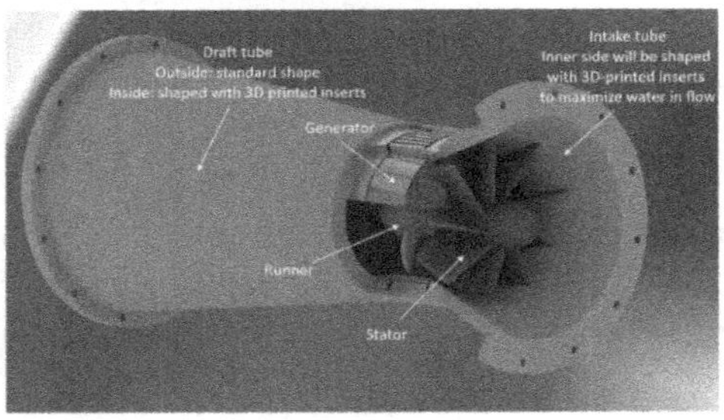	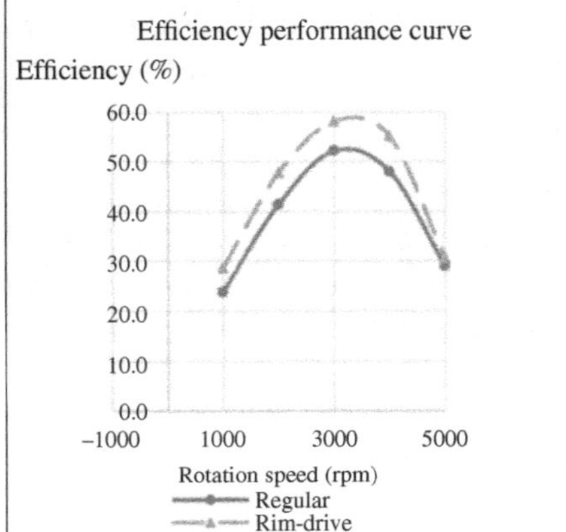
Overview with draft tube and intake	Efficiency performance comparison between regular and rim-driven turbines

Figure 7.35 Rim-driven turbine design.

pump-turbine systems that do not offer frequency adjustment while in pumping mode. Additionally, during the turbine part load, operation at peak efficiency cannot be granted. Although adjustable-speed units have a variety of other operational and performance characteristics that make them very desirable for power systems with a large share of baseload generation or, at the other end of the spectrum, for power systems with a high penetration of variable renewable energy (VRE), this type is very rare or even none in operation in the United States. Therefore, this project puts high advantages on the existing PSH power plants as it aims to use rim-driven hydro turbines to recover energy and, hence, offers innovative technology to enhance performance and reduce installation space.

Key Technical Risks

Key resources and risks required to implement the proposed recommendations are (1) initial cost, (2) legal issues associated with free stream turbine installation, (3) maintenance issues related to free stream turbines, and (4) project needs to achieve long-duration energy storage (LDES). The proposed solution will enlarge the components' life by 100%, but the maintenance plan must be accompanied well.

Energy Impact

The energy impact has the main advantage, which has a highly site-specific in their performance, costs, and effects. Therefore, focusing on the processes that lead to sustainable systems is crucial, not just on PSH performance and cost indicators. Several value propositions are:

Great Savings: PSH has several advantages: the capability to store large amounts of energy for long periods of time; the ability to provide large quantities of flexible, dispatchable generating capacity; a very long economic lifetime (100 years or more); and a long cycle life without significant degradation of performance. Typically, if adequately maintained, PSH infrastructure is designed for 80 or more years of operation. The power units (pump turbines and motor generators) may last 40–50 years with the rim-driven unit. The proposed recommendations can achieve significant savings in electricity consumption, natural gas consumption, operating cost, and financial resources that can be earned by selling extra electricity to the grid.

Benefits for Power Plants: Conventional electric power plants and natural gas production companies can achieve a demand reduction, mitigating the need for expansion in the power plants' capacities, which prevents the increase of the cost per unit of energy produced. Developing hybrid systems: coupling PSH with batteries, floating solar PV, heat storage, and desalination can provide additional services with reduced costs and environmental impacts.

Social and Environmental Benefits: Concerning environmental issues, fish, and debris are the main concerns surrounding turbine equipment in rivers and streams. Trash rack structures prevent debris and fish from moving downstream through the turbine. Compared to other schemes, the proposed concept has unique features: the rim design minimizes the need for sealing and isolating the water from the outside (important for drinking water, for instance). Furthermore, the possible use of a hubless design will avoid water basin interruption and allow fish and other life to navigate through with minimized risk. Because of the VRE, a reduction of CO_2 emissions associated with conventional electricity production will also be achieved.

Green Hydrogen Production: The integration of an electrolysis station with a pumped storage hydropower system for green hydrogen production. The green hydrogen produced from this novel solution offers huge potential for further development in the United States. Water recycling methods can be utilized to generate green hydrogen and oxygen products, especially in industrial facilities such as distilleries and wastewater treatment plants. Green hydrogen will be utilized by distilleries to replace fossil fuels in boilers, and unused green hydrogen from both operations will be sold on the market.

References

1 Renewable Energy Policy Network for the 21st Century, Renewables 2018 Global Status Report, Paris, France, REN21 Secretariat, 2018, http://www.ren21.net/gsr-2018/, Accessed 8 Jul. 2019.

2 Prasad, S., "Energy Efficiency, *Sources* and Sustainability," *ASME. J. Energy Resour. Technol.*, Vol. 132, No. 2, pp. 020301–020302, 2010. https://doi.org/10.1115/1.4001684.

3 World Energy Resources: Hydropower 2016, World Energy Council, London, United Kingdom, 2016, https://www.worldenergy.org/data/resources/resource/hydropower/, Accessed 8 Jul. 2019.

4 Wong, K. V., "Engineering Solutions to the Greenhouse Gases Generated by Hydroelectric Plants," *ASME. J. Energy Resour. Technol.*, Vol. 136, No. 2, pp. 024701–024703, 2014. https://doi.org/10.1115/1.4027260.

5 U.S. Energy Information Administration (EIA), Report: Annual Energy Production and Consumption by *Source*, U.S. Energy Information Administration, Washington, DC, USA, 25 Jun. 2019, https://www.eia.gov/electricity/monthly/, Accessed 8 Jul. 2019.

6 World Energy Resources, Report, 2016, World Energy Council, London, United Kingdom, 2016, https://www.worldenergy.org/wp-content/uploads/2016/10/World-Energy-Resources-Full-report-2016.10.03.pdf, Accessed 8 Jul. 2019.

7 2017 Hydropower Market Report, Oak Ridge National Laboratory U.S. Department of Energy, TN, USA, Apr. 2018, https://www.energy.gov/sites/prod/files/2018/04/f51/Hydropower%20Market%20Report.pdf, Accessed 8 Jul. 2019.

8 Derakhshan, S., and Kasaeian, N., "Optimization, Numerical, and Experimental Study of a Propeller Pump as Turbine," *ASME. J. Energy Resour. Technol.*, Vol. 136, No. 1, pp. 012005–012007, 2014. https://doi.org/10.1115/1.4026312.

9 Yang, C. S., "Performance of the Vertical Turbine Pumps as Hydraulic Turbines," *Proceedings of symposium on performance characteristics of Hydraulic Turbines and Pumps*, American Society of Mechanical Engineers, Boston, MA, Dec. 1983.

10 Wilbur, L. C. (Ed.), *Handbook of Energy System Engineering*, John Wiley & Sons, New York, 1985.

11 Gulliver, J. S., and Arndt, R. E., *Hydropower Engineering Handbook*, McGraw-Hill, New York, 1991.

12 Schweiger, F., and Gregor, J., "Developments in the Design of Water Turbines", *International Water Power & Dam Construction*, Reed Business Publishing, Sutton, Surrey, UK, May 1989, pp. 23–26.

13 Warnick, C. C., et al., *Hydropower Engineering*, Prentice-Hall, New York, 1984.

14 Moody, L. F., and Zowski, T., "Hydraulic Machinery," in *Handbook of Applied Hydraulics*, Davis, C.V., and Sorensen, K.E. (Eds), McGraw-Hill, New York, 1969. pp. 26–36.

15 "New Austrian Stamps," The Sun. No. 1765, Sydney, 24 Jan. 1937, p. 13. Retrieved 10 Mar. 2017.

16 Abeykoon, C., and Hantsch, T., "Design and Analysis of a Kaplan Turbine Runner Wheel," *Proceedings of the 3rd World Congress on Mechanical, Chemical, and Material Engineering*, Rome, Italy, Paper No. HTFF 151, 2017. https://doi.org/10.11159/htff17.151.

17 Jost, D., Skerlavaj, A., and Lipej, A., "Improvement of Efficiency Prediction for a Kaplan Turbine with Advanced Turbulence Models," *J. Mech. Eng.*, Vol. 60, No. 2, pp. 124–134, 2014. https://doi.org/10.5545/sv-jme.2013.1222.

18 Abbas, A., Qandil, M., Al-Haddad, M., and Amano, R. S., "Investigation of Horizontal Micro Kaplan Hydro Turbine Performance Using Multidisciplinary Design Optimization," *Trans. ASME J. Energy Resources Technology*, Vol. 142, No. 5, 052101, pp. 8, Paper No: JERT-19-1465, May 2020. https://doi.org/10.1115/1.4045821, Published Online: 20 Dec. 2019.

19 Permana, E., and Rudianto, Y., "Design and Velocity Distribution of Runner Blade of Kaplan Turbine Using CFD (Computer Fluid Dynamic) for Small Hydroelectric Power Plant," *IOP Conf. Ser.: Mater. Sci. and Eng.*, Vol. 180, No. 1, pp. 012269, 2017. https://doi.org/10.1088/1757-899X/180/1/012269.

20 Singh, S. K., "A Novel Concept for Non-Linear Multidisciplinary Aerodynamic Design Optimization," *Aerosp. Sci. Technol.*, Vol. 70, pp. 626–635, 2017. https://doi.org/10.1016/j.ast.2017.08.043.

21 Ujwala, M., Chowdary, P. C. K., and Naik, L. S., "Design and Analysis of Low Head, Lightweight Kaplan Turbine," *Int. Ref. J. Eng. Sci.*, Vol. 6, No. 2, pp. 17–25, 2017. ISSN: 2319-183X.

22 Wang, L., Xiong, C., Hu, J., Wang, X., and Qiu, Z., "Sequential Multidisciplinary Design Optimization and Reliability Analysis Under Interval Uncertainty," *Aerosp. Sci. Technol.*, Vol. 80, pp. 508–519, 2018. https://doi.org/10.1016/j.ast.2018.07.029.

23 Yao, W., Chen, X., Luo, W., van Tooren, M., and Guo, J., "Review of Uncertainty-Based Multidisciplinary Design Optimization Methods for Aerospace Vehicles," *Prog. Aerosp. Sci.*, Vol. 47, No. 6, pp. 450–479, 2011. https://doi.org/10.1016/j.paerosci.2011.05.001.

24 Zhang, M., Gou, W., Li, L., Wang, X., and Yue, Z., "Multidisciplinary Design and Optimization of the Twin-Web Turbine Disk," *Struct. Multidisc. Optim.*, Vol. 53, No. 5, pp. 1129–1141, 2016. https://doi.org/10.1007/s00158-015-1373-2.

25 Yen, Y., ElGammal, T., Amano R.S., Millevote, J., Mueller, R.J., and Lequesne, B., "Numerical Optimization on Micro Kaplan Hydro Turbine System," *Fluids Engineering Division Summer Meeting*, Washington, DC, 2016. https://doi.org/10.1115/FEDSM2016-7575.

26 Wong, K. V., "Engineering Solutions to the Greenhouse Gases Generated by Hydroelectric Plants," *ASME. J. Energy Resour. Technol.*, Vol. 136, No. 2, pp. 024701–024703, 2014. https://doi.org/10.1115/1.4027260.

27 Prasad, S., "Energy Efficiency, *Sources* and Sustainability," *ASME. J. Energy Resour. Technol.*, Vol. 132, No. 2, pp. 020301–020302, 2010. https://doi.org/10.1115/1.4001684.

28 Renewable Energy Policy Network for the 21st Century, Renewables 2017 Global Status Report, Paris, France, REN21 Secretariat, 2017, http://www.ren21.net/wp-content/uploads/2017/06/17-8399_GSR_2017_Full_Report_0621_Opt.pdf, Accessed 10 Aug. 2018.

29 Hamududu, B., and Killingtveit, A., "Assessing Climate Change Impacts on Global Hydropower," *Energies*, Vol. 5, No. 2, pp. 305–322, 2012. https://doi.org/10.3390/en5020305.

30 World Energy Council, World Energy Resources: 2013 Survey, World Energy Council, London, England, 2013, https://www.worldenergy.org/wp-content/uploads/2013/09/Complete_WER_2013_Survey.pdf, Accessed 10 Aug. 2018.

31 International Energy Agency (IEA), Key World Energy Statistics 2016 Report, Paris, France, 2017, https://www.iea.org/publications/freepublications/publication/KeyWorld2017.pdf, Accessed 10 Aug. 2018.

32 Derakhshan, S., and Kasaeian, N., "Optimization, Numerical, and Experimental Study of a Propeller Pump as Turbine," *ASME. J. Energy Resour. Technol.*, Vol. 136, No. 1, pp. 012005–012007, 2014. https://doi.org/10.1115/1.4026312.

33 Bijukchhe, V., "Comparison of Experimental Results of Horizontal Kaplan Turbine with Computational Fluid Dynamics," M.Sc. thesis, University of Iowa, Iowa City, IA, 2012, https://iro.uiowa.edu/esploro/outputs/graduate/Comparison-of-experimental-results-of-horizontal/9983777237702771 (accessed 1 May 2025).

34 ElGammal, T., Yen, Y., Amano R.S., Millevote, J., Mueller, R.J., and Lequesne, B., "Numerical Investigations on Intake Tube Design of Micro Kaplan Hydro-Turbine System," *Fluids Engineering Division Summer Meeting*, Washington, DC, 2016. https://doi.org/10.1115/FEDSM2016-7569.

35 Krovel O., Nilssen R., Skaar S.E., Lovli E., and Sandoy N., "Design of An Integrated 100 kW Permanent Magnet Synchronous Machine in A Prototype Thruster for Ship Propulsion," *Proceedings of 2004 International Conference on Electric Machines*, Beijing, China, 2004.

36 Abbas, A., Qandil, M., Al-Haddad, M., Saravani, M. S., and Amano, R., "Utilization of Hydro-turbines in Wastewater Treatment Plants," *ASME. J. Energy Resour. Technol.*, Vol. 144, No. 6, pp. 062011–062015, 2019. https://doi.org/10.1115/1.4042969.

37 Gorban, A. N., Gorlov, A. M., and Silantyev, V.M., "Limits of the Turbine Efficiency for Free Fluid Flow," *ASME. J. Energy Resour. Technol.*, Vol. 123, No. 4, pp. 311–317, 2001. https://doi.org/10.1115/1.1414137.

38 Dixon, S. L., and Hall, C. A., *Fluid Mechanics and Thermodynamics of Turbomachinery*, 7th ed., Butterworth-Heinemann, Waltham, MA, 2013.

39 Morrison, G., Yin, W., Agarwal, R., and Patil, A., "Development of Modified Affinity Law for Centrifugal Pump to Predict the Effect of Viscosity," *ASME. J. Energy Resour. Technol.*, Vol. 140, No. 9, pp. 092005–092009, 2018. https://doi.org/10.1115/1.4039874.

Problems

7.1 A Pelton wheel turbine is designed for a water head H of 100 ft. The jet diameter d is 6 in., nozzle efficiency η_n is 0.94, and maximum overall efficiency $\eta_{o/a}$ is 0.9, with a rotating speed N of 120 rpm. If the relative flow exiting from the bucket is such that $\beta_3 = 90°$ and $CV_b = 0.95$, determine the design flow rate, shaft power, torque, wheel diameter, and hydraulic efficiency.

7.2 A hydraulic turbine is to be designed to produce 36,000 hp at 95 rpm under 54 -ft water head. A model of the turbine is to be designed to produce 50 hp at 550 rpm. If the overall efficiency is estimated to be 88%, determine (a) the diameter of the model, (b) the diameter of the prototype, and (c) the type of turbine.

7.3 The mixed-flow hydraulic turbine shown is used to convert hydraulic energy of water at 25-m high to mechanical energy. The stator velocity coefficient is estimated to be 70%, and the overall efficiency of the turbine is 65%. If the impeller with inlet blade angle of 35° is operating at 120 rpm to obtain the best efficiency condition without swirl at discharge, determine the flow rate in (in m^3/s) and the output shaft power (in kW) (draw the velocity diagram at impeller inlet).

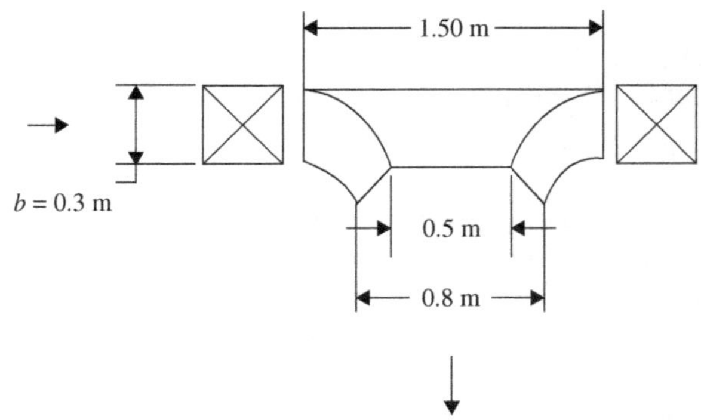

7.4 In problem 7.3, determine the hydraulic efficiency, maximum utilization factor ε_m, and required impeller blade angle at discharge, if the blade height there is also 0.3 m.

7.5 A hydraulic turbine is designed to produce 25,000 hp at 90 rpm under 150 ft of head. Laboratory facilities are available to provide 25 ft of head and to absorb 50 hp from the model turbine. The peak efficiency of both model and prototype turbines is estimated to be 85%. Determine the appropriate model test speed, scale ratio, and volumetric flow rate.

7.6 Preliminary calculations for a hydraulic power generation site show a net head of 2350 ft and a water flow rate of 75 ft^3/s are available. Select an appropriate turbine type and its size to run at (a) 450 rpm and (b) 600 rpm.

7.7 Conditions at the inlet to the nozzle of a Pelton wheel are $p = 4.81$ MPa (gage) and $V = 6.10$ m/s. The jet diameter $d = 20$ cm and the nozzle efficiency $\eta_n = 0.96$. The wheel diameter $D = 2.45$ m. If the turbine efficiency is 86% at this condition, calculate (a) the power output, (b) the normal operating speed, (c) the approximate runaway speed, (d) the torque at normal operating speed, and (e) the approximate torque at zero speed.

7.8 In English units, the common definition of specific speed for a hydraulic turbine is $N_s = NP_s^{0.5}/H^{1.25}$ (rpm $hp^{0.5}/ft^{1.25}$). Develop a conversion value between this definition and a truly nondimensional one. Evaluate the specific speed of an impulse turbine operating at 400 rpm under a net head of 1190 ft with 86% efficiency when supplied by a single jet of 6 in. diameter. Also estimate the wheel diameter.

7.9 Study on a hydropower site indicates that an effective net head of 24 m with a flow rate of 95 m^3/s are available. Determine the type of turbine to be installed and its rated output power, rotating speed, and possible size.

7.10 A Francis turbine of 2 m diameter rotating at 120 rpm has been tested to produce a shaft power of 205 MW under a head of 72 m and a flow rate of 320 m^3/s. If a larger turbine is to be scaled up for a higher head of 120 m while rotating at the same speed, determine the diameter of this new turbine and its corresponding flow rate, shaft power, and efficiency.

8

Wind Turbines

8.1 Introduction to Wind Power

Wind power is a well-proven and cost-effective technology. It is expected to be the main way the industry responds to the government's targets; so, it will become an important source of electricity in years to come. In this book, we have focused on basic and some advanced topics needed to understand the current situation in wind power engineering. This first section provides introductory materials for wind energy.

Wind as an energy source can be traced back some two thousand years to Persian windmills. Until the Industrial Revolution and the advent of the steam engine, windmills ranked second only to wood as an energy source. Wind power is also a form of renewable energy derived from solar energy. As portions of the earth are heated by the sun, air will rise due to buoyancy force. The air in the cool regions with higher pressure then rushes to fill this low-pressure area, creating wind power.

As the clean-energy industry emerges from the initial stage caused by the global economic downturn, it is entering a new stage of rapid business change. The worldwide energy demand is expected to double by 2030 and triple by 2050, when fossil fuels will account for no more than two-thirds of all energy consumed, compared with 80% of the energy consumed today. Traditional fossil sources such as oil, gas, and coal are not renewable and cause pollution by releasing huge quantities of carbon dioxide and other pollutants into the atmosphere, damaging the environment in many ways, from acid rain to climate change. To help combat these problems, many states in the United States are seeking ways to use renewable energy sources, such as wind, solar, and biomass. Along with its environmental and cost benefits, renewable energy is a rapidly growing industry with vast economic growth and job creation potential. The U.S. Secretary of Agriculture has identified wind, solar, and biomass as key factors for advancing the U.S. economy.

The wind resource in the United States is vast. Using today's technology, wind power is theoretically enough to supply all the electricity needed in the United States. Wind energy has recently become the world's fastest-growing source of renewable energy. The U.S. Department of Energy (DOE) expects wind energy to contribute 20% of the U.S. electricity supply by 2030. As a result, there has been a revived interest in wind turbines because they are emissions-free and the wind is renewable and cost-free; however, the amount of electricity generated and obtained by wind energy conversion systems is still unsteady, relatively expensive, and difficult to integrate into traditional electricity systems because of the variation in wind source and unresolved energy storage issues. On a large scale, spatial variability describes the fact that there are many different climatic regions in the world, some much windier than others. These regions are largely dictated by the latitude, affecting the insolation amount. Within any one climatic region, there is a great deal of variation on a smaller scale, largely dictated by physical geography—the proportion of land and sea, the size of land masses, and the presence of mountains or plains, for example. The resource map of wind energy in the U.S. Ref. 1 indicates the vast majority of available wind is very unsteady; strong wind zones are concentrated in certain regions and not uniformly distributed throughout the nation, making wind power collection more difficult. Conversely, the easy-to-collect wind energy is primarily

Fundamentals of Turbomachinery: Theory and Applications, Second Edition. Ryoichi S. Amano and William W. Peng.

Companion website: www.wiley.com/go/amano/turbomachinery2e

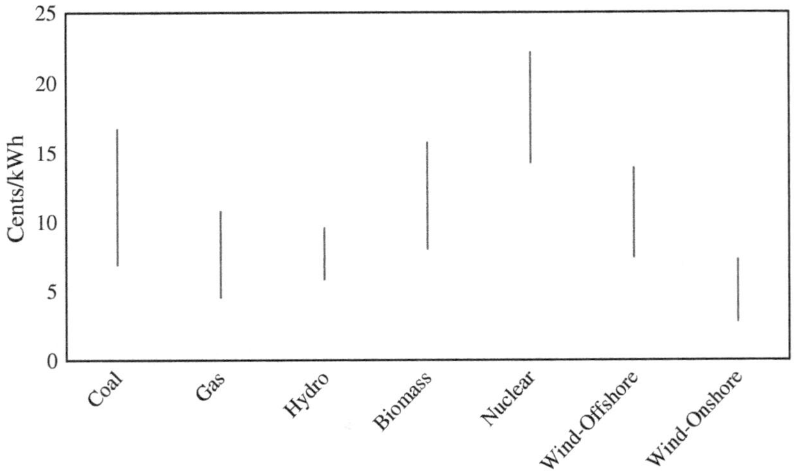

Figure 8.1 Predicted energy cost.[1]

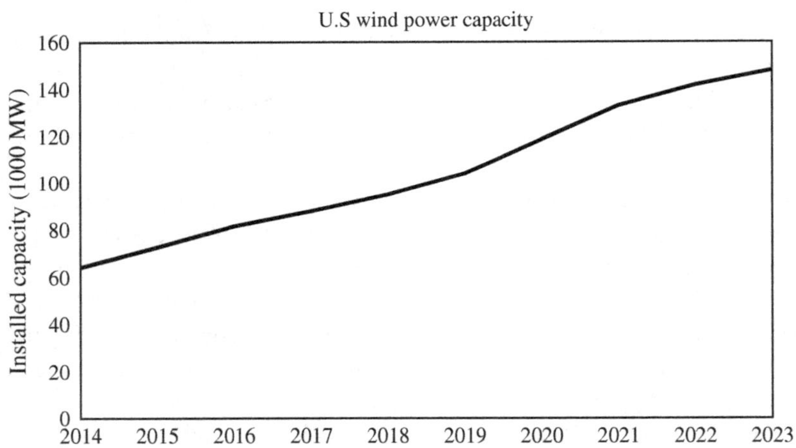

Figure 8.2 U.S. and global wind power capacity. (*Source:* IRENA (2024), *Renewable capacity statistics 2024*, International Renewable Energy Agency, Abu Dhabi.)

confined to remote locations, making electricity distribution difficult. The costs (Cent/kWh) of major energy sources are shown in (Figure 8.1).

The annual average available wind power in the United States and the world is shown in Figure 8.2.[2] The wind power density is related to the mean wind speed as $P/A \sim \rho V^3$. In reality, the wind speed at any location varies. Its variation over time is idealized with a frequency function called the Rayleigh speed distribution.

Wind availability varies according to season and geographical location. Public acceptance of wind energy conversion schemes is necessary for the design of any widespread usage of wind energy. The environmental impact of wind energy conversion systems is small compared with that of conventional electric power generating plants. One problem frequently brought up is noise, including acoustic noise in human hearing and interference with communication electromagnetic signals. The concern that migrating birds might fly into a wind turbine or tower should also be considered. These problems can be mitigated by proper siting of wind turbines using cable transmission, directing antennas, and nonmetallic materials for turbine blades.

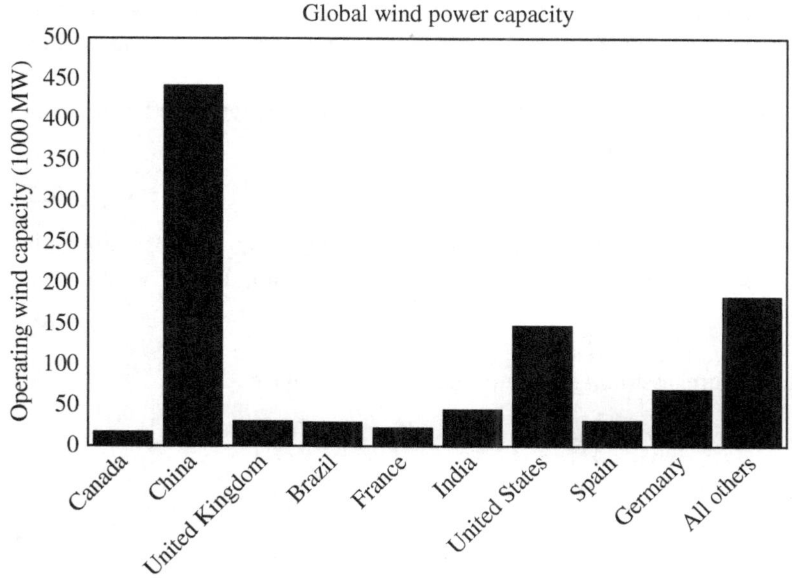

Figure 8.3 Wind farms on the hill and offshore.

The National Climatic Center in Asheville, North Carolina, collects and organizes wind speed distribution data in the continental United States.

In general, two possible siting locations for wind energy conversion systems are on hills or offshore, as shown in Figure 8.3. The ultimate selection of a wind energy conversion system is a lengthy procedure that is more or less based on intuition and good judgment rather than on an exact set of guidelines. Extensive analysis of meteorological data and wind characteristics is necessary. It includes the variation of wind velocity magnitude, direction concerning time, and elevation from ground level.

The most economical application of wind turbines to generate electricity is in groups of large machines (660 kW and up), called wind power plants or wind farms. Each plant can range from a few megawatts to hundreds of megawatts. Small wind turbines (from 250 W to 50 kW) are the ideal power source for remote sites without a utility system. However, due to their variability, these wind turbines should be used in hybrid or with storage systems, such as wind/photovoltaic/diesel or wind/compressed air/hydro-pumped storage. Successful application of wind turbines to harness wind energy depends on knowledge of several disciplines, including atmospheric fluid mechanics, aerodynamics, structure analysis, and electrical and control engineering.

8.2 Actuator Theory

To extract energy from a wind stream with a wind turbine, the first question to ask is, how much wind power is available? Or what is the maximum power that can be extracted with the best wind turbine of a certain size? This question can be answered with actuator theory.

As shown in Figure 8.4, the wind turbine is replaced with an actuator (disk) of the same size in the wind stream. The actuator causes a divergence of streamlines and a deceleration of the flow from an upstream speed of V to a downstream speed of $V - u$. The extreme

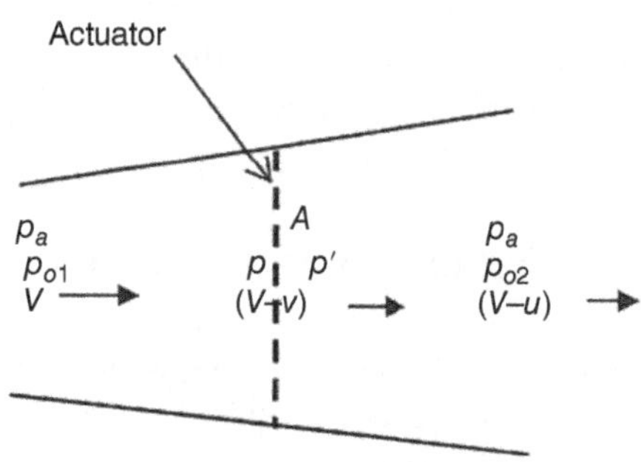

Figure 8.4 Actuator in a wind stream.

case of zero $(V - u)$ is impossible because that implies stagnant air downstream. Around the actuator, the flow speed is $V - v$. Hence, the mass flow rate of the stream is $m = \rho(V - v)A$. The magnitudes of u and v are related and will be shown.

The drag force on the actuator by the stream can be related to the axial momentum change from far upstream to far downstream where the static pressures are the same, that is,

$$F_d = \dot{m}[V - (V - u)] = \dot{m}u = \rho(V - v)Au. \tag{8.1}$$

It can also be expressed in terms of the static pressure difference immediately upstream and downstream of the actuator, that is, $F_d = (p - p')A$, since the axial-flow velocity across the actuator is constant. Hence, we have

$$p - p' = \rho(V - v)u. \tag{8.2}$$

Also, from the definition of stagnation pressure, assuming incompressible fluid, we have

$$p_{01} = p_a + \frac{1}{2}\rho V^2 = p + \frac{1}{2}\rho(V - v)^2, \qquad p_{02} = p_a + \frac{1}{2}\rho(V - u)^2 = p' + \frac{1}{2}\rho(V - v)^2.$$

Hence, the static pressure difference can also be expressed as

$$p - p' = \rho V u - \frac{1}{2}\rho u^2 = \rho u\left(V - \frac{1}{2}u\right). \tag{8.3}$$

Comparing Eqs. (8.2) and (8.3), it is concluded that $u = 2v$. Substituting this back into Eq. (8.1) results in $F_d = 2\rho Av(V - v)$. The wind turbine efficiency η is $\eta = P_s/(F_d V)$, where P_s is the shaft power of the actuator (ideal wind turbine) and $F_d V$ is the thrust power available from the stream. The shaft power is equal to the rate of kinetic energy change of flow through the turbine if the flow friction is negligible. Hence, we have

$$P_s = \dot{m}\left[\frac{1}{2}V^2 - \frac{1}{2}(V - u)^2\right] = \dot{m}u\left(V - \frac{1}{2}u\right) = F_d(V - v).$$

The efficiency can then be expressed as $\eta = 1 - v/V$, and the shaft power can be related to the efficiency as

$$P_s = \eta F_d V = 2\rho Av(V - v)\eta V = 2\rho AV^3\left(\frac{v}{V}\right)\left(1 - \frac{v}{V}\right)\eta = 2\rho AV^3\eta^2(1 - \eta).$$

To determine the maximum shaft power for a set of given ρ, A, and V, the above expression can be differentiated with respect to η and be set to zero such that

$$\frac{dP_s}{d\eta} = 2\rho AV^3(2\eta - 3\eta^2) = 0, \qquad \eta_{\max} = \frac{2}{3}.$$

Hence, the corresponding maximum shaft power is

$$P_{s,\max} = \frac{8}{27}\rho AV^3. \tag{8.4}$$

Or we can express $P_s = F_d(V - v) = \rho(V - v)Au(V - v) = 2\rho Av(V - v)^2$ and set $dP_s/d(v/V) = 0$ for the maximum power condition. Then, we also obtain $v_{\max} = V/3$ and $P_{s,\max} = 2\rho A(V/3)(2V/3)^2 = (8/27)\rho AV^3$. This result is frequently used to estimate the upper limit of the power attainable from a wind turbine with a cross-sectional area A in a wind stream of flow velocity V and air density ρ.

8.3 Types of Wind Turbines

Many types of wind turbines have been devised. They can be classified according to the orientation of their axis of rotation with respect to the wind stream.

Horizontal-axis rotors can be either upwind or downwind, as shown in Figure 8.5. The rotor can be designed with different numbers of blades, ranging from the one-blade rotor with a counterweight up to a rotor of 50 or

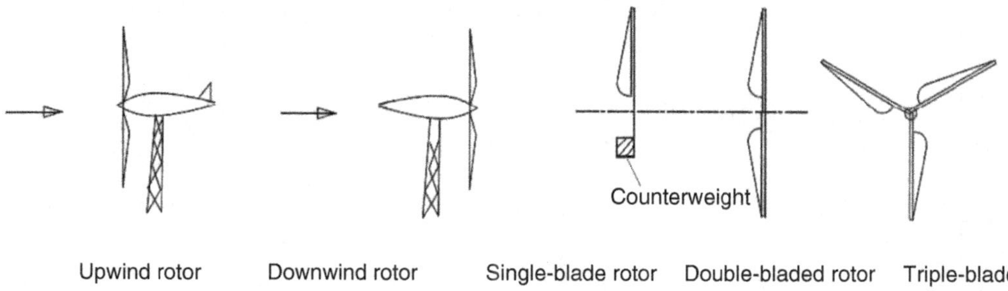

Figure 8.5 Horizontal-axis wind turbine rotors.

more blades such as the bicycle design, which is more rigid in structure. The ratio of the projected area of the rotor blades (on a plane perpendicular to its axis of rotation) to the swept area of the rotor is known as *solidity*:

$$\text{Solidity} = \frac{\text{blade area} \times \text{number of blades}}{\text{area swept by the rotor}}.$$

It can range from 0.01 to 0.7. This is related to the design rotating speed to be discussed in the next section.

The down-wind design can naturally align the axis with the wind direction. During a strong or gusty wind, the blades can be easily furled or coned downstream, as shown in Figure 8.6, to protect the blades from excessive stresses. This is not practical for upwind design. So, for medium and large units, the downwind design is preferred. However, there is a problem called "tower shadow" in the downwind design. The tower will obstruct the lower part of the wind, causing the blades to be subjected to an unsteady force when they pass the tower. Hence, a special tower design is needed to minimize its obstruction to the wind. For the small units, the upwind design is employed. A tail vane is used to align with the direction of the wind. Slight coning is acceptable during gusty wind.

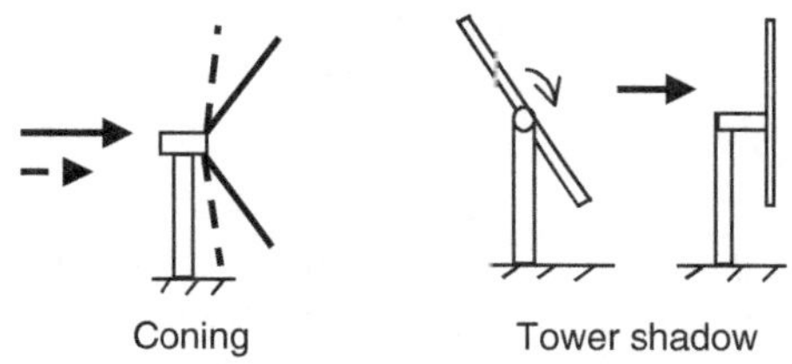

Figure 8.6 Coning and tower shadow for downwind design of a horizontal-axis wind turbine.

In the field, the rotor speed is variable because of the changing wind velocity. However, the electrical generator requires a constant speed to generate electricity with constant frequency. Hence, a transmission or speed regulator is required between the rotor and generator, as shown in Figure 8.7.

Vertical-axis rotors, for which the axis of rotation is perpendicular to both the earth's surface and the wind stream, are shown in Figure 8.8. These types of rotors have a major advantage over horizontal-axis rotors. They do not have to be turned into the wind stream as the wind changes direction. This reduces the system design complexity and decreases the gyro forces on the rotor. But they are lower in rotational speeds, starting torque, and shaft power output. Also, the system is dynamically unstable and sensitive to the off-design condition.

The Savonius (or S) rotor is a drag-type device. The drag force on one blade is greater than the other, and hence, the rotor will rotate. Its rotor tip speed is slower than the wind speed. The lower rotating speed and power output make generating electricity impractical. But its simplicity and self-starting characteristics with relatively high torque at low speed make it ideal for starting other machines or driving irrigation water pumps.

The Darrieus is a lift type with airfoil-shaped blades. However, it has low torque at low wind speeds and is not self-starting. So, it requires external power to start, either an electrical motor or a Savonius turbine. Some variations of vertical-axis wind turbines have been envisioned recently; two examples are shown in Figure 8.9.[3] The H-rotor turbine is installed on an abandoned industrial chimney.

Detailed operating principles of lift-type machines, both horizontal and vertical axes, will be discussed in the next section.

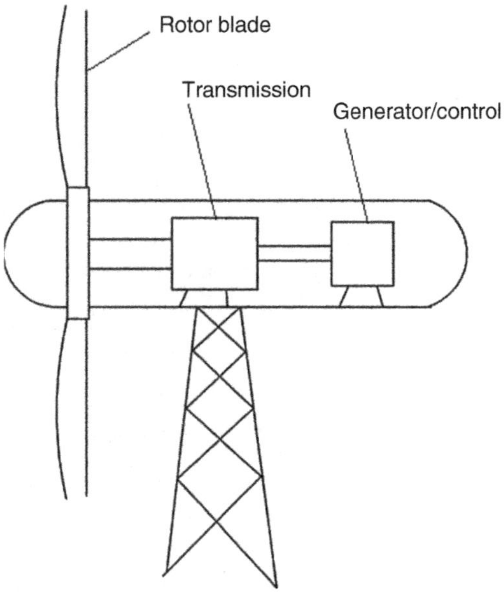

Figure 8.7 Components inside a horizontal-axis wind turbine.

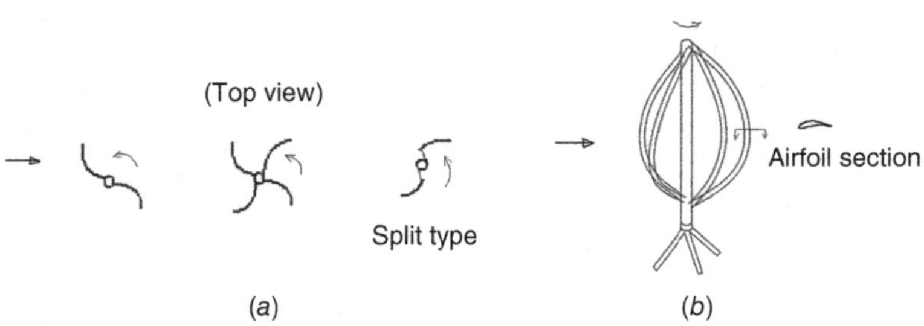

Figure 8.8 Vertical-axis wind turbines. (*a*) Savonius turbines (drag type). (*b*) Darrieus turbine (lift type).

8.4 Wind Turbine Characteristics and Preliminary Design Analysis

It was shown in Section 2.2 that wind turbine performance can be expressed in terms of two nondimensional parameters, that is, power coefficient defined as $C_p = P_s/(\rho V^3 A/2)$ and rotor tip speed ratio $\lambda = r_t\omega/V$. The characteristic curves for various types of wind turbines are shown in Figure 8.10. They are similar to the basic performance curves for the gas/steam and hydraulic turbines discussed in previous chapters. The peak power coefficient occurs at different tip speed/wind velocity ratios for the different types of turbines, as shown. The numbers of blades (Z_b) and the rotor tip speed ratio ($r_t\omega/V$) at the maximum power coefficient for the horizontal-axis turbines can be related approximately as $Z_b \approx 12/(r_t\omega/V)$, based on the test data given in Ref. 5. Hence, the impeller with less blades will be used for the higher rotating speed machine.

Velocity diagrams upstream and downstream of a horizontal-axis propeller-type turbine can be drawn as shown in Figure 8.11 for a radial position r along the blades. Since the airflow is not confined, its effective velocity throughout the blades is reduced by a factor a, called the axial interference factor (it was obtained to be $\frac{1}{3}$ under the maximum-efficiency condition from actuator theory, discussed in Section 8.2). Its actual magnitude and the deviation of relative flow velocity from the blade trailing edge have to be determined from empirical correlation. A theoretical treatment based on the combined actuator theory and airfoil aerodynamics results in the relationship

(a)

(b)

Figure 8.9 Variation of vertical-axis wind turbines. (a) Anemometer-type turbine. (b) H-rotor turbine installed on top of abandoned industrial chimney. (Courtesy of Mechanical Engineering magazine, Vol. 126/No. 6, June 2004, pp. 28–31; © the American Society of Mechanical Engineers, New York.)

shown in Figure 8.12,[4] where $\varepsilon = C_d/C_L$ and $\phi = 90° - \beta_l$. The interference relative flow angle can also be expressed in the form of $\tan \beta_l = r\omega(1 + a')/[V(1 - a)]$, where a' is the tangential interference factor.

From airfoil data, such as those shown in Figure 3.40 or Appendix B, the lift and drag coefficients can be utilized to compute the forces exerted on the rotor. Hence, the axial and tangential components of the force per unit length are given as

$$F_a = \frac{\rho W_m^2 c(C_L \sin \beta_m + C_d \cos \beta_m)}{2g_c}, \qquad F_u = \frac{\rho W_m^2 c(C_L \cos \beta_m - C_d \sin \beta_m)}{2g_c},$$

where c is the blade chord and β_m is the average relative flow angle, defined as $\tan \beta_m = (\tan \beta_{f1} + \tan \beta_{f2})/2$. The second term of F_u is subtractive, contrary to that of Eq. (3.15) for the axial-flow fan. Then, the torque on the rotor is computed by integrating $F_u r$ from hub to tip, that is, $\tau = Z_b \int F_u r \, dr$, which can be evaluated numerically. The turbine shaft power follows from $P_s = \tau\omega$. For a preliminary design, chord c is suggested as $c = 4\pi r \cos^2 \beta_m/(Z_b C_L \sin \beta_m)$ by Ref. 6. An iteration process is generally needed for the optimum design. Expressed in terms of the power coefficient $C_p = P_s/(\pi r_t^2 \rho V^3/2)$ versus the tip speed ratio $\lambda = r_t\omega/V$, the optimum design of the propeller wind turbines are shown in Figure 8.13.[4] It is shown that the maximum C_{pm} is equal to $16/27 = 0.592$ for $\varepsilon = 0$ as predicted from the actuator theory in Section 8.2.

For the Darrieus-type vertical-axis rotor, the airfoil-shaped blades move along a circular path, as indicated in Figure 8.14. This motion results in a continually changing angle of attack. However, the lift force on each blade always has a component in the direction of its motion. It can be explained from the velocity diagrams shown. It should be realized that the absolute flow velocity $(1 - a)V$ is assumed to be constant in both magnitude and direction, and the peripheral velocity U is constant at any angular position. A lift vector is sketched to be perpendicular to the relative velocity vector, **W** in each diagram. It can be shown that a tangential component of the lift vector exists in any position except when the blade is at the 3 and 9 o'clock positions, where all three velocities are lined up and the angle of attack is zero. Otherwise, a net torque is always produced, regardless of the blade's position, if the drag force is negligible.

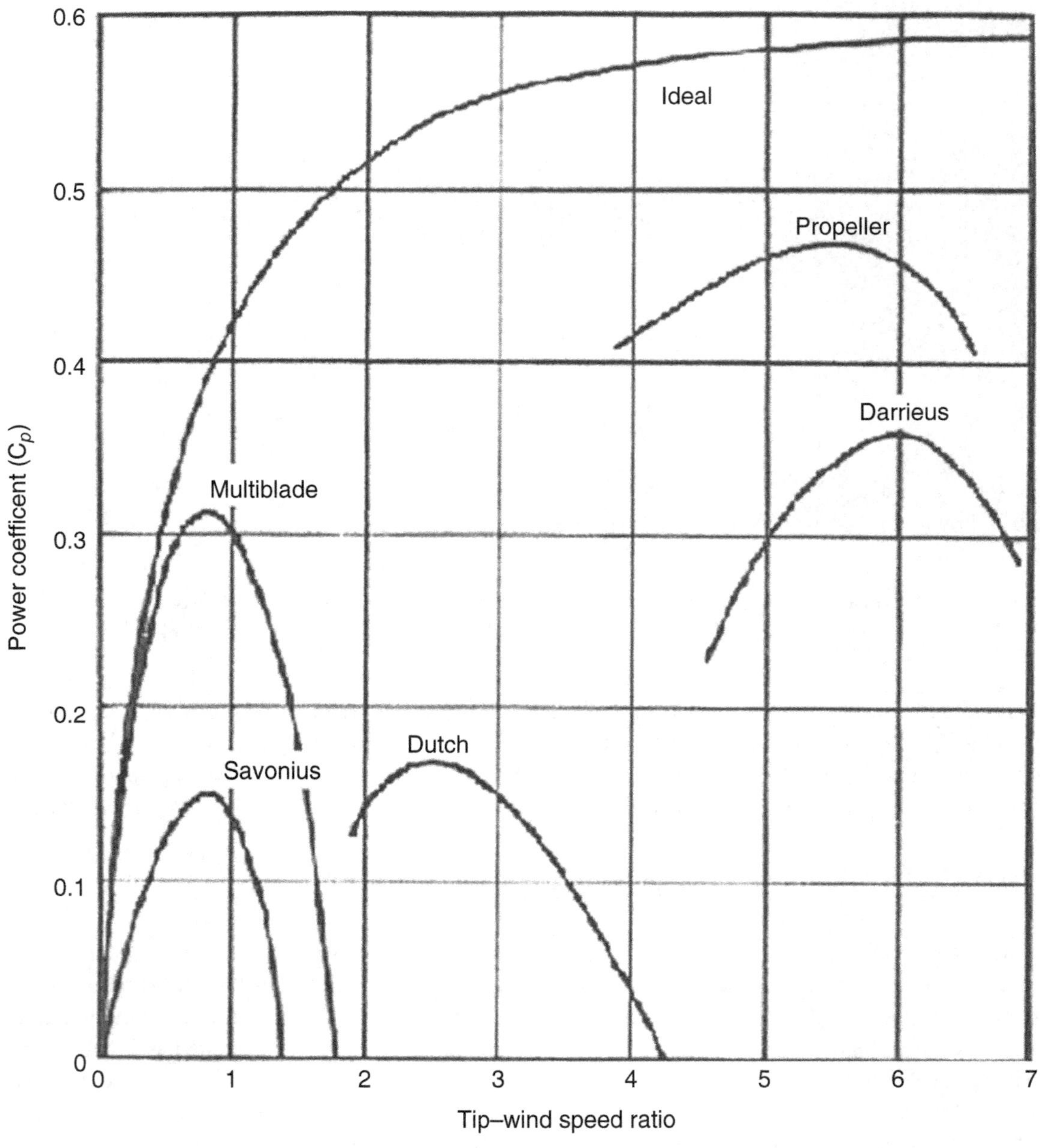

Figure 8.10 Typical performance curves for various types of wind turbines. (*Source:* Wortman[4]/Butterworth-Heinemann.)

The flow pattern is extremely complicated and different in each of the four quadrants. No rigorous analysis is available, except for some simplified analyses given in Refs. 6–8.

In designing a wind turbine, serious consideration should be placed on structure integrity. The blade materials and construction techniques include laminated wood, riveted stainless steel, and aluminum. Solid laminated wood can be used for small blades, whereas large blades can be built with a laminated wood spar and plywood ribs. Construction with composite materials and filament wound techniques are generally favored due to the absence of welded or riveted joints and their resistance to fatigue and corrosion. Composite blades also can better

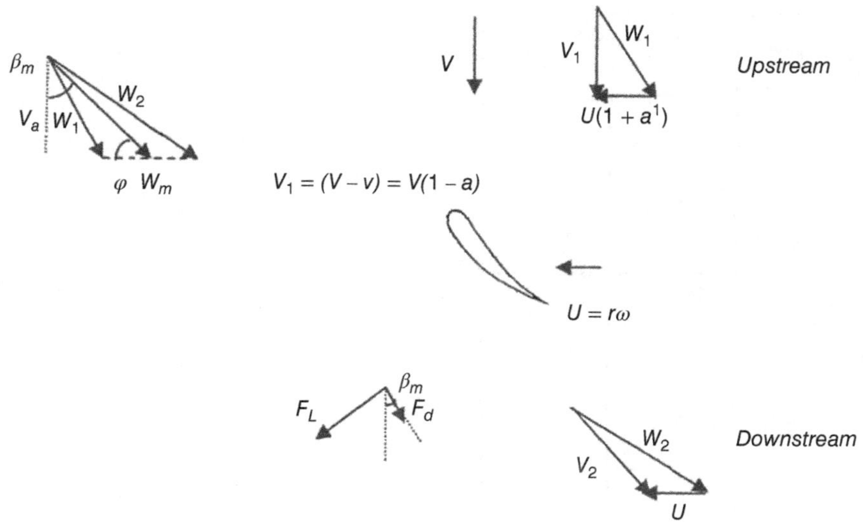

Figure 8.11 Velocity diagrams for a horizontal-axis wind turbine at mean radius r_m.

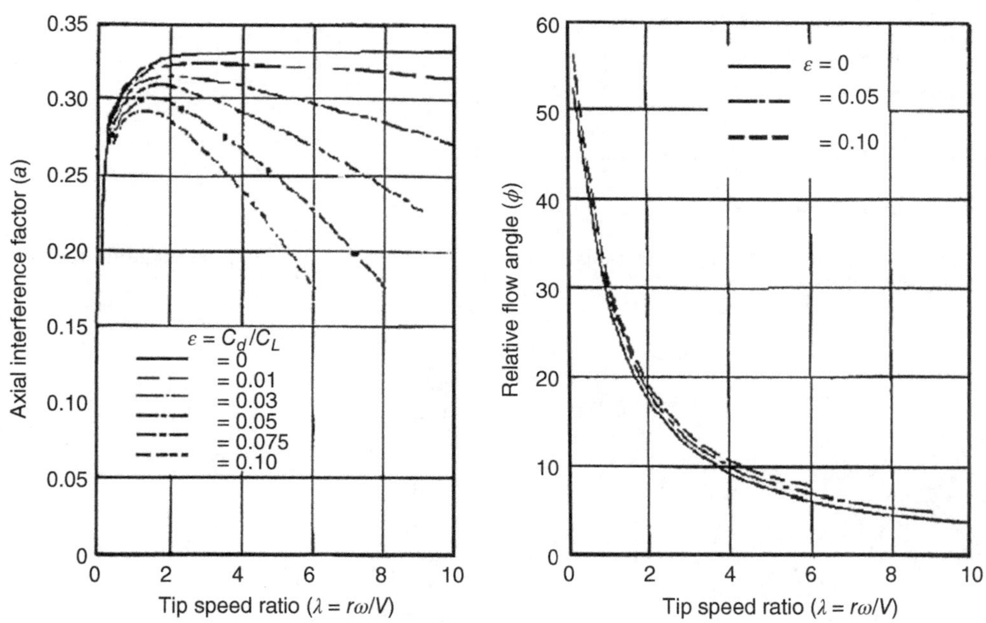

Figure 8.12 Variation of axial interference factor and relative flow angle with tangential tip speed ratio. (*Source:* Wortman[4]/Butterworth-Heinemann.)

resist the damage from foreign objects, are easier to repair, and have less impact to electromagnetic (EM) wave signals.

The detailed setup and procedure for field performance test of wind turbines are specified in Ref. 9. It includes the measurements of wind speed, direction, and ambient air temperature and pressure on a meteorological tower, in addition to the wind turbine power output during a given period of time.

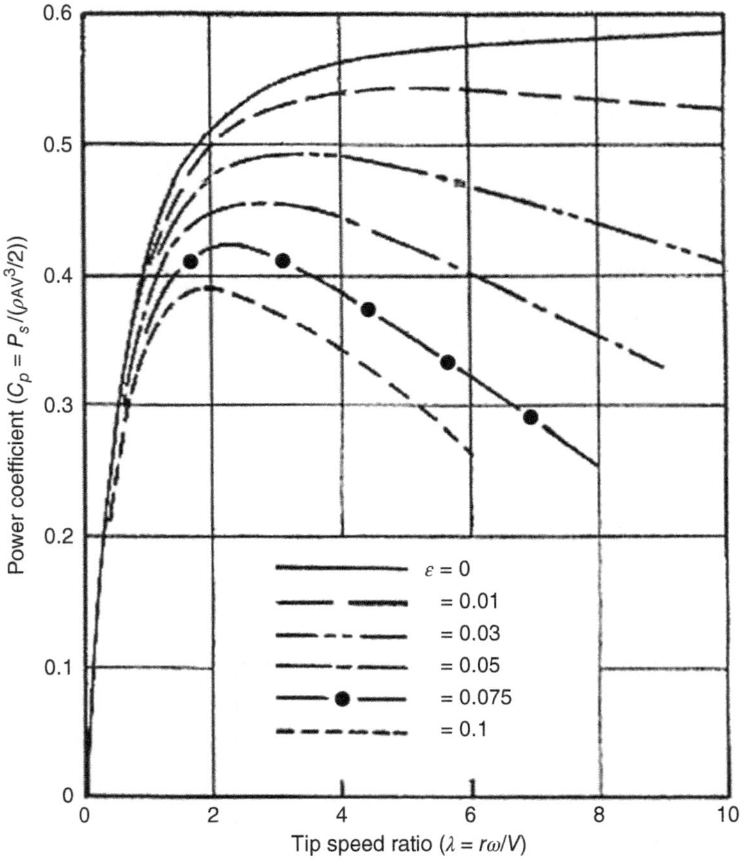

Figure 8.13 Variation of power coefficient versus tip speed ratio for propeller wind turbines with different drag/lift ratio of blades. (*Source:* Wortman[4]/Butterworth-Heinemann.)

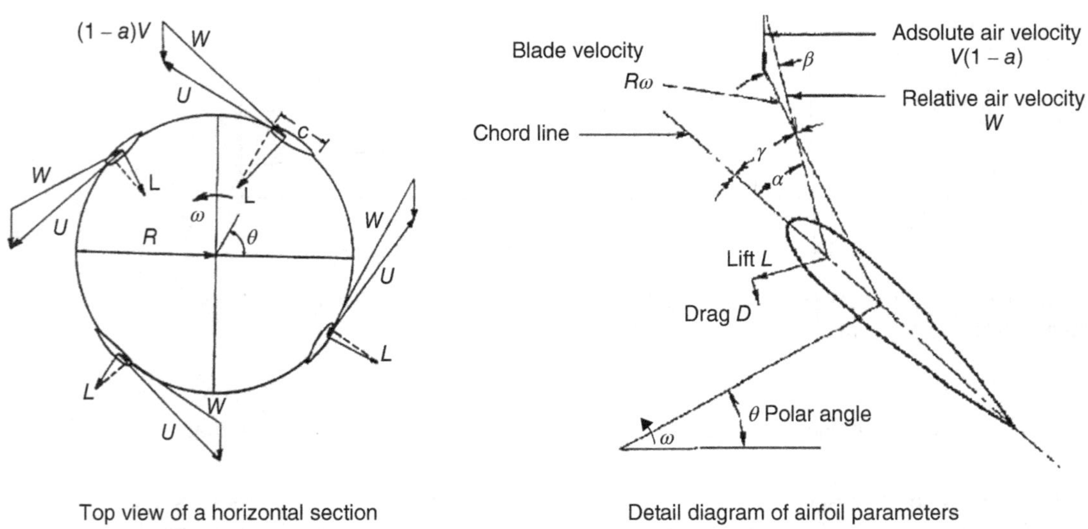

Figure 8.14 Velocity diagrams for a vertical-axis turbine.

Example 8.1 A double-blade wind turbine is to be designed to generate electrical power of 1.5 MW under wind speed of 13 m/s. The estimated generator and transmission efficiencies are 96% and 94%, respectively. Determine the rotor diameter and rotating speed under the rated condition.

Solution

From Figure 8.10, the maximum C_p is read as 0.47, corresponding to $\lambda = R\omega/V = 5.3$ on the curve for the propeller. Hence, we have

$$P_s = \frac{P_e}{\eta_g \eta_m} = \frac{1.5}{0.96 \times 0.94} = 1.662\ \text{MW} = 1.662 \times 10^6\ \text{W}$$
$$= C_p\left(\frac{1}{2}\rho A V^3\right) = 0.47\left(\frac{1}{2} \times 1.222 \times A \times 13^3\right) = 630.9\,A,$$

or $A = 2634.7\ \text{m}^2 = r_t^2\pi$, or $r_t = 28.9$ m, or $D = 57.8$ m. Also, from $r_t\omega/V = 12/Z_b$, or $\omega = b \times V/r_t = 6 \times 13/28.9 = 2.7\ \text{rad/s} = N\pi/30$, we have $N = 25.8$ rpm.

Example 8.2 A horizontal wind turbine is designed to produce power from a wind stream of 40 mph. The preliminary specifications are given as $N = 80$ rpm, $r_t = 15$ ft. and $r_h = 1$ ft; the selected airfoil data at the mean radius are $C_L = 0.28$, $C_d = 0.015$, chord $c = 1.2$ ft, and exit blade angle $\beta_2 = 65°$ concerning the axial direction. Neglecting the airflow velocity reduction, that is, $a = 0$, and the relative airflow deviation at exit, determine the optimal number of blades, the shaft power based on the mean radius calculations, and the maximum possible shaft power based on actuator theory.

Solution

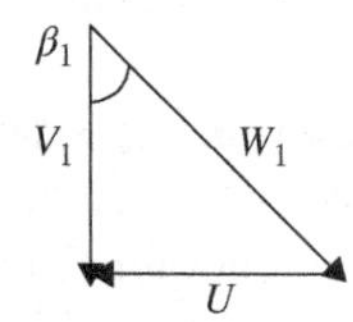

From $V = 40 \times 5280/3600 = 58.67$ ft/s, $\omega = N\pi/30 = 80\pi/30 = 8.38$ rad/s. Hence, the tip velocity ratio $r_t\omega/V = 15 \times 8.38/58.67 = 2.14$, and the optimal number of blades is obtained from $Z_b = 12/2.14 \cong 5$.

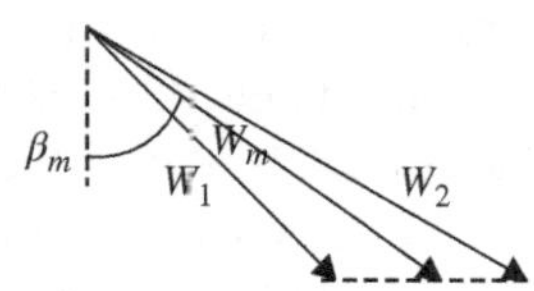

The mean radius can be obtained as $r_m = [(r_t^2 + r_h^2)/2]^{0.5} = [(15^2 + 1.0^2)/2]^{0.5} = 10.63$ ft, and hence, the blade peripheral velocity at the mean radius is $U_m = r_m\omega = 10.63 \times 8.38 = 89.1$ ft/s. Assuming $V_1 = V$, the relative flow angle at the inlet is $\beta_1 = \tan^{-1}(U_m/V_1) = \tan^{-1}(89.1/58.67) = 56.6°$. Hence, the mean relative flow angle can be obtained from $\tan\beta_m = 0.5(\tan\beta_1 + \tan\beta_2) = 1.831$ or $\beta_m = 61.36°$. Also, the relative flow velocity can be obtained as $W_m = V_1/\cos\beta_m = 58.67/\cos 61.36° = 122.4$ ft/s.

The tangential force is calculated from

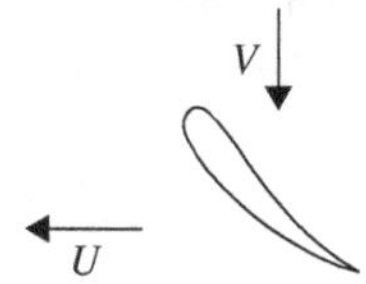

$$F_{um} = \frac{\rho W_m^2 c(C_L \cos\beta_m - C_d \sin\beta_m)}{2g_c}$$
$$= \frac{0.0763 \times 122.4^2 \times 1.2 \times (0.28\cos 61.36° - 0.015\sin 61.36°)}{2 \times 32.2}$$
$$= 2.58\ \text{lb}_f/\text{ft}.$$

Hence, $P_s \approx r_m F_{um}\,\Delta r\,\omega Z_b = 10.63 \times 2.58 \times 14 \times 8.38 \times 5/550 = 29.2$ hp. A more accurate calculation can be obtained by dividing the blades into more sections. From actuator theory, the maximum possible shaft power should be

$$P_{s,\max} = \left(\frac{8}{27}\right)\left(\frac{\rho}{g_c}\right)AV^3 = \frac{(8/27)(0.0763/32.2) \times \pi 15^2 \times (58.67)^3}{550} = 182.3\ \text{hp}.$$

Example 8.3

Determine the tangential force acting on a unit section of the blade of a vertical wind turbine when it is at the angular position shown. The relevant data are as follows:

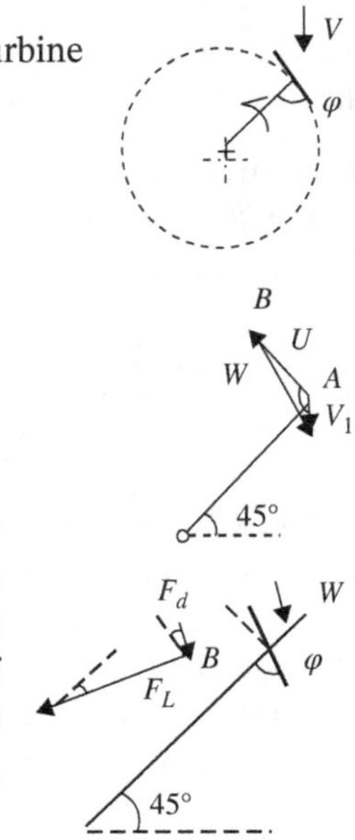

a) Wind velocity $V = 40$ mph, interference factor $a = 0.27$
b) Turbine diameter $D = 50$ ft, rotating speed $N = 60$ rpm
c) Airfoil section NACA 4312 with C_L, C_d given in Figure 3.40
d) Chord length $c = 1.2$ ft, setting angle with respect to the radius, $\varphi = 85°$

Solution

Here, $V = 40$ mph $= 58.7$ ft/s, so $V_1 = V(1 - a) = 42.8$ ft/s, $U = (N\pi 30)(D/2) = 6.283 \times 25 = 157.1$ ft/s. From the velocity triangle, we have $A = 90 + 45 = 135°$, and from the law of cosine, we have $W = (U^2 + V_1^2 - 2UV_1 \cos A)^{0.5} = 189.8$ ft/s, and from the law of sine, we have $\sin B = V_1(\sin A)/W = 0.159$ or $B = 9.2°$, and hence, the angle of attack $\alpha = B - (90° - \varphi) = 9.2 - (90\text{–}85) = 4.2°$. From Figure 3.40, we obtain $C_L = 0.58$ and $C_d = 0.027$. Hence, the tangential force is obtained as $F_u = \rho W^2 c(C_L \sin B - C_d \cos B)/2g_c = 0.0763 \times 189.8^2 \times 1.2 \times 0.066/(2 \times 32.2) = 3.38\ \text{lb}_f/\text{ft}$.

8.5 Variable-Speed Performance of Wind Turbines

Constant-speed operation of a wind turbine results in a simple system design and will be able to avoid the critical speeds which may destroy the turbine. But constant-speed operation means that the maximum power coefficient is achievable only at one wind velocity. A lower value has to be accepted at the other wind velocities.

The effects of wind velocity on turbine output power can be illustrated better by plotting the performance curves in terms of the dimensional parameters. The propeller curve in Figure 8.10 can be replotted for the rotor diameter of 20 m into curves of shaft power P_s versus rotating speed N for a few different wind velocities, as shown in Figure 8.15. It is clear that a wind turbine can be operated at the maximum output power only at one rotating speed for one wind velocity. For example, the maximum power for $V = 14$ m/s is more than five times that for $V = 8$ m/s but at different rotating speeds.

The maximum C_p conditions for different V can be related by eliminating V from $P_m = (1/2)C_{pm}\rho AV^3$ and $\lambda_m V = r_t\omega_m$. Or we have

$$P_m = \frac{1}{2}C_{pm}\rho A\left(\frac{r_t\omega_m}{\lambda_m}\right)^3 = \frac{C_{pm}\rho A r_t^3}{2\lambda_m^3}\left(\frac{N_m\pi}{30}\right)^3 = kN_m^3,$$

where k is a constant. In other words, the constant C_p contours on a $P_s - N$ diagram is in cubical functions. The load curve passing through the maximum C_p is also shown in Figure 8.15. This type of characteristic is actually achievable if the load is a pump or blower, as discussed in earlier chapters. But the shaft power required to drive an electrical generator connected to a fixed resistance varies as the square of the rotational speed. And other types of load characteristics are also possible for some machinery. Hence, a transmission and/or some control system will generally be required in wind turbine applications.

The torque characteristics of wind turbines are also important, especially at low rotating speeds. The torque produced by a wind turbine must be greater than the opposing or tare torque due to friction of the load for the turbine to accelerate from stationary to some steady-state speed. Otherwise, an external power source will be needed. Typical torque curves of three types of wind turbines (Darrieus, propeller, and Savonius) are shown in Figure 8.16. The Darrieus type of vertical-axis turbine has minimum starting torque, as mentioned earlier.

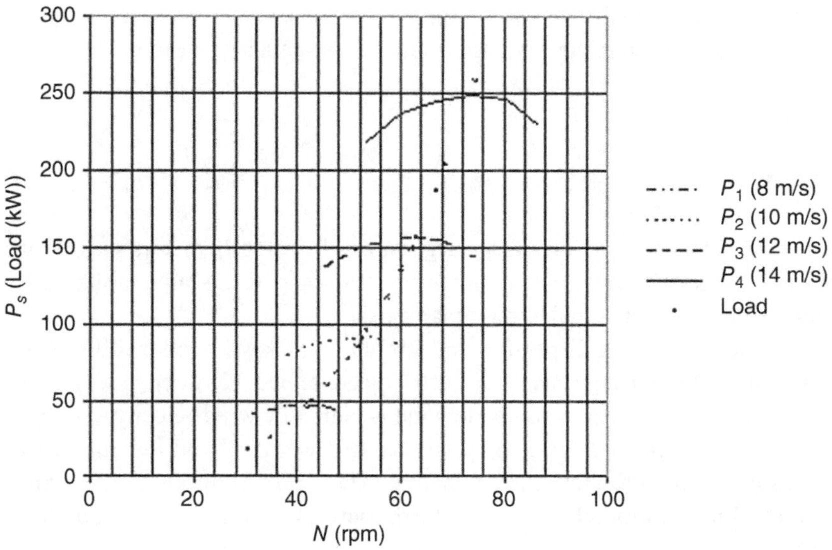

Figure 8.15 Variable-speed operation of wind turbine.

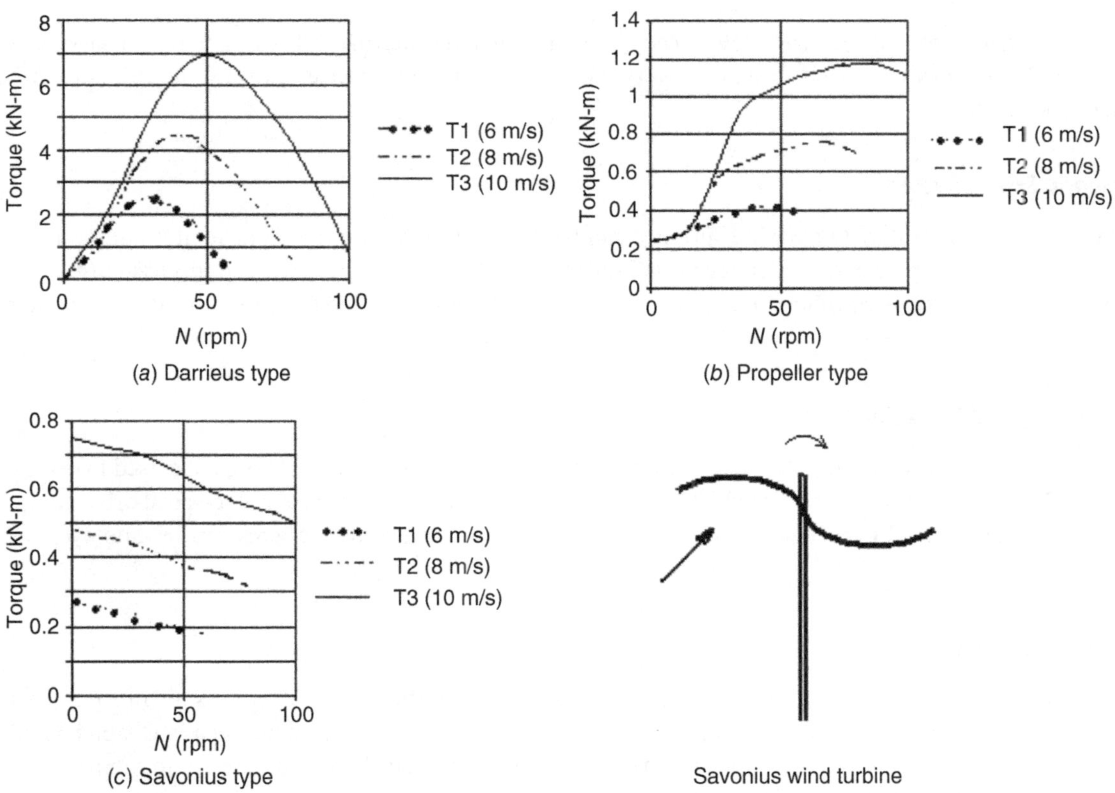

Figure 8.16 Torque characteristics curves of wind turbines under different wind speeds.

The Savonius type with drag elements, as shown, has the maximum torque at zero speed and drops quickly as the rotating speed picks up. Hence, it can be used to start the other turbines. The propeller type can start by itself with low torque.

8.6 Wind Turbine Applications

Commercially available wind turbines range from a few kilowatts to megawatts. The rotor diameters range from around 10 to 200 ft.[10] Generally, large units of 100 kW and greater are for wind farms, whereas small units under 100 kW are for irrigation, isolated communities, or individual home or farm usage.

The main problems in wind power application are its dispersion and variability. A key issue for utility-scale wind plants that must be resolved is transmission line capacity. The electricity generated from vast and sparsely populated areas must be transmitted to large cities where electricity demand is high. The wind velocity at a given location changes direction and magnitude constantly, resulting in variations in power level and the output frequency of an AC generator. A mechanism is needed to align the horizontal-axis wind turbines to the direction of the wind. Some low-starting torque machines, especially the vertical-axis Darrius turbines, require auxiliary power to accelerate them to a minimum speed. Hence, most wind turbines require other devices, increasing the total cost. However, a few applications are currently economically feasible in some situations.

8.6.1 Grid-Intertie System

Electricity is generated from a direct-drive, low-speed, permanent-magnet alternator. The output power varying in both voltage and frequency with wind speed is converted with a synchronous inverter into grid-quality 240-V AC, single-phase, 60-Hz power.

8.6.2 Hybrid Power System

In remote areas without grid systems, wind turbines can be used with a backup diesel generator and optional photovoltaic (PV). The DC power generated is charged and stored in a battery string. An inverter is used to convert it to the AC power of the required voltage and frequency, whereas DC power is used for communication and other purposes.

8.6.3 Pumping and Irrigation Systems

The three-phase AC power with variable voltage and frequency produced by wind turbines can be used to drive a standard three-phase induction motor directly. The system works like a conventional variable-speed drive pump. If a certain flow rate or head is needed, some control valves and multiple-pump systems can be employed, as discussed in Chapter 3.

8.6.4 Other Systems

Wind turbines may power the hydrogen-making process, drive air conditioners, or compress air in some heat engines or the compressed-air energy storage (CAES) with a modified gas turbine as mentioned in Chapter 5 and the small unit shown in Figure 8.17.[11] The power from wind turbines is used to drive a compressor to compress the air. The compressed air is stored in the air cylinders to be used when wind power is low. The thermal storage can also serve as a combustion chamber to supplement the power output with fuel if needed. A control mechanism is required to manipulate the system to match the variable wind power and load. Although the overall efficiency of the systems may be low, they provide some alternatives for special situations.

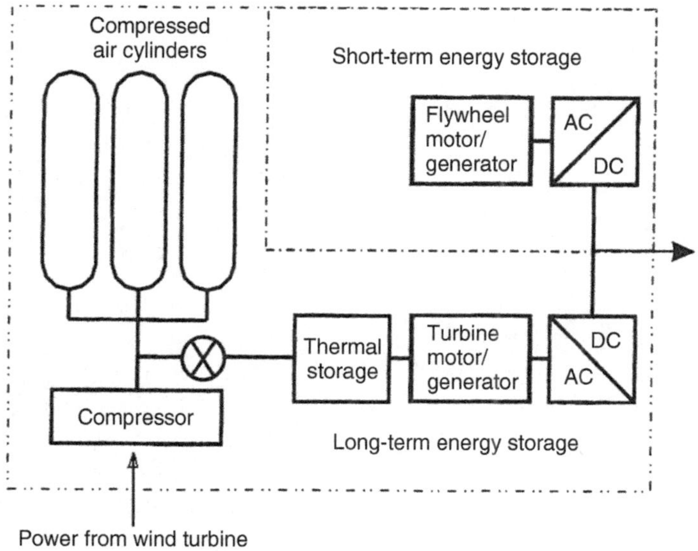

Figure 8.17 Small-scale energy storage with compressed air and flywheel for wind turbines. (*Source:* Reproduced from Reproduced from Cogeneration and Distributed Generation Journal, Vol. 20, No. 2, Spring 2003/with permission of Fairmont Press.)

8.7 Icing Problem

8.7.1 Cold Climate

Wind turbines in cold climates are exposed to icing conditions and low temperatures outside the design limits of standard wind turbines. Standard turbines operating in such extreme environments are prone to production losses and increased loads, which in turn will cause a risk of premature mechanical failure and financial losses. Although exact numbers are hard to assess, some 60 GW of installed capacity is in cold climate (CC) areas in Scandinavia, North America, Europe, and Asia. Narrowing this assessment only to regions with a high likelihood of extended CC operation, Scandinavia, Switzerland, Canada, and northern parts of the United States and China, wind turbines to the power of 20 GW were conservatively estimated to be installed in CC areas in 2011. Additionally, microclimates with similar conditions are found in more temperate areas, such as central and southern Europe, China, Japan, and many parts of the United States, and locations in the southern hemisphere, such as Australia, New Zealand, and southern South America. Recent interest in offshore wind power development introduces further demands for knowledge in dealing with icing, as turbines installed in the shallow waters in northern Europe and in the coast of New England in the United States also face icing conditions.

Sites within CC constitute a vast wind energy production potential. Because fewer temperate sites are available and the offshore wind development faces higher costs than expected, large wind energy projects in CC become tempting and are likely to be implemented. Increased experience and knowledge, combined with improvements in technology directed to be used within CC, have enabled such projects to become more competitive when compared with those onshore with low wind resources and those offshore to be built at higher costs.

Wind turbines in cold climates refer to sites that may experience significant time or frequency of either icing events or low temperatures outside the operational limits of standard wind turbines. Apart from lower energy production, or even no production, which directly influences a wind farm's economy, there are legal issues, such as ice throw and increased noise, fatigue loading, and operations and maintenance aspects that need to be considered.[2] Areas where periods with temperatures below the operational limits of standard wind turbines occur are defined

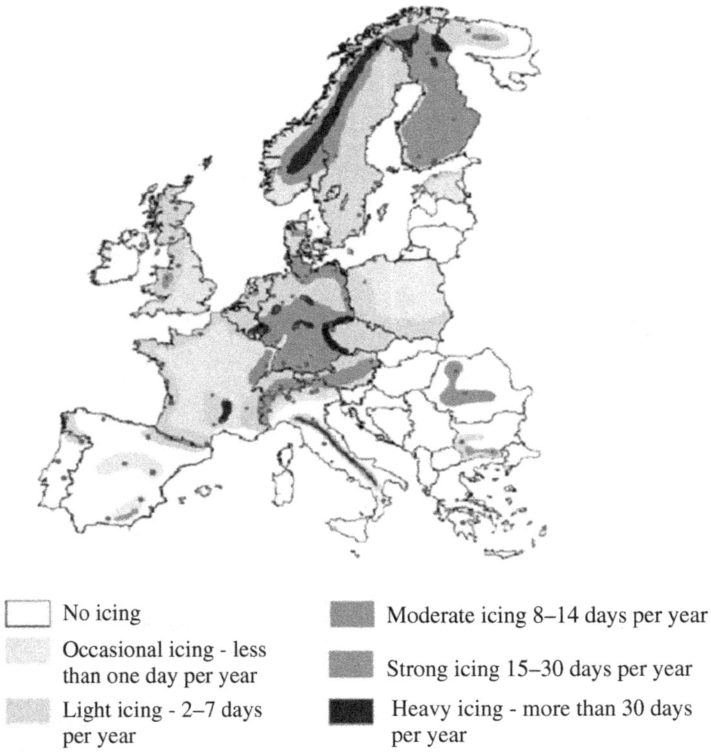

Figure 8.18 Regions with cold climates. (*Source:* Adapted from.)

as low-temperature climates (LTC), and areas with icing events are defined as icing climates (IC). A region can be in an LTC or in an IC or both while they are still all denoted CC sites. In Figure 8.18, a description of different CC regions is given.

8.7.2 Low-temperature Climate

Clear sky radiation during the winter typically causes extremely low temperatures, which are often associated with high-pressure zones. Furthermore, a site's elevation plays an important role.

An LTC can have different effects on a wind energy project:

- Materials used in turbines can be affected by low temperatures
- Maintenance work at low temperatures is more time-consuming
- Cold start of wind turbine can be more difficult at low temperatures
- Lubricants can lose their viscosity
- Internal heat of wind turbines reduces energy production

8.7.3 Icing Climate

Atmospheric icing is defined as the accretion of ice or snow on structures that are exposed to the atmosphere. There are two different types of atmospheric icing: in-cloud icing (rime ice or glaze ice) and precipitation icing (freezing rain, wet snow). The different forms of atmospheric icing are as follows:

- Rime Ice: The wind transported super-cooled liquid water droplets from clouds or fog. When they hit the surface, they freeze. Rime icing is common at high altitudes and at low temperatures. The most severe rime icing event occurs on freely exposed mountains or hilltops where air is forced upwards and consequently cooled. Depending on droplet size and air temperature during the icing event, rime ice can form structures of different density and strength, which leads to a division into two types of rime ice, hard and soft rime ice. Hard rime ice is opaque, usually white, and it adheres firmly on surfaces, making it very difficult to remove. The density of hard rime ice ranges between 600 and 900 kg/m^3 according to ISO 12494. Soft rime ice is thin ice with needles and flakes. The growth of soft ice starts usually at a small point and grows triangularly into the windward direction. The density of it is between 200 and 600 kg/m^3 according to ISO 12494, and it can be more easily removed. In Figure 8.19, there is an example how rime ice is formed.[12]

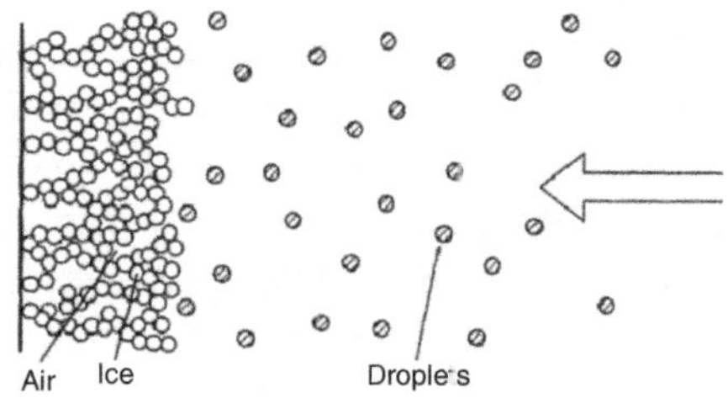

Figure 8.19 Rime ice. (*Source:* Ref. 3/The American Society of Mechanical Engineers.)

- Glaze Ice: it is caused by freezing rain, freezing drizzle, or wet in-cloud icing and forms a smooth, transparent, and homogeneous ice layer with a strong adhesion on the structure. Traditionally, glaze ice forms at temperatures between 0°C and −4°C. Part of the water droplets freeze upon impact and remainder run along the surface before freezing, forming a profile shape of high-density clear ice. Its density is around 900 kg/m^3 according to ISO 12494. Freezing rain or drizzle occurs when warm air melts snow crystals and forms rain droplets, which afterwards fall through a freezing air layer near the ground. Wet in-cloud icing occurs when the surface temperature is near 0°C, and the water droplets which hit the surface do not freeze completely. A layer of liquid water forms, which may flow around the object and freeze also on the leeward side. (Figure 8.20) shows the formation of glaze ice and the difference with rime ice.[12]

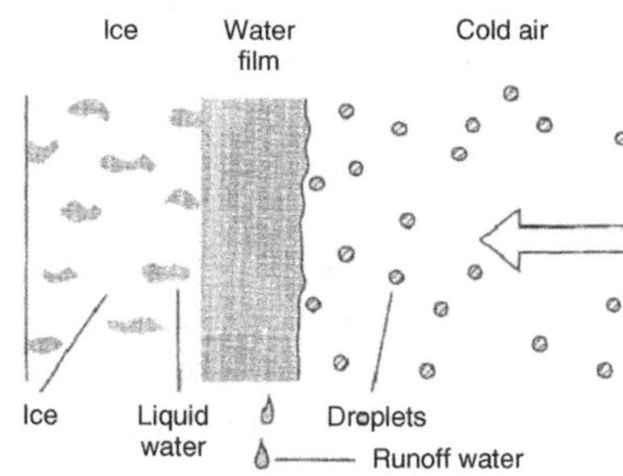

Figure 8.20 Glaze ice. (*Source:* Adapted from Baring-Gould et al.[12])

- Wet Snow: Partly melted snow crystals with high liquid water content become sticky and are able to adhere on the surface of an object. Wet snow accretion therefore occurs when the air temperature is between 0°C and 3°C. The typical density is 300–600 kg/m^3. The wet snow will freeze when the wet snow accretion is followed by a temperature decrease.[12]
- Hoarfrost: It is formed when water vapor sublimes to ice. This happens at very low temperatures and will form ice with very low density, which is normally not considered to cause any significant load on a structure.

8.7.4 Effects of Icing on Wind Energy

Climatic conditions with turbulence, gusts, icing, and lightning strikes in cold regions will affect wind turbine performance and cause wind farms to idle on windy days. Icing, in particular, affects wind assessment and wind farm operation. The following problems are related to icing and cold climates.

- Measurement Errors: During the assessment phase, the anemometers, wind vanes, and temperature sensors can be affected by ice. At icing conditions, wind speed errors can be as high as 30%. Another study identifies a maximum error of 40% for an ice-free anemometer and 60% for a standard anemometer during icing events.[13]

- Power Losses: Ice accretion changes the shape and roughness of the blade aerofoil (consequently affecting their aerodynamic characteristics) and introduces measurement errors from turbine instruments (wrong wind speed or direction, which affects yaw and power controls). Small amounts of ice on the leading edge of aerofoils significantly reduce the aerodynamic properties of the blade and the resulting power production. Power losses may vary from 0.005% to 50% of the annual production, depending on icing intensity and its duration on the site, wind turbine models, and the evaluation methodology.[13]
- Overproduction: Higher air density related to low temperatures and aerofoil modifications can lead to wind turbine overproduction, which has been recorded at up to 16%.
- Mechanical Failures: Ice accretion will increase the load on the blades and on the tower structure, causing high amplitude vibrations and/or resonance as well as mass imbalance between blades. Operation at low temperatures affects oil viscosity and changes the dimensions and mechanical properties of different components of the wind turbine. This results in possible overheating and higher fatigue charges on components.
- Electrical Failure: Snow infiltration in the nacelle and extreme temperature lead to condensation in the electronics.[13]
- Safety Hazard: The pieces of ice shed from wind turbines can be quite large and definitely not insignificant. It is likely that something will occur eventually if preventive measures are not implemented. Though the risks are greatest for service personnel who must approach the wind turbine, others can be at risk when the wind turbine is located near a road or recreation area.[13]

8.7.5 Ice Accretion

Although the estimation of icing on wind turbines and other structures is very important, no undisputed model has so far been developed. Due to the nonlinear relationships between the parameters affecting ice accretion, most models include empirical formulas, and none can be solved analytically. The most widely used model today is Makkonen's algorithm for ice accretion. The Makkonen model[14] was originally developed for modeling ice accretion on overhead power lines and, therefore, models ice accretion on a cylinder resembling a power line. Now, the Makkonen model will be explained, and it is based on Ref. 9.[9]

Irrespective of what kind of particles, liquid or solid, cause ice on an object, the accumulation rate will be dependent on the flux density, in other words, the cross-sectional area of the object, the velocity of the particles, and the mass concentration of particles in the air. Because not all particles will collide with the object and not all particles that do collide will stick or accrete, three efficiency coefficients need to be introduced:

$$\frac{dM}{dt} = \alpha_1 \alpha_2 \alpha_3 wvA \tag{8.5}$$

where α_1 is the collision efficiency, α_2 is the sticking efficiency, α_3 is the accretion efficiency, w is the mass concentration, v is the particle velocity relative to the object, and A is the cross-sectional area relative to the particle velocity. The Makkonen equation can easily be used to model ice load and ice accretion rate. It is common to give the ice load and ice accretion rate in kg/m and kg/m/hr, respectively. The latter refers to per meter of the cylinder used in the model. The ice accretion rate is sometimes presented in terms of active ice hours, which refers to hours with ice accretion rate above a predefined threshold, commonly 10 g/m/hr.

8.7.5.1 The Collision Coefficient, α_1

The collision coefficient, α_1, is the ratio of the flux density of particles hitting the object to the total flux density. α_1 decreases from 1.0 when particles are forced around the object instead of hitting it. Small droplets, large cross-sections, and low wind velocity reduce α_1. The droplet trajectory is determined by the aerodynamic drag and the inertia. For a small droplet, the inertia will be small, and it will tend to follow the streamlines around the object instead of hitting it. If the inertia and drag forces are known, the collision coefficient can be calculated theoretically by numerical calculation of the droplet trajectories. Because this is computationally expensive, Makkonen suggests the simplification to assume a cylindrical object and to use an empirical equation for α_1,

$$\alpha_1 = A - 0.028 - C(B - 0.0454) \tag{8.6}$$

where

$$A = 1.066K^{0.00616}e^{-1.103K^{-0.688}} \tag{8.7}$$

$$B = 3.641K^{0.498}e^{-1.497K^{-0.694}} \tag{8.8}$$

$$C = 0.00637(\phi - 100)^{0.381} \tag{8.9}$$

where K and ϕ are dimensionless parameters

$$\phi = \frac{Re_d^2}{K} \tag{8.10}$$

where Re_d is the Reynolds number based on the free stream velocity and given by

$$Re_d = \frac{\rho_a dv}{\mu} \tag{8.11}$$

and

$$K = \frac{\rho_w d^2 v}{9\mu D} \tag{8.12}$$

where d is the droplet diameter, D is the cylinder diameter, ρw is the water density, μ is the dynamic viscosity of air, and ρ_a is the air density.

Because α_1 depends on the droplet size, one can safely assume $\alpha_1 = 1.0$ for wet snow and freezing rain as long as the structure studied is not extremely large. This means that α_1 normally needs to be calculated only for occasions with in-cloud icing with small droplets.

8.7.5.2 The Sticking Coefficient, α_2

The coefficient α_2 represents the efficiency of collection of those particles that hit the object. It means that α_2 is the ratio of the flux density of the particles that stick to the object to the flux density of the particles that hit the object. The sticking efficiency α_2 is reduced from unity when the particles bounce from the surface. The particles here are considered to stick when they are permanently collected, or their residence time on the surface is sufficient to affect the icing rate due to, for example, heat exchange with the surface. For liquid droplets, it can normally be assumed that they do not bounce, and hence, α_2 equals 1.0. As soon as there is a liquid layer on the surface, combined with favorable conditions, basically all droplets will stick to the surface. For ice and snow particles, the ratio of particles that bounce off the object can be a significant share. For solid particles, such as dry snow, α_2 is close to or equal to 0.

There is presently no detailed theory for the sticking efficiency of wet snow. The available approximation methods of α_2 are empirical equations based on laboratory simulations and some field observations. The best first approximation for α_2 for cylindrical shapes is probably

$$\alpha_2 = \frac{1}{v} \tag{8.13}$$

where the wind speed, v, is in m/s. When $v < 1.0$ m/s, $\alpha_2 = 1.0$. The air temperature and humidity also affect α_2, but there are presently not enough consistent data to take them into account, despite several experimental studies on this problem. However, as pointed out above, $\alpha_2 > 0$ only when the snow particle surface is wet. Thus, snow does not accrete effectively when the wet-bulb temperature is below 0°C. This criterion is very important because it allows determination of the duration of wet-snow events by climatic weather data.

8.7.5.3 The Accretion Coefficient, α_3

The accretion efficiency coefficient, α_3, is the ratio of icing to the flux density of droplets sticking to the object's surface. During dry growth, α_3 equals 1.0 because there is no run-off. During wet growth and glaze icing, there is a liquid film on the adhesion surface, and α_3 will be less than 1.0. The freezing rate will then be determined by the rate at which latent heat from freezing can be transported away from the surface. A heating balance over the surface of the object can be used to determine the portion that will freeze.

$$Q_f + Q_v = Q_c + Q_e + Q_l + Q_s \tag{8.14}$$

where Q_f is the latent heat released during freezing, Q_v is the frictional heating of air, Q_c is the loss of sensible heat to air, Q_e is the heat loss due to evaporation, Q_l is the heat loss in warming the impinging supercooled water to the freezing temperature, and Q_s is the heat loss due to radiation.

The terms of the heat-balance equation can be parametrized using the meteorological and structural variables.

The heat released in the freezing is transferred from the ice–water interface through the liquid water into the air, and, consequently, there is a negative temperature gradient ahead of the growing ice. This kind of supercooling results in a dendritic growth morphology, and, consequently, some liquid water is trapped within the spray ice matrix. Because the unfrozen water can be entrapped without releasing any latent heat, the term Q_f in Eq. (8.14) is

$$Q_f = (1 - \lambda)\alpha_3 F L_f \tag{8.15}$$

where λ is the liquid fraction of the accretion, and F is the flux density of water to the surface ($F = \alpha_1\alpha_2 wv$). Attempts to determine the liquid fraction, λ, have been made both theoretically and experimentally. These studies suggest that λ is rather insensitive to the growth conditions and that a value of λ around 0.3 is a reasonable first approximation.

Another important consequence of the existence of porous ice is that the heat conduction into the ice does not need to be considered in the heat balance because the mixture of ice and water is at the uniform temperature of 0°C and, therefore, involves no heat conduction through it.

The kinetic heating of air, Q_v, is a relatively small term (except at aircraft speeds), but because it is easily parametrized by

$$Q_v = \frac{hrv_w^2}{2C_{pa}} \tag{8.16}$$

it is usually included in the heat balance. Kinetic heating of the droplets is insignificant and is ignored. Here, h is the convective heat-transfer coefficient, r is the recovery factor for viscous heating ($r = 0.79$ for a cylinder), vw is the wind speed, and C_{pa} is the specific heat of air.

The convective heat transfer is

$$Q_c = h(t_s - t_a) \tag{8.17}$$

where t_s is the temperature of the icing surface ($t_s = 0$°C for pure water), and t_a is the air temperature.

The evaporative heat transfer is parametrized as

$$Q_e = \frac{h\psi L_e(e_s - e_a)}{C_p p} \tag{8.18}$$

where ψ is the ratio of the molecular weights of dry air and water vapor ($\psi = 0.622$), L_e is the latent heat of vaporization, e_s is the saturation water vapor pressure over the accretion surface, e_a is the ambient vapor pressure in the airstream, and p is the air pressure. Here, e_s is a constant (6.17 mbar) and e_a is a function of the temperature and relative humidity of the ambient air.

The term Q_l refers to the temperature difference between the impinging droplets and the surface of the icing object:

$$Q_l = FC_w(t_s - t_d) \tag{8.19}$$

where C_w is the specific heat of water, and t_d is the temperature of the droplets at impact.

For cloud droplets, $t_d = t_a$ can be assumed because of the small terminal velocity. This assumption must usually be made also for supercooled raindrops, although they are not fully adjusted to the temperature at the surface layer during their fall.

The heat loss due to long-wave radiation may be parametrized as

$$Q_s = \sigma a(t_s - t_a) \tag{8.20}$$

where σ is the Stefan–Boltzmann constant (5.67 × 10–8 W/m^2/K^4), and a is the radiation linearization constant (8.1 × 107 K^3). This equation only takes into account long-wave radiation and assumes emissivities of unity for both the icing surface and the environment. Short-wave radiation of the sun is neglected because, in practice, atmospheric icing always occurs in cloudy weather. Using the parametrization of Eqs. (8.15)–(8.20) in the heat-balance Eq. (8.14) and solving the accretion efficiency result in the following equation:

$$\alpha_3 = \frac{1}{F(1-\lambda)}\left[(h + 6a)(t_s - t_a) + \frac{h\psi L_e}{C_{pa}p}(e_s - e_a)\frac{hrv_w^2}{2C_{pa}} + FC_w(t_s - t_d)\right] \tag{8.21}$$

8.7.5.4 Limitations of the Model

Although Makkonen's model for ice accumulation is widely used, it is limited because it models ice accretion on a cylinder. To translate the ice load on an ideal cylinder to a complex structure such as a wind turbine is computationally expensive and so far, not possible for the general case. To model ice accumulation on a wind turbine, one needs to break down the structure into small elements, still considering shadowing effects from other elements and ice-growing elements together to form a single element.

One of the more practical problems is that the liquid water content, LWC, of the air, the cloud droplet density and the median volume droplet size, MVD, in the air are required values. These parameters are not routinely measured, nor does there exist a satisfactorily way to measure them. The average droplet diameter has to be approximated, and the average number of droplets in cloud has to be known. The liquid water content can be modeled, but the result has a high uncertainty. There is an uncertainty in the model when α_1, the collision efficiency, is small because α_1 depends on the droplet diameter and errors are introduced in estimating MVD. Also, the effects of droplet trajectory angle on the collision efficiency are poorly known, which means that the effects of wind direction and rotational speed of a wind turbine are uncertain. Another uncertainty regarding α_1 concerns that low velocities are introducing uncertainties in the Reynolds number and flow patterns.

Modeling ice accretion also requires modeling of the ice density because the icing rate depends on the dimensions of the iced object. The density of the ice can be modeled using ballistic models, but for rime ice, best-fit equations have been developed in wind-tunnel experiments.

$$\rho_{\text{ice}} = 0.378 + 0.425\log(R) - 0.823\log(R^2) \tag{8.22}$$

where R is the Macklin parameter, a dimensionless density parameter. It is given by

$$R = -\frac{v_o d_m}{2t_s} \tag{8.23}$$

where v_o is the droplet impact speed, d_m is the droplet median diameter, and t_s is the surface temperature. t_s can be approximated as the air temperature. To solve the equation, the impact speed must first be calculated from the heat balance equation. A difficulty with solving the heat balance equation is the convective heat transfer coefficient. Even assuming a simple structure, such as a cylinder, to determine the heat transfer coefficient is complicated.

Another limitation of the model is that it considers only accumulation of ice. Ice will be removed only due to the sign of the heat balance. Melting is not included either shedding of ice nor sublimation.

8.7.6 Conclusions

The topic of this chapter is new, and it is important nowadays because of the benefits of cold climates for wind turbines. This means that there is not much data available about icing events. Thus, it has been difficult to obtain the information needed. This chapter resumes part of the field of ice accretion in wind turbines. It shows different icing events, the weather variables of importance, and the effects of ice accretion.

Also, systems of anti-icing and de-icing are described. As it is a relatively new field, not much work has been done on such systems. Accordingly, a lot of work is needed to optimize these processes. Many existing systems are based on those used in aircraft, as the field presents similar challenges, though encountered in different contexts. The warm air de-icing system is described in detail, based on information provided by ENERCON. The data highlights two locations with icing events and clearly demonstrates the significant impact of de-icing: the presence of the system results in a 50.4% increase in power production compared to a turbine without de-icing capability.

There is a general physical model for ice accretion, but it has some limitations that have to be corrected. One of them is that there are some coefficients in the formula that are based on cylinder shapes. Some programs transform these shapes to more complex structures like blades, but it would be better to reformulate the model.

The simulations show differences in the dependence on ice accretion of the different variables. The ambient temperature is the most important factor to take into account for icing events. The wind speed is the second most important variable. Also, if the rotational speed is added, the relative wind speed in different parts of the blade would make this variable even more important. LWC is also important, but depending on the range of its value, it does not really affect the ice accretion. The same is true for MVD, if the values of this variable are low, the ice thicknesses of each case are really different, but after 100 μm, the effect of it seems to be not important. Finally, the effect of AOA on ice thickness seems to be insignificant. However, AOA affects the ice shape, so it is a variable that must be taken into account depending on which kind of de-icing systems is installed in the blade turbine, for example, heating resistance that focus on a specified part of the blade surface.

References

1 Lazard. (2024). *Lazard's Levelized Cost of Energy Analysis – Version 17.0.* Retrieved from https://www.lazard.com/media/xemfey0k/lazards-lcoeplus-june-2024-_vf.pdf.

2 IRENA. (2024). Renewable capacity statistics 2024, International Renewable Energy Agency, Abu Dhabi.

3 ASME Mechanical Engineering, American Society of Mechanical Engineers, New York, Jun. 2004.

4 Wortman, A., Introduction to Wind Engineering, Butterworth Publishers, Woburn, MA, 1983.

5 Bossel, V., Energie vom Wind, Deutsche Gesellschaft für Sonnen-energie, Munich, Germany, 1974.

6 Ashley, H., "Some Contributions to Aerodynamic Theory for Vertical Axis Wind Turbines," *AIAA J. Energy*, Vol. 2, No. 2, pp. 113–119, Mar. 1978, American Institute of Aeronautics & Astronautics, New York.

7 Wilson, R. E., Walker, S. N., and Lissaman, P., "Aerodynamics of the Darrieus Rotor," *AIAA J. Aircraft*, Vol. 15, No. 12, pp. 1023–1024, Dec. 1976, American Institute of Aeronautics & Astronautics, New York.

8 Sheldahl, R. E., Klimas, P. C., and Feltz, L. V., "Aerodynamics Performance of a 5-M Diameter Darrieus Turbine," *AIAA J. Energy*, Vol. 4, No. 5, pp. 227–232, Sept. 1980, American Institute of Aeronautics & Astronautics, New York.

9 ASME PTC Committee No. 42, *Wind Turbine Performance Test Codes*, American Society of Mechanical Engineers, New York, 1988.

10 GE Energy, www.gepower.com; Bergey Windpower Co., www.bergey.com.

11 Sears, J. R., "Thermal and Compressed Air Storage: The Next Generation of Energy Storage Technology," *Cogen. Distributed Gen. J.*, Vol. 20, No. 2, pp. 19–33, Spring 2003, The Fairmont Press, Lilburn, GA.

12 Baring-Gould, I., Cattin, R., Durstewitz, M., Hulkkonen, M., Krenn, A., Laakso, T., Lacroix, A., Peltola, E., Ronsten, G., Tallhaug, L., and Wallenius, T., "Wind Energy Projects in Cold Climates," International Energy Agency, 2011.

13 Ilinca, A. (2011). *Analysis and Mitigation of Icing Effects on Wind Turbines.* In I. Al-Bahadly (Ed.), Wind Turbines (pp. 177–214). Rijeka, Croatia.

14 Makkonen, L., "Models for the Growth of Rime, Glaze, Icicles and Wet Snow on Structures," *Phil. Trans. R. Soc. Lond.*, 358, pp. 2913–1939, 2000.

Problems

8.1 A horizontal-axis wind turbine has a blade diameter of 7 m and delivers power of 15 kW when rotating at 180 rpm under a wind speed of 50 km/hr. What is its power coefficient and its probable number of blades? What is the maximum power that can be extracted at this site with the same size wind turbine?

8.2 A four-blade horizontal-axis wind turbine of 30 ft diameter is to be installed at a location with an average wind velocity of 40 mph. Determine the maximum power deliverable and the probable rotating speed for maximum efficiency.

8.3 A multiblade farm windmill with $D = 1$ m is to develop full power at wind velocity of 40 km/hr. Calculate the rotating speed for optimum power generation. Estimate the power output.

8.4 Determine the tangential forces on the blade of the vertical wind turbine in Example 8.3 when its angular position is at the other three quadrants as shown.

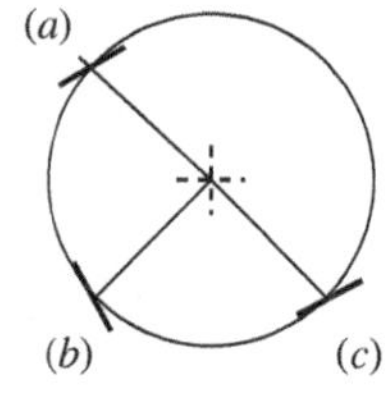

8.5 Repeat the calculations of Example 8.3 to include the axial and tangential interference factor.

8.6 A propeller wind turbine with a radius of 15 m is needed to produce 780 hp in a wind velocity of 50 km/hr while rotating at 40 rpm. A model is to be designed and tested with the air velocity of 12 m/s and rotating speed of 200 rpm. What is the model radius and its expected power output'?

8.7 Determine the maximum possible axial force acting on the prototype turbine described in problem 8.6.

8.8 A vertical-axis wind turbine with the dimensions shown rotating at 40 rpm under a wind speed of 4 m/s. If it generates a shaft power of 4 kW, estimate the downstream average wake flow angle with respect to the mean wind direction.

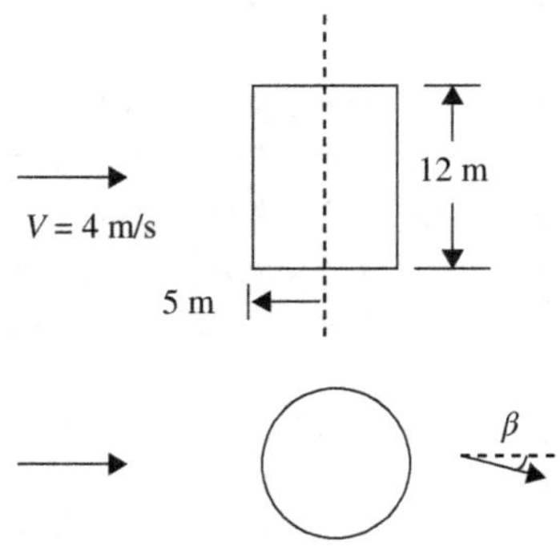

9

Artificial Intelligence as a New Frontier in Turbomachinery Design and Applications

9.1 Introduction

Turbomachines play a crucial role in converting renewable, chemical, or potential energy into propulsion, mechanical force, or electrical power. As the need for efficiency, reliability, compactness, and affordability grows, it is imperative to improve the accuracy of design methodologies. These enhancements are essential to accommodate the emergence of new materials, fluids, and fuels, alongside advances in manufacturing techniques and technologies. This chapter explores recent developments in design strategies that leverage the combined strengths of artificial intelligence (AI) and precise simulation techniques. These strategies benefit from integrating design information from various sources and improving design validation accuracy. They are effective in aerodynamic and thermal design, as well as in the critical areas of materials and fluid mechanics. In the future, the integrated use of machine learning (ML) with detailed simulation processes could allow researchers to perform reliable virtual experiments during the design stages, thus reducing time and risk.

The quest for innovative energy conversion and propulsion systems through turbomachines demands advanced design methodologies that enhance precision and open new design possibilities. Traditional design approaches have often relied on company-specific correlations, especially in the early design stages. However, these correlations can be inaccurate, necessitating significant safety margins. Classic examples of such conventional techniques are found in the work of Wisler et al.,[1] on compressors and turbines, and in the work of Craig and Cox.[2] These safety margins, designed to account for predictive uncertainties as highlighted in these studies, are expected to decrease as designs become more complex. This evolution is supported by advanced tools such as computational fluid dynamics (CFD), mechanical structural mechanics analysis, and chemical analysis.

The vast volume and diversity of data now available—from component testing, field data, and multi-domain simulations present a remarkable opportunity for innovation. Data-driven design methods, powered by AI, excel at analyzing these extensive datasets to reveal hidden patterns, thus driving innovation in turbomachinery research. This potential is evident in studies ranging from wind turbine aerodynamics to focused research on impingement jet cooling, a technique for rapid surface cooling, as discussed by Amano and Sundén.[3] These works provide innovative insights relevant to both wind and gas turbine sectors. Advances in computational methods, as explored by Amano and Sundén,[4] and the pioneering heat transfer research by Amano et al.[5] directly influence turbomachinery design. Studies on topics such as turbulent jet heat transfer by Wang et al.,[6] blade sweep efficiency by Xu and Amano,[7–12] cooling channel optimization by Nourin and Amano,[13–15] buoyancy effects by Saravani et al.,[16] innovative temperature control by Gupta et al.,[17] advanced wall modeling by Rodi et al.,[18] and ribbed passage configurations by Kumar and Amano[19, 20] illustrate the broad scope of active research areas.

In some cases, the design process includes multi-objective optimization algorithms, as described by Astrua.[21] These advanced techniques often rely on simplified models to save computational resources but can introduce uncertainties, necessitating additional safety margins and potentially limiting extensive design optimization. Over the years, turbomachinery design has evolved significantly, leveraging numerous data sources, including scaled

Fundamentals of Turbomachinery: Theory and Applications, Second Edition. Ryoichi S. Amano and William W. Peng.

Companion website: www.wiley.com/go/amano/turbomachinery2e

component tests, comprehensive engine field data, and digital simulations spanning various domains, such as aerodynamics, combustion, and rotor dynamics. Navigating such diverse and extensive datasets, with varying resolutions, accuracy levels, and operating conditions, is challenging but also paves the way for using data-driven design methods, such as ML, as discussed by Michelassi et al.[22] As Brunton et al.[23] elaborate, ML, a branch of AI, uses sophisticated algorithms to identify data patterns. This discussion will focus on three key areas where data-driven design methods are enhancing turbomachinery design: accelerating the development of new materials and manufacturing processes through sequential learning, simultaneous optimization of materials and component performance, and integrating data-driven models to better understand turbulence effects. Finally, this chapter highlights the importance of data management, as data is crucial for ML algorithms to function effectively. Addressing this aspect is essential to fully harness the potential of AI-based design methodologies.

9.2 Machine Learning

Scientists and engineers frequently utilize Reynolds-averaged Navier–Stokes (RANS) models in machinery with fluid flow because more detailed methods such as large eddy simulation (LES) and direct numerical simulation (DNS) require substantial computational resources, particularly for large engines and specific flow conditions. While CFD has facilitated the design of these machines, RANS models exhibit limitations, especially in capturing turbulence effects, as highlighted by Laskowski[24] and Sandberg and Michelassi.[25] These inaccuracies can affect various components of a gas turbine. For example, RANS models are often insufficient for tasks requiring detailed design decisions, such as in compressors, combustion chambers, or cooling systems in high-pressure turbines, as noted by Tucker[26] and Cozzi.[27] The behavior of fluids within machinery varies significantly between components. For instance, the repetitive interaction of flows with walls in compressors and turbines necessitates detailed analysis, whereas combustion chambers, characterized by consistent behavior and complex chemical reactions, may require less detailed analysis.

When applied under appropriate conditions and with sufficient grid resolution, LES and DNS have demonstrated their effectiveness in accurately interpreting flow physics across various turbomachinery components. Ling[28] provided insights into axial compressors, showing how LES revealed energy loss mechanisms that deviated significantly from unsteady Reynolds-averaged Navier–Stokes (URANS) predictions. Traditional RANS and URANS methods often fall short in predicting mixing dynamics and subsequent heat release in combustion chambers.

In contrast, LES, as documented by Michelassi et al.,[29] successfully forecasted both stable and unstable operational states within annular industrial gas turbine combustors. Simulating high-pressure turbines presents challenges due to the demanding environment of elevated pressures, temperatures, and transonic speeds, resulting in high Reynolds and Mach numbers. Consequently, DNS is typically confined to smaller geometric sections. Wheeler et al.[30] stated that disturbances on the suction side are undetected by conventional Reynolds-averaged techniques, which impact the computation of the aerodynamic throat area.

Conversely, low-pressure turbines are more amenable to simulations due to their moderate Reynolds and Mach numbers, typically below 3×10^5 and 0.6, respectively. Multiple studies have explored simulations of single airfoils up to 2D-stage and 3D-stage DNS and LES.[25, 31] Notably, DNS and LES can replace traditional experimental approaches for gas turbine components, offering critical insights into design trajectories based on key design variables. Michelassi et al.[32] explored the interplay between flow coefficient and reduced frequency, while Pichler et al.[33] focused on the impact of stator-rotor gaps. Additionally, Pichler et al.[34] provided a detailed analysis of the inaccuracies inherent in URANS models, offering insights into their limitations.

Traditional turbulence models were initially developed for planar-free shear flows and attached boundary layers, yet their application has extended beyond these parameters, as noted by Wilcox[35] and Patel et al.[36] Despite their broader use, Reynolds-averaged models provide reasonable accuracy but may be inadequate for performing existing gas turbine designs or exploring innovative design spaces. The limitations of URANS have motivated researchers to control detailed data from DNS and LES to improve these more computationally efficient methods. However, despite numerous models introduced over the years, advancements in turbulence modeling have been limited in recent decades. Efforts to enhance Reynolds-averaged strategies through high-fidelity CFD, such as complex full-Reynolds stress models, have not yielded the anticipated results, as illustrated by Parneix et al.,[37] who desired to refine fundamental terms using DNS. Rodi et al.[18] advanced two-equation modeling near walls using highly detailed simulations of basic flow types, yet these improvements did not easily apply to more complex situations.

In engineering and physics, there is an interesting link between how turbulence moves and how heat is spread through turbulence, a concept explored in depth by Wissink et al.[38] The relationship is crucial for understanding and improving systems that control heat and airflow, such as jet engines. To improve these models, researchers often turn to ML, which allows scientists to create highly accurate and customized models for specific situations. A remarkable example is a study by Sandberg et al.[39] They used a technique called gene expression programming (GEP) to improve how we predict heat movement in a part of a jet engine that helps manage the engine's temperature. The researchers used advanced simulations to gather data. Then, using GEP, they developed a new formula that helps in calculating a key number called the turbulent Prandtl number, which is essential for understanding how turbulence affects heat diffusion. The beauty of their work is that it keeps the traditional way of looking at heat transfer (Fourier's law) but adds a layer of precision and customization through new equations:

$$\overline{U_i'T'} = -\frac{\mu_t}{p_{r_t}(I)}\frac{d\overline{T}}{dx_i} \tag{9.1}$$

In the above equation, the turbulent heat flux is represented as $\overline{U_i'T'}$. Within the representation, μ_t symbolizes the turbulent viscosity and $\frac{d\overline{T}}{dx_i}$ stands for the spatial gradient of the time-averaged temperature. Lumley[40] introduced an expression, $p_{r_t}(I)$, for the turbulent Prandtl number that is dependent on core turbulence invariants, denoted by I. The formula has been proven to significantly improve alignment with observed heat transfer data, without necessitating excessive computational resources compared to standard RANS computations. Wissink et al.[31] devised groundbreaking foundational law leveraging consistent LES flow fields. However, if the new model alters the flow field as predicted by RANS, a feedback mechanism becomes essential during the training phase. The mechanism evaluates the new foundational law by initiating a RANS simulation with the updated closure and comparing it against the original LES flow domain. Figure 9.1 details the procedure, where each new model produced by GEP is tested through an assessment. Incorporating a feedback loop from RANS into the GEP training addresses certain limitations identified by Weatheritt,[41] Akolekar, and Sandberg.[42] Specifically, when rotor–stator interactions are unsteady and historical impacts are pronounced, a training regimen without direct RANS feedback might not always align perfectly with LES outcomes. Figure 9.2 demonstrates challenges, revealing that machine learning models, developed using various strategies, still exhibit noticeable deviations from LES.

Enhancing the performance and trustworthiness of ML models combined with RANS is a significant step forward in the field. Zhao et al.[43] explored this area deeply. They created a new, complex rule specifically for analyzing turbulent flows, which are especially important in studying the flow behind turbines, often referred to as turbine wakes, as shown in Figure 9.3. Understanding turbine wakes is essential for evaluating how turbines perform under varying conditions and designing them more effectively. However, accurately predicting these wakes has always been a challenge with RANS.

The advent of deep neural networks (DNNs) presents an exciting frontier. Ling et al.[44] demonstrated that DNNs have the potential to revolutionize our predictions of fluid flow, especially within the context of RANS models.

Building upon this foundation, Milani et al.[45] applied this innovative ML approach to the film cooling flow, which is critical in engineering applications such as maintaining the thermal management of aircraft engine components. Their findings indicated that a universal model for predicting all types of fluid flows might be unattainable

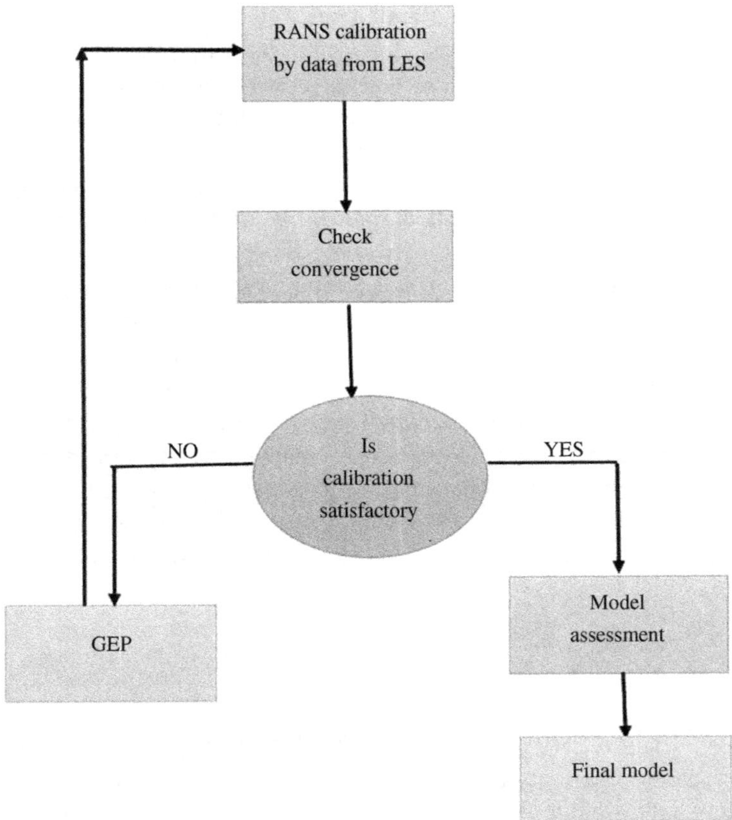

Figure 9.1 Integrating gene expression programming (GEP) with CFD.

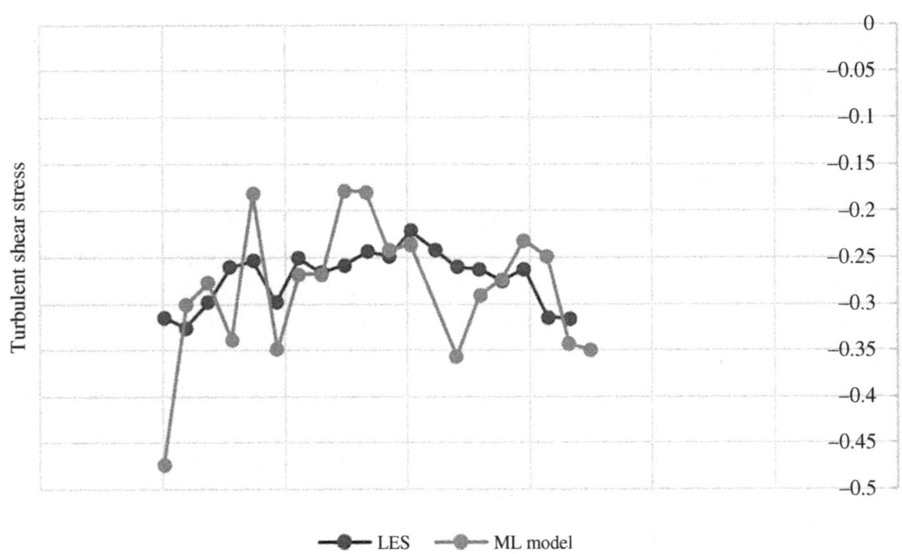

Figure 9.2 Assessing turbulence shear stress around a turbine blade.

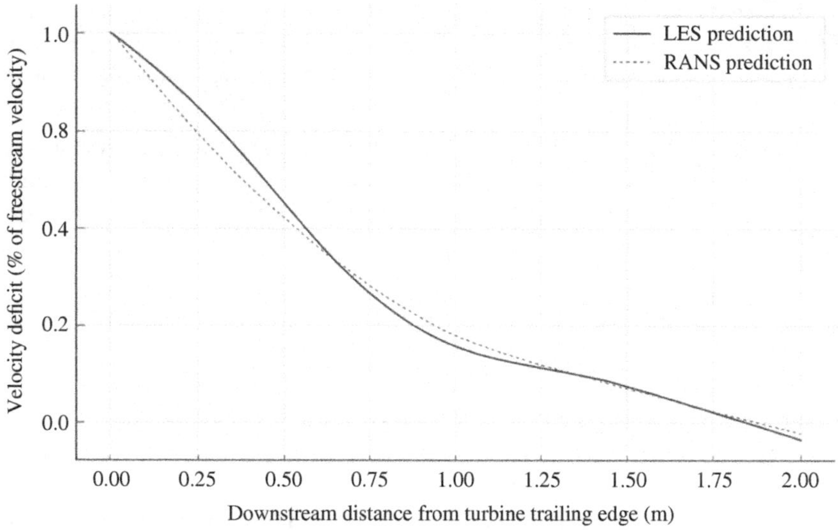

Figure 9.3 Predictions of exit wake in low-pressure turbines.

due to the distinct characteristics inherent to each flow type. Instead, they proposed that highly accurate models could be developed by concentrating on specific types of flows.

Focusing on film cooling flow, their research examined model performance under various conditions, including alterations in the speed and direction of the cooling flow and modifications to the design of cooling holes. The results were promising, demonstrating that models tailored to specific flow types could achieve both precise predictions and computational efficiency. This advancement is particularly beneficial for the design of machinery such as turbines, as it facilitates faster and more informed design decisions.

Weatheritt et al.[46] introduced an innovative method for enhancing the simulation of heat effects from a jet interacting with another flow using computational techniques. This work addresses a significant challenge in integrating ML methods with traditional physics-based simulations, namely, ensuring that these ML methods adhere to fundamental physical principles. A pivotal aspect is maintaining the fidelity of these principles within ML models.

For instance, Ling et al.[47] emphasized the importance of incorporating Galilean invariance into their models. Galilean invariance ensures that a model's predictions remain consistent, regardless of the observer's frame of reference or motion with the flow. Without directly incorporating this principle, models would require substantially more data to learn these rules accurately, and even then, they might not fully capture them. This highlights the necessity of developing ML models that inherently respect established physical laws.

Another critical challenge is the interpretability of ML models, understanding the rationale behind their predictions. Accurate predictions alone are insufficient if the decision-making process remains opaque, especially as models become increasingly complex. In engineering, particularly in designing systems with complex fluid dynamics like turbines, numerous simulations are conducted to predict performance under various conditions. Visualization tools have proven invaluable in deciphering this extensive data. One such tool is t-distributed Stochastic Neighbor Embedding (t-SNE), which aids engineers and scientists in visualizing how well their ML models perform in untrained scenarios. *t*-SNE effectively identifies when models are making reliable predictions versus when they might be extrapolating based on their training data. Milani et al.[48] adapted *t*-SNE to gain insights into turbomachinery, specifically film cooling explores their adaptability to new cooling structures, and air-blowing rates not directly included in their training data. This approach also helped them identify the most critical aspects of their model for predicting cooling efficiency in various machinery components.

Similarly, Wu[49] utilized *t*-SNE to enhance the understanding of their ML model's decision-making processes. This tool enabled them to visualize how the model categorizes distinct types of flow, facilitating the evaluation of the model's performance on unseen flow types. By observing how different flow patterns cluster, researchers can assess the model's ability to make reliable predictions in new situations.

While tools like GEP and t-SNE have significantly advanced data and model visualization, there remains a notable gap in user comprehension and trust. Engineers and designers often struggle to fully understand and trust these complex models. A more intuitive approach could enhance their ability to prototype and grasp the intricacies of these systems swiftly.

9.3 Data Governance

In this review, we have emphasized the transformative potential of AI in the future of gas turbine engines and similar machinery. A significant challenge in realizing this potential lies in the effective management of data. For AI to function optimally, it requires access to high-quality data.

The design process of turbomachinery generates vast amounts of data across multiple stages from conceptual design through final testing. This data encompasses design specifications, simulation outcomes, performance metrics, and operational parameters. Implementing effective data governance requires a structured approach across several key dimensions:

1) Data Collection and Validation
   - Implementation of standardized data collection protocols
   - Real-time data validation systems
   - Quality assurance checkpoints at each stage of data acquisition
   - Sensor calibration and maintenance protocols
2) Data Storage and Security
   - Secure cloud-based storage systems
   - Regular backup procedures
   - Access control mechanisms
   - Data encryption protocols
   - Compliance with industry standards and regulations
3) Data Processing and Organization
   - Standardized data formatting
   - Metadata tagging for easy retrieval
   - Data cleaning and preprocessing procedures
   - Version control systems
4) Data Integration and Accessibility
   - Unified data platforms
   - Cross-departmental data-sharing protocols
   - Real-time data access capabilities
5) Data Quality Management
   - Regular data quality assessments
   - Error detection and correction procedures
   - Data completeness monitoring
   - Accuracy verification protocols

To implement these frameworks effectively, organizations should:

- Establish clear data governance policies
- Define roles and responsibilities for data management
- Implement training programs for personnel
- Maintain documentation of data management processes

By implementing these structured data management practices, organizations can ensure that AI systems have access to high-quality, reliable data for learning and decision-making. This foundation is essential for leveraging AI effectively in turbomachinery design and optimization.

9.3.1 Validation—Validation Data Repository

The Validation Data Repository (VDR) is a vital resource that contains detailed information on various engine components, serving as a crucial foundation for ML applications in turbomachinery design. This data spans from targeted tests on specific parts like combustion chambers, axial compressors, and turbines to broader systems such as cooling and purge mechanisms. The repository is enriched with data from engine certifications and real-world engine performance assessments, providing the extensive dataset necessary for training robust ML models.

The primary use of this individual component test data is to train ML algorithms that can predict the performance of each engine part at the initial design stages. ML models, particularly deep neural networks and ensemble learning techniques, can process this historical data to identify complex patterns and relationships between design parameters and performance outcomes. However, the true value of the VDR lies in its comprehensive engine data, which enables more sophisticated ML applications. Although this data might vary in detail and precision, it is essential for training integrated AI models that can simulate the interactions between different engine components. For example, while detailed measurements such as combustion discharge temperatures or extensive pressure maps might not always be available, ML algorithms can learn from the available data points to interpolate and predict missing values.

The collective data from the VDR powers various ML applications, from predictive maintenance to design optimization. By training on data gathered from various engines operating under different conditions, AI models can learn to predict engine performance, identify potential failures, and suggest design improvements. This ML-driven approach has been successfully demonstrated by Ramprasad et al.,[50] who developed AI models for industrial gas turbines, and Pichler et al.,[51] who applied ML techniques to analyze aero-derivative turbines. The extensive dataset of VDR enables AI systems to make accurate predictions and recommendations, ultimately advancing engine technology through data-driven insights and optimization.

9.3.2 Conceptual Design—Conceptual Design Repository (CnDR)

In the conceptual design stage, engineers create a basic one-dimensional (1D) model. This model is designed to give early predictions about an engine's performance, size, and potential costs. To build this model, engineers rely on proprietary correlations, special formulas, or rules that summarize the knowledge gained from past engine designs, experiments on smaller parts or subsystems, and extensive tests on complete engines. These correlations are a way to use what has been learned in the past to make educated guesses about modern designs. It is important to understand that the data used to create these correlations might not be detailed or precise. This means the model can give a good overview of an engine's performance but might not be able to explain exactly why the engine behaves in certain ways under specific conditions. While these models are good for broad assessment, they do not delve deeply into the physics that drive engine behavior.

The accuracy of these predictive correlations can also vary. According to the research by Wisler et al.[1] and Craig and Cox,[2] the reliability of these predictions usually falls within a range of ±1–5% of actual performance, at least within the scope of the design parameters they were created for. This variability highlights uncertainty in early design stages, pointing out that while these early models are incredibly useful for getting an approximation on engine performance and costs, they need to be refined and tested further as the design process progresses.

9.3.3 Comprehensive Design Repository (CpDR)

In this phase, engineering design takes a deep dive into the details, utilizing advanced 3D modeling techniques. These techniques are based on complex mathematical equations known as partial differential equations. CFD and finite element analysis (FEA) are the most common methods used. These methods allow engineers to create detailed simulations of how fluids flow and how structures respond to various forces, providing insights into the complex behaviors of engine components.

However, these detailed simulations often involve approximations of physical phenomena rather than exact calculations. This approach was chosen to reduce the enormous computational power that would be required. A common choice is to use the RANS method instead of the more precise but computationally demanding DNS. These approximations help understand the flow within the engine and identify where losses occur, but they might not always accurately rank or quantify these losses, as pointed out by Tucker.[26]

Nonetheless, LES has shown promise in reducing design uncertainties through limited simulations on specific components or subsystems. For instance, Sandberg and Michelassi[25] used LES to study combustion dynamics in heavy-duty combustors, and Andreini et al.[52] explored hot streak phenomena in combustor-turbine interactions. Such high-precision simulations are valuable for identifying key performance improvement areas and focusing on specific design aspects. Lengani et al.[53] further demonstrated this using LES to examine interactions between low-pressure turbine stators and rotors. By simulating real operational conditions and paying close attention to geometric details, these "virtual tests" can offer insights that surpass traditional experimental methods in accuracy and detail. When it comes to understanding turbine performance, challenges arise from the complexity of the data, especially in distinguishing between different types of losses. As illustrated in Figure 9.4, losses are often categorized as profile and end-wall losses. The interaction between the turbine's moving parts (like stators and rotors) and other factors contributes to these losses, complicating the measurement of changing flows within the turbine. The distinction between profile and end-wall losses is not always clear-cut, as shown by Michelassi and Ling[54] who used both LES and RANS to study how losses are generated across the span of a low-pressure turbine, which highlights the limitations of traditional design methods and RANS models in capturing complex loss mechanisms. They suggested adding a new dimension to the design process, termed "virtual testing," which advocates for high-fidelity CFD simulations to explore new design territories or clarify specific flow behaviors. Ideally, virtual testing should enhance traditional RANS models and influence the development of design correlations, empowering designers to make more informed decisions early in the design lifecycle.

Michelassi et al.[55] used a special computer program to see how changing certain parts of a turbine affects how well it works. They focused on two main things: how fast the air moves through the turbine and how often the turbine blades encounter air pockets. First, they looked at the flow coefficient, which is a fancy way of saying how fast the air moves compared to how fast the turbine spins. The flow coefficient is defined below as

$$\varphi = \frac{\text{Axial speed}}{\text{Revolution speed}} \tag{9.2}$$

and the reduced frequency as

$$F_{\text{Red}} = \frac{(\text{Incoming wakes passing frequency}) \times (\text{Airfoil axial chord})}{\text{Airfoil exit velocity}} \tag{9.3}$$

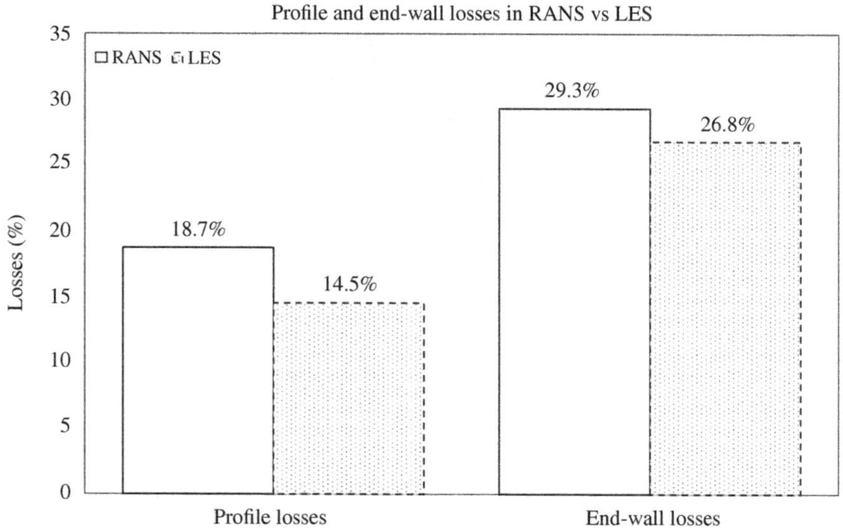

Figure 9.4 Profile and end-wall losses illustration.

Their experiments showed that these two things can cause the turbine to lose more energy than expected, which happens because the blades do not interact well with the air coming from the blades in front of them. They also found out how these air pockets move, which is important for understanding how the turbine works.

Pichler et al.[34] investigated how changing the distance between the stator and rotor rows affects turbine's performance. Conventional wisdom suggests that increasing this distance reduces efficiency. However, Pichler's research offers a fresh perspective, showing that a larger gap might improve performance by allowing for better wake decay and reducing losses in the rotor, which challenges traditional design assumptions and opens new possibilities for turbine optimization.

Sandberg and Michelassi[25] used a special computer program to predict when a gas turbine might start to have combustion problems. Knowing how the pressure inside the turbine changes is important because this information helps make the turbine run smoothly and safely under different conditions. Their research in Figure 9.5 proved that the computer could accurately predict these pressure changes, despite some slight differences. This accuracy is super helpful for making turbines that are both safe and work well.

Handling lots of diverse types of data is a big challenge in engineering. Michelassi[29] talked about how important it is to ensure that the tools and models being used for design are accurate and mentioned the importance of keeping all the data well-organized and easy to find, so it can be used properly, even if there is not much of it. Making sure the data can be easily used and understood by others in research and development is a key factor, especially if it is going to help train computer models to learn on their own.

When it comes to using ML and AI, the old systems for keeping track of laboratory data might not be enough. We need a new way to manage data that makes it easy to share and use across different projects and teams which includes giving everything consistent names, ensuring data fit together in different ways, and linking up all kinds of data, so computer models can use it easily. For example, if you want to compare two experiments or mix data from different tests and simulations, you need to organize the data in a certain way which helps AI models understand the data better, make good comparisons, and help with decision-making. So, if a company wants to use AI well, they must look at how they manage their data first. Improving how data is handled is a crucial step. This makes sure that AI models have the best data to work with, leading to better designs for turbines and other innovations.

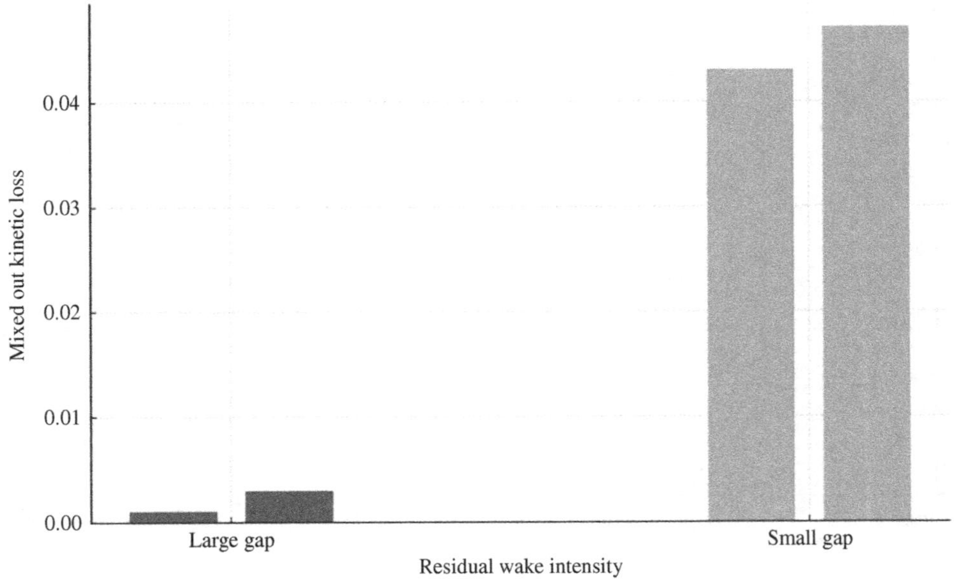

Figure 9.5 Changes in rotor losses with the distance from the stator.

9.4 Revolutionizing Turbomachinery with Machine Learning

In the energy world, getting the most out of thermodynamic cycles means operating at very high pressures and temperatures, which can stress out the parts of stationery and moving machinery, possibly forcing designers to make changes that might lower performance, make the machine unavailable, or raise costs. Recently, there have been some cool innovations in aviation with new materials. For instance, GE Aviation introduced composite fan blades in the GE90–115B engine to make it lighter, allow for bigger blades without adding stress, and handle intense conditions better. There has also been a push to use ceramic matrix composites (CMCs) in the hottest parts of gas turbines. CMCs are great because they do not need as much cooling in super-hot environments, unlike older materials that cannot take the heat.

Researchers like Fink et al.[56] have been working on creating superalloys that can handle high temperature conditions without needing too much Rhenium, which is pricey and not easy to find. Patsias[57] investigated using special coatings to dampen vibrations as a cheaper alternative to more expensive options. Another big leap forward has been in additive manufacturing (3D printing), highlighted by Attallah et al.,[58] which helps to make complex parts that are right for the job. However, special materials are needed to work with the printing process. Much research is being done into new kinds of coatings, composites, and alloys for turbine parts.

Designing gas turbines is still a tough nut to crack. They must deal with extreme heat, last a long time without breaking down, and resist corrosion and wear. Finding new materials to do all this is a long, costly journey of trial and error. Here, ML comes in to change the game. It starts with training a ML model with all the data we already have. This model learns the relationship between what materials are made of, how they are made, and how well they perform. Then, it can help pick out the best new materials to try, using criteria like how well they are expected to perform or how certain we are about those expectations. Researchers can then pick which experiments to do next, combining their know-how with the model's advice. The results of those experiments make the model even smarter, hopefully cutting down on the number of tests needed to find the right materials.

This approach, illustrated in Figure 9.6, blends the smarts of scientists with the power of data to find new materials faster and more efficiently than ever before.

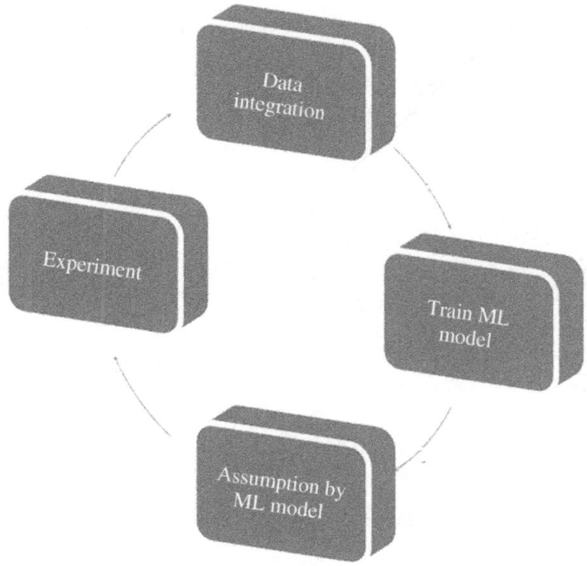

Figure 9.6 Diagram of the step-by-step learning procedure.

Recent advancements in material science, particularly in turbomachinery, demonstrate how ML and additive manufacturing transform material discovery and development. Ling[28] has successfully used ML techniques to design novel materials like corrosion-resistant aluminum alloys, high-strength nickel superalloys, and shape memory alloys. Similarly, Gaultois et al.[59] identified new thermoelectric materials with superior properties. These examples show how ML expedites the discovery process, uncovering unconventional materials overlooked by traditional methods.

Additive manufacturing (3D printing) introduces the ability to create complex parts with tailored characteristics. However, ensuring consistent material properties post-production remains a challenge. Martin et al.[60] demonstrated that data-driven methods effectively address this issue, pinpointing optimal powder and nanoparticle combinations for alloys used in additive manufacturing which highlights the synergy between ML and additive manufacturing for efficient material development.

Transfer learning, a ML subset, is particularly beneficial. Researchers need less experimental data by applying knowledge from existing datasets to new models. This is evident in the work of Hutchinson,[61] where transfer learning significantly improves accuracy in predicting material properties like the experimental band gap. This approach accelerates material development while reducing costs.

Overall, the convergence of ML and additive manufacturing signals a revolutionary era in material science. These technologies streamline discovery, push the boundaries of material design, and hold immense potential for innovations in turbomachinery and other fields. As they mature, they promise custom-tailored materials that could change industry practices and open new research avenues.

9.4.1 Integrated Optimization of Materials and Component Performance

Traditionally, turbomachinery design involved selecting materials from a limited pool of existing options or investing significant resources in developing custom materials. However, data-driven models are now transforming this process, enabling the prediction of material properties and the exploration of novel materials tailored to specific performance requirements. This integrated approach optimizes components for performance, manufacturability, and testability.

A prime example is the work of Peerless et al.,[62] where a ML model was used to simultaneously evaluate numerous material options for a turbine blade exposed to persistent organic pollutants (POPs). Their study predicted

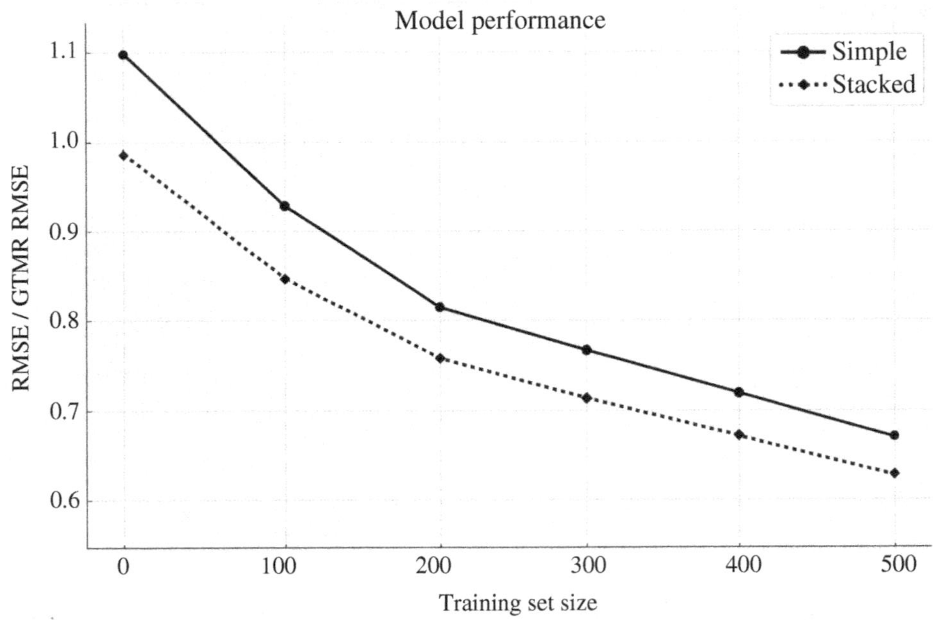

Figure 9.7 Comparison of error rates of stacked and basic models for forecasting experimental band gap.

trade-offs between material performance and biodegradability, providing insights into the balance between performance and environmental considerations.

This integrated materials-component optimization shifts design philosophy, potentially significantly impacting high-temperature applications where material properties play a critical role. By concurrently optimizing the design of components like turbine blades and the materials from which they are constructed, engineers can push the envelope toward higher-performing and more robust engines.

Beyond predicting the performance of existing or conceptual materials, ML models can guide the development of completely novel alloys. These models can identify patterns and correlations that traditional methods often lack by analyzing vast datasets of material properties and processing conditions, which paves the way for discovering materials with unprecedented combinations of strength, thermal conductivity, corrosion resistance, and other desirable attributes. Such tailored materials could unlock significant performance gains in critical turbomachinery components, enabling more efficient, powerful, and durable engines in aerospace and power generation industries.

Figure 9.7 illustrates the operation of two models: the **stacked model** and the **basic model**. The stacked model, a ML approach that predicts sequential phases by leveraging intermediate outputs, first estimates the **density functional theory (DFT) band gap** based on the chemical composition. The DFT band gap is a theoretical calculation of a material's electronic properties derived from quantum mechanical principles. The model then uses the predicted DFT band gap along with the chemical formula to estimate the experimental band gap. In contrast, the basic model directly predicts the experimental band gap from the chemical formula, bypassing any intermediate calculations or theoretical predictions.

9.5 Conclusions

ML has the potential to change how we design turbomachinery. The discussion has highlighted three key areas where its influence will be strongest:

Faster Development of High-Performance Materials: Sequential learning, designed for experiments, helps ML algorithms guide researchers toward the best materials and settings.

Combined Improvement of Materials and Components: The ability to improve both materials and component performance simultaneously promises better parts that are more efficient and long lasting.

Data-Driven Models for System Prediction: ML allows us to build models that accurately predict how complex turbomachinery systems will work, which leads to significant improvements in design and operation.

Importantly, the success of these data-driven methods depends on combining them with the basic principles of turbomachinery design which is not about replacing scientific knowledge with mysterious AI; it is about making that knowledge even stronger with data insights. Since data on turbomachinery can be limited, it is even more important to include expert knowledge within ML systems to keep them on track. The future of innovation in turbomachinery lies in this smart combination of traditional understanding and data-driven insights. Promising areas for future research include:

Including Expert Knowledge: Finding ways to incorporate the rules and principles of science directly into ML models.

Understanding ML Predictions: Developing methods to explain the reasoning behind ML predictions, helping designers make the best decisions.

The main goal remains clear: to reduce the number of design cycles, lower uncertainty, improve the design process, and confidently explore new design possibilities. By embracing this potential, ML will change the future of turbomachinery.

Example 9.1 List the potential future uses of AI in turbomachinery design. Discuss how these AI applications can enhance the design process and improve turbomachinery performance.

Solution

Future Uses of AI in turbomachinery design:

Predictive Maintenance: AI can analyze operational data to predict when components will fail, allowing for timely maintenance and reducing downtime.

Design Optimization: AI algorithms can explore a vast design space to identify the most efficient and effective designs faster than traditional methods.

Material Discovery: ML models can predict the properties of new materials, accelerating the development of high-performance components.

Performance Monitoring: AI can continuously monitor the performance of turbomachinery in real time, adjusting operational parameters to maintain optimal efficiency.

Automated Testing: AI can automate the testing process, quickly identifying design flaws and suggesting improvements.

Flow Simulation: AI can enhance CFD simulations, providing more accurate predictions of fluid behavior with less computational effort.

Example 9.2 ***Using ML to Predict Optimal Blade Shapes in Gas Turbine Design***

In the development of gas turbine blades, engineers traditionally design a blade profile, test it under various conditions, analyze the results, and then refine the design. This approach is often time-consuming and resource intensive. Now, a gas turbine design team is utilizing ML to streamline this process and enhance both the performance and durability of the blades.

Solution
Realistic Application of Machine Learning

1) **Data Collection and Training**: The design team has collected extensive data from previous blade tests, including measurements of lift, drag, surface pressure distributions, and material stress under operational conditions. This data feeds an ML model, which learns how certain profile shapes correlate with high efficiency and balanced stress distribution.
2) **Profile Prediction**: Once trained, the ML model generates blade profile suggestions. For example, it might suggest a profile with a specific curvature and thickness distribution based on its prediction of how these shapes will perform. The model considers both aerodynamic performance (like achieving maximum lift with minimum drag) and mechanical aspects (like reducing areas prone to high stress).
3) **Design Testing and Validation**: Engineers can then take the ML-suggested profiles and test them in a digital simulation environment, such as a CFD simulation. In this virtual setting, they can evaluate airflow patterns, lift, and drag, as well as identify areas of mechanical stress without needing to manufacture a physical prototype for each test.
4) **Iterative Improvement**: The ML model continuously refines its predictions based on feedback from simulation results. For example, if a suggested profile causes high stress at certain points, the model learns to adjust its shape suggestions to distribute stress more evenly across the blade. This iterative improvement cycle allows the team to reach an optimized design faster than with manual adjustments.

How Machine Learning Impacts Blade Design

- **Aerodynamic Efficiency**: By predicting the best curvature and thickness combinations, the ML model enables the blade to achieve higher aerodynamic efficiency.
- **Material Durability**: The ML model can predict profiles that avoid high-stress regions that might lead to material fatigue. This prolongs blade life and reduces maintenance intervals, lowering the long-term costs of turbine operation.

Benefits of ML-Driven Design Optimization
ML accelerates the blade design process and allows engineers to explore a broader range of shapes and profiles in contrast to traditional methods. This approach helps identify innovative designs that might otherwise be overlooked, yielding more efficient and resilient blades under operational stresses.

References

1 Wisler, D. C., Koch, C. C., and Smith, L. H. Jr., "Preliminary Design Study of Advanced Multistage Axial Flow Core Compressors," 31 Jan. 1977.

2 Craig, H. R. M., and Cox, H. J. A., "Performance Estimation of Axial Flow Turbines," *Proc. Inst. Mech. Eng.*, Vol. 185, pp. 407–424, 1970. https://doi.org/10.1243/PIME_PROC_1970_185_048_02.

3 Amano, R. S., and Sundén, B., Impingement Jet Cooling in Gas Turbines, 2024, https://www.witpress.com/books/978-1-84564-906-7, Accessed 25 Nov. 2024.

4 Amano, R. S., and Sundén, B., Computational Fluid Dynamics and Heat Transfer, 2024, https://www.witpress.com/books/978-1-84564-144-3.

5 Amano, R. S., Arakawa, H., and Suga, K., "Turbulent Heat Transfer in a Two-Pass Cooling Channel by Several Wall Turbulence Models," *Int. J. Heat Mass Transf.*, Vol. 77, pp. 406–418, 2014. https://doi.org/10.1016/j.ijheatmasstransfer.2014.04.066.

6 Wang, Q., Chen, Y.-T., and Sundén, B. (Eds.), *Emerging Topics in Heat Transfer: Enhancement and Heat Exchangers*, WIT Press, Southampton, 2014.

7 Xu, C., and Amano, R. S., "Development of a Low Flow Coefficient Single Stage Centrifugal Compressor," *Int. J. Comput. Met. Eng. Sci. Mech.*, Vol. 10, pp. 282–289, 2009. https://doi.org/10.1080/15502280902939460.

8 Xu, C., and Amano, R. S., "On the Development of Turbomachine Blade Aerodynamic Design System," *Int. J. Comput. Met. Eng. Sci. Mech.*, Vol. 10, pp. 186–196, 2009. https://doi.org/10.1080/15502280902795052.

9 Xu, C., and Amano, R. S., "The Development of Centrifugal Compressor Impeller," *ASME Turbo Expo 2008: Power Land, Sea, Air*, pp. 539–544, 2009. https://doi.org/10.1115/GT2008-50200.

10 Xu, C., and Amano, R. S., "Computational Analysis of Scroll Tongue Shapes to Compressor Performance by Using Different Turbulence Models," *Int. J. Comput. Met. Eng. Sci. Mech.*, Vol. 11, pp. 85–99, 2010. https://doi.org/10.1080/15502280903563517.

11 Xu, C., and Amano, R. S., "Study of the Flow in Centrifugal Compressor," *Int. J. Fluid Mach. Syst.*, Vol. 3, pp. 260–270, 2010. https://doi.org/10.5293/IJFMS.2010.3.3.260.

12 Xu, C., and Amano, R. S., "Empirical Design Considerations for Industrial Centrifugal Compressors," *Int. J. Rotating Mach.*, Vol. 2012, pp. 1–15, 2012. https://doi.org/10.1155/2012/184061.

13 Nourin, F. N., and Amano, R. S., "Experimental Study on Flow Behavior and Heat Transfer Enhancement with Distinct Dimpled Gas Turbine Blade Internal Cooling Channel," *J. Energy Resour. Technol.*, 2022. https://doi.org/10.1115/1.4052035.

14 Nourin, F. N., and Amano, R. S., "Heat Transfer Augmentation with Multiple Jet Impingement Cooling on Dimpled Surface for Gas Turbine Blades," *J. Energy Resour. Technol.*, 2023. https://doi.org/10.1115/1.4054465.

15 Nourin, F. N., Blum, B. L., and Amano, R. S., "Exploring the Effects of Guide Vane on Dimpled Cooling Channel of Gas Turbine Blade," *J. Energy Resour. Technol.*, 2023. https://doi.org/10.1115/1.4056334.

16 Saravani, M. S., Amano, R. S., DiPasquale, N. J., and Halmo, J. W., "Turning Guide Vane Effect on Internal Cooling of Two-Passage Channel with Parallel Ribs," *J. Energy Resour. Technol.*, 2020. https://doi.org/10.1115/1.4046731.

17 Gupta, A., Ibrahim, M. S., and Amano, R. S., "Experimental Study of Novel Passive Control Methods to Improve Combustor Exit Temperature Uniformity," *Heat Mass Transf.*, Vol. 51, pp. 23–32, 2015. https://doi.org/10.1007/s00231-014-1357-6.

18 Rodi, W., Mansour, N. N., and Michelassi, V., "One-Equation Near-Wall Turbulence Modeling with the Aid of Direct Simulation Data," *J. Fluids Eng.*, Vol. 115, pp. 196–205, 1993. https://doi.org/10.1115/1.2910124.

19 Kumar, S., and Amano, R. S., "Experimental Investigation of Heat Transfer and Flow Using V and Broken V Ribs Within Gas Turbine Blade Cooling Passage," *Heat Mass Transf.* Vol. 51, pp. 631–647, 2015. https://doi.org/10.1007/s00231-014-1436-8.

20 Kumar, S., Amano, R. S., and Lucci, J. M., "Numerical Simulations of Heat Transfer Distribution of a Two-Pass Square Channel with V-rib Turbulator and Bleed Holes," *Heat Mass Transf.*, Vol. 49, pp. 1141–1158, 2013. https://doi.org/10.1007/s00231-013-1156-5.

21 Astrua, P., Piola, S., Silingardi, A., and Bonzani, F., "Multi-objective Constrained Aero-Mechanical Optimization of Axial Compressor Transonic Blade," Copenhagen, Denmark: s.n.

22 Michelassi, V., Allegorico, C., Cioncolini, S., Graziano, A., Tognarelli, L., and Sepe, M., "Machine Learning in Gas Turbines," *Mech. Eng.*, Vol. 140, pp. S54–S55, 2018. https://doi.org/10.1115/1.2018-SEP5.

23 Brunton, S. L., Noack, B. R., and Koumoutsakos, P., "Machine Learning for Fluid Mechanics," *Annu. Rev. Fluid Mech.*, Vol. 52, pp. 477–508, 2020. https://doi.org/10.1146/annurev-fluid-010719-060214.

24 Laskowski, G., Kopriva, J., Michelassi, V., Shankaran, S., Paliath, U., Bhaskaran, R., Wang, Q., Talnikar, C., Wang, Z. J., and Jia, F., "Future Directions of High Fidelity CFD for Aerothermal Turbomchinery Analysis and Design," Washington, DC: s.n, 2016.

25 Sandberg, R. D., and Michelassi, V., "The Current State of High-Fidelity Simulations for Main Gas Path Turbomachinery Components and Their Industrial Impact," *Flow Turbulence Combust.* Vol. 102, pp. 797–848, 2019. https://doi.org/10.1007/s10494-019-00013-3.

26 Tucker, P. G., "Computation of Unsteady Turbomachinery Flows: Part 1—Progress and Challenges," *Prog. Aerosp. Sci.*, Vol. 47, pp. 522–545, 2011. https://doi.org/10.1016/j.paerosci.2011.06.004.

27 Cozzi, L., Rubechini, F., Marconcini, M., Arnone, A., Astrua, P., Schneider, A., and Silingardi, A., "Facing the Challenges in CFD Modelling of Multistage Axial Compressors," Charlotte, NC: s.n, 2017.

28 Ling, J., Antono, E., Bajaj, S., Paradiso, S., Hutchinson, M., Meredig, B., and Gibbons, B. M., "Machine Learning for Alloy Composition and Process Optimization," Oslo, Norway: s.n, GT2018-75207, 2018.

29 Michelassi, V., "Turbomachinery Research and Design: The Role of DNS and LES in Industry," Berlin, Germany: s.n, 2018.

30 Wheeler, A. P.S., Sandberg, R. D., Sandham, N. D., Pichler, R., Michelassi, V., and Laskowski, G., "Direct Numerical Simulations of a High-Pressure Turbine Vane," *J. Turbomach.*, 2016. https://doi.org/10.1115/1.4032435.

31 Michelassi, V., Wissink, J. G., and Rodi, W., "Direct Numerical Simulation, Large Eddy Simulation and Unsteady Reynolds-Averaged Navier—Stokes Simulations of Periodic Unsteady Flow in a Low-Pressure Turbine Cascade: A Comparison," *Proc. Inst. Mech. Eng. A J. Power Energy.*, Vol. 217, pp. 403–411, 2003. https://doi.org/10.1243/095765003322315469.

32 Michelassi, V., Chen, L.-W., Pichler, R., and Sandberg, R. D., "Compressible Direct Numerical Simulation of Low-Pressure Turbines—Part II: Effect of Inflow Disturbances," *J. Turbomach.*, 2015. https://doi.org/10.1115/1.4029126.

33 Pichler, R., Michelassi, V., Sandberg, R., and Ong, J., "Highly Resolved Large Eddy Simulation Study of Gap Size Effect on Low-Pressure Turbine Stage," *J. Turbomach.*, 2018. https://doi.org/10.1115/1.4038178.

34 Pichler, R., Sandberg, R. D., Michelassi, V., and Bhaskaran, R., "Investigation of the Accuracy of RANS Models to Predict the Flow Through a Low-Pressure Turbine," *J. Turbomach.*, 2016. https://doi.org/10.1115/1.4033507.

35 Wilcox, D. C., "Reassessment of the Scale-Determining Equation for Advanced Turbulence Models," *AIAA J.*, Vol. 26, pp. 1299–1310, 1988. https://doi.org/10.2514/3.10041.

36 Patel, V. C., Rodi, W., and Scheuerer, G., "Turbulence Models for Near-Wall and Low Reynolds Number Flows—A Review," *AIAA J.*, Vol. 23, pp. 1308–1319, 1985. https://doi.org/10.2514/3.9086.

37 Parneix, S., Laurence, D., and Durbin, P. A., "A Procedure for Using DNS Databases," *J. Fluids Eng.*, Vol. 120, pp. 40–47, 1998. https://doi.org/10.1115/1.2819658.

38 Wissink, J. G., Michelassi, V., and Rodi, W., "Heat Transfer in a Laminar Separation Bubble Affected by Oscillating External Flow," *Int. J. Heat Fluid Flow*, Vol. 25, pp. 729–740, 2004. https://doi.org/10.1016/j.ijheatfluidflow.2004.04.005.

39 Sandberg, R. D., Tan, R., Weatheritt, J., Ooi, A., Haghiri, A., Michelassi, V., and Laskowski, G., "Applying Machine Learnt Explicit Algebraic Stress and Scalar Flux Models to a Fundamental Trailing Edge Slot," *J. Turbomach.*, 2018. https://doi.org/10.1115/1.4041268.

40 Lumley, J. L., "Toward a Turbulent Constitutive Relation," *J. Fluid Mech.*, Vol. 41, pp. 413–434, 1970. https://doi.org/10.1017/S0022112070000678.

41 Weatheritt, J., Sandberg, R. D., Ling, J., Saez, G., and Bodart, J., "A Comparative Study of Contrasting Machine Learning Frameworks Applied to RANS Modeling of Jets in Crossflow," Charlotte, NC: s.n, 2017.

42 Akolekar, H. D., Sandberg, R. D., Hutchins, N., Michelassi, V., and Laskowski, G., "Machine-Learnt Turbulence Closures for Low-Pressure Turbines with Unsteady Inflow Conditions," *J. Turbomach.*, 2019. https://doi.org/10.1115/1.4043907.

43 Zhao, Y., Akolekar, H. D., Weatheritt, J., Michelassi, V., and Sandberg, R. D., "RANS Turbulence Model Development Using CFD-Driven Machine Learning," *J. Comput. Phys.*, Vol. 411, pp. 109413, 2020. https://doi.org/10.1016/j.jcp.2020.109413.

44 Ling, J., Kurzawski, A., and Templeton, J., "Reynolds Averaged Turbulence Modelling Using Deep Neural Networks with Embedded Invariance," *J. Fluid Mech.*, Vol. 807, pp. 155–166, 2016. https://doi.org/10.1017/jfm.2016.615.

45 Milani, P. M., Ling, J., Saez-Mischlich, G., Bodart, J., and Eaton, J. K., "A Machine Learning Approach for Determining the Turbulent Diffusivity in Film Cooling Flows," *J. Turbomach.* 2018. https://doi.org/10.1115/1.4038275.

46 Weatheritt, J., Zhao, Y., Sandberg, R. D., Mizukami, S., and Tanimoto, K., "Data-Driven Scalar-Flux Model Development with Application to Jet in Cross Flow," *Int. J. Heat Mass Transf.*, Vol. 147, pp. 118931, 2020. https://doi.org/10.1016/j.ijheatmasstransfer.2019.118931.

47 Ling, J., Jones, R., and Templeton, J., "Machine Learning Strategies for Systems with Invariance Properties," *J. Comput. Phys.*, Vol. 318, pp. 22–35, 2016. https://doi.org/10.1016/j.jcp.2016.05.003.

48 Milani, P. M., Ling, J., and Eaton, J. K., "Generalization of Machine-Learned Turbulent Heat Flux Models Applied to Film Cooling Flows," *J. Turbomach.* 2020. https://doi.org/10.1115/1.4045389.

49 Wu, J., Wang, J.X., Xiao, H., and Ling, J., "Visualization of High Dimensional Turbulence Simulation Data Using t-SNE," Grapevine, TX: AIAA, 2017.

50 Ramprasad, R., Batra, R., Pilania, G., Mannodi-Kanakkithodi, A., and Kim, C., "Machine Learning in Materials Informatics: Recent Applications and Prospects," *NPJ Comput Mater.*, 2017. https://doi.org/10.1038/s41524-017-0056-5.

51 Pichler, R., Zhao, Y., Sandberg, R., Michelassi, V., Pacciani, R., Marconcini, M., and Arnone, A , "Large-Eddy Simulation and RANS Analysis of the End-Wall Flow in a Linear Low-Pressure Turbine Cascade, Part I: Flow and Secondary Vorticity Fields Under Varying Inlet Condition," *J. Turbomach.*, 2019. https://doi.org/10.1115/1.4045080.

52 Andreini, A., Bacci, T., Insinna, M., Mazzei, L., and Salvadori, S., "Hybrid RANS-LES Modeling of the Aerothermal Field in an Annular Hot Streak Generator for the Study of Combustor–Turbine Interaction," *J. Eng. Gas Turbines Power*, 2017. https://doi.org/10.1115/1.4034358.

53 Lengani, D., Simoni, D., Pichler, R., Sandberg, R. D., Michelassi, V., and Bertini, F., "On the Identification and Decomposition of the Unsteady Losses in a Turbine Cascade," *J. Turbomach.*, 2019. https://doi.org/10.1115/1.4042164.

54 Michelassi, V., and Ling, J., "Challenges and Opportunities for Artificial Intelligence and High-Fidelity Simulations in Turbomachinery Applications: A Perspective," *J. Glob. Power Propuls. Soc.*, Vol. 2021, pp. 1–14, 2021. https://doi.org/10.33737/jgpps/135173.

55 Michelassi, V., Chen, L., Pichler, R., Sandberg, R., and Bhaskaran, R., "High-Fidelity Simulations of Low-Pressure Turbines: Effect of Flow Coefficient and Reduced Frequency on Losses," *J. Turbomach.*, 2016. https://doi.org/10.1115/1.4033266.

56 Fink, P. J., Miller, J. L., and Konitzer, D. G., "Rhenium Reduction—Alloy Design Using an Economically Strategic Element," *JOM*, Vol. 62, pp. 55–57, 2010. https://doi.org/10.1007/s11837-010-0012-z.

57 Patsias, S., "Advanced Ceramic Coatings and Interfaces III," Ceramics Engineering and Science Proceedings: Ceramic Damping Coatings: Evaluating their Effectiveness and Predicting Added Damping," 2008, pp. 19.

58 Attallah, M. M., Jennings, R., Wang, X., and Carter, L. N., "Additive Manufacturing of Ni-based Superalloys: The Outstanding Issues," *MRS Bull.*, Vol. 41, pp. 758–764, 2016. https://doi.org/10.1557/mrs.2016.211.

59 Gaultois, M. W., Oliynyk, A. O., Mar, A., Sparks, T. D., Mulholland, G. J., and Meredig, B., "Perspective: Web-Based Machine Learning Models for Real-Time Screening of Thermoelectric Materials Properties," *APL Mater.*, 2016. https://doi.org/10.1063/1.4952607.

60 Martin, J. H., Yahata, B. D., Hundley, J. M., Mayer, J. A., Schaedler, T. A., and Pollock, T. M., "3D Printing of High-Strength Aluminium Alloys," *Nature*, Vol. 549, pp. 365–369, 2017. https://doi.org/10.1038/nature23894.

61 Hutchinson, M., Antono, E., Gibbons, B. M., Paradiso, S., Ling, J., and Meredig, B., "Overcoming Data Scarcity with Transfer Learning," Long Beach, CA: s.n, 2017.

62 Peerless, J. S., Sevgen, E., Edkins, S. D., Koeller, J., Kim, E., Kim, Y., et al., "Design Space Visualization for Guiding Investments in Biodegradable and Sustainably *Source*d Materials," *MRS Commun.*, Vol. 10, pp. 18–24, 2020. https://doi.org/10.1557/mrc.2020.5.

Problems

9.1 Describe three primary functions of turbomachinery in energy systems. Explain how each function benefits from improvements in design accuracy and material selection.

9.2 List two traditional design methodologies used in turbomachinery and two ways AI could improve upon these methods. Compare the limitations of traditional methods with the potential benefits offered by AI.

9.3 Why are safety margins necessary in early turbomachinery design? How does AI help reduce these margins, and what impact could this have on the efficiency and cost of turbomachinery?

9.4 Explain the difference between RANS models and LES in the context of turbomachinery design. Identify one advantage and one limitation of each.

9.5 Using the example of impingement jet cooling in turbines, describe how data-driven methods could enhance cooling efficiency. How does ML provide insights into optimizing this process?

9.6 Consider that a ML model is predicting the properties of a new alloy for high-temperature turbine applications. What data might this model require to make accurate predictions? How can AI accelerate material discovery for turbomachinery components?

9.7 Imagine a blade design team is trying to optimize a blade's camber line for higher aerodynamic efficiency and lower stress. Describe two ways a ML model could assist in predicting and refining the optimal blade shape.

9.8 Discuss how turbulence impacts heat transfer in turbomachinery components. How does ML help improve turbulence modeling to predict heat transfer more accurately?

9.9 Why is data quality important for AI models in turbomachinery design? Identify three essential aspects of data governance and describe how each helps ensure that AI models produce reliable results.

9.10 What is *t*-SNE, and how is it used in ML for turbomachinery? Describe a scenario in which engineers could use *t*-SNE to assess model reliability on new cooling profiles.

9.11 Define multi-objective optimization in turbomachinery design and provide an example of how AI could balance two competing objectives, such as durability and efficiency, when selecting materials for turbine blades.

9.12 List three potential applications of AI in turbomachinery that have not yet been widely adopted. For each application, describe the potential impact on design accuracy, material selection, or maintenance.

10

Computational Fluid Dynamics in Turbomachinery

10.1 Introduction

Computational fluid dynamics (CFD) is integral to the design and analysis of turbomachinery, encompassing devices, such as compressors, turbines, pumps, and fans. These machines facilitate energy transfer between rotating components and the interacting fluid, resulting in complex, turbulent flow patterns. CFD allows engineers to simulate these intricate flows, offering critical insights that drive improvements in design, efficiency, and reliability. Through virtual testing and performance evaluation, CFD enables the development of optimized, cost-effective, and robust designs, significantly minimizing reliance on extensive physical testing.

CFD facilitates the simulation of fluid behavior, offering detailed insights into flow patterns, pressure gradients, and temperature distributions across a range of operating conditions. This capability has significantly advanced turbomachinery design by providing performance data that would otherwise require substantial time and experimental resources to obtain. CFD is applied in various turbomachinery applications, including aerodynamic analysis, thermal management, unsteady and transient flow analysis, as well as cavitation and noise prediction.

10.2 CFD Application Advances in Turbomachinery

Early attempts (before 1960), predicting the complex fluid flow in any turbomachine was limited to analytical and experimental methods only. Researchers and engineers derived simplified mathematical models and empirical correlations to help them understand the flow behavior, but they could not tackle the complex 3D and turbulent flow that often represents the turbomachines. During 1960s, developed numerical methods (finite difference, finite element, and finite volume) were used to approximate the solution of Navier–Stokes equations; however, applications were limited to simple 2D linear flow due to the limitation in computational power.[1] In the 1970s, the first applications of CFD in turbomachinery were conducted through a simplified 2D model, such as air flow analysis in compressors and turbines, which provided early insights into velocity patterns and pressure distributions. However, these models could not be utilized to design practical turbomachinery components because of their assumptions regarding turbulence.[1] The development of computational resources and turbulence models, such as the Reynolds-Averaged Navier–Stokes (RANS) equations in the 1980s, significantly improved the accuracy of 3D turbomachinery modeling. Introducing turbulence models like k-epsilon and k-omega empowered the simulation of more complex, high-speed turbulent flows of turbomachinery. This advancement marked a key step in CFD's capability to predict the real-world performance of compressors, turbines, and fans, thereby contributing to improved efficiency in turbomachinery design.[2] A breakthrough in the 1990s was the introduction of unsteady flow modeling, which facilitated the simulation of transient effects, interactions between stators and rotors, and blade wake phenomena. This advancement helped in predicting the dynamic instabilities leading to performance losses.[3] The 1990s and 2000s brought significant advancements in both

Fundamentals of Turbomachinery: Theory and Applications, Second Edition. Ryoichi S. Amano and William W. Peng.

Companion website: www.wiley.com/go/amano/turbomachinery2e

hardware, such as parallel computing, increased memory capacity, and faster processors, and in CFD algorithms. High-resolution techniques, including large eddy simulation (LES) and detached eddy simulation (DES), became increasingly prominent, allowing for more accurate predictions of the unsteady and complex flow patterns typical of turbomachinery. Both commercial (e.g., ANSYS Fluent, STAR-CCM+) and open-source (e.g., OpenFOAM, SU2) CFD software have been developed to support turbomachinery design and analysis. Commercial tools are typically more user-friendly and widely adopted in industry, facilitating seamless integration into engineering workflows, while open-source platforms offer greater flexibility for customization and research applications. After 2010, optimization tools were incorporated into CFD software packages, enhancing the design development cycle by providing designers with a more efficient, automated, and streamlined workflow. Nowadays, machine learning (ML) and artificial intelligence (AI) are being integrated into CFD applications to improve and accelerate the design optimization process. Furthermore, the concept of real-time processing is being developed in conjunction with CFD to enable real-time modeling in turbomachinery.

10.3 Fundamentals of CFD Methodologies in Turbomachinery

In general, there are three primary approaches to solving fluid dynamics and heat transfer problems. The first is the theoretical or analytical approach, which provides exact solutions to the governing equations but is feasible only for a limited range of cases. The second is the experimental approach, which involves testing the real scenario under controlled conditions; however, it often requires considerable time to construct and troubleshoot the experimental setup and can be costly. The third approach is CFD, which is a field integrating fluid mechanics, mathematics, and computer science fields to solve any fluid flow dynamics and heat transfer. Fluid mechanics principles are employed to derive mathematical equations (partial differential equations, known as governing equations) that describe fluid behavior. These equations are then transformed into a set of algebraic equations through mathematical techniques and solved using numerical methods via advanced computer programs or commercial software packages. It was initially believed that CFD would replace all theoretical and experimental techniques; however, this was not entirely accurate. At that time, both experimental and analytical or theoretical methods are expected to remain valuable, serving as distinct analytical tools for validation and would continue to coexist alongside CFD.

For first-time users, there are two ways to introduce CFD. One approach involves diving into the complex partial differential equations governing fluid flow and their numerical representations, which can often feel overwhelming for beginners. The other approach focuses on introducing key features and workflows commonly available in most commercial and shared online CFD software. In this chapter, we will focus primarily on the second approach and how it can be implemented for a turbomachine, incorporating detailed aspects of the first approach where relevant. Most CFD codes and graphical user interface (GUI) are made user-friendly to enter the problem geometry, parameters, and visualize results.; Therefore, it includes preprocessors, physics solver, and postprocessing sections. Several processes lie under the preprocessing stage, such as building the geometry, specifying boundary conditions, and mesh generation. The physics solver stage then involves selecting the transport equations that govern the physics of the simulated problem. Finally, the postprocessing stage offers tools for visualizing solution progress through plots and monitors, along with scenes that display physical properties like temperature, pressure, and velocity. The following sections will provide some key aspects and workflow steps when using CFD for modeling fluid dynamics in turbomachines, but for more details about the CFD methodology, please refer to books about CFD.[4]

10.3.1 Preprocessing

Conducting any CFD analysis necessitates the creation of the fluid flow region, which can either be modeled using CAD software or extracted from the space between the internal and external solid walls of a turbomachine. This geometry is commonly referred to as the computational domain. The computational domain can represent an internal (most turbomachines) or external flow (wind and hydro turbines) problem where the bounding boundaries will determine its type. Any computational domain is bounded by boundary conditions (BCs) which define the behavior of the fluid at the boundaries and acts as a constraint to ensure accurate and realistic solution. Inlet and outlet BCs are the main types that specify the fluid conditions entering and exiting the computational domain, such as velocity, temperature, pressure, mass flow rate, and fluid phase volume fraction. Wall boundaries are specified at the interfaces between solids and fluids and can be constrained by heat transfer or slip/nonslip conditions. The symmetric BC empowers the modeling of partial domains when symmetric axis or plane is located, on the other hand, the periodic BC allows the modeling of repeating or cyclic geometries (such as infinitely long pipe).

CFD codes employ numerical methods to solve the governing equations of fluid flow (Navier–Stokes equations), which necessitates discretizing the computational domain into small cells, or mesh (finite volume). This process, known in the CFD field as meshing, is crucial for controlling solution accuracy. In general, increasing the number of cells enhances solution accuracy but also raises computational costs. Thus, achieving the optimal balance between computational expense and accuracy level is essential. Other factors may affect the solution accuracy, such as mesh type, mesh quality, the boundary layer mesh size, and the size of the first cell on the wall, and the numerical method order of accuracy. Accurate simulation of the boundary layer effects requires fine cells near the wall, sometimes called prism layers or inflation layers, the size of these cells shall satisfy the requirements of the turbulence simulation technique used in the transport equations (will be discussed in the next section). Meshes are commonly classified based on cell connectivity into structured, unstructured, and hybrid types. They can also be categorized by element type, including hexahedral, polyhedral, and tetrahedral cells. Each mesh type impacts solution accuracy and computational speed, making careful selection essential based on the specific application, prior experience, and relevant literature. Polyhedral meshes are highly versatile and are particularly well-suited for complex computational domains.

10.3.2 Physics Solver

The physics solver section in any CFD code is responsible for specifying the physics models that govern the solution within the computational domain. The accuracy and validity of the results heavily depend on selecting appropriate models that accurately represent the physical phenomena occurring in the domain. In general, CFD codes solve the fundamental governing equations of fluid dynamics, specifically the conservation equations for mass, momentum, and energy. The continuity equation states that the mass flow rate entering the control volume equals the mass flow rate exiting the control volume in addition to the mass accumulation within that control volume, which can be stated mathematically as follows:

$$\frac{\partial \rho}{\partial t} + \nabla \cdot (\rho \mathrm{v}) = 0, \tag{10.1}$$

where ρ is the density and v is the continuum velocity.

The momentum equation, also known as Navier–Stokes equation, expresses momentum conservation by equating the sum of forces acting on the fluid to the rate of change of momentum within the fluid. Mathematically, it is expressed as

$$\frac{\partial (\rho \mathrm{v})}{\partial t} + \nabla \cdot (\rho \mathrm{v} \otimes \mathrm{v}) = \nabla \cdot \sigma + f_b, \tag{10.2}$$

where f_b is the resultant of body forces (such as gravity and centrifugal forces) per unit volume acting on the control volume and σ is the stress tensor.

While the energy equation represents energy conservation and is expressed mathematically as follows:

$$\frac{\partial (\rho E)}{\partial t} + \nabla \cdot (\rho E \mathrm{v}) = \nabla (\mathrm{v} \cdot \sigma) + f_b \cdot \mathrm{v} - \nabla \cdot q + S_E, \tag{10.3}$$

where E is the total energy per unit mass, q is the heat flux, and S_E is an energy source per unit volume.

The equation of state is used to complete the governing equation system by linking pressure, temperature, and density, typically through models such as constant density, polynomial density, ideal gas, or real gas formulations. In addition to the primary governing equations, simulating physics in turbomachines necessitates accounting for turbulence, rotational and Coriolis forces, compressibility effects, and, in certain cases, phase change models, such as those for cooling fluid transitions or cavitation.

Turbulence is present in most turbomachinery applications, making turbulence modeling essential for accurately predicting fluid dynamics. However, this modeling process is complex and challenging. The following discussion provides a brief introduction to turbulence and explores various methodologies for predicting it. The turbulence represents irregular fluctuations of the velocity and pressure over time. These fluctuations occur at small scales and high frequencies, demanding significant computational resources to resolve fully, which is called the direct numerical solution (DNS). The DNS is unfeasible for complex geometry (like turbomachines) and high Re flows due to its enormous computational power demands. Therefore, another less expensive methodology, turbulence modeling, is used to solve using time averaged quantities and approximate the small fluctuations around the average values. The Navier–Stokes equations are time-averaged and reformulated as the Reynolds-Averaged Navier–Stokes (RANS) equations, which focus on solving for the average flow characteristics while modeling all turbulence scales. Time averaging is performed by breaking down each solution variable into an average component ($\overline{\phi}$) and a fluctuating component (ϕ').

$$\phi = \overline{\phi} + \phi', \tag{10.4}$$

where ϕ represents any fluid property like velocity, pressure, species concentration, and energy.

The RANS equations are the same as the original Navier–Stokes equations but with an additional term called the stress tensor. The stress tensor can be modeled using two approaches: eddy viscosity models and Reynolds stress transport models. The eddy viscosity models solve additional transport equations to determine a parameter called turbulent eddy viscosity to model the stress tensor. The eddy viscosity models include Spalart–Allmaras model, k-epsilon model, and k-omega model.[5–11] On the other hand, the Reynolds stress transport model directly solves the components of the stress tensor, providing a more detailed and accurate description of the turbulence.[12, 13] The RANS models are widely used in industry, but they struggle with complex separation problems and unsteady flows. Therefore, the large eddy simulation (LES) technique is employed to predict turbulence in transient problems, where large turbulence scales are directly resolved, while smaller scales are modeled using sub-grid scale models. This method is less computationally demanding than direct numerical simulation but still requires significantly more computational power than RANS models, as it necessitates fine meshes near the walls (the mesh size shall be less than the Taylor microscale and Kolmogorov scale). The need for a hybrid approach led to the development of the Detached Eddy Simulation (DES) method, which combines elements of both LES and RANS techniques. The DES approach models turbulence using RANS equations near the walls where fine mesh is required and solves the large-scale turbulence far from walls in the fluid core using the LES approach. For more background information about the LES and DES, see Refs. 14, 15.

Aside from the external walls bounding the computational domain for almost all CFD applications, turbomachines have internal rotating walls (rotors) and sometimes stationary walls (stators). These internal walls add complexity to the solution since walls often introduce vorticities to the flow which necessitates accurate prediction of the flow in the boundary layer. The boundary layer is defined as the layer close to the wall where the flow velocity changes from zero (first layer touches the wall) to 99% of the free stream velocity. The turbulent boundary layer consists of the outer and inner layers where the turbulent effects dominate the outer layer. On the other hand, the inner layer is divided into three sub-layers: viscous, buffer, and log sublayers. The viscous sublayer is the region adjacent to the wall, dominated by viscous effects. Beyond this lies the log layer, located at the outer edge of the inner layer, where turbulent and viscous effects exert equal influence. Between these two layers, the buffer layer acts as a transitional zone. The limits of each layer are determined by the y^+ value which is a dimensionless parameter representing the distance from the centroid of the first cell in contact with the wall from the wall.

$$y^+ = \frac{yU_T}{\nu}, \tag{10.5}$$

where y is the distance from the first cell centroid to the wall, U_T is the friction velocity at the wall, and the ν is the kinematic viscosity.

In the viscous sublayer, y^+ values are less than 5. The buffer layer has y^+ values ranging from 5 to 30, while the log layer spans from 30 to 500, in the outer layer, y^+ values exceed 500. The discretization of the computational domain to ensure appropriate y^+ values near the walls as recommended by the utilized turbulence model is a key factor in acquiring accurate results. Most commercial CFD codes offer three primary wall treatment approaches: Low y^+, High y^+, and All y^+. Additionally, some codes provide an option for a two-layer All y^+ approach. Whenever the computational domain walls have a y^+ less than 1, the low y^+ wall treatment approach shall be used, while the high y^+ approach shall be used for computational domains with y^+ higher than 30. The All y^+ and two-layer All y^+ are using blended wall functions for a wide range of near wall y^+ values.

Modeling turbomachines is inherently complex, primarily due to the presence of rotating components, such as the impeller or rotor. The complexity of the model depends on the specific objectives of the simulation—whether it aims to optimize blade design, analyze individual components, or assess the machine's impact on its surroundings. If the intention is to study the effect of the turbomachine on its surrounding, some commercial CFD codes gives an option to use simplified approach (which do not need the detailed geometry of the turbomachine) to model the turbomachine blades by adding a source term for the volumetric body forces to the momentum equation, or by specifying a pressure jump derived from the turbomachine performance curve. More details about the available models used in this simplified approach can be found in each software help document. If the purpose of the CFD study is optimization of the blade or study the internal flow within the turbomachine itself, it is necessary to have the whole detailed geometry. The rotational motion of the rotor can be simulated using various approaches, including Moving Reference Frame (referred to as Rotating Reference Frame in STAR-CCM+), Rigid Body Motion, Overset Mesh, and Dynamic Fluid-Body Interaction (also known as Six Degrees of Freedom or FSI models). Please note that the terminology may vary between CFD software; the terms used here are based on Siemens STAR-CCM+, with their counterparts in ANSYS Fluent indicated where applicable. In steady-state simulations, rotational motion can be modeled by rotating the reference frame while keeping the mesh stationary. This approach is efficient for obtaining approximate performance curves for turbomachines, as it reduces computational effort and time. For transient simulations, rotor motion can be represented by moving the vertices of the rotor domain during the simulation. The rotation rate, which can be determined from experimental data, is then specified in the CFD code and can be described using rigid body motion. In cases where the rotation rate is not directly available, commercial CFD codes can determine the rotation rate based on the forces exerted by the fluid dynamics. This process is referred to as dynamic fluid–body interaction or fluid–structure interaction. Another technique, known as morphing or dynamic fluid–body interaction morphing, is employed to deform the mesh based on the forces calculated from the fluid dynamics. However, this approach is not applicable to turbomachinery. The reason is that it is designed for small mesh deformations—such as those occurring during valve openings—that do not significantly impact the mesh quality, allowing for successful solution convergence. In contrast, turbomachinery involves high-speed rotations that can cause substantial mesh distortion, compromising the integrity of the solution.

Using either commercial or in-house CFD codes necessitates the understanding of the underlying techniques utilized in the numerical solver. Any numerical solution for a given case begins with the initialization of the computational domain, followed by the selection of appropriate numerical methods for solving the problem. Next, relevant physical properties are selected for being monitored throughout the solution process. The transport equations are then solved using a computational solver, which involves iterative progress through a feedback loop. This loop continues until the solution converges and meets the established convergence criteria. Domain initialization can be arbitrary and has default values in most commercial codes but having close initial conditions to the real problem property values eases the model convergence and shortens the solution time. The transport partial differential equations are approximated by finite volume method, where algebraic equations are approximated using the finite difference for the finite volume cells. For further details on the various interpolation schemes and pressure–velocity coupling methods available for controlling the solution, please refer to Reference.[4] The convergence of a CFD model entails that the numerical solution becomes stable with the solution iteration progression. The stability of the solution can be determined from the solution residuals which are the relative change of the solution among the current and previous iterations. Usually, a decreasing residual toward zero below the convergence criteria means a solution convergence. Physical quantities such as velocity, temperature, power, and lift and drag forces are commonly used to monitor convergence. A specific parameter can be chosen to track either at a particular location or as an average over a surface. The stabilization of the monitored parameter over solution time signals that the model has converged. Additionally, mass conservation can be monitored between the system inlets and outlets. Meshing plays a significant role in the solution accuracy and discretization errors; users are advised to conduct a systematic mesh independence study followed by model validation to ensure a high-fidelity CFD model.

10.3.3 Post-Processing

Post-processing refers to the analysis and visualization of the results obtained from a CFD model. This process is conducted to interpret the physical phenomena that have occurred during the simulation and to extract key parameters that describe the solution. A high-fidelity CFD model is of limited value without proper post-processing, as the insights gained from the simulation cannot be fully understood or utilized. Post-processing includes reporting maximum, minimum, average values of physical properties (forces, moment, temperature, velocity, and pressure) at certain locations, surface, or volume. Scalar functions such as pressure and temperature can be visualized using contour or scalar scenes to highlight the intensity of these scalar functions in a certain section of the computational domain. The direction of the flow or any other vector function can be visualized using the vector scenes, it is also useful in determining the location of high turbulence and vortices. Some CFD software offers the ability to display 3D representations of a specific scalar function within the computational domain, either for a constant value or over a defined range. This visualization is known as iso-surfaces. Examples through figures will be provided in the next sections.

10.4 Applications of CFD in Turbomachinery Design and Analysis

In this section, several practical examples will be presented to demonstrate how the methodologies discussed earlier are applied in real-world turbomachinery scenarios. These examples will illustrate the connection between theoretical approaches and their implementation in the design, analysis, and optimization of turbomachinery components, such as turbines, compressors, and pumps. All the below discussed examples were built and simulated in Siemens STAR-CCM+ software, however, other CFD software packages could be used for the same purpose.

10.4.1 Wind Turbine

Problem Description: Study the effect of new airfoils designs characterized with high lift to drag ratios on the aerodynamics performance and power output of residential-scale wind turbines.

Preprocessing: The first step involves creating the CAD model of the wind turbine, which can be done using any CAD software, such as SolidWorks or CREO. Alternatively, the geometry can be constructed directly within STAR-CCM+, or using ANSYS SpaceClaim or ANSYS DesignModeler, depending on the preferred software environment. Either way, the model will be imported and ready in your parts directory for next steps as shown in Figure 10.1. This surface geometry represents the solid outer surface; however, CFD codes model fluid around solid or bounded by solid. Therefore, the fluid (Air) around the wind turbine shall be extracted. This case lies under the external flow type; hence a block of air shall be created, and the wind turbine geometry shall be subtracted from it. The moving reference frame is used in this example to model the rotational motion of the turbine since we are interested in the overall turbine performance when using the new airfoils in terms of output power. The computational domain is divided into two distinct sections. The first section is the moving region, represented by the rotating cylinder, which simulates the rotation of the turbine rotor. The second section comprises the surrounding air around the rotor. Both sections are illustrated in Figure 10.2. Previous studies suggested that to minimize the influence of boundary layers near the domain walls on simulation results, the ratio of the blade swept area to the cross-sectional area of the computational domain should be kept below 30%. In addition, to accurately capture the wake region and the subsequent velocity recovery downstream of the wind turbine, the domain was extended to a length of 10.5 times the rotor diameter downstream of the turbine. Upstream of the turbine, the domain length was limited to 1.5 times the rotor diameter. Additionally, the rotating cylinder representing the turbine's rotor was modeled with a radius of 1.01 times the rotor diameter and a length of 1.1 times the hub height. In the simulation, the inlet velocity and outlet pressure boundary conditions were applied upstream and downstream of the wind turbine, respectively. The ground (bottom wall) was modeled as a no-slip wall, while the surrounding walls, representing the environment, were assigned slip walls boundary conditions. In this example, the meshing tool in the STARCCM+ software (or ANSYS Meshing, if ANSYS was used) was used to discretize the domain. Figure 10.3 illustrates three distinct meshing domains. The first is the rotating (moving) domain, which represents the turbine rotor, where a fine mesh is used to capture the detailed fluid dynamics within this complex volume. A smaller, stationary cuboid mesh extends till the outlet boundary condition from the rotor to ensure accurate resolution of the wake region, where velocity components undergo significant variation. The remaining computational domain, located further from both the rotating rotor and the wake region, uses a coarser mesh due to the reduced complexity in these areas. The prism layer (inflation layers) option was applied to generate prism layers on all the walls of the fluid domain, including the tower and nacelle

Figure 10.1 Horizontal axis wind turbine geometry.

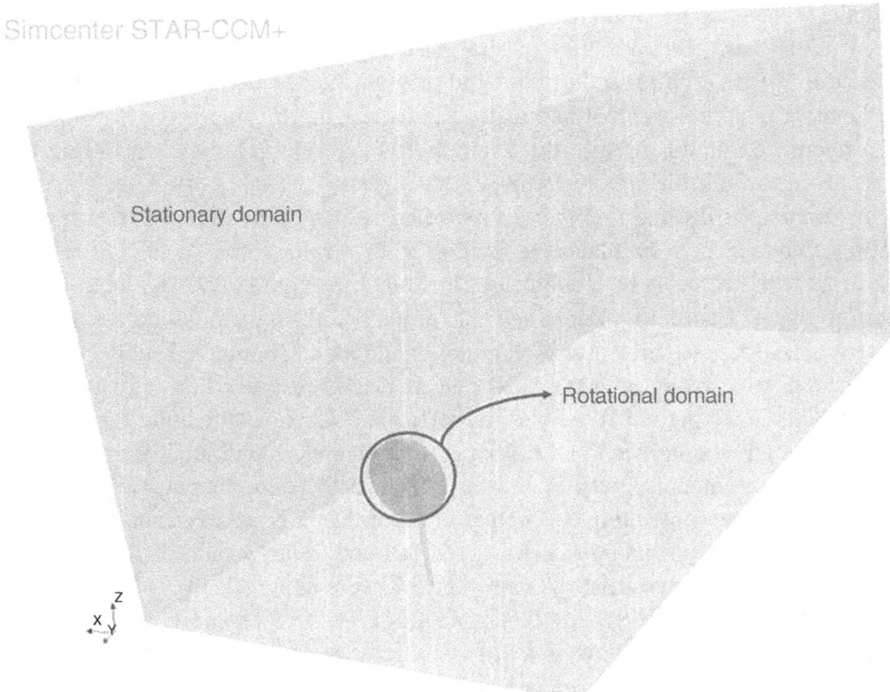

Figure 10.2 Computation domain geometry.

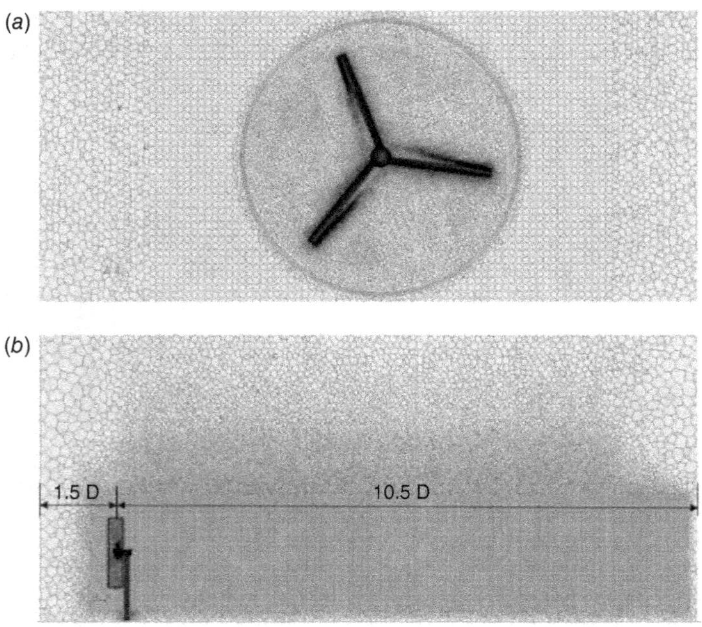

Figure 10.3 Mesh details of the computational domain (*a*) transverse sectional plane cuts through rotor blades, (*b*) longitudinal vertical sectional plane at wake centerline.[16]

walls. This approach was implemented to accurately resolve the physical equations near the walls and account for the effects of boundary layers in the solution.

Physics Solver: The physics in the computational domain is solved based on the selected models you specify within the continua section of STARCCM+ software (or Setup section if ANSYS was used), therefore, selecting the physical models representing the current problem is crucial. In this example, the computational domain is three dimensional, and hence, a 3D steady-state model is used. There is no heat addition, and the air density is constant; therefore, the constant gas density model was selected. The flow around the turbine blade considering its definition for airfoil profile was determined to be turbulent, so the RANS equations were solved along with the k-epsilon turbulence model. In addition, the gravity model was selected to mimic the gravity force. A rotational rate was specified for a rotational reference frame approximated from experiments with the same airfoil profile.

Post-Processing: Before running the simulation, it is essential to set up various reports, monitors, plots, and scenes. Key reports to establish prior to solving the model include those that will be averaged over a specified number of iterations or time steps. For steady-state simulations, especially when a parameter exhibits slight fluctuations, it is advisable to create a monitor for that parameter and average it over an appropriate number of iterations to ensure reliable results. In this example, we are interested in the average power output from the turbine and the velocity and pressure distribution in the wake region after the turbine. To get the power, you need to calculate the moment around the rotation axis for the turbine blades through the moment report within STARCCM+ and multiply it by the rotational speed through an expression report. *Any numerical model must be tested using various mesh sizes until the results become independent of the mesh resolution (called, Mesh Independence Study). Subsequently, the model should be validated against experimental data before being applied in further analyses.* Mesh independence study and validation were performed for the model in the current example as shown in Table 10.1. The table presents a comparison of the average error percentage for each grid resolution. A cell counts of 6.5 million was chosen to strike a balance between result accuracy (with an average error of 5% in axial velocity measurements) and simulation time. It is worth noting that the maximum deviation between the 6.5 M and 7.8 M grids is approximately 0.25%.[16]

Contour scenes can be created for pressure distribution around airfoils, axial velocity, and vorticity scenes for better understanding of the increase or decrease in the acquired power. For example, the absolute pressure distribution around both root ($r/R = 0.2$) and tip ($r/R = 0.9$) sections of the blades generated using NACA 64(3)618, NACA 6412, and GOE 447, respectively, are shown in Figure 10.4. The counterclockwise direction of rotation is evident since, in the six parts of the figure, the high-pressure side of the blade is to the left of the airfoils, while the low-pressure side to the right of the airfoils. In addition, higher pitch angles at the root are well-demonstrated in parts (*a*), (*c*), and (*e*), while the pitch angles are almost zero at blade tip in parts (*b*), (*d*), and (*f*). Another example of the contour surfaces is shown in Figure 10.5, which presents absolute axial velocity distribution and vorticity magnitude distribution over a horizontal plane cut through the wake body. More detailed information about wind turbines' performance and blade modifications can be found in references.[16–20]

Table 10.1 Axial Velocity Error Percentages for Different Grid Counts.

Case No.	Grid Count (Millions)	Average Error % at $Z/D = 0.25$	Average Error % at $Z/D = 0.5$
1	2.8	7.82	8.46
2	4.7	7.51	8.26
3	6.5	4.77	5.03
4	7.8	4.72	4.79

Source: Ref. 16.

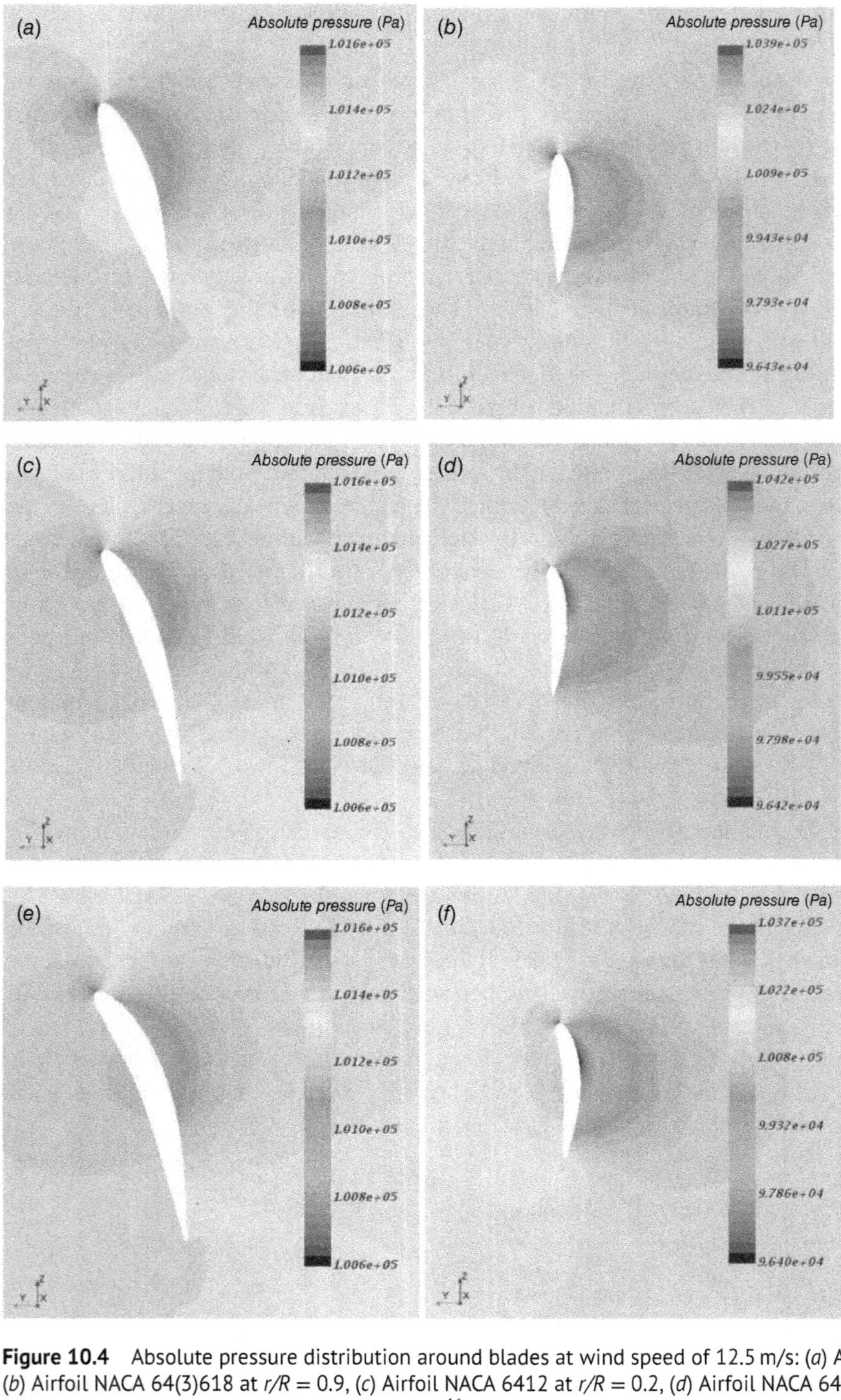

Figure 10.4 Absolute pressure distribution around blades at wind speed of 12.5 m/s: (*a*) Airfoil NACA 64(3)618 at $r/R = 0.2$, (*b*) Airfoil NACA 64(3)618 at $r/R = 0.9$, (*c*) Airfoil NACA 6412 at $r/R = 0.2$, (*d*) Airfoil NACA 6412 at $r/R = 0.9$, (*e*) Airfoil GOE 447 at $r/R = 0.2$, and (*f*) Airfoil GOE 447 at $r/R = 0.9$.[16]

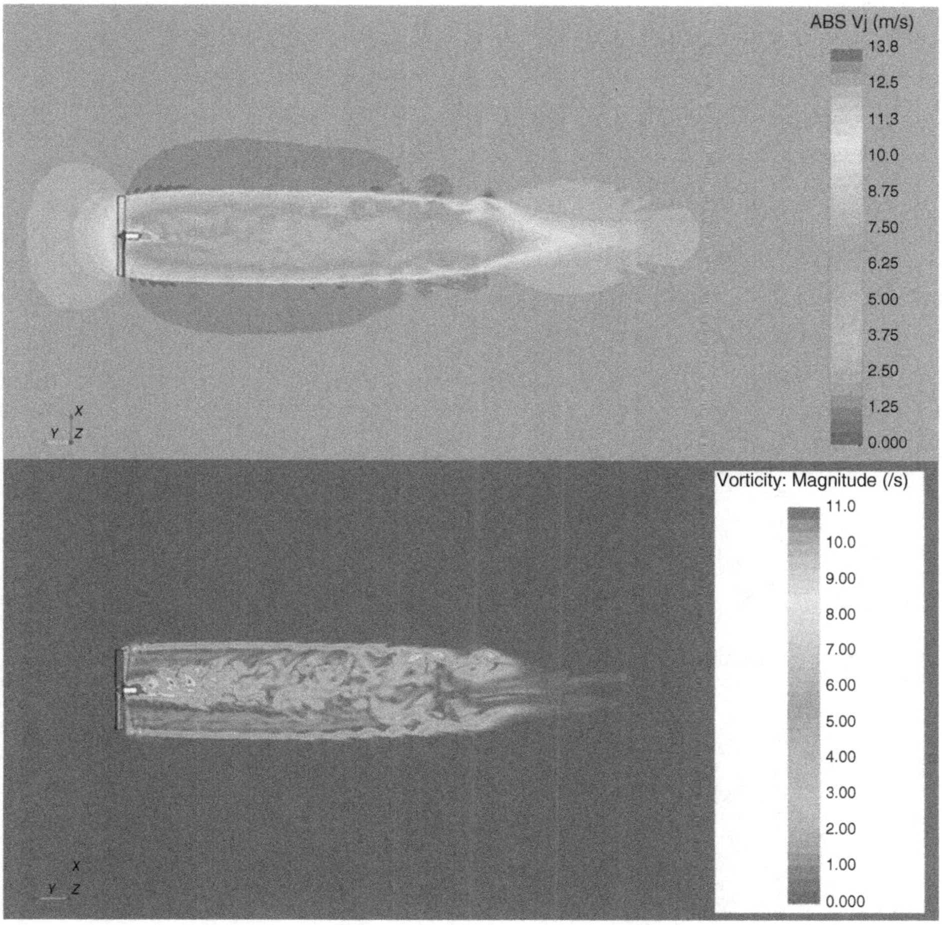

Figure 10.5 Absolute axial velocity distribution and vorticity magnitude distribution over a horizontal plane cut through the wake body at each model's hub height.[16]

10.4.2 Hydro-Turbine

Problem Description: Investigate the number of rotor blades' influence on the hydro-turbine performance in terms of power output and pressure drop.

Preprocessing: As explained before, the turbine stator and rotors shall be created using any CAD software, such as SolidWorks or Creo. Alternatively, the geometry can be constructed directly within STAR-CCM+, or using ANSYS SpaceClaim or ANSYS Design Modeler, depending on the preferred software environment. Following the same procedures as Section 10.4.1. the model will be imported and ready in your parts directory for next steps as shown in Figure 10.6. This surface geometry represents the solid outer surface; however, CFD codes model fluid around solid or bounded by solid. Therefore, the fluid (water) entrapped between the hydro-turbine and the external pipe shall be extracted. The moving reference frame is used in this example to model the rotational motion of the turbine since we are interested in the overall turbine performance in terms of power output and pressure drop. The computational domain is divided into two distinct sections. The first section is the moving region, represented by the rotating cylinders, which simulates the rotation of the turbine rotor, and the shaft connected to it. The second section involves the flow of water through the pipe, starting before the stator, moving through the stator, circulating around the rotor rotational section, and continuing past the rotor to the pipe outlet. Both

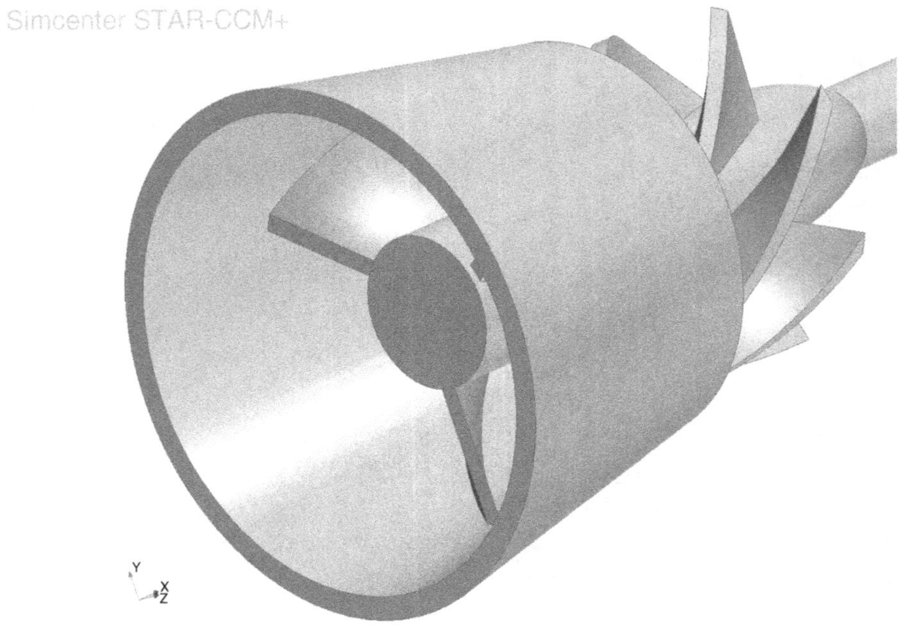

Figure 10.6 Hydro-turbine stator and rotor geometry.

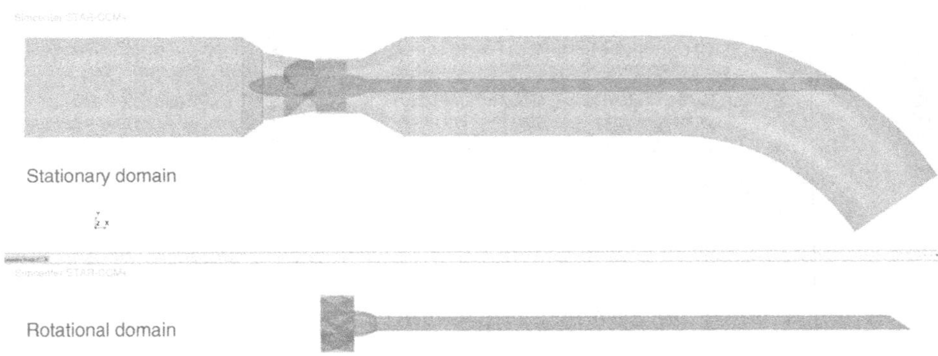

Figure 10.7 Stationary and rotational sections of the computational domain.

sections are illustrated in Figure 10.7. In the simulation, the inlet velocity and outlet pressure boundary conditions were applied upstream and downstream of the wind turbine, respectively, as shown in Figure 10.8. Internal interfaces are used to connect different regions (rotational and non-rotational), while the rest of the walls are modeled as no-slip walls boundary conditions. In this example, the meshing tool in the STARCCM+ software (or ANSYS Meshing, if ANSYS was used) was used to discretize the domain. Figure 10.9 illustrates two distinct meshing sizes. Coarse mesh is used far from the stator and rotor blades, while a fine mesh is used to capture the detailed fluid dynamics within the complex volume around the stator and rotor blades. The prism layer (inflation layers) option was applied to generate prism layers on all the walls of the fluid domain, including the rotor and stator. This approach was implemented to accurately resolve the physical equations near the walls and account for the effects of boundary layers in the solution.

Physics Solver: Like the previous example, selecting the physical models representing the current problem is crucial. In this example, the computational domain is three-dimensional, and hence, a 3D steady-state model is used. There is no additional heat, and the water density is approximately constant; therefore, the constant density

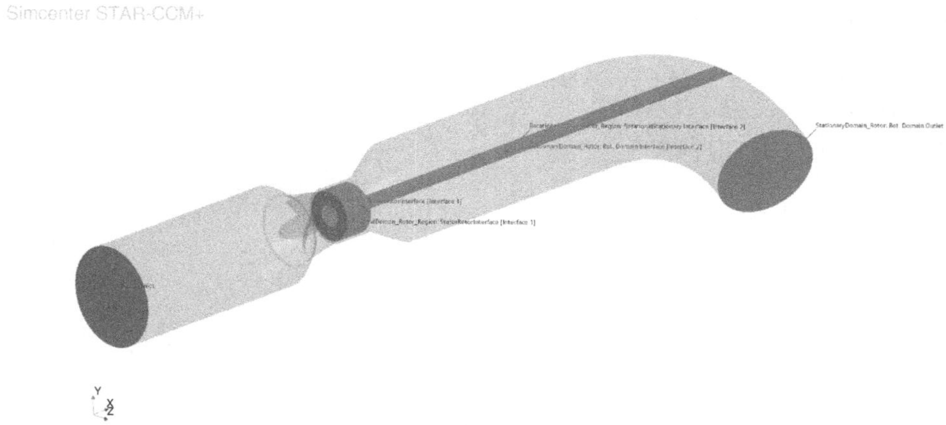

Figure 10.8 Interfaces between different regions, in addition to the inlet and outlet BCs.

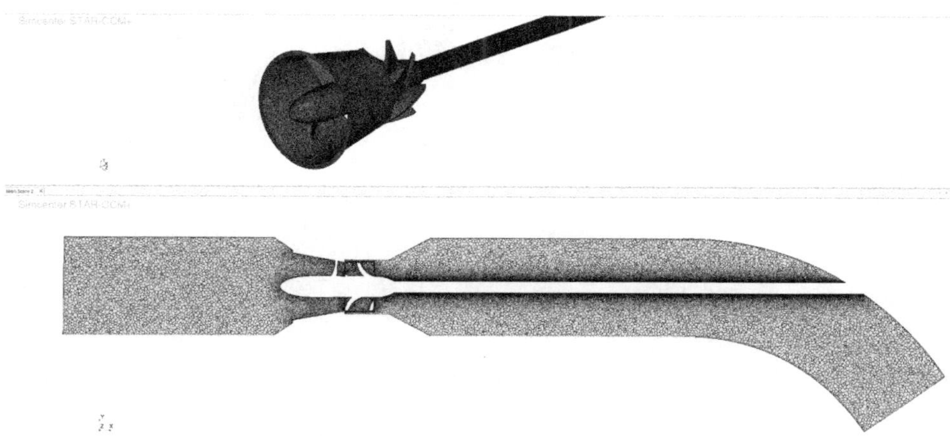

Figure 10.9 Mesh details of the computational domain.

model was selected. The flow around the turbine blades was determined to be turbulent, so the RANS equations were solved along with the k-epsilon turbulence model. In addition, the gravity model was selected to mimic the gravity force. A rotational rate was specified for a rotational reference frame approximated from experiments.

Post-Processing: Before running the simulation, it is essential to set up various reports, monitors, plots, and scenes. Key reports to establish prior to solving the model include those that will be averaged over a specified number of iterations or time steps. For steady-state simulations, especially when a parameter exhibits slight fluctuations, it is advisable to create a monitor for that parameter and average it over an appropriate number of iterations to ensure reliable results. In this example, we are interested in the average power output and the pressure drop of the turbine. To get the power, you need to calculate the moment around the rotation axis for the turbine blades through the moment report within STARCCM+ and multiply it by the rotational speed through an expression report. *Any numerical model must be tested using various mesh sizes until the results become independent of the mesh resolution (called, Mesh Independence Study). Subsequently, the model should be validated against experimental data before being applied in further analyses.* A mesh independence study was conducted by assessing the spatial discretization error using Richardson extrapolation. As shown in Table 10.2, a cell size of 4 mm yields an error percentage of less than 1% compared to the predicted exact power value, demonstrating that the results are independent of the mesh size.

Table 10.2 Spatial Discretization Error Using Richardson-Extrapolation.

No.	Description	Mesh Size	Percentage Error
1	Coarse	8	3.45%
2	Medium	4	0.98%
3	Fine	2	0.31%

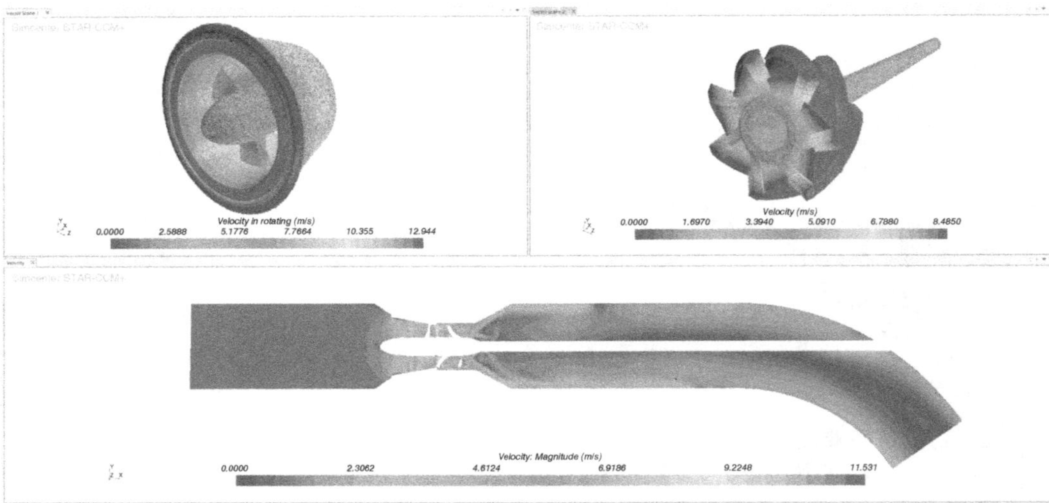

Figure 10.10 Velocity scalar and vector scenes around the turbine rotor and stator.

Validation against experiments was also performed for the current model, and the error level was less than 8%, except for only one case, it was 13%. Contour scenes can be created for velocity distribution in a plane cutting horizontally through the computational domain as shown in Figure 10.10. Additionally, in this example, vector scenes were created to show the flow direction and the velocity intensity on the stator and rotor walls as shown in the figure.

10.4.3 Gas-Turbine Blade Cooling

Problem Description: Study the effectiveness of different passive heat transfer augmentation techniques on the gas turbine blades internal cooling in terms of the increase of the heat transfer coefficient and pressure drop.

Preprocessing: The internal cooling of the gas turbine blades can be acquired by various techniques, such as internal convection channel cooling, jet impingement, film cooling. In this example, the internal channel and jet impingement cooling will be investigated. Same steps as previous examples will be followed. The internal cooling channels (original and modified) shall be created using any CAD software, such as SolidWorks or Creo. Alternatively, the geometry can be constructed directly within STAR-CCM+, or using ANSYS SpaceClaim or ANSYS DesignModeler, depending on the preferred software environment. Then, the model shall be imported in the parts directory for next steps as shown in Figure 10.11. The first CAD models are for the flat and dimpled internal cooling channels, while the other CAD models are for the jet impingement cooling with and without diverters. The moving reference frame is used to model the rotational motion of the internal cooling channel (geometry (*a*)). The

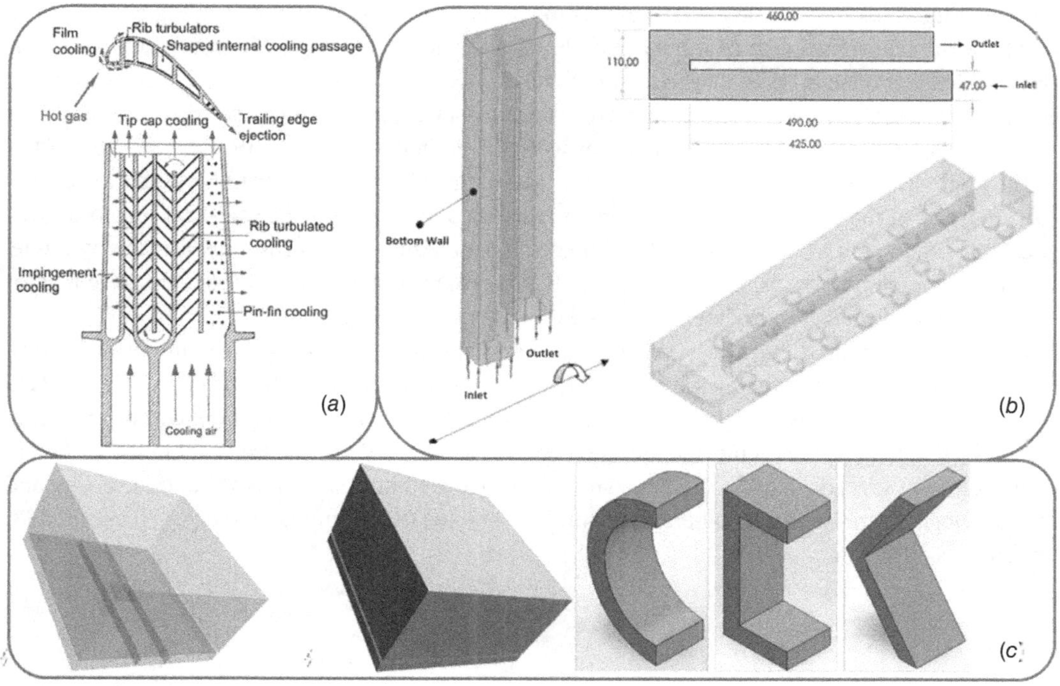

Figure 10.11 Internal and external cooling of the gas turbine blade, (*a*) gas turbine blade cooling techniques, ([21]/NETL (National Energy Technology Laboratory)/Public domain.) (*b*) U-bend internal cooling channel geometries, original and modified, (Adapted from[22, 23]) (*c*) jet impingement cooling geometry with selected diverters for heat transfer enhancement. ([24]/The American Society of Mechanical Engineers.)

computational domain in both cooling techniques is composed of one region, either rotational or stationary. Both computational domains are illustrated in Figure 10.11. In each simulation, the inlet velocity or inlet mass flow rate and outlet pressure boundary conditions were applied to drive the flow through the domain. The computational domains were discretized with a fine mesh in regions exhibiting complex flow turbulence and in areas of particular interest, while a coarser mesh was used in all other regions, as depicted in Figure 10.12. The prism layer (inflation layers) option was applied to generate prism layers on all the walls of the fluid domain to accurately resolve the physical equations near the walls and account for the effects of boundary layers in the solution.

Physics Solver: Like the previous example, selecting the physical models representing the current problem is crucial. Both models utilize a three-dimensional computational domain. Specifically, a steady-state 3D model is

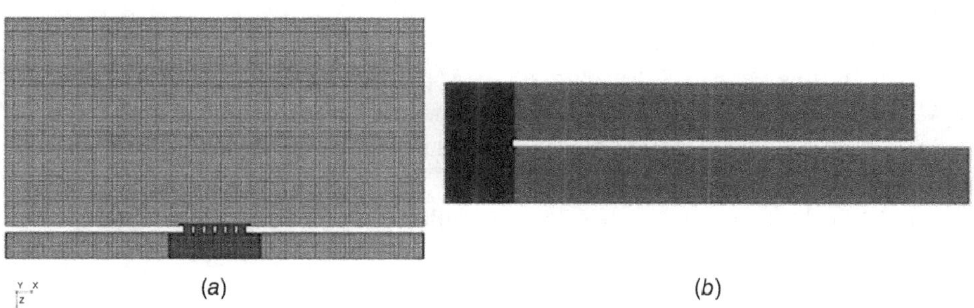

Figure 10.12 Mesh cross-sectional view: (*a*) Jet Impingement cooling model,[24] (*b*) internal cooling channel model.[22]

applied for jet impingement cooling, while a transient 3D model is used for the internal cooling channel. The transient approach for the internal cooling channel is required due to the use of large eddy simulation (LES) turbulence modeling, which is incompatible with the assumptions of a steady-state analysis. There is heat addition, and the air density will be variable with its temperature; therefore, the ideal gas model was selected. The flow was identified as turbulent, and therefore, the RANS equations were solved in conjunction with the k-epsilon turbulence model for the jet impingement cooling scenario. In contrast, large eddy simulation (LES) methodology was employed for the internal cooling channel to resolve turbulence in large eddies, while the wall-adapting local eddy (WALE) sub-grid scale model was utilized to model turbulence in smaller eddies. In addition, the gravity model was selected to mimic the gravity force. A rotational rate for the internal cooling channel model was specified for a rotational reference frame similar to experiments for model validation.

Post-Processing: The same procedures and recommendations as outlined in the previous example apply, with a particular emphasis on the importance of conducting a Mesh Independence Study. Specifically, it is essential that any numerical model be tested using a range of mesh sizes until the results no longer depend on the mesh resolution. Once mesh independence is achieved, the model must be validated against experimental data before being used for further analyses. For details on the mesh independence study, as well as validation results and visualizations of the velocity, pressure, and temperature contours obtained from numerically solving these models, please refer to references.[22–24]

References

1 Anderson, J. D. *Computational Fluid Dynamics: The Basics with Applications*. McGraw-Hill, New York, 1995.

2 Versteeg, H. K., & Malalasekera, W. *An Introduction to Computational Fluid Dynamics: The Finite Volume Method*. Pearson Education, London, 2007.

3 Denton, J. D. "The Calculation of Three-Dimensional Viscous Flow Through Multistage Turbomachines," *J. Turbomach.*, Vol. 115, No. 4, pp. 621–639, 1993.

4 Tu, J., Liu, C., and Yeoh, G.-H. *Computational Fluid Dynamics: A Practical Approach*. Butterworth-Heinemann, San Diego, CA, 2018.

5 Spalart, P.R. "Strategies for Turbulence Modelling and Simulations," *Int. J. Heat Fluid Flow*, Vol. 21, pp. 252–263, 2000.

6 Jones, W.P., and Launder, B.E. "The Prediction of Laminarization with a Two-Equation Modelof Turbulence," *Int. J. Heat Mass Transf.*, Vol. 15, pp. 301–314, 1972.

7 Launder, B.E., and Sharma, B.I. "Application of the Energy Dissipation Model of Turbulence to the Calculation of Flow Near a Spinning Disc," *Lett. Heat Mass Transf.*, Vol. 1, No. 2, pp 131–138, 1974.

8 Lien, F.S., Chen, W.L., and Leschziner, M.A. "Low-Reynolds Number Eddy-Viscosity Modelling Based on Non-Linear Stress-Strain/Vorticity Relations," Proceedings of the 3rd Symposium on Engineering Turbulence Modelling and Measurements, Crete, Greece, 27–29 May 1996.

9 Rodi, W. "Experience with Two-Layer Models Combining the k-e Model with a One-Equation Model Near the Wall," 29th Aerospace Sciences Meeting, January 7–10, Reno, NV, AIAA 91-0216, 1991.

10 Wilcox, D.C. *Turbulence Modeling for CFD*, 2nd ed., DCW Industries, Inc., La Cañada Flintridge, 1998.

11 Menter, F.R. "Two-Equation Eddy-Viscosity Turbulence Modeling for Engineering Applications," *AIAA J.*, Vol. 32, No, 8, pp. 1598–1605, 1994.

12 Sarkar, S., and Balakrishnan, L."Application of a Reynolds-Stress Turbulence Model to the Compressible Shear Layer," ICASE Report 90-18, NASA CR 182002, 1990.

13 Speziale, C.G., Sarkar, S., and Gatski, T.B. "Modelling the Pressure-Strain Correlation of Turbulence: An Invariant Dynamical Systems Approach," *J. Fluid Mech.*, Vol. 227, pp. 245–272, 1991.

14 Tucker, P. G. *Advanced Computational Fluid Dynamics*. Cambridge University Press, Cambridge, 2016.

15 Spalart, P.R., Jou, W.H., Strelets, M. and Allmaras, S.R. "Comments on the Feasibility of LES Forwings, and on a Hybrid RANS/LES Approach," in *1st AFOSR Int. Conf. on DNS/LES Aug, 4–8 1997. Advances in DNS/LES*, Liu, C. and Liu, Z. (Eds), Greyden Press, Columbus, OH, 1997.

16 Hasan, A. S., Abousabae, M., Salem, A. R., and Amano, R. S. "Study of Aerodynamic Performance and Power Output for Residential-Scale Wind Turbines." *ASME. J. Energy Resour. Technol.* Vol. 143, No. 1, pp. 011302, 2021. https://doi.org/10.1115/1.4047602.

17 Hasan, A. S., Abousabae, M., Al Hamad, S., and Amano, R. S. "Experimental and Numerical Investigation of Vortex Generators and Winglets in Horizontal Axis Wind Turbine Blade Design." *ASME. J. Energy Resour. Technol.*, Vol. 145, No. 1, pp. 011301, 2023.

18 Hasan, A. S., Abousabae, M., Al Hamad, S., and Amano, R. S. "Experimental and Numerical Investigation of Tubercles and Winglets Horizontal Axis Wind Turbine Blade Design." *ASME. J. Energy Resour. Technol.*, Vol. 145, No. 1, pp. 011302, 2023.

19 Al Hamad, S, Abousabae, M, Hasan, A, Habash, O, & Amano, R.S. "Effect of Winglet Blade on the Performance of Small-Scale Horizontal Axis Wind Turbine." Proceedings of the ASME Turbo Expo 2023: Turbomachinery Technical Conference and Exposition. Volume 14: Wind Energy. Boston, Massachusetts, USA. June 26–30, 2023. V014T37A002. ASME. https://doi.org/10.1115/GT2023-101643

20 Al Hamad, S, Abousabae, M, Shaker, O, & Amano, R.S. "Experimental and Simulation Investigation of J-Shaped and Kammtail Virtual Airfoils in Small-Scale Horizontal Axis Wind Turbines." Proceedings of the ASME Turbo Expo 2024: Turbomachinery Technical Conference and Exposition. Volume 13: Heat Transfer: General Interest/Additive Manufacturing Impacts on Heat Transfer; Wind Energy. London, United Kingdom. June 24–28, 2024. V013T37A003. ASME. https://doi.org/10.1115/GT2024-121013.

21 Smith, L., Karim, H, Etemad, S., Pfefferle, W., and Cunha, F.J., *Gas Turbine Handbook*, U.S. Department of Energy-National Energy Technology Laboratory (NETL), Morgantown, WV, pp. 389–410, 2006.

22 Saravani, M. S., DiPasquale, N. J., Beyhaghi, S., Amano, R. S. "Heat Transfer in Internal Cooling Channels of Gas Turbine Blades: Buoyancy and Density Ratio Effects," *ASME. J. Energy Resour. Technol.*, Vol. 141, No. 11, pp. 112001, 2019.

23 Nourin, F. N., Blum, B. L., and Amano, R. S. "Exploring the Effects of Guide Vane on Dimpled Cooling Channel of Gas Turbine Blade." *ASME. J. Energy Resour. Technol.*, Vol. 145, No. 5, pp. 051702, 2023.

24 Salem, A. R., Nourin, F. N., Abousabae, M., and Amano, R. S. "Experimental and Numerical Study of Jet Impingement Cooling for Improved Gas Turbine Blade Internal Cooling With In-Line and Staggered Nozzle Arrays." *ASME. J. Energy Resour. Technol.*, Vol. 143, No. 1, pp. 012103, 2021.

Problems

10.1 What are the three primary approaches to solving fluid dynamics and heat transfer problems?

10.2 What are the key stages of CFD workflow in commercial software?

10.3 What is the computational domain in CFD?

10.4 What factors affect the accuracy of CFD simulations?

10.5 What is the role of the physics solver in CFD?

10.6 What is the Reynolds-Averaged Navier–Stokes (RANS) equation used in CFD?

10.7 How is turbulence predicted in CFD?

10.8 What is the importance of the boundary layer in turbomachines?

10.9 What are the different approaches for modeling rotor motion in turbomachines?

10.10 What is mesh independence study, and why is it important?

10.11 What is post-processing in CFD, and what does it involve?

10.12 Which turbulence model is typically employed during the early stages of design, and what are the reasons for its selection?

10.13 What is the difference between the Eulerian and the Lagrangian fluid motion description?

A

Review on Thermodynamics and Compressible Flow

A.1 Thermodynamic Properties

In a compressible fluid medium of pure substance, such as gas, vapor, or steam, a thermodynamic state can be defined with any two thermodynamic properties, such as temperature and pressure. The other properties, such as density, specific enthalpy, internal energy, and entropy, can be determined from these two chosen properties in a function, so that

$$h = f_1(p, T), \qquad u = f_2(p, T), \qquad v = \frac{1}{\rho} = f_3(p, T), \qquad s = f_4(p, T).$$

These are the so-called equations of state. In the simplest case of an ideal gas, when the density is relatively low and the intermolecular interaction force is weak, the equation of state can be expressed in the simple form $p = \rho RT$. The gas constant is $R = 53.3\,\text{ft-lb}_\text{f}/(\text{R-lb}_\text{m})$ for air. Also, some other properties can be defined, such as the specific heats with constant volume or constant pressure, $C_v = \delta q/\delta T)_v = du/dT$, $C_p = \delta q/\delta T)_p = dh/dT$, and $k = C_p/C_v$, where the derivative is taken with v or p kept as constant. With $p = \rho RT$ and $h = u + pv$, we obtain $C_v = R/(k-1)$ and $C_p = kR/(k-1)$. The entropy change is defined as $\Delta s = \int \delta q/T_{\text{rev}}$, where the integration is carried out along a reversible process. The energy equation for a closed system in the differential form can be written as $T\,ds = \delta q = du + p\,dv = dh - dp/\rho$, since $h = u + pv$.

When a fluid medium cannot be treated as an ideal gas, empirical formulas, charts, or tables (e.g., the steam Mollier diagrams shown in Figure A.1[1]) have to be employed to relate or to determine these thermodynamic properties. More details can be found in textbooks on thermodynamics or other engineering handbooks.

A.2 Thermodynamic Processes

A thermodynamic process is defined as a sequence of thermodynamic states. Two processes important to the study of turbomachinery are compression and expansion processes, corresponding to the processes taking place in the blower or compressor and the gas or steam turbine. To study the thermodynamic process, it is convenient to use thermodynamic diagrams, such as the T–s or p–v diagrams shown in Figure A.2. A state is represented by a point, and a process is represented by a line or curve on these diagrams. According to the second law of thermodynamics, the entropy will always increase in an actual adiabatic process (without heat transfer), as shown in the process 1–2 in the T–s diagram. For convenience of analysis, an ideal process with constant entropy, called an isentropic process, is defined. This process can be realized if there is no heat transfer, no friction, and no other dissipative mechanism in the process. Under this ideal condition,

$$T\,ds = dh - v\,dp = 0.$$

Fundamentals of Turbomachinery: Theory and Applications, Second Edition. Ryoichi S. Amano and William W. Peng.

Companion website: www.wiley.com/go/amano/turbomachinery2e

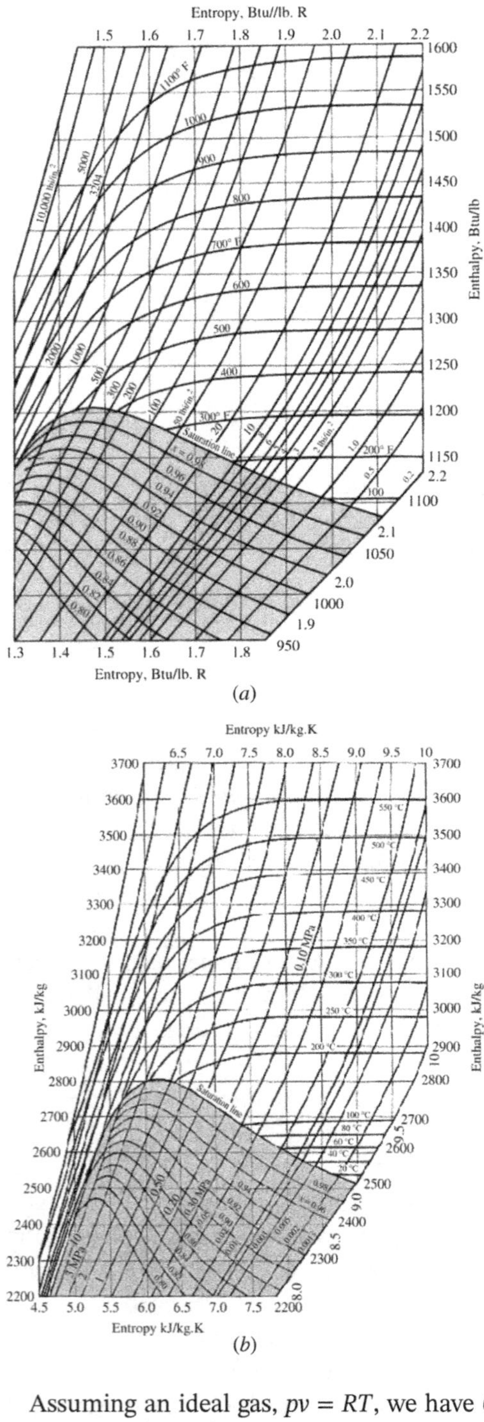

Figure A.1 Mollier diagrams for steam. (*a*) English units and (*b*) SI units (Reprinted by permission from Jones, J. B. & Hawkins, G. A., *Engineering Thermodynamics*, 2nd ed., John Wiley & Sons, Inc., New York, 1986.)

Assuming an ideal gas, $pv = RT$, we have $C_p\, dT/T - R\, dp/p = 0$. With constant specific heats, the isentropic equations result in

$$\frac{T_{s2}}{T_1} = \left(\frac{p_{s2}}{p_1}\right)^{(k-1)/k} = \left(\frac{\rho_{s2}}{\rho_1}\right)^{(k-1)} = \left(\frac{v_1}{v_{s2}}\right)^{(k-1)}, \tag{A.1}$$

which are important to the study of compression and expansion processes in turbomachines.

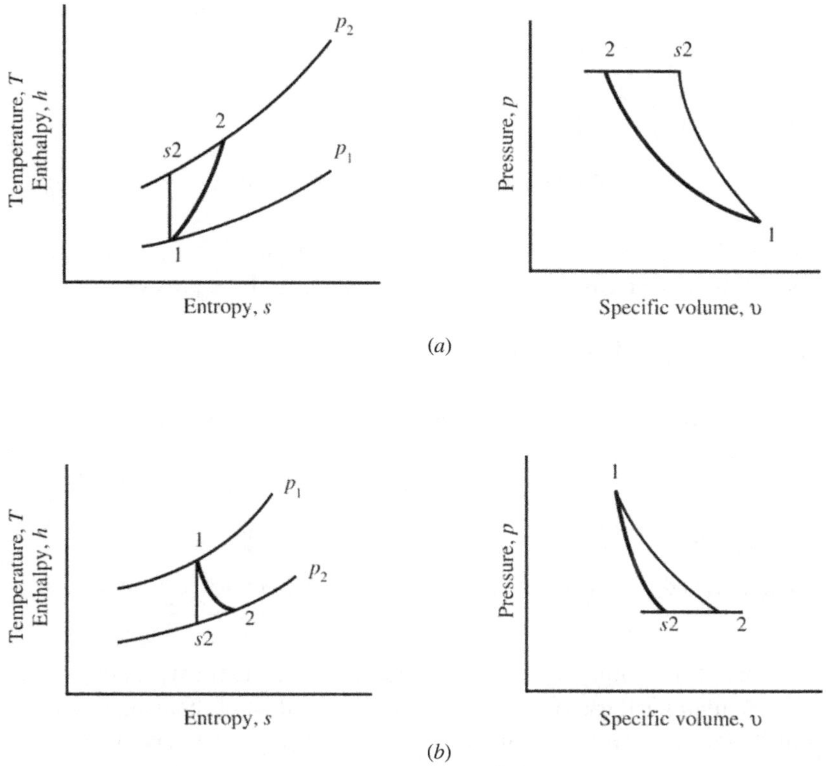

Figure A.2 Thermodynamic diagrams for compression and expansion processes. (*a*) Compression processes from 1 to 2 and (b) expansion processes from 1 to 2.

The work done on or by the fluid per unit mass for the above isentropic process can be derived as follows:

(a) In a closed system, such as the gas in a cylinder–piston system, the work done by unit mass of fluid defined as positive is $w/m = \int p\,dv$. Equation (A.1) can also be written as $p = cv^{-k}$, and hence

$$\frac{w}{m} = \int_1^{S_2} cv^{-k}\,dv = -\frac{c}{(k-1)[1/(v^{k-1})]_{v_1}^{v_{s2}}} = \left(\frac{c}{k-1}\right)\left(\frac{1}{v_1^{k-1}} - \frac{1}{v_{s2}^{k-1}}\right)$$
$$= \left(-\frac{1}{k-1}\right)(p_{s2}v_{s2} - p_1v_1) = -\frac{RT_1}{(k-1)\left[(p_{s2}/p_1)^{(k-1)/k} - 1\right]^{-1}}. \quad \text{(A.2)}$$

(b) In an open system, such as flow through a centrifugal compressor, the work done on unit mass of fluid defined as negative is $-w/m = -\int_1^2 p\,dv + (p_{s2}v_{s2} - p_1v_1)$, where $(p_{s2}v_{s2} - p_1v_1)$ is the so-called flowing work at the inlet and outlet. Hence we have

$$-\frac{w}{m} = \frac{k}{k-1}(p_{s2}v_{s2} - p_1v_1) = C_pT_1\left[\left(\frac{p_{s2}}{p_1}\right)^{(k-1)/k} - 1\right]. \quad \text{(A.3)}$$

This expression can also be derived from the general energy equation (first law of thermodynamics) for an open system, that is,

$$\frac{q}{m} + h_1 + \frac{V_1^2}{2} + gZ_1 = h_{s2} + \frac{V_{s2}^2}{2} + gZ_2 + \frac{w}{m}.$$

For an adiabatic process, $q = 0$, and neglecting the difference of the kinetic energy and gravitational potential energy at the inlet and outlet, we have

$$-\frac{w}{m} = h_{s2} - h_1 = C_p(T_{s2} - T_1) = C_p T_1 \left(\frac{T_{s2}}{T_1} - 1\right).$$

With the isentropic relationship in Equation (A.1), the above expression can be written as $-w/m = C_p T_1 [(p_{s2}/p_1)^{(k-1)/k} - 1]$, which is identical to Equation (A.3).

Another process occasionally encountered for turbomachines is the isothermal process, where the temperature is kept constant with heat transfer. For an ideal gas, $pv =$ const. and the work can be expressed as

$$-\frac{w}{m} = -\int_1^2 p\, dv = -c\int_1^2 \frac{dv}{v} = -c \ \ln\left(\frac{v_2}{v_1}\right) = p_1 v_1 \ \ln\left(\frac{v_1}{v_2}\right) = p_1 v_1 \ \ln\left(\frac{p_2}{p_1}\right).$$

This expression is valid for either closed or open systems, since the flow work at the inlet and outlet is equal. The corresponding entropy change will be

$$\Delta s = \int_1^2 ds = -\int_1^2 \frac{v\, dp}{T} = -R\int_1^2 \frac{dp}{p} = R \ \ln\left(\frac{p_1}{p_2}\right), \tag{A.4}$$

and the amount of heat to be removed or added will be $\delta q = T\, ds$, depending on whether it is a compression or expansion process.

In the actual processes, where some degree of heat transfer and friction occurs, a general formula $pv^n = c$, or a so-called polytropic process can apply, when an ideal gas is assumed. With the test data of the actual process available, the exponent n can be determined with curve fitting such that $n = \ln(p_1/p_2)/[\ln(v_2/v_1)]$. For air at room temperature, we have $1 < n < 1.394$. The extreme case of $n = 1.394$ is the isentropic process, and $n = 1.0$ is the isothermal process just mentioned.

A.3 Other Thermoflow Properties

In dealing with a flowing system, sometimes it is more convenient to use the so-called stagnation state, defined by stagnation properties (sometimes called total properties in the literature). The stagnation state is defined for a point in a flowing system as the thermodynamic state when the flow at that point is brought to rest through an isentropic process. The stagnation state is related to the flowing state (defined in terms of the static properties) through the adiabatic energy equation and the isentropic equation. For an ideal gas, the equations can be expressed as

$$h_{01} = h_1 + \frac{1}{2}V_1^2 \quad \text{or} \quad C_p T_{01} = C_p T_1 + \frac{1}{2}V_1^2,$$

$$\frac{T_{01}}{T_1} = \left(\frac{p_{01}}{p_1}\right)^{(k-1)/k} = \left(\frac{\rho_{01}}{\rho_1}\right)^{k-1} = \left(\frac{v_1}{v_{01}}\right)^{k-1},$$

where the subscript 0 designates the stagnation properties. The stagnation and flowing states of a point in a flowing system can also be represented with two points in the h–s or T–s diagram, as shown in Figure A.3. It is important to emphasize that the points 1 and 01 in this diagram refer to the same points in a flowing system, while the points 1 and $s2$ in Figure A.2 refer to two end states of an ideal process or two points in an ideal steady-state flowing system.

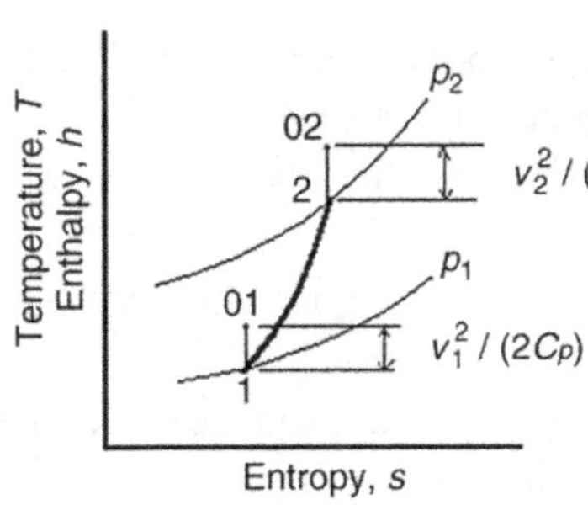

Figure A.3 Relationship between stagnation and static properties of fluid.

In a compressible fluid medium, a disturbance propagates with a finite speed, that is, the speed of sound (or acoustic speed). Its magnitude is a function of local thermodynamic properties. Hence the speed of sound is also considered as a thermodynamic property and can be expressed in the form $a = \sqrt{(\partial p/\partial \rho)_s}$, with the subscript s meaning the derivative is taken with the entropy being kept constant. For an ideal gas, using the isentropic equations of Section A.1, it can also be expressed as $a = \sqrt{kp/\rho} = \sqrt{kRT}$.

In high-speed compressible flow, the local flow velocity can be expressed in a nondimensional parameter, the Mach number, defined as $M = V/a$. If the flow velocity is less than the speed of sound, $V < a$ (or $M < 1$), it is subsonic flow. Otherwise, it is supersonic flow. In supersonic flow, it is always likely that the flow will change to subsonic through a shock wave, which incurs significant energy loss. These topics will be discussed in later sections. The compressors rotating at high speed always encounter this problem, similar to the cavitation problems in liquid pumps.

A.4 Adiabatic Efficiencies of Flow Components

Due to friction and other irreversible processes in the flow, fluid entropy always increases from inlet to outlet in the flow components, such as a nozzle, diffuser, compressor, and turbine. The adiabatic efficiencies, assuming no heat transfer, of these components are defined according to their functions.

A.4.1 Diffuser

A diffuser is used to convert the kinetic energy at the inlet to pressure (a form of thermal energy) at the outlet, as shown in Figure A.4*a*. The diffuser adiabatic efficiency is defined as the fraction of available kinetic energy at the inlet being converted to the pressure rise, or

$$\eta_d = \frac{h_{s02} - h_1}{V_1^2/2} = \frac{h_{s02} - h_1}{h_{01} - h_1}.$$

From the energy equation, we have $h_{01} = h_1 + V_1^2/2$, and hence the maximum attainable outlet pressure is p_{01} with zero velocity, while the actual outlet static pressure is p_2 and stagnation pressure is p_{02}.

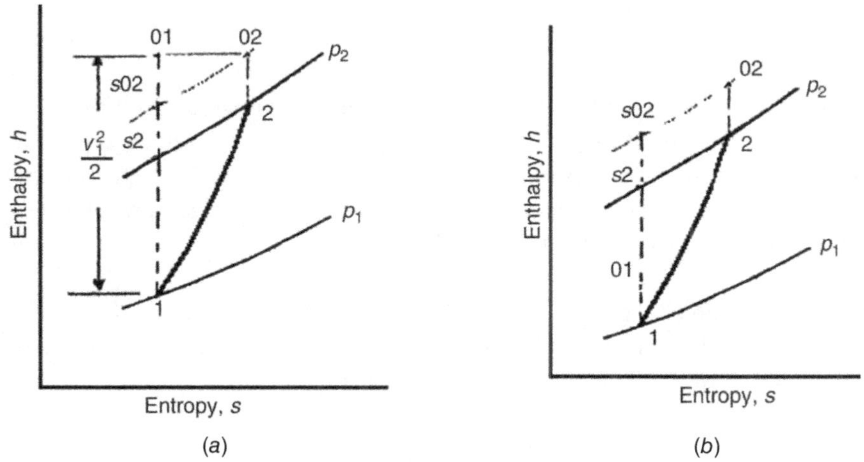

Figure A.4 Processes through (*a*) diffuser and (*b*) compressor.

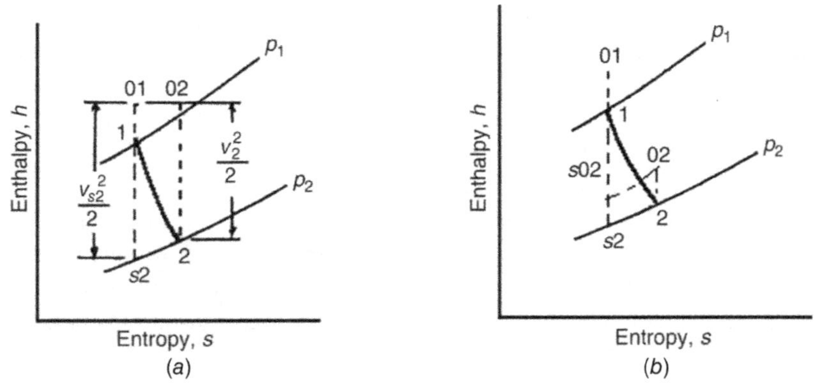

Figure A.5 Processes through (*a*) nozzle and (*b*) turbine.

A.4.2 Nozzle

A nozzle is used to convert fluid thermal/pressure energy (static enthalpy) into kinetic energy with a given set of inlet and outlet pressures. Nozzle efficiency is defined as the actual kinetic energy at the outlet divided by the maximum attainable kinetic energy under an isentropic process, as shown in Figure A.5*a*, that is,

$$\eta_n = \frac{h_{01} - h_2}{h_{01} - h_{s2}} = \frac{V_2^2}{V_{s2}^2} = \mathrm{CV}^2,$$

where CV is the velocity coefficient of a nozzle. From the energy equation, we have $h_{01} = h_{02} = h_2 + V_{s2}^2/2 = h_{s2} + V_{s2}^2/2$, where $V_2 < V_{s2}$ and $CV < 1$ since $h_2 > h_{s2}$.

A.4.3 Compressor and Turbine

For a compressor or turbine, the energy equation of flow is given as $h_{01} = h_{02} + w/m$, where the flow velocities at the inlet and outlet are approximately the same and are combined into the stagnation enthalpies. Work done by the turbine is treated as positive, and that on the compressor is treated as negative. Hence we have

$$\frac{w_t}{m} = h_{01} - h_{02}, \quad -\frac{w_c}{m} = h_{02} - h_{01}.$$

When the inlet and outlet pressures are fixed, the minimum required work on the compressor and the maximum output work by the turbine can be attained through the isentropic process, as shown in Figures A.4*b* and A.5*b*. Hence the total-to-total efficiencies are defined as $\eta_{\mathrm{tt},c} = (h_{s02} - h_{01})/(h_{02} - h_{01})$ for a compressor and $\eta_{\mathrm{tt},t} = (h_{01} - h_{02})/(h_{01} - h_{s02})$ for a turbine. In some literature, the differences in inlet and outlet flow velocities are neglected, and hence the adiabatic efficiencies are defined as $\eta_c = (h_{s2} - h_1)/(h_2 - h_1)$ for a compressor and $\eta_t = (h_1 - h_2)/(h_1 - h_{s2})$ for a turbine.

It should be mentioned here that there is no change of total enthalpy across the diffuser or nozzle, that is, $h_{02} = h_{01}$, since there is no energy or work being transferred into or out of the fluid. But significant change of kinetic energy occurs. Across the compressor and turbine, total enthalpies change.

Example A.1 Airflow at 120 kPa and 25 °C is being compressed to 600 kPa and 230 °C through a centrifugal compressor. Determine the adiabatic efficiency of the compressor and the input power required per unit mass.

Solution

Assuming steady-state flow, an adiabatic process, and negligible difference between the inlet/outlet flow velocities, we have

$$T_{s2} = T_1\left(\frac{p_2}{p_1}\right)^{(k-1)/k} = (25+273)\left(\frac{600}{120}\right)^{(1.4-1)/1.4} = 298\times 1.584 = 472\ \text{K}.$$

With constant C_p assumed, we have

$$\eta_c = \frac{T_{s2}-T_1}{T_2-T_1} = \frac{472-298}{503-298} = 0.848 = 84.8\%.$$

Then

$$-\frac{w_c}{m} = C_pT_1\left(\frac{T_2}{T_1}-1\right) = 1.004\times 298\left(\frac{503}{298}-1\right) = 205.8\ \text{kJ/kg}.$$

Example A.2 In a processing plant, carbon dioxide is being compressed from 14.7 psia and 30 °F to 40 psia through a compressor with flow rate 150 cfm measured at the inlet. Determine the input power required if the compressor is well insulated and its efficiency is 87%. Also determine the minimum power required if the compressor is being cooled to keep the gas temperature constant during the compression process.

Solution

The gas density at the inlet can be calculated as

$$\rho_1 = \frac{p_1}{RT_1} = \frac{14.7\times 144}{35.11\times(30+460)} = 0.123\ \text{lb}_\text{m}/\text{ft}^3.$$

Hence we have a mass flow rate $m = 0.123\times 150/60 = 0.307\ \text{lb}_\text{m}/\text{s}$. Neglecting the mechanical and leakage losses, from $h_2 = h_1 + (h_{s2}-h_1)/\eta_c$ and assuming ideal gas with constant C_p, we obtain

$$T_2 = T_1\left(1+\frac{(p_2/p_1)^{(k-1)/k}-1}{\eta_c}\right)$$
$$= 490\left(1+\frac{(40/14.7)^{(1.29-1)/1.29}-1}{0.87}\right) = 632\ \text{R}.$$

Hence we have

$$-\frac{w_c}{m} = C_p(T_2-T_1) = 0.2007\,(632-490) = 28.5\ \text{Btu/lb}_\text{m},$$
$$-w_c = \frac{0.307\times 28.5\times 778}{550} = 12.4\ \text{hp}.$$

For an isothermal process, we have

$$-\frac{w_c}{m} = p_1v_1\ \ln\left(\frac{p_2}{p_1}\right) = \left(14.7\times\frac{144}{0.123}\right)\ \ln\left(\frac{40}{14.7}\right)$$
$$= 17,227\ \text{lb}_\text{f}\text{-ft/lb}_\text{m}.$$

Hence $-w_c = 0.307\times 17{,}227/550 = 9.6$ hp.

Example A.3 Steam of 250 psia and 1000 °F enters a steam turbine at a flow rate of 1.2 lb_m/s and discharges to the condenser at 10 psia. If the adiabatic efficiency is estimated to be 85%, determine the possible output horsepower.

Solution

From the steam Mollier diagram, we have $h_1 = 1528$ Btu/lb$_m$, $s_1 = 1.82$ Btu/(lb$_m$-R) at 250 psia and 1000 °F. Hence at $p_2 = 10$ psia, we have $h_{s2} = 1165$ Btu/lb$_m$, where $s_{s2} = s_1$. From $\eta_t = (h_1 - h_2)/(h_1 - h_{s2})$, we have

$$h_2 = 1528 - 0.85(1528 - 1165) = 1219.5 \text{ Btu/lb}_m,$$

$$w_t = m(h_1 - h_2) = 1.2(1528 - 1219.5) = 370.3 \text{ Btu/s} = \frac{370.3 \times 778}{550} = 523.7 \text{ hp}.$$

A.5 Thermodynamic Cycles

A working fluid medium may undergo a series of processes and return to its initial state, such as those in a power-producing system or a refrigeration system. These series of processes make a complete thermodynamic cycle. The basic ideal cycle is called a Carnot cycle and consists of isothermal heat addition, isothermal heat rejection, isentropic expansion, and isentropic compression processes, as shown in Figure A.6. In power-producing systems, the processes proceed in the direction of 1–2–3–4–1; in refrigeration systems, they proceed in the direction of 1–4–3–2–1. The heat addition and rejection processes take place in a heat exchanger or combustion chamber or simply through exhausting and intaking between the engine and the ambient atmosphere (for open systems). Expansion and compression processes take place in the turbine, compressor, or simple expansion valve, nozzle, and diffuser. In an internal combustion engine, all of these processes take place in a cylinder.

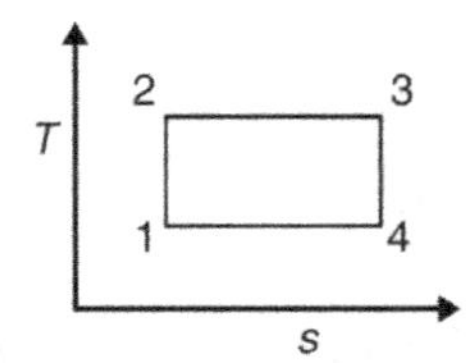

Figure A.6 Carnot cycle.

The thermal efficiency of a heat engine cycle is defined as the ratio of the net power produced to the heat addition from a high-temperature reservoir or from combustion, that is,

$$\eta_{\text{th,c}} = \frac{w}{q_h} = \frac{q_h - q_L}{q_h} = 1 - \frac{q_L}{q_h} = 1 - \frac{T_L}{T_h},$$

where q_h is the heat addition from the high-temperature reservoir of T_h, q_L is the heat rejection to the low-temperature reservoir of T_L, and $q_L/q_h = T_L/T_h$ for a Carnot cycle.

Typical values are as follows: for an aircraft gas turbine engine, $T_h = 2300$ K, $T_L = 217$ K, and $\eta_{\text{th,c}} = 0.90$; for an automotive or stationary gas turbine engine, $T_h = 1700$ K, $T_L = 300$ K, and $\eta_{\text{th,c}} = 0.82$; in a steam turbine power plant, $\eta_{\text{th,c}} \cong 0.70$.

In practice, however, the Carnot cycle cannot be achieved due to irreversibility and other difficulties, such as heat transfer in the components and pumping or expanding two-phase mixtures. Instead, the Brayton cycle and Rankine cycle, as shown in Figure A.7, are used in gas turbine engines and steam power plants or as a refrigeration system in the reversed processes.

A.6 Thermodynamic Cycles of Gas Turbine Engines

Gas turbine engines are extensively used in modern high-speed aircraft propulsion systems and are becoming more common in shaft power-producing systems to generate electricity and to drive some ground or marine vehicles and other machinery. The detailed construction and performance characteristics of the compressor and turbine components are discussed in Chapters 4 and 5–. The different systems and related thermodynamics cycles will be elaborated on in this section.

The gas turbine is a heat engine that converts the chemical energy of the fuel to mechanical energy in either shaft power or thrust force power form. Compared with reciprocating engines, the gas turbine is better in balancing and has less rubbing and seal problems. Compared with the steam turbine, it is more compact, especially for the

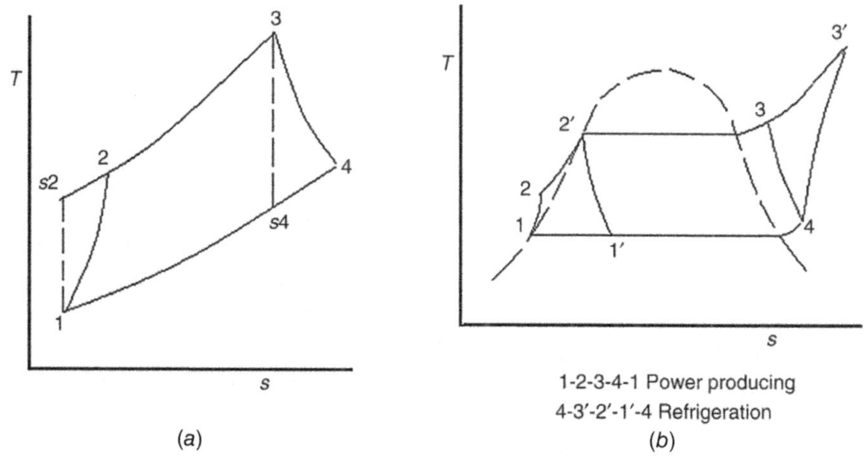

Figure A.7 Thermodynamic cycles for gas turbine, steam power plant, and refrigeration plant. (*a*) Brayton cycle with air or gas and (*b*) Rankine cycle with vapor.

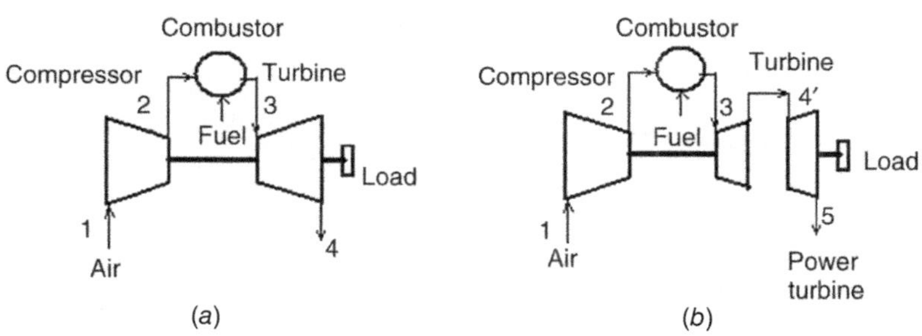

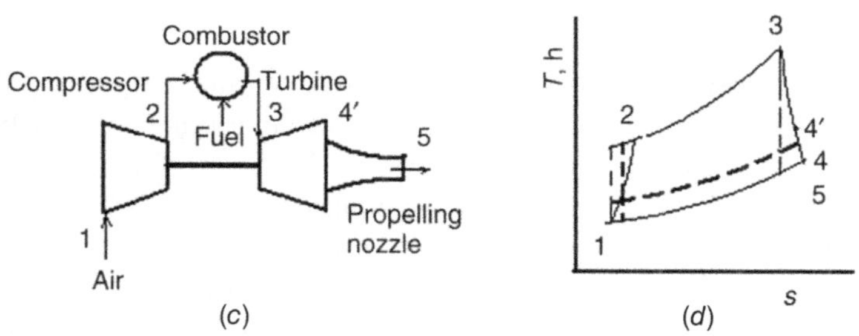

Figure A.8 Gas turbines with basic open cycles. (*a*) single stage turbine, (*b*) two-stage turbine, (*c*) single stage turbine with propelling nozzle, and (*d*) T-S diagram.

open-cycle engine. Since the combustion product gas is used as the working medium and expands through the turbine directly, no bulky boiler is required. And the gas is exhausted into the ambient atmosphere without a condenser, hence no cooling water is required. Basic open-cycle gas turbines consisting of three main components and the T–s diagram are shown in Figure A.8. Air is drawn from the ambient into the compressor and compressed to a higher pressure. Then it enters the combustion chamber, where the fuel, either in liquid or gaseous form, is introduced and burnt. The resulting high-temperature, high-pressure product gas is then expanded through the turbine, and mechanical shaft power is produced. Part of this power is used to drive the compressor. The useful output power is the shaft power produced by either the same turbine or a separate turbine (A.8*a* and *b*). In a jet engine, it will be the thrust produced by a propelling nozzle, expanding the product gas into a high-speed jet (A.8*c*).

The primary parameters affecting gas turbine performance in terms of power output and efficiency are

(a) component efficiencies of the compressor, η_c, combustor, η_b, turbine, η_t, and nozzle, η_n.
(b) maximum temperature at the turbine inlet, T_{03}; and
(c) compression ratio across the compressor, $r = p_{02}/p_{01}$.

The steady-state energy equation or the first law of thermodynamics applied to the unit mass flow in an open-system component can be written as $q = (h_2 - h_1) + \frac{1}{2}\left(V_2^2 - V_1^2\right) + w$, where positive q is the heat transfer into the flow, positive w is power output from the impeller, and h, V are enthalpy and flow velocity at the inlet and outlet of the component. Neglecting the change of kinetic energy and the mechanical losses in the component as illustrated in the T–s diagram of A.8d, it can be written as $w_{12} = -(h_2 - h_1) = -C_p(T_2 - T_1)$ for a compressor, $q_{23} = (h_3 - h_2) = C_p(T_3 - T_2)$ for a combustor, and $w_{34} = (h_3 - h_4) = C_p(T_3 - T_4)$ for a turbine. So, with the friction losses and gas properties change neglected, that is, assuming isentropic compression and expansion, the cycle efficiency and specific work can be expressed as

$$\begin{aligned}\eta &= \frac{w_{34} + w_{12}}{q_{23}} = \frac{C_p(T_3 - T_4) - C_p(T_2 - T_1)}{C_p(T_3 - T_2)} = 1 - \frac{T_4 - T_1}{T_3 - T_2} \\ &= 1 - \frac{T_1(T_4/T_1 - 1)}{T_2(T_3/T_2 - 1)} = 1 - \frac{T_1}{T_2} \\ &= 1 - \left(\frac{1}{r}\right)^{(k-1)/k},\end{aligned} \tag{A.5}$$

where $k = C_p/C_v$, $r = p_2/p_1 = p_3/p_4$, and $T_2/T_1 = T_3/T_4 = r^{(k-1)/k}$, and so $T_4/T_1 = T_3/T_2$, and

$$w = C_p(T_3 - T_4) - C_p(T_2 - T_1),$$

or

$$\begin{aligned}\frac{w}{C_pT_1} &= \frac{1}{T_1}[(T_3 - T_4) - (T_2 - T_1)] = \frac{T_3}{T_1}\left(1 - \frac{T_4}{T_3}\right) - \left(\frac{T_2}{T_1} - 1\right) \\ &= t\left(1 - \frac{1}{r^{(k-1)/k}}\right) - (r^{(k-1)/k} - 1),\end{aligned} \tag{A.6}$$

where $t = T_3/T_1$. So, in general, we have $\eta = f(r, k)$ and $w/(C_pT_1) = f(t, r, k)$. The specific work output w is a measure of the plant size for a given output power and t is an indication of the turbine metallurgical limit, which is around 5 or 6 for a typical air-cooled turbine.

As mentioned earlier, to produce a net useful power, both compression and combustion are needed, that is, $r > 1$ and $T_3 > T_2$. This can be illustrated from the T–s diagram in Figure A.8d. As r or T_3/T_2 approaches 1, the cycle becomes that shown with the heavy dashed lines, and the net output power, which is represented by the net enclosed area of the T–s diagram, approaches zero. Also from Equation (A.6), for $w = 0$, we have $t(1-1/x) = (x - 1)$, where $x = r^{(k-1)/k}$, or $(x - 1)(x - t) = 0$. Hence we have $x = 1$ and $x = t$, or $r = t^{k/(k-1)}$ and $r = 1$, that is, $T_3 = T_2$ and $p_2 = p_1$.

To find the condition of maximum w under constant t, we can differentiate Equation (A.6) with respect to $r^{(k-1)/k}$ or x, and set it to zero, that is,

$$\frac{d}{dx}\left(\frac{w}{C_pT_1}\right) = \frac{t}{x^2} - 1 = 0.$$

So we have $x = \sqrt{t}$ or $r_{opt}^{(k-1)/k} = \sqrt{t}$. Since $r^{(k-1)/k} = T_2/T_1 = T_3/T_4$, it is equivalent to have

$$\frac{T_2}{T_1}\frac{T_3}{T_4} = t = \frac{T_3}{T_1},$$

or $T_2 = T_4$ for the maximum output power with a given T_3/T_1 value.

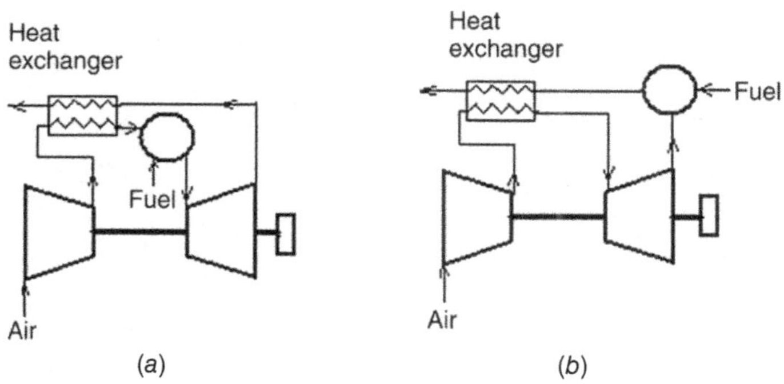

Figure A.9 Open-cycle gas turbines with heat exchanger. (*a*) Upstream of the combustor and (*b*) downstream of the combustor.

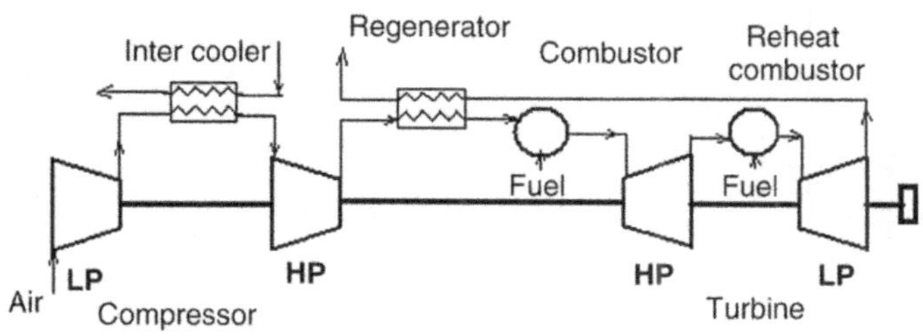

Figure A.10 Complex plant with intercooling, heat exchange, and reheat.

To improve the overall efficiency and power output, the basic gas turbine system can be modified by

(a) adding an intercooler between compressor stages,
(b) reheating between turbine stages, and/or
(c) using turbine exhaust gas to preheat the compressed gas before it enters the combustor.

A heat exchanger can be used to extract heat from the exhaust gas to preheat the compressed air, as shown in Figure A.9*a*. If dirty fuel is used, the combustor can be located downstream of the turbine, as shown in Figure A.9*b*. A more complex cycle with intercooling between compressor stages, reheating between turbine stages, and heat exchanger between exhaust gas and compressed air is shown in Figure A.10. The overall efficiency is higher, but the advantage of compactness is lost.

In order to satisfy some application requirements, the gas turbine can be designed as a closed system or be combined with a steam cycle. In a closed system, some special gas can be used as the working fluid medium, and heat is transferred to and rejected out of the working fluid with the closed-type heat exchangers, as shown in Figure A.11. It is similar to the steam turbine cycle, except no condenser is needed since gas instead of vapor is used as the working medium.

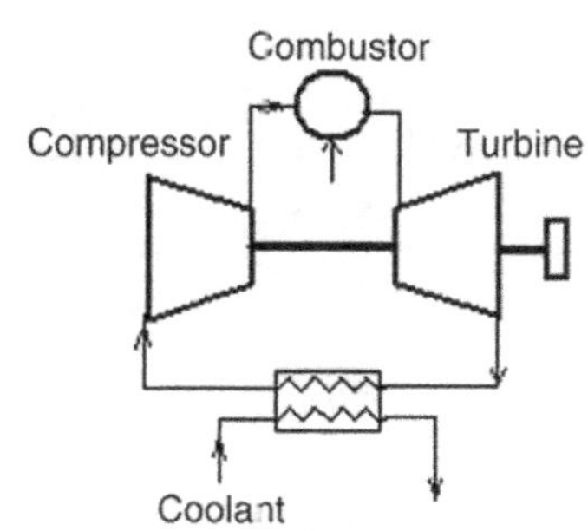

Figure A.11 Basic closed cycle.

In the combined gas and steam turbine cycle, as shown in Figure A.12,[2] the high-temperature exhaust gas from the gas turbine is used to generate steam in a boiler or steam generator to drive the steam turbine. Sometimes, the generated steam or part of it is used for space heating or other processes. Then the system is called cogeneration or combined heat and power. With improved, gas turbine technology more advantages of this combined cycle compared with the conventional steam power plant can be realized, especially when using gasified coal as the fuel source, as mentioned in Chapter 5.

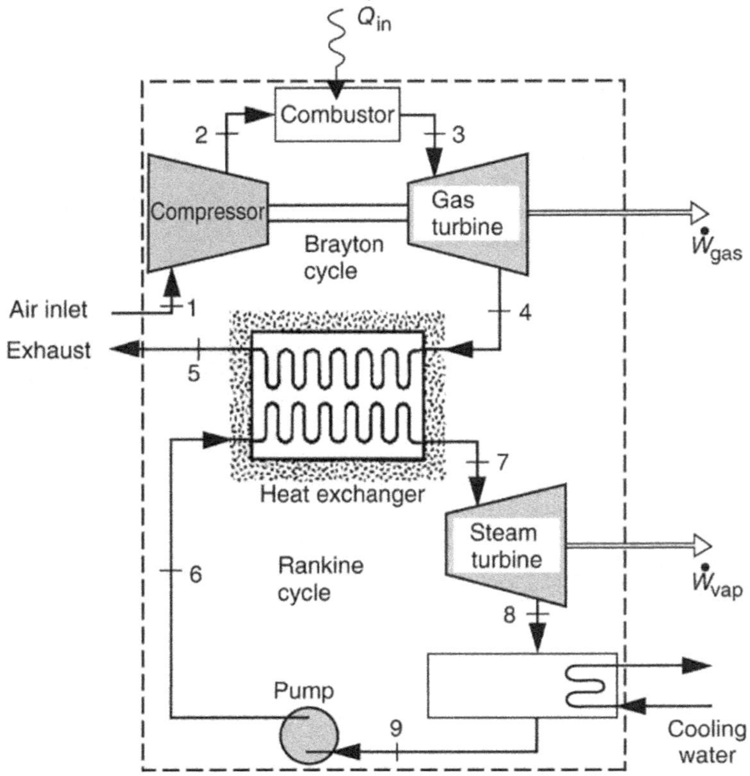

Figure A.12 Combined gas and steam turbine cycles. (Reprinted by permission from Moran, M. J. & Shapiro, H. N., *Fundamentals of Engineering Thermodynamics*, 3rd ed., John Wiley & Sons, Inc., New York, 1995.)

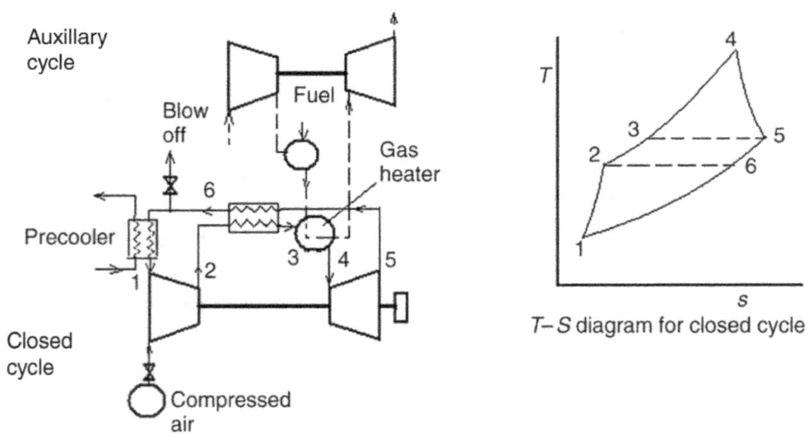

Figure A.13 Closed cycle with an auxiliary cycle.

In another application, a closed cycle is designed as shown in Figure A.13. Fuel is burnt in the combustor or boiler of an auxiliary cycle, then heat is transferred to the gas heater of the closed cycle. Its advantages are that higher system pressure can be employed to reduce the plant size and the power output can be adjusted by changing the system pressure. Also, no filter is needed to prevent erosion in the main closed loop and any gas can be used as the working fluid medium. The disadvantage is that it needs an external heating system and a precooler, hence the system becomes more bulky.

A.6.1 Aircraft Propulsion

Applications of gas turbine to aircraft propulsion are primarily to produce the thrust with a high-speed jet or with a combination of jet and propeller driven by a turbine. The main part of the gas turbine can be considered as a high-temperature, high-pressure gas generator. The gas exhausted from the turbine is expanded through a propelling nozzle, resulting in a high-speed jet to produce the thrust force. The process through the nozzle is designated from 4′ to 5 in the T–s diagram of Figure A.8. The thrust force and propulsive efficiency are important parameters. Three basic types of aircraft gas turbine engines are shown in Figure 5.1. In a turbojet, all the thrust is produced by the jet issued from a propelling nozzle. In a turbofan, the jet is formed by mixing the core flow from the nozzle and the bypass flow through the outer annulus from the fan upstream. The noise level is lower in a turbofan since the jet velocity is lower while the mass flow rate is higher for the same amount of thrust. In a turboprop, the thrust is produced from the propeller and jet. Its propulsive efficiency is higher, but it is limited to subsonic flight.

The T–s diagram for a turbojet with frictional effects is shown in Figure A.14; the dashed lines are for the ideal processes. Since the power produced by the turbine is equal to that required by the compressor, we have $T_{03} - T_{04} = T_{02} - T_{01}$. The thrust force produced by the jet in flight can be written as

$$F = m(V_j - V_a) + A_j(p_j - p_a), \tag{A.7}$$

where m is the air mass flow rate, V_a is the forward speed, and V_j is the jet flow velocity relative to the engine (Figure A.15). The first term is called momentum thrust and the second term is pressure thrust. Velocity V_j can be calculated as $V_j = [2C_p(T_{04} - T_5)]^{0.5}$ if the expansion through the nozzle is unchoked, and the pressure at the nozzle exit plane is equal to the ambient pressure. If the expansion is choked, we have $p_j > p_a$ and V_j has to be calculated for the condition of unit Mach number, that is, $M_j = 1$. From $V_j = a_5 = (kRT_5)^{0.5}$ and $V_j^2/2 = C_p(T_{04} - T_5)$, we have $T_5 = 2T_{04}/(k + 1)$. Then from $\eta_n = (T_{04} - T_5)/(T_{04} - T_{s5})$, we have $T_{s5} = T_{04} - (T_{04} - T_5)/\eta_n$ and $p_5 = p_{04}(T_{s5}/T_{04})^{k/(k-1)}$.

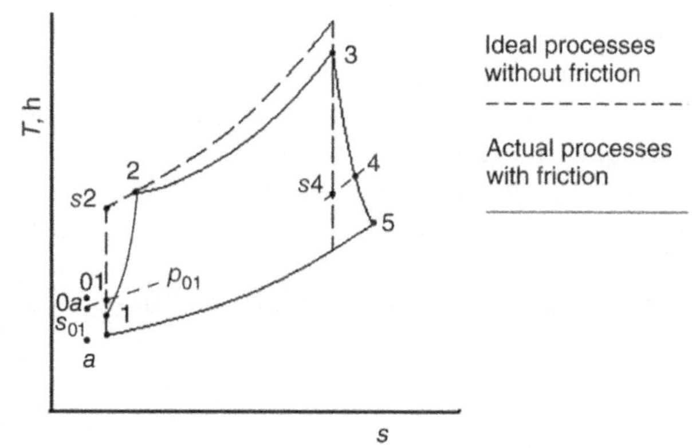

Figure A.14 T–s diagram for a basic turbojet with frictional effects.

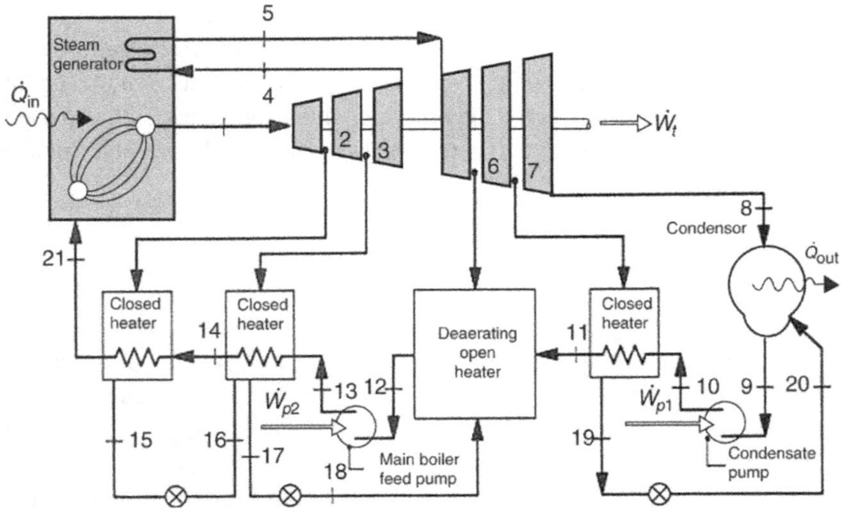

Figure A.15 Variations of Mach number and flow velocity in a channel with variable cross-sectional area.

For the unchoked condition, the propulsive efficiency can be expressed as

$$\eta_p = \frac{mV_a(V_j - V_a)}{m\left[V_a(V_j - V_a) + (V_j - V_a)^2/2\right]} = \frac{2}{2 + (V_j - V_a)/V_a} = \frac{2}{1 + V_j/V_a}. \tag{A.8}$$

From Equations (A.7) and (A.8) neglecting the pressure thrust, it is shown that F is maximum but $\eta_p = 0$ with $V_a = 0$ and η_p is maximum but $F = 0$ with $V_j/V_a = 1$. The optimum operation will be obtained when V_j is moderately greater than V_a. In terms of the nondimensional parameters, they can be plotted as shown below. Since the optimum range of V_j/V_a is relatively narrow, the choice of engine type for application is mainly determined by the flight speed. In other words, the turbojet is for the highest speed, the turbofan for the medium speed, and the turboprop for the lower speed.

Example A.4 A basic open-cycle gas turbine operates with air entering the compressor at 14.7 psia and 80 °F. The pressure ratio is 6, and air exits the combustor at 1650 °F. Determine the specific work output per unit mass of air and the thermal efficiency of the cycle if the compressor and turbine adiabatic efficiencies are 82 and 87%, respectively.

Solution

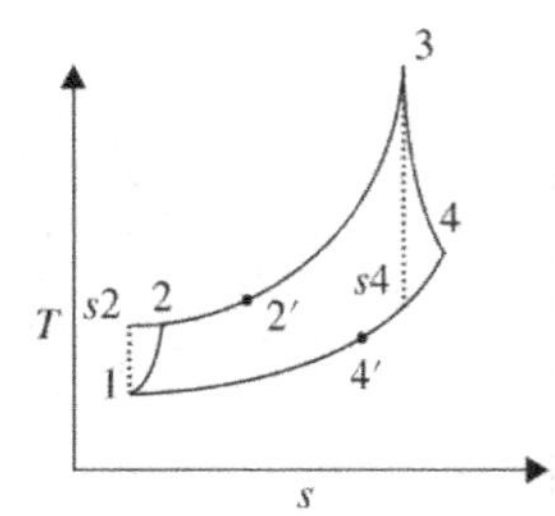

From $T_{s2}/T_1 = T_3/T_{s4} = (p_2/p_1)^{(k-1)/k}$, we have $T_{s2} = 901$ R, $T_{s4} = 1264.6$ R. From $\eta_c = (T_{s2} - T_1)/(T_2 - T_1)$, we have $T_2 = 980.2$ R, and hence $w_c = C_p(T_2 - T_1) = 149.4$ hp./(lb$_m$/s). From $\eta_t = (T_3 - T_4)/(T_3 - T_{s4})$, we have $T_4 = 1374.5$ R, and hence $w_t = 249.7$ hp./(lb$_m$/s). So we have $w = w_t - w_c = 100.2$ hp./(lb$_m$/s), $q_{in} = C_p(T_3 - T_2) = 383.5$ hp./(lb$_m$/s), and $\eta_{th} = w/q_{in} = 0.26 = 26\%$.

Example A.5 A heat exchanger of 70% effectiveness is added in the gas turbine cycle in the above example as shown in Figure A.9a. Determine the improved thermal efficiency.

Solution

From $\eta_{hx} = (T_{2'} - T_2)/(T_4 - T_{4'})$, where $T_{4'} = T_2$, we have $T_{2'} = T_2 + \eta_{hx}(T_4 - T_2) = 980.2 + 0.7(1374.5–980.2) = 1256.2\,\text{R}$. Hence $q_{in} = C_p(T_3 - T_{2'}) = 0.339(2110–1256.2) = 289.4\,\text{hp./(lb}_\text{m}\text{/s)}$ and $\eta_{th} = 99.8/289.4 = 0.346 = 34.6\%$.

Example A.6 A turbojet aircraft flies at 650 mph in the atmosphere of 10.5 psia and 50 °F. The compressor pressure ratio is 5, and the turbine inlet temperature is 1350 °F. Assume the adiabatic efficiencies of intake, compressor, turbine, and nozzle are $\eta_i = 0.90$, $\eta_c = 0.80$, $\eta_t = 0.87$, and $\eta_n = 0.92$, respectively, and the pressure drop in the combustor is negligible. Determine the jet velocity and thrust if the air mass flow rate is 60 lb_m/s.

Solution

Referring to Figure A.14, with $V_a = 650\,\text{mph} = 953\,\text{ft./s}$, we have $T_{01} = T_{0a} = T_a + V_a^2/2C_p = 510 + 953^2/(2 \times 0.24 \times 778 \times 32.2) = 585.5\,\text{R}$. From $\eta_i = (T_{s01} - T_a)/(T_{01} - T_a)$, we have $T_{s01} = T_a + \eta_i(T_{01} - T_a) = 577.9\,\text{R}$ and $p_{01} = p_a(T_{s01}/T_a)^{k/(k-1)} = 10.5(577.9/510)^{3.5} = 16.3$ psia. Hence we have $p_{03} = p_{02} = 5 \times 16.3 = 81.5$ psia and $T_{s02} = T_{01}(5)^{0.2857} = 927.3\,\text{R}$. From $\eta_c = (T_{s02} - T_{01})/(T_{02} - T_{01})$, we have $T_{02} = T_{01} + (T_{s02} - T_{01})/\eta_c = 1012.7\,\text{R}$. For the turbojet, if $w_c = w_t$ is assumed, we have $T_{04} = T_{03} - (T_{02} - T_{01}) = 1810 - (1012.7–585.5) = 1382.8\,\text{R}$. From $\eta_t = (T_{03} - T_{04})/(T_{03} - T_{s04})$, we have $T_{s04} = T_{03} - (T_{03} - T_{04})/\eta_t = 1318.9\,\text{R}$, and $p_{04} = p_{03}(T_{s04}/T_{03})^{3.5} = 26.9$ psia.

To check whether the flow is choked at the nozzle exit, we calculate the choking condition from $T_c = 2\,T_{04}/(k+1) = 2 \times 1382.8/(1.4+1) = 1152.3\,\text{R}$, $T_{sc} = T_{04} - (T_{04} - T_c)/\eta_n = 1382.8 - (1382.8–1152.3)/0.92 = 1132.3\,\text{R}$, and $p_c = p_{04}(T_{sc}/T_{04}) = 26.9(1132.3/1382.8)^{3.5} = 13.36$ psia. Since $p_c > p_a$, the flow is choked at the nozzle exit plane, and hence we have $T_5 = T_c = 1152.3\,\text{R}$, $p_5 = p_c = 13.36$ psia, and $V_j = (\text{kRT}_5)^{0.5} = (1.4 \times 53.33 \times 32.2 \times 1152.3)^{0.5} = 1664.4\,\text{ft./s}$. From $\rho_5 = p_5/RT_5 = 13.36 \times 144/(53.33 \times 1152.3) = 0.0313\,\text{lb}_\text{m}/\text{ft}^3$, and $m = \rho VA$, we have $A_5 = 60/(0.0313 \times 1664.4) = 1.15\,\text{ft}^2$, and hence

$$F = m(V_j - V_a) + A_5(p_5 - p_a) = \frac{60(1664.4 - 953)}{32.2} + 1.15(13.36 - 10.5) \times 144$$
$$= 1325.6 + 473.6 = 1799.2\,\text{lb}_\text{f}.$$

Example A.7 A steam power plant operating on a Rankine cycle is combined with the gas turbine of Example A.4. The exhaust gas from the gas turbine is used to generate steam as shown in Figure A.12. The boiler efficiency is 75% with a water/gas mass flow ratio of 0.10. The steam turbine and pump efficiencies are 84% and 80%, respectively. The condenser pressure and pump discharge pressure are 2 and 100 psia, respectively. Determine the thermal efficiency of the combined cycle.

Solution

Referring to Figure A.12 and assuming pure saturated liquid at the condenser outlet, we have $h_9 = h_f = 93\,\text{Btu/lb}_\text{m}$ from the steam table. With $p_8 = p_9 = 2$ psia and $p_6 = 100$ psia, we have $h_6 = h_9 + \Delta p/\eta_p\rho = 93 + (100–2) \times 144/[0.8(62.4 \times 778)] = 93.4\,\text{Btu/lb}_\text{m}$. Assuming $T_6 \cong T_9 = 585$ R with the definition of boiler efficiency, $\eta_b = (h_7 - h_6)m_w/[C_p(T_4 - T_5)m_g]$, we have $h_7 = 0.75 \times 0.24(1374.5–585)/0.10 + 93.4 = 1514.5\,\text{Btu/lb}_\text{m}$. Assuming $p_6 \cong p_7 = 100$ psia, we obtain $s_{8s} = s_7 = 1.908\,\text{Btu/lb}_\text{m}\text{R}$ and $h_{8s} = 1120\,\text{Btu/lb}_\text{m}$ from the steam table. From the definition of turbine efficiency, $\eta_t = (h_7 - h_8)/(h_7 - h_{8s})$, we have $h_8 = 1514.5–0.84 \times (1514.5–1120) = 1183\,\text{Btu/lb}_\text{m}$, or $w_t = h_7 - h_8 = 331.4\,\text{Btu/lb}_\text{m}$, $w_p = 0.4\,\text{Btu/lb}_\text{m}$.

Hence the additional power output from the steam power plant per unit mass flow rate of gas through the gas turbine is $\Delta w = (w_t - w_p)m_w/m_g = 331 \times 0.1 = 33.1\,\text{Btu/lb}_\text{m} = 46.8\,\text{hp./(lb}_\text{m}\text{/s)}$ and the new thermal efficiency $\eta_{th} = (99.8 + 46.8)/382.6 = 38.3\%$.

A.7 Rankine Cycles with Vapor

The Rankine cycle with vapor as the working medium for power producing or refrigeration is shown in Figure A.7*b*. The power-producing cycle proceeds as 1–2–3–4–1, with steam as a common medium. Other media are used for some special applications. The process from 1 to 2 is through a pump requiring power input w_p. The process from 2 to 3 is through the boiler or steam generator, requiring heat transfer in, q_i. The process from 3 to 4 is through the steam turbine producing power output w_t. while that from 4 to 1 is through the condenser rejecting heat out, q_0.

Hence the thermal efficiency is defined as $\eta_{\text{th}} = (w_t - w_p)/q_i = 1 - q_0/q_i$, where $w_t - w_p = q_i - q_0$ from the conservation of energy.

A detailed schematic of the more sophisticated system as shown in Figure A.16 is designed to improve overall efficiency. Both steam reheating and feedwater regeneration are implemented. Detailed discussions are available in thermodynamics texts.

The refrigeration cycle in Figure A.7*b* proceeds as 4–3′–2′–1′–4. The process from 4 to 3′ is through a vapor compressor, requiring power input w_i. The process from 3′ to 2′ is through a condenser rejecting heat out, q_0. The process from 2′ to 1′ can be a simple throttle valve to produce pressure drop (hence $h_{1'} = h_{2'}$) or it can be a small turbine to produce some power output w_0 with pressure drop (then $h_{1'} < h_{2'}$). The process from 1′ to 4 is through an evaporator to absorb heat in, q_i, to cool the enclosed space air or chilled water. For a refrigeration system, the parameter concerned is called the coefficient of performance, defined as $\beta = q_i/w_i$.

Since the medium in the condenser and evaporator is a two-phase saturated mixture, the processes are constant pressure, and the enthalpy through a throttle valve is constant, the Mollier diagram for the refrigerants is commonly drawn as a *p*–*h* diagram. A detailed schematic of the heating, ventilation, and air conditioning (HVAC) system is shown in Figure 1.5.

Example A.8 A basic refrigeration system is operated with R-134A refrigerant with the following conditions:

1) Evaporator temperature is kept at 0 °F.
2) Condenser pressure is kept at 200 psia.
3) Refrigerant mass flow rate is 10 lb_m/s.

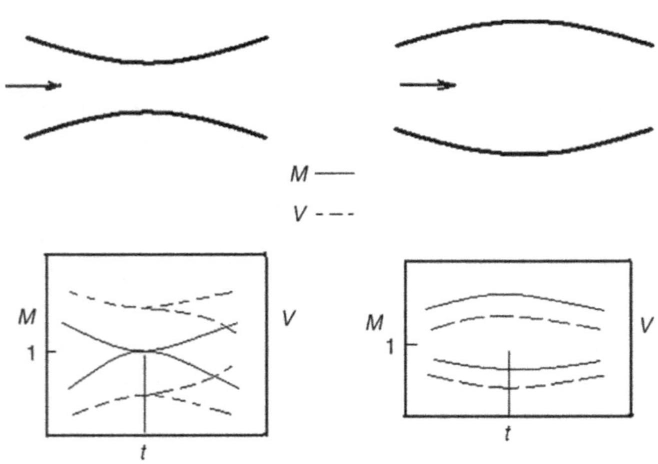

Figure A.16 Schematic diagram of the Rankine cycle with reheating and regeneration. (Reprinted by permission from Moran, M. J. & Shapiro, H. N., *Fundamentals of Engineering Thermodynamics*, 3rd ed., John Wiley & Sons, Inc., New York, 1995.)

Determine the coefficient of performance β, compressor power required P_c, and refrigeration capacity Q assuming no superheat in the evaporator outlet, no subcooling in the condenser outlet, and the compressor efficiency is 75%.

Solution

The T–s and p–h diagrams are shown below for convenience, where 4–$s3$ is the isentropic process to be used for the definition of compressor efficiency.

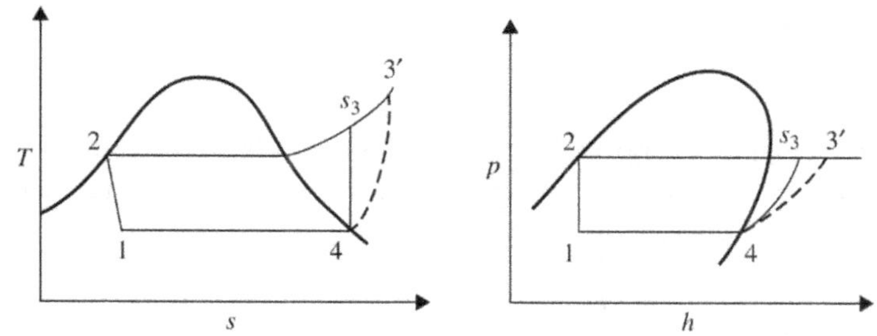

From the p–h diagram for R134A from a thermodynamics text, we have $h_4 = 104\,\text{Btu/lb}_\text{m}$, $h_2 = h_1 = 53\,\text{Btu/lb}_\text{m}$, $p_4 = p_1 = 20$ psia, $s_4 = s_{s3} = 0.226\,\text{Btu/lb}_\text{m}$-R, and $h_{s3} = 122\,\text{Btu/lb}_\text{m}$.

Hence from $\eta_c = (h_{s3} - h_4)/(h_{3'} - h_4)$, we have $h_{3'} = h_4 + (h_{s3} - h_4)/\eta_c = 104 + (122–104)/0.75 = 128\,\text{Btu/lb}_\text{m}$, or the compressor work required per unit mass is $w_i = h_{3'} - h_4 = 128–104 = 24\,\text{Btu/lb}_\text{m}$, the heat absorbed by the evaporator per unit mass is $q_i = h_4 - h_1 = 104–53 = 51\,\text{Btu/lb}_\text{m}$, the coefficient of performance is $\beta = q_i/w_i = 51/24 = 2.1$, total compressor power required is $P_c = mw_i = 10 \times 24 = 240$ Btu/s = 340 hp., and refrigeration capacity is $Q_r = mq_i = 10 \times 51 = 510$ Btu/s = 153 tons (1 refrigeration ton = 12,000 Btu/h).

A.8 Compressible Flow Through a Channel

The compressible flow through a channel is governed by

(a) Conservation equation of mass,
(b) Conservation equation of momentum,
(c) Conservation equation of energy, and
(d) Equation of state.

In general, there are three momentum equations since momentum is a vector parameter. Hence there are a total of six equations to solve for six unknowns, ρ, p, T, V_x, V_y, and V_z. If the flow is turbulent or a two-phase mixture, there will be more unknowns and equations. These equations can be solved numerically with a computer. Its application to turbomachine design and analysis is briefly discussed in Chapter 10.

Some basic physics will be presented in terms of quasi-one-dimensional flow through a channel with variable cross-sectional area. It is applicable to the flow analysis in a nozzle, diffuser, or radial-flow impeller channel. The governing equations based on the assumptions of adiabatic and inviscid (hence isentropic) flow are as follows:

Mass: $\rho VA = \text{const.}$,

$$\frac{d\rho}{\rho} + \frac{dA}{A} + \frac{dV}{V} = 0. \tag{A.9}$$

Energy: $C_pT + V^2/2 = \text{const.}$,

$$C_p dT + V\,dV = 0. \tag{A.10}$$

State: $p = \rho RT$,

$$\frac{dp}{p} = \frac{d\rho}{\rho} + \frac{dT}{T}. \tag{A.11}$$

Isentropic: $p/\rho^k = \text{const.}$,

$$\frac{dp}{p} = k\frac{d\rho}{\rho}. \tag{A.12}$$

Definitions: $M = V/a$, $a^2 = kRT$,

$$\frac{dM}{M} = \frac{dV}{V} - \frac{da}{a} = \frac{dV}{V} - \frac{1}{2}\frac{dT}{T}. \tag{A.13}$$

Notice that the momentum equation is not included, since it is redundant for isentropic flow, and the differential forms are obtained from the algebraic forms by taking the logarithm on both sides of the equations and then taking the differential on each term. Equation (A.10) can be rewritten as, or

$$C_p\frac{dT}{T} = \frac{V^2 dV}{TV} = -\frac{kRV^2}{kRT}\frac{dV}{V} = -kRM^2\frac{dV}{V},$$

or

$$\frac{dT}{T} = -(k-1)M^2\frac{dV}{V}.$$

From (A.11) and (A.12) we obtain

$$(k-1)\frac{d\rho}{\rho} = \frac{dT}{T}.$$

Hence

$$\frac{d\rho}{\rho} = -M^2\frac{dV}{V}. \tag{A.14}$$

Combined with (A.10), we finally have

$$\frac{dA}{A} = (M^2 - 1)\frac{dV}{V} = C_1\frac{dV}{V}. \tag{A.15}$$

Also, from (A.13), we have

$$\frac{dM}{M} = \left(1 + \frac{k-1}{2}M^2\right)\frac{dV}{V}$$

and

$$\frac{dA}{A} = \frac{M^2 - 1}{1 + [(K-1)/2]M^2}\frac{dM}{M} = C_2\frac{dM}{M}. \tag{A.16}$$

Equations (A.15) and (A.16) have some important and interesting implications.

The effects of cross-sectional area change on flow are different for subsonic and supersonic flow:

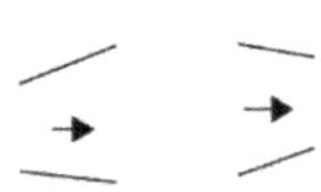

For $M < 1$ (subsonic flow, $C_1 < 0$, $C_2 < 0$), in the divergent channel, that is, $dA > 0$, both V and M decrease in the flow direction, and in the convergent channel, that is, $dA < 0$, both V and M increase. The trend is the same as that in incompressible flow.

For $M > 1$ (supersonic flow, $C_1 > 0$, $C_2 > 0$), in the divergent channel, both V and M increase, and in the convergent channel, they decrease, just opposite to subsonic or incompressible flow.

When $dA = 0$, that is, at the throat of a convergent-divergent channel or at the maximum cross-sectional area, the Mach number can be either unity or the extreme (maximum or minimum), as shown from Equation (A.16). The velocity has to be extreme from Equation (A.15) and is shown in Figure A.16.

Detailed flow characteristics will be analyzed for two types of flow passages: convergent nozzle and convergent-divergent nozzle.

A.8.1 Convergent Nozzle

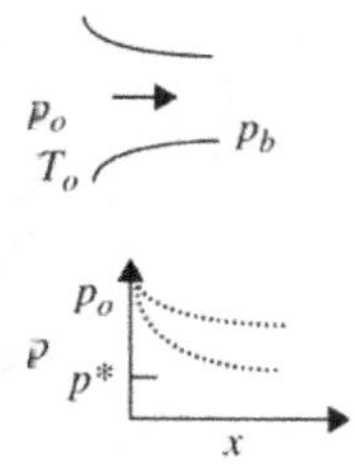

If the reservoir condition is fixed at p_0 and T_0 and the back pressure p_b decreases, the pressure through the channel will change as shown until $p_b = p^*$, where $p^* = p_0[2/(k+1)]^{k/(k-1)}$ is the critical pressure. The Mach number at the exit will be unity. But the flow inside the nozzle is always subsonic. Further decrease in p_b will not affect the flow upstream, and the flow is choked. But the expansion waves develop near the exit. Hence in the range of $p_0 > p_b > p^*$, $p_e = p_b$, and the flow velocity at the exit can be calculated from the isentropic flow equations,

$$V_e^2 = 2C_pT_0\left(1 - \frac{T_b}{T_0}\right) = \frac{2kRT_0}{k-1}\left(1 - \frac{T_b}{T_0}\right)$$
$$= \left\{\left(\frac{2k}{k-1}\right)\left(\frac{p_0}{\rho_0}\right)\left[1 - \left(\frac{p_b}{p_0}\right)^{(k-1)/k}\right]\right\}$$

and the mass flow rate is given as

$$\dot{m} = \rho_b V_e A_e = \rho_0 A_e\left(\frac{\rho_b}{\rho_0}\right)V_e = \rho_0 A_e\left(\frac{p_b}{p_0}\right)^{1/k}V_e.$$

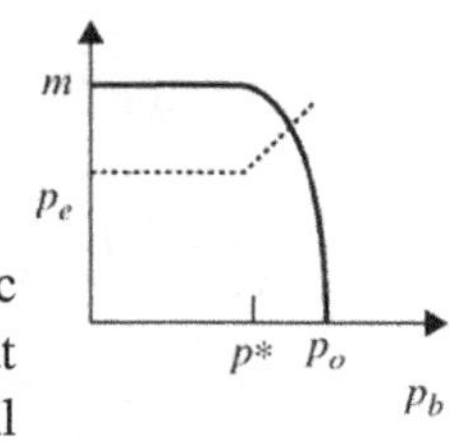

When $p_b < p^*$, $p_e = p^* = p_0[2/(k+1)]^{k/(k-1)}$, and $V_e = a^* = \sqrt{2/(k+1)}a_0 = \sqrt{2kRT_0/(k+1)} = \sqrt{[2k/(k+1)]p_0/\rho_0}$, $\dot{m} = \rho^* V_e A_e =$ const. The variations of p_e and $\dot{m}$ with respect to p_b are shown qualitatively in the diagram.

If the back pressure is fixed and reservoir pressure increases, the flow will be isentropic and subsonic all the way until $p_0 = p^*[(k+1)/2]^{k/(k-1)}$, where $p^* = p_b$. Beyond that, flow at the exit will be at Mach 1, that is, $M_e = 1$, and $\rho_e = \rho^*$, $a_e = a^*$. But the mass flow rate will keep increasing since we have

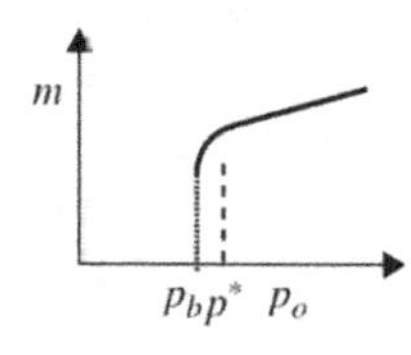

$$\dot{m} = \rho_e V_e A_e = \rho^* a^* A_e = \rho_0 a_0\left(\frac{\rho^*}{\rho_0}\right)\left(\frac{a^*}{a_0}\right)A_e \sim \rho_0 a_0 A_e \sim \frac{p_0}{\sqrt{T_0}},$$

where ρ_0 increases with p_0. In other words, the mass flow rate will be directly proportional to the reservoir pressure p_0, as illustrated in the following diagram.

A.8.2 Convergent-Divergent Nozzle

There are four possible flow regimes in the convergent-divergent nozzle when the back pressure decreases relative to the reservoir pressure, as shown in Figure A.17.

1) In the flow regime a to d, we have $p_0 > p_b > p_{b,crit} > p^*$, the flow is isentropic and subsonic all the way. Hence p_b and p_t can be related with the isentropic flow equation,

$$\left(\frac{p_t}{p_0}\right)^{1/k}\left[1 - \left(\frac{p_t}{p_0}\right)^{(k-1)/k}\right]^{1/2} = \left(\frac{A_e}{A^*}\right)\left(\frac{p_b}{p_0}\right)^{1/k}\left[1 - \left(\frac{p_b}{p_0}\right)^{(k-1)/k}\right]^{1/2}, \qquad \text{(A.17)}$$

where p_t, is the pressure at the throat.

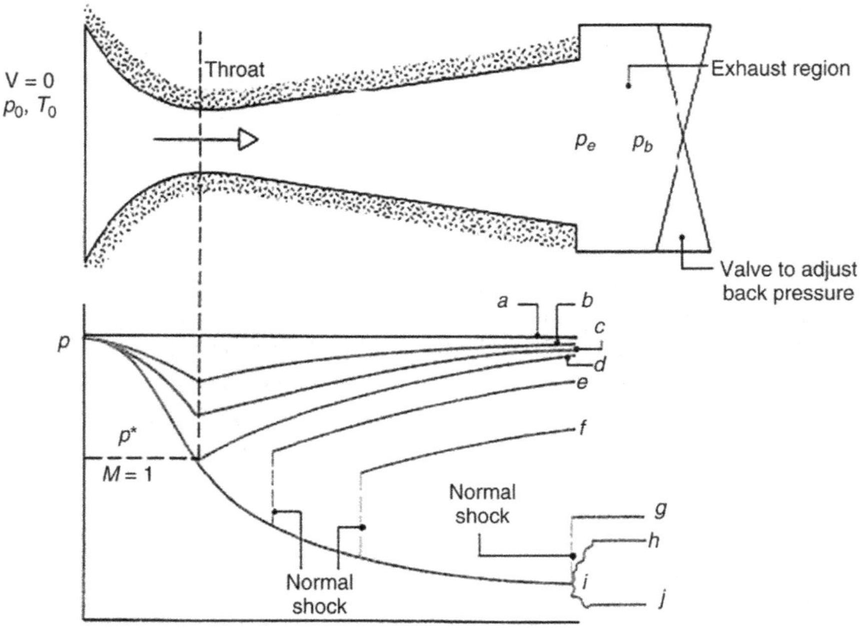

Figure A.17 Flow regimes in the convergent-divergent channel (Reprinted by permission from Moran, M. J. & Shapiro, H. N., *Fundamentals of Engineering Thermodynamics*, 3rd ed., John Wiley & Sons, Inc., New York, 1995.)

And $p_{b\,crt}$ is defined as the value of p_b when M_t just reaches 1, or $p_t = p^*$. It can be evaluated by setting $p_t/p_0 = p^*/p_0 = [2/(k+1)]^{k/k-1}$ in the above equation, that is,

$$\left(\frac{A_e}{A^*}\right)\left(\frac{p_{b,\text{crit}}}{p_o}\right)^{1/k}\left[1-\left(\frac{p_{b,\text{crit}}}{p_0}\right)^{(k-1)/k}\right]^{1/2} = \left(\frac{k-1}{k+1}\right)^{1/2}\left(\frac{2}{k+1}\right)^{1/(k-1)}. \tag{A.18}$$

2) If p_b keeps decreasing, the supersonic flow will expand from the throat into the divergent section with decreasing pressure and a shock wave occurs somewhere to bring the pressure up to match the back pressure at the exit plane. This is the flow regime *d* to *g* until the shock wave reaches the exit plane as shown in Figure A.17. The shock wave will be discussed in the next two sections.
   At this point, the pressure p_e upstream of the shock wave can be related to A_e/A^* through the isentropic flow equation, which is the second solution of Equation (A.18) in the form

$$\left(\frac{p_e}{p_0}\right)^{2/k} - \left(\frac{p_e}{p_0}\right)^{(k+1)/k} = \left(\frac{2}{k+1}\right)^{(k+1)/k-1}\left(\frac{k-1}{2}\right)\left(\frac{A^*}{A^e}\right)^2. \tag{A.19}$$

   And the Mach number M_e upstream of the shock wave can be determined as

$$\frac{1}{M_e}\left\{\frac{1+[(k-1)/2]M_e^2}{(k+1)/2}\right\}^{(k+1)/[2(k-1)]} = \frac{A_e}{A^*}. \tag{A.20}$$

   This means that for each A_e/A^* there are two Mach numbers, one for subsonic, the other for supersonic, depending on the magnitude of back pressure. The back pressure, designated as p_{b3}, is related to p_e, through the normal shock wave equation, to be discussed in the next section.
3) As p_b keeps decreasing from p_{bg}, the shock wave moves downstream of the exit plane and the flow becomes "overexpanded." It is called overexpanded because $p_e < p_b$ and a series of oblique shock waves and expansion waves are required to recover pressure back to p_b.

When $p_b = p_i$, the "design point" is reached and there is no shock wave anywhere. This pressure, designated as $p_b = p_i = p_e$, can be determined from Equation (A.19).

4) If p_b keeps decreasing, the flow becomes "underexpanded" and expansion waves occur at the exit plane to reduce the pressure from p_i to p_b.

A.9 Normal Shock Waves

A shock wave is a finite discontinuity in compressible flow and can be either stationary or moving with respect to the observer's reference frame. Its formation is due to the accumulation of a series of weak waves (small disturbances), either compression or expansion waves, as explained below:

(a) If the piston has a sudden motion to the right with a finite velocity V, which can be approximated as a series of small steps of increase of velocity, ΔV. These small steps of increase of velocity will create a series of compression weak waves propagating to the right, as shown in Figure A.18*a*. The first wave will travel at speed a_1 related to the temperature T_1 of the gas medium undisturbed. But the second wave behind will travel at speed a_2, which is greater than a_1, since the gas ahead of the second wave has been compressed by the first wave. The same process will happen to the third, fourth waves, and so on $(\ldots a_4 > a_3 > a_2 > a_1)$. The waves behind will catch up to the wave ahead and accumulate into a strong wave, that is, a shock wave.

(b) If the gas in the channel is initially stationary and the back pressure p_b starts to decrease, it will create a series of expansion waves propagating upstream and causing the gas flow velocity to increase, as shown in Figure A.18*b*. After the flow velocity at the throat reaches Mach 1, the expansion waves due to further decrease of the back pressure cannot propagate upstream beyond the throat because the flow will carry the waves back downstream. The flow downstream of the throat becomes supersonic. In other words, the expansion waves can propagate upstream only in the subsonic flow region, not in the supersonic flow region, and will accumulate between these two regions to form a shock wave.

Shock waves can be normal, oblique, or curved with respect to the flow direction due to interference with other shock waves or expansion waves in the two- or three-dimensional.

A.9.1 Stationary Normal Shock Waves

Since a shock wave is a discontinuity in an otherwise continuous gas medium, the flow properties across the shock wave have to be derived from the finite forms of mass, momentum, and energy equations (differential forms are not applicable and the detailed process inside the shock wave has to be studied with a statistical method).

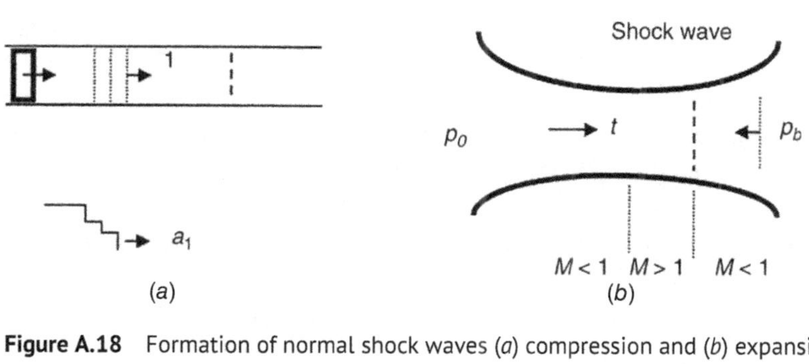

Figure A.18 Formation of normal shock waves (*a*) compression and (*b*) expansion.

To relate the flow properties upstream and downstream of a stationary normal shock wave, we apply the conservation equations to the control volume enclosing the shock wave:

$$\rho_1 V_1 = \rho_2 V_2, \tag{A.21}$$

$$p_1 - p_2 = \begin{cases} \rho_1 V_1 (V_2 - V_1), \\ \rho_2 V_2 (V_2 - V_1). \end{cases} \tag{A.22}$$

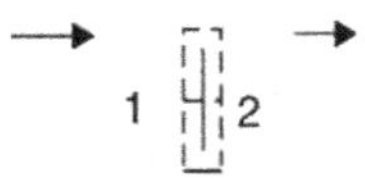

(A.23)

Since the control volume is very thin, the friction effect on the side surface is neglected in the above momentum equation.

Neglecting the heat transfer into or out of the control volume, the energy equation can be written as

$$\frac{1}{2}V_1^2 + C_p T_1 = \frac{1}{2}V_2^2 + C_p T_2 = C_p T_0. \tag{A.24}$$

Or Equations (A.22) and (A.23) can be combined into

$$V_2^2 - V_1^2 = (p_1 - p_2)\left(\frac{1}{\rho_1} + \frac{1}{\rho_2}\right). \tag{A.25}$$

It is interesting to compare this equation with the Bernoulli equation for constant ρ.

With $P = \rho RT$, Equation (A.24) becomes

$$V_2^2 - V_1^2 = \left(\frac{2k}{k-1}\right)\left(\frac{p_1}{\rho_1} - \frac{p_2}{\rho_2}\right). \tag{A.26}$$

Combining Equations (A.25) and (A.26) and using (A.24), we obtain

$$\frac{V_1}{V_2} = \frac{\rho_2}{\rho_1} = \frac{\left(\frac{k+1}{k-1}\right)\frac{p_2}{p_1} + 1}{\left(\frac{k+1}{k-1}\right) + \frac{p_2}{p_1}}. \tag{A.27}$$

Also with $P = \rho RT$, we have

$$\frac{T_2}{T_1} = \frac{p_2}{p_1}\frac{\rho_1}{\rho_2} = \frac{\left(\frac{k+1}{k-1}\right) + \frac{p_2}{p_1}}{\left(\frac{k+1}{k-1}\right) + \frac{p_1}{p_2}}. \tag{A.28}$$

These relations, called Rankine–Hugoniot relations, show the velocities, densities, and temperature ratio across the shock wave, as a function of the shock wave strength (p_2/p_1), which can also be related to the upstream Mach number M_1, to be discussed next.

The fact that p_2/p_1 should be greater than 1 across a shock wave can be ascertained from the entropy consideration. From $s_2 - s_1 = C_p \ln(T_2/T_1) - R\ln(p_2/p_1)$ and Equation (A.28), $\Delta s/R = (s_2 - s_1)/R$ can be expressed in terms of p_2/p_1 and plotted as shown. Since entropy always increases in the adiabatic process, that is, $\Delta s/R > 0$, it is only possible that $P_2/P_1 > 1$ across a shock wave.

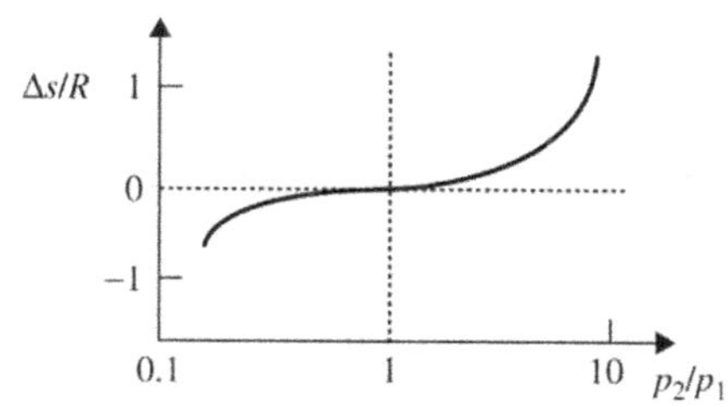

Also, across a shock wave, we have $T_{01} = T_{02}$ since it is an adiabatic process, but $p_{01} \neq p_{02}$ because it is a non-isentropic process. Or, from the entropy equation,

$$s_{02} - s_{01} = C_p \ln\left(\frac{T_{02}}{T_{01}}\right) - R \ln\left(\frac{p_{02}}{p_{01}}\right) = -R \ln\left(\frac{p_{02}}{p_{01}}\right).$$

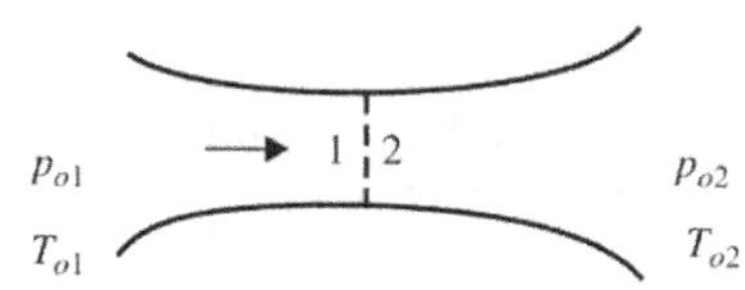

If the flow is set up with two reservoirs as shown, it is also clear from the intuition that $p_{02} < p_{01}$ for a flow to be possible.

In engineering analysis, it is more convenient to express the flow properties ratio across a shock wave in terms of the upstream Mach number M_1. From the same set of mass, momentum, energy, and state equations, the following expressions can be obtained:

$$\frac{p_2}{p_1} = \frac{2kM_1^2 - (k-1)}{k+1}, \tag{A.29}$$

$$\frac{\rho_2}{\rho_1} = \frac{(k+1)M_1^2}{2 + (k-1)M_1^2}, \tag{A.30}$$

$$\frac{T_2}{T_1} = \frac{p_2}{p_1}\frac{\rho_1}{\rho_2} = \frac{\left[2kM_1^2 - (k-1)\right]\left[2 + (k-1)M_1^2\right]}{(k+1)^2 M_1^2}, \tag{A.31}$$

$$M_2^2 = \frac{(k-1)M_1^2 + 2}{2kM_1^2 - (k-1)}. \tag{A.32}$$

Since the stagnation and static pressure ratios can be related to isentropic equations, p_{02}/p_{01} can also be expressed in terms of M_1,

$$\frac{p_{02}}{p_{01}} = \frac{p_{02}/p_2}{p_{01}/p_1} x \frac{p_2}{p_1} = f(M_1)$$

So M_1 is the only parameter required to determine the properties ratios across a shock wave.

Again the entropy change $(s_2 - s_1)$, if expressed in terms of M_1 and plotted as shown, demonstrates that the Mach number upstream of a shock wave has to be greater than 1 $(M_1 > 1)$.

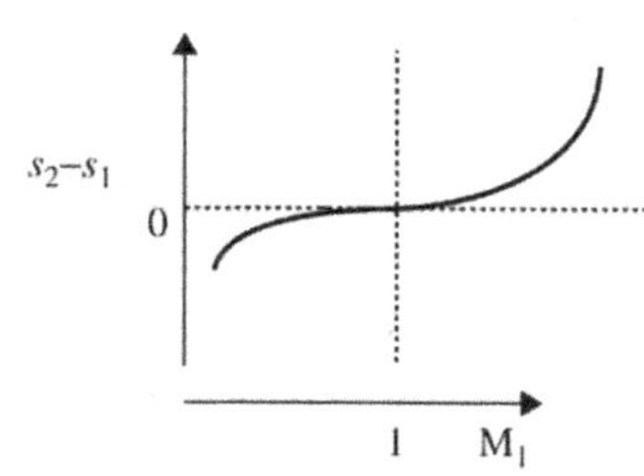

Also, by rearranging Equation (A.32) into

$$M_2^2 = \frac{M_1^2 + 2/(k-1)}{[2k/(k-1)]M_1^2 - 1},$$

it can be shown that the denominator is greater than the numerator for $M_1 > 1$. Hence it is concluded that M_2 is always less than 1. This means that across a shock wave flow changes from supersonic to subsonic (it cannot be the other way around).

In the limiting case, when M_1 approaches 1 (say $1 < M_1 < 1.1$), the flow across a shock wave will become isentropic, and the weak shock wave will become an acoustic wave.

A.9.2 Moving Shock Waves

Assume the shock wave (s.w.) is moving to the right at U_s into a stationary gas medium at p_1, T_1 and induces the gas to move at V and compresses it to p_2, T_2. Converting the flow into a frame with a stationary shock wave, the gas will flow from right to left as shown. So with respect to the stationary shock wave, the parameters should be

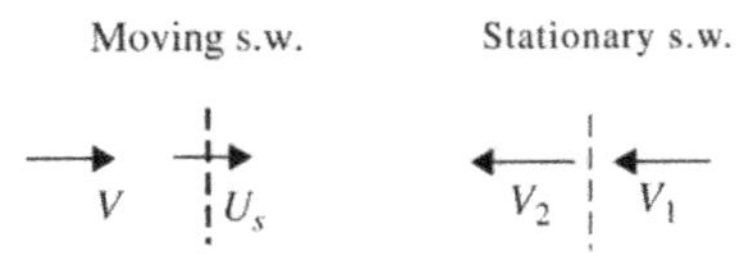

$$V_1 = U_s, \qquad V_2 = U_s - V,$$

or

$$M_1 = \frac{U_s}{a_1} \equiv M_s, \qquad M_2 = \frac{U_s - V}{a_2} = \left(\frac{U_s}{a_1}\right) = \left(\frac{a_1}{a_2}\right) - \frac{V}{a_2} = \frac{M_s a_1}{a_2} - M_2'.$$

Substituting M_1, M_2 into the stationary shock wave equations derived in the previous section, we obtain the equations for a moving shock wave in terms of M_s and M_2' as follows:

$$\frac{p_2}{p_1} = \frac{2kM_s^2 - (k-1)}{k-1}, \qquad \frac{\rho_2}{\rho_1} = \frac{(k+1)M_s^2}{2 + (k-1)M_s^2}$$

$$\frac{T_2}{T_1} = \frac{p_2}{p_1}\frac{\rho_1}{\rho_2} = \frac{\left[2kM_s^2 - (k-1)\right]\left[2 + (k-1)M_s^2\right]}{(k+1)^2 M_s^2}.$$

$$\left(M_s \frac{a_1}{a_2} - M_2'\right)^2 = \frac{(k-1)M_s^2 + 2}{2kM_s^2 - (k-l)}$$

From the last equation, M_2' can be obtained in terms of M_s, and $V = M_2' a_2$.

A.10 Oblique Shock Waves

Oblique shock waves can occur in flows over a sharp wedge or a sharp corner as shown. Flow can be divided into normal and parallel components. The parallel components do not change across the shock wave, and the normal components can be related to the normal shock wave equations derived. The first problem on an oblique shock wave is to determine the relationship between the shock wave angle β and the deflection angle δ with respect to the upstream Mach number M_1. The flow velocity components across an oblique shock wave are depicted as shown.

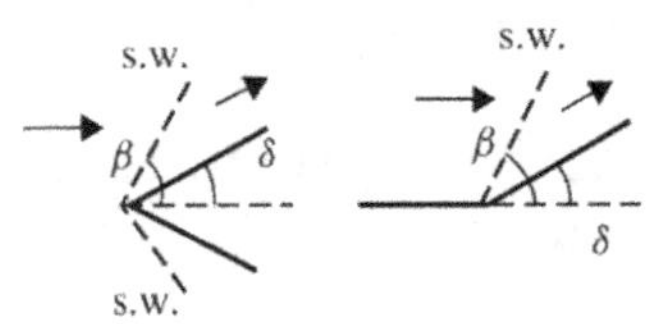

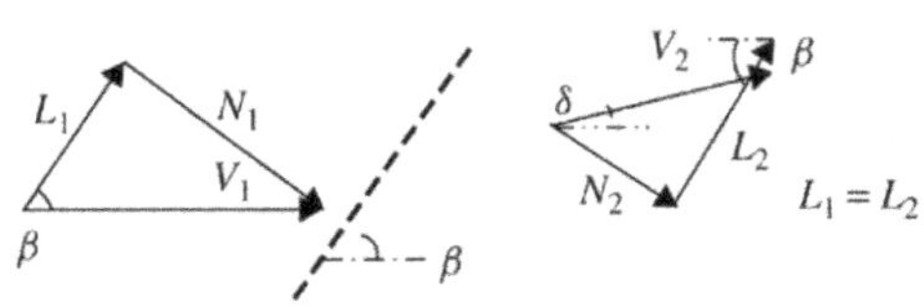

It is shown that $N_1 = V_1 \sin\beta$, $N_2 = V_2 \sin(\beta - \delta)$ or $M_{n1} = M_1 \sin\beta$, $M_{n2} = M_2 \sin(\beta - \delta)$, and they can be related through the normal shock wave equations such that

$$\frac{p_2}{p_1} = \frac{2kM_1^2 \sin^2\beta - (k-1)}{k+1} \tag{A.33}$$

$$\frac{\rho_2}{\rho_1} = \frac{(k+1)M_1^2 \sin^2\beta}{2 + (k-1)M_1^2 \sin^2\beta}, \tag{A.34}$$

$$\frac{T_2}{T_1} = \frac{p_2}{p_1}\frac{\rho_1}{\rho_2} = \frac{\left[2kM_1^2\sin^2\beta - (k-1)\right]\left[2+(k-1)M_1^2\sin^2\beta\right]}{(k+1)^2 M_1^2\sin^2\beta}, \tag{A.35}$$

$$M_2^2\sin^2(\beta-\delta) = \frac{M_1^2\sin^2\beta + 2/(k-1)}{2\gamma M_1^2\sin^2\beta/(k-1) - 1}, \tag{A.36}$$

$$\frac{p_{02}}{p_{01}} = \frac{p_{02}/p_2}{p_{01}/p_1} x \frac{p_2}{p_1},$$

where the stagnation and static pressure ratios can be obtained from the isentropic equations. Since $M_{n1} \geq 1$ as proved from the entropy consideration earlier, it is necessary that $M_1 \sin\beta \geq 1$. That means $\sin^{-1}(1/M_1) \leq \beta \leq 90°$.

When $\beta = 90°$, it is a normal shock wave; when $\beta = \sin^{-1}(1/M_1) < 90°$, it becomes a Mach wave, where the object diminishes to a point or δ approaches zero, and $p_2/p_1 = 1$.

From $\tan\beta = N_1/L_1$, $\tan(\beta-\delta) = N_2/L_2$, and $\rho_1 N_1 = \rho_2 N_2$ [with $L_1 = L_2$ and Equation (A.34)], an important relationship between M_1, β, and δ can be obtained as

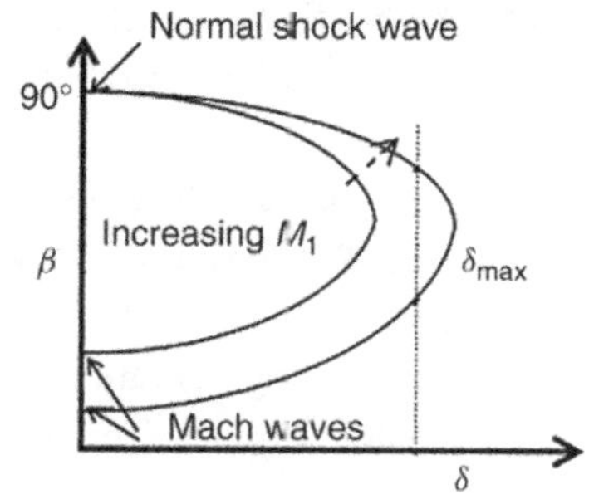

Figure A.19 Relationship between oblique shock wave parameters.

$$\tan\delta = \frac{2\cot\beta\left(M_1^2\sin^2\beta - 1\right)}{2 + M_1^2(k+\cos 2\beta)}. \tag{A.37}$$

It is plotted in Figure A.19.

For each δ and M_1, there are two solutions for β, one for the strong shock wave with $M_2 < 1$ and the other for the weak shock wave with $M_2 > 1$. Which one actually occurs depends on the downstream condition. When $\delta = 0$, it will be either normal shock waves or Mach waves. When $\delta > \delta_{max}$, the shock wave becomes detached, or for a given δ, the shock wave changes from detached to attached as M_1 increases since δ_{max} increases with M_1.

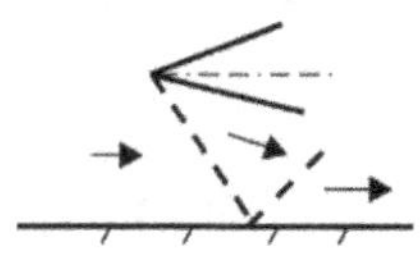

When a shock wave is incident on a flat wall, a second shock wave has to be reflected in order to turn the flow next to the wall back to the parallel direction as shown.

Examples of shock wave interactions occur in the following two cases:

(a) In flow along a concave wall consisting of several angular changes, the oblique shock waves generated at the changes tend to converge and coalesce into a single stronger shock wave, as shown in Figure A.20*a*.
(b) The oblique shock waves *a* and *b* initiated at the opposite concave walls will interact to generate two transmitted shock waves *c* and *d*, as shown in Figure A.20*b*. And in regions 4 and 5, the flow velocity directions and pressures should be the same along a slip-line. But the density, flow velocity magnitude, and entropy can be different, since the strength of shock waves *a*, *c* and *b*, *d* could be different.

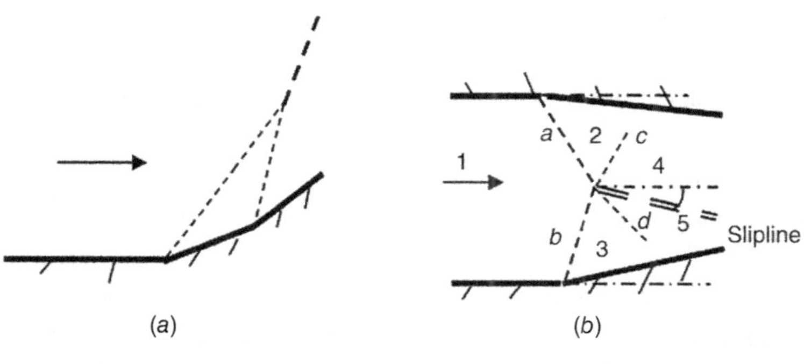

Figure A.20 Interaction of two oblique shock waves (*a*) single shock wave along a concave wall and (*b*) two shock waves at opposite concave walls.

Problems

A.1 Air expands from a reservoir of 2 MPa, 290 K to a back pressure of 100 kPa through a nozzle. Determine the exhaust air velocity.

A.2 Air in a flow passage has a temperature of 150 °C, a pressure of 300 kPa, and a velocity of 120 m/s. Determine the stagnation temperature, pressure, and Mach number at that point.

A.3 Air enters a diffuser of area ratio 1.5 at 100 kPa, 30 °C with a velocity of 150 m/s, and the exit temperature is 40 °C. If heat loss of 0.4 kJ/kg occurs, find the exit air velocity and static pressure.

A.4 Air entering a compressor at 100 kPa, 15 °C is compressed to 1.0 MPa. If the compressor efficiency is 80%, what is the power input required per unit mass flow rate of air?

A.5 A compressor draws air from an atmosphere of 101 kPa and 25 °C. At the inlet the static pressure is measured to be 70 kPa, and at the outlet the flow velocity is measured to be 20 m/s. If the static discharge pressure ratio is 9 and the total-to-total efficiency is 87%, what is the static temperature at the outlet and the shaft power required per unit mass flow rate of air?

A.6 Air is being compressed from 95 kPa and 25 °C to 700 kPa at an inlet volumetric flow rate of 8 m^3/min with a polytropic process (n = 1.30). Cooling water of 14 kg/min through the compressor jacket has a temperature rise of 5 °C. If air velocity at the inlet is negligible but is 120 m/s at discharge, determine the input power required.

A.7 Air enters a turbine at 3 MPa, 500 °C, and 70 m/s and expands through a pressure ratio of 10 : 1. The air exhausts at a velocity of 140 m/s. If the turbine efficiency is 85%, determine the work output per unit mass flow rate of air.

A.8 A nozzle is tested with air entering at 1.2 MPa, 1500 K, 12 m/s, and the discharge is measured as 600 kPa, 1280 K. What is the exit flow velocity, nozzle efficiency, and nozzle inlet/outlet diameter ratio?

A.9 The turbine in a jet engine receives air at 1280 K, 1.8 MPa and exhausts to a nozzle at 35 psia with negligible velocity. If the ambient atmosphere pressure is 101 kPa and the turbine and nozzle efficiencies are 87% and 95%, respectively, determine the nozzle inlet temperature and the air velocity at the nozzle exit.

A.10 A gas turbine has a compressor inlet condition of 100 kPa, 35 °C, 30 m/s and discharges at 700 kPa, 290 °C, 10 m/s. Fuel enters the combustion chamber and raises the air temperature to 930 °C before entering the turbine with a negligible velocity. The turbine discharge temperature is 435 °C, 95 kPa, and 50 m/s. Determine (a) the compressor and turbine total-to-total efficiencies and (b) the cycle thermal efficiency.

A.11 A gas turbine power plant is designed to deliver 8000 hp. to drive an electric generator. The ambient condition is 25 °C and 101 kPa, pressure ratio is 6 : 1, the adiabatic efficiencies of compressor and turbine are 85 and 89%, respectively, and the inlet temperature of the turbine is 700 °C. Determine the required airflow rate and the cycle thermal efficiency.

A.12 The gas turbine in Problem A.19 is installed with a scale-down turbine, just enough to produce the power to drive the compressor, and a propelling nozzle with the efficiency of 90%. Determine the nozzle exit plane area and the static thrust when tested at the given ambient condition mass flow rate.

A.13 A steam turbine is tested with steam of 1.5 MPa and 540 °C at the inlet. If the condenser pressure is maintained at 70 kPa and the turbine exit temperature is measured as 180 °C, estimate the turbine efficiency and its shaft power output per unit steam mass flow rate.

A.14 A vapor compression refrigeration cycle with refrigerant R134a as the working fluid operates with an evaporator at 70 kPa, 3 °C superheated and a condenser at 650 kPa. A centrifugal compressor of 82% adiabatic efficiency and a simple throttle valve are employed. Determine the refrigeration capacity in tons (1 ton = 12,000 Btu/h) and the shaft power required to drive the compressor if the mass flow rate of the refrigerant is 0.12 lb./s.

A.15 Consider an ideal refrigeration cycle that uses R134A as the working fluid. The temperature of the refrigerant in the evaporator is 20 °F, and in the condenser it is 140 °F. The refrigerant is circulated at the rate of 0.15 lb_m/s. Determine the coefficient of performance and the refrigeration capacity of the plant.

B

Airfoil Characteristics

Because airfoil sections are extensively used in axial-flow machines, their aerodynamic characteristics and the details of some NACA airfoils will be discussed in this Appendix.

B.1 Flow Over Objects

The complexity of the flow pattern over an object depends on the flow velocity and the object geometry. To simplify the analysis, the flow field can be divided into two regions. The viscous effect is considered only in the boundary layer next to the object surface. Outside the boundary layer, the fluid is considered as inviscid potential flow. At low flow velocity, it can even be assumed as incompressible fluid.

The boundary layer can be either laminar or turbulent. It can transit from laminar to turbulent, and be separated from the surface, when the flow velocity gradient at the surface changes from positive to negative, as shown above. The occurrences of boundary layer transition and separation depend on the Reynolds number, pressure gradient, surface roughness, object geometry, and its orientation with respect to the mean flow direction. If the pressure decreases in the flow direction, there is no possibility for the boundary layer to separate. If it increases, a condition called adverse pressure gradient, the boundary layer can separate, causing adverse effects, due to changes of flow pattern and pressure distribution. The turbulent boundary layer can resist the separation better than the laminar boundary layer. Hence enhancing the boundary layer transition is a way to delay boundary layer separation, but the viscous friction due to the laminar layer is less than that due to the turbulent boundary layer.

The resultant force on an object submerged in the flow is due to the unbalanced pressure distribution and the viscous friction on its surface. In an airfoil, the resultant force can be decomposed into lift and drag force components. The lift is normal to the mainstream flow direction and is primarily due to the pressure distribution. The drag is parallel to the mainstream flow direction and is due to the combination of pressure distribution and viscous friction. In boundary layer separation (called stall), a significant decrease of lift and increase of drag result from the distorted pressure distribution.

The lifting characteristics are mainly related to the pressure distribution and are negligibly affected by viscosity. If the angle of attack α, camber z_c/c, and blade thickness $t_{\max}/c$ are small, they can be predicted with the thin airfoil theory. For low velocity of potential flow, the airfoil is replaced with its mean line, and the flow pattern is determined such that one streamline is coincident with the mean line. For subsonic flow, a correction factor $1/\sqrt{1-M_\infty^2}$ is applied; for supersonic flow, the linearized theory is applicable. It is more difficult to analyze the flows associated with the turbulent boundary layer, boundary layer transition and separation. For details on these subjects, the readers should refer to texts on aerodynamics.

The airfoil as a streamlined body is designed to maximize the lift–drag ratio (L/D) and to delay boundary layer separation. In general, the overall lift and drag forces as a function of angle of attack are determined from laboratory tests.

Fundamentals of Turbomachinery: Theory and Applications, Second Edition. Ryoichi S. Amano and William W. Peng.

Companion website: www.wiley.com/go/amano/turbomachinery2e

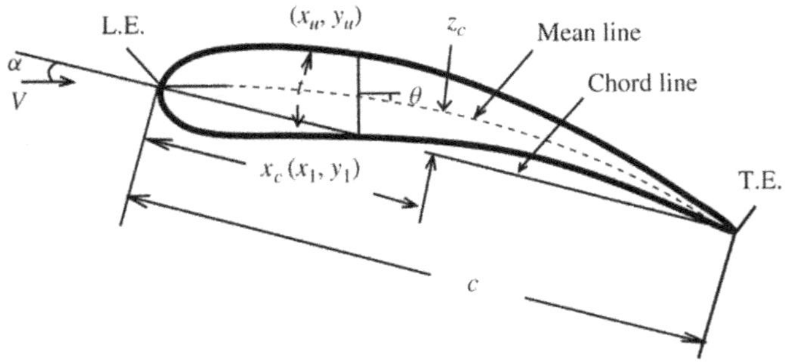

Figure B.1 Typical airfoil geometry.

B.2 Airfoil Geometry

The airfoil is composed of a thickness envelope wrapped around a mean camber line. The mean line lies halfway between the upper and lower surfaces of the airfoil and intersects the chord line at the leading edge (L.E.) and trailing edge (T.E.), as shown in Figure 3.39 and again in Figure B.1, where the angle of attack α is also shown.

The chord c is used as a reference length. All the other dimensions are expressed in terms of its fractions. The mean line and thickness distributions are specified for different airfoil types. Then the upper and lower surfaces can be determined with the expressions $x_u = x - y_t \sin\theta$, $y_u = y_m + y_t \cos\theta$, $x_l = x + y_t \sin\theta$, and $y_l = y_m - y_t \cos\theta$, where $x \cong x_m$, $y_t \cong t/2$, $\tan\theta = dy_m/dx_m$, where (x_m, y_m) are the points along the mean line.

B.3 NACA Airfoil Specifications

There are several series of NACA airfoils. Only the two most commonly cited will be discussed here. Other types of airfoils, including those developed elsewhere, are available in the literature:

1) The four-digit series is designated as NACA *mpxy*, where m is the maximum camber of the mean line (z_c) in percent of chord c, p is the position of maximum camber (x_c) in tenths of c, and xy represents the maximum thickness ($t_{\max}$) in percent of c. The mean line is expressed by two parabolic functions such that $dy_m/dx = 0$ at $x = p$ for both functions, that is,

$$y_{mf} = \frac{m}{p^2}(2px - x^2),$$

for forward portion of the maximum ordinate and

$$y_{ma} = \frac{m}{(1-p)^2}[(1-2p) + 2px - x^2],$$

for aft portion of the maximum ordinate. The thickness ordinate is given as

$$y_t = \frac{t_{\max}}{0.20}(0.2969\sqrt{x} - 0.1260x - 0.3516x^2 + 0.2843x^3 - 0.1015x^4),$$

and the radius of the arc at the leading edge is given as $r_l = 1.1019t_{\max}^2$.

2) The five-digit series, designated as NACA *mpqxy*, was designed for the extreme forward position of the maximum camber, where m is the maximum camber of the mean line (z_c) in percent of c and the design lift coefficient (at maximum L/D) $C_L = 0.15$ m, pq are two times of the maximum camber position ($2x_c$) in percent of c, and xy represents the maximum thickness ($t_{\max}$) in percent of c.

The mean line is expressed as

$$y_{mf} = \frac{1}{6}k[x^3 - 3nx^2 + n^2(3-n)x] \qquad \text{for } x = 0, \ldots, n,$$

$$y_{ma} = \frac{1}{6}kn^3(1-x) \qquad \text{for } x = n, \ldots, 1,$$

where k and n values are given for different x_c. For the design value $C_L = 0.3$, they are as follows:

x_c	0.05	0.10	0.15	0.20	0.25
n	0.058	0.126	0.2025	0.290	0.391
k	361.4	51.64	15.957	6.643	3.230

The expression for y_t is the same as that for the four-digit series given above.

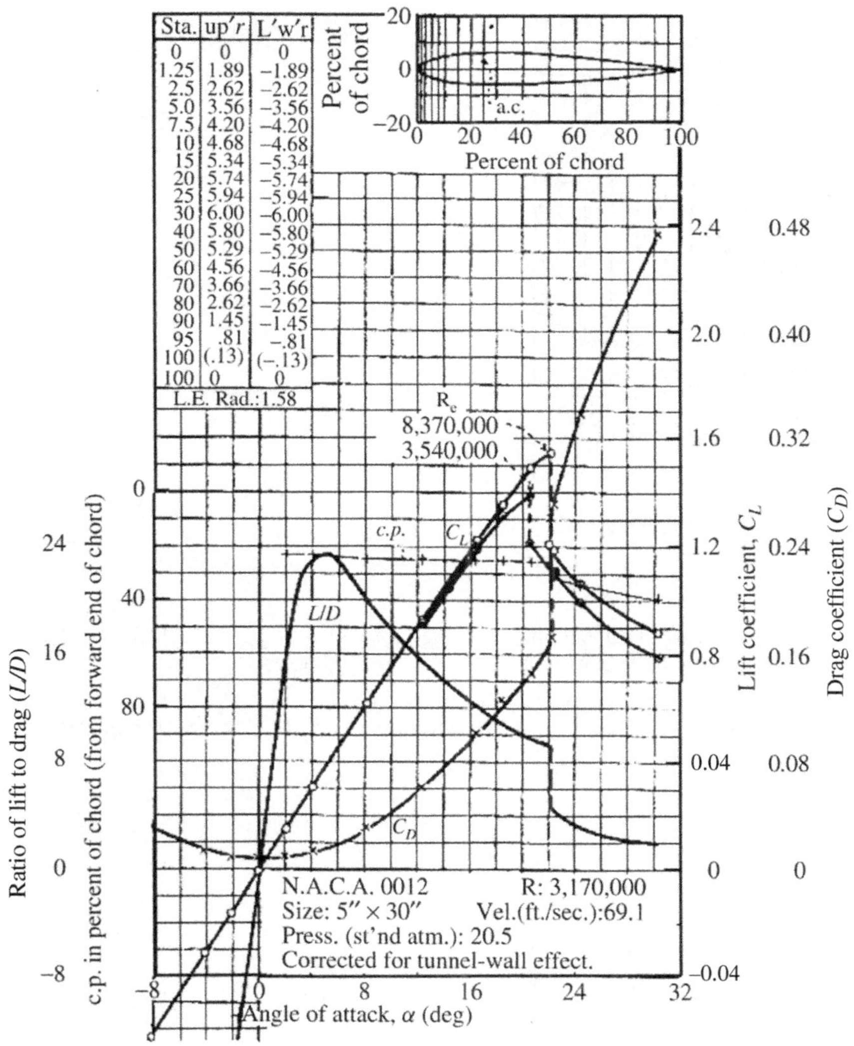

Figure B.2 NACA 0012 airfoil data (four digits with straight camber line, upper and lower surfaces are symmetric, $t_{max} = 0.12c$). (*Source*: Rice, M. S., *Handbook of Airfoil Sections for Light Aircraft*, Aviation Publications, Milwaukee, WI, 1971).

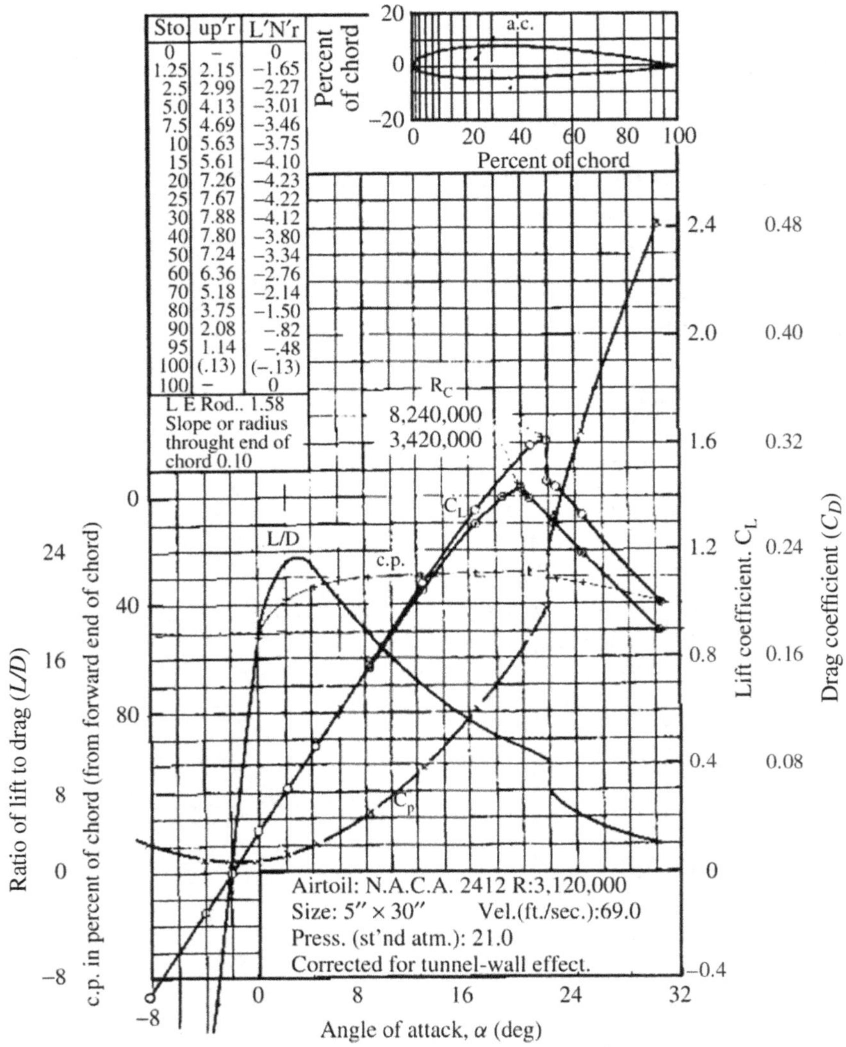

Figure B.3 NACA 2412 airfoil data (four digits with curved camber line, $z_c = 0.02c$ at $x_c = 0.4c$, $t_{max} = 0.12c$). (*Source*: Rice, M. S., *Handbook of Airfoil Sections for Light Aircraft*, Aviation Publications, Milwaukee, WI, 1971).

Some NACA airfoil data are given in Figures B.2, B.3, B.4.[5] In addition to L/D, C_L, and C_d, the center of pressure (c.p.) versus the angle of attack is also given:

$$\text{c.p.} = \frac{\oint xp\,dx}{\oint p\,dx}.$$

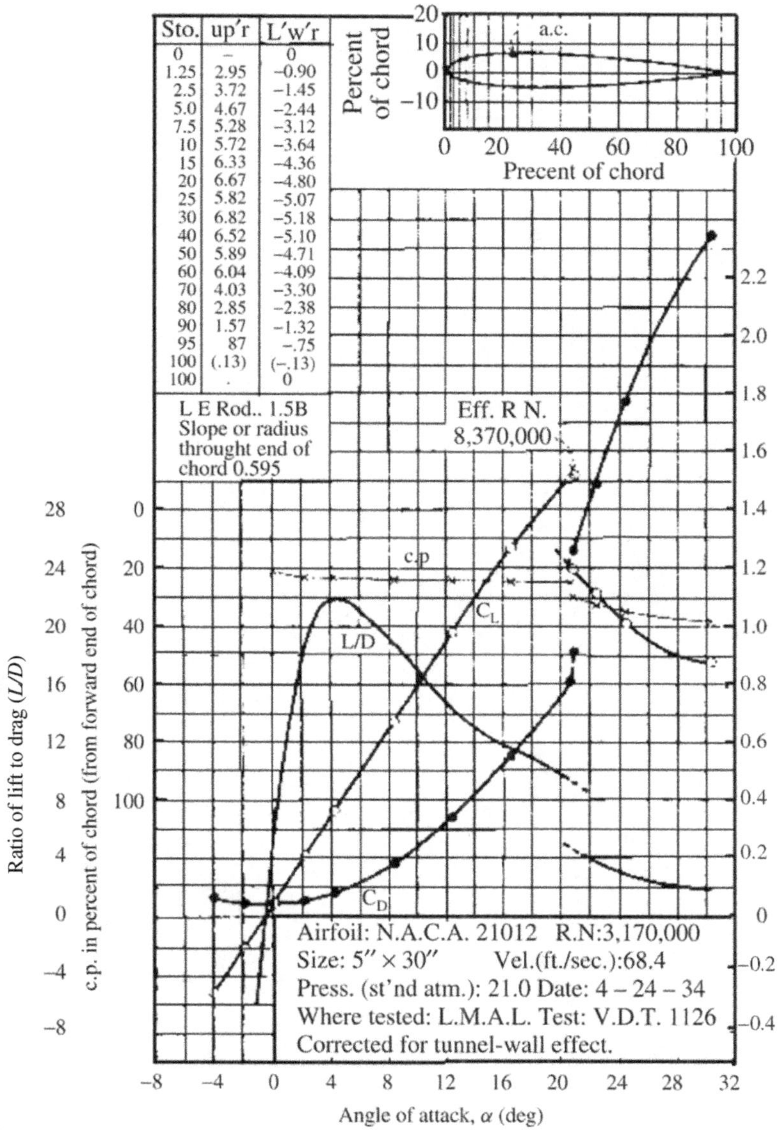

Sto.	up'r	L'w'r
0	–	0
1.25	2.95	−0.90
2.5	3.72	−1.45
5.0	4.67	−2.44
7.5	5.28	−3.12
10	5.72	−3.64
15	6.33	−4.36
20	6.67	−4.80
25	5.82	−5.07
30	6.82	−5.18
40	6.52	−5.10
50	5.89	−4.71
60	6.04	−4.09
70	4.03	−3.30
80	2.85	−2.38
90	1.57	−1.32
95	87	−.75
100	(.13)	(−.13)
100	.	0

Figure B.4 NACA 21012 airfoil data (five digits with curved camber line, $z_c = 0.02c$ at $x_c = (10/2)\%\ c = 0.05c$, $C_{L,m} = 0.15 \times 2 = 0.3$ at maximum L/D, $t_{max} = 0.12c$). (*Source*: Rice, M. S., *Handbook of Airfoil Sections for Light Aircraft*, Aviation Publications, Milwaukee, WI, 1971).

C

Conversion Factors and Some Constants

Length	1 in. = 2.54 cm, 1 ft = 0.3048 m, 1 mile = 1.609 km = 5280 ft
Area	1 acre = 43,560 ft^2
Volume	1 U.S. gallon = 231 $in.^3$ = 0.003785 m^3
Velocity	1 ft/s = 0.3048 m/s, 1 mph = 0.447 m/s
Mass	1 lb_m = 0.4536 kg, 1 slug = 32.174 lb_m
Density	1 lb_m/ft^3 = 16.02 kg/m^3
Flow rate	1 gpm = 0.227 m^3/h = 0.00223 ft^3/s 1 cfm = 1.699 m^3/h
	1 ft^3/s = 0.0283 m^3/s
Force	1 lb_f = 4.448 N
Temperature	°F = 1.8 °C + 32, R = °F + 460, K = °C + 273
Pressure	1 psi = 6.895 kPa, 1 ft of H_2O = 2.989 kPa, 1 in. of Hg = 3.386 kPa
	1 bar = 100.0 kPa (1 Pa = 1 N/m^2)
Energy	1 ft-lb_f = 1.356 J, 1 Btu = 778 ft-lb_f = 1.055 kJ (1 J = 1 N-m)
Power	1 hp = 0.7457 kW = 550 ft-lb_f/s = 2545 Btu/h, 1 W = 0.737 ft-lb_f/s
Viscosity	1 ft^2/s = 0.0929 m^2/s, 1 lb_f-s/ft^2 = 47.88 N-s/m^2(kg/m-s)
	(1 stoke = 1 cm^2/s, 1 poise = 0.1 N-s/m^2)
Specific energy	1 ft-lb_f/lb_m = 2.989 J/kg
Specific heat capacity	1 ft-lb_f/(lb_m °F) = 5.38 J/(kg °C), 1 Btu/(lb_m °F) = 4.188 kJ/(kg °C)
Universal gas constant	R_o = 8314.3 J/(kg-mol K) = 1.98592 Btu/(lb_m-mol R) = 1545 ft-lb_f/(lb_m-mol R)

Fundamentals of Turbomachinery: Theory and Applications, Second Edition. Ryoichi S. Amano and William W. Peng.

Companion website: www.wiley.com/go/amano/turbomachinery2e

D

Thermodynamic Properties of Air, Water, Superheated Steam, and CO_2 at STP (14.7 psia, 60 °F)

	Air	Water	Steam (Superheated)	CO_2
Molecular weight	28.98	18	18.0	44
R (J/kg K)	287.0	—	461.4	189.1
(ft-1bf/1bm R)	53.3	—	85.78	35.1
C_p (J/kg K)	1004	—	~2000	837
(Btu/lbm R)	0.24	—	~0.48	0.20
k (C_p/C_v)	1.4	—	~1.3	~1.3
μ (N-s/m^2)	1.8×10^{-5}	1.0×10^{-3}	—	1.5×10^{-5}
(lbf-s/ft^2)	3.7×10^{-7}	2.1×10^{-5}	—	3.1×10^{-7}
Density at STP (kg/m^3)	1.22	998	~0.5	1.85
(lbm/ft^3)	0.0762	62.4	~0.03	0.115

Fundamentals of Turbomachinery: Theory and Applications, Second Edition Ryoichi S. Amano and William W. Peng.

Companion website: www.wiley.com/go/amano/turbomachinery2e

E

Internal Diameter of Standard Pipe of Schedule 40

Nominal Size	Inside Diameter (in.)	Nominal Size	Inside Diameter (in.)
2	2.067	4	4.026
$2\frac{1}{2}$	2.469	5	5.047
3	3.068	6	6.065
$3\frac{1}{2}$	3.548	8	7.981

Fundamentals of Turbomachinery: Theory and Applications, Second Edition. Ryoichi S. Amano and William W. Peng.

Companion website: www.wiley.com/go/amano/turbomachinery2e

Index

Fundamentals of Turbomachinery: Theory and Applications, Second Edition. Ryoichi S. Amano and William W. Peng.

Companion website: www.wiley.com/go/amano/turbomachinery2e